Scientific Computation

Editorial Board

J.-J. Chattot, Davis, CA, USA
P. Colella, Berkeley, CA, USA
R. Glowinski, Houston, TX, USA
M. Holt, Berkeley, CA, USA
Y. Hussaini, Tallahassee, FL, USA
P. Joly, Le Chesnay, France
H. B. Keller, Pasadena, CA, USA
D. I. Meiron, Pasadena, CA, USA
O. Pironneau, Paris, France
A. Quarteroni, Lausanne, Switzerland
J. Rappaz, Lausanne, Switzerland
R. Rosner, Chicago, IL, USA
J. H. Seinfeld, Pasadena, CA, USA
A. Szepessy, Stockholm, Sweden
M. F. Wheeler, Austin, TX, USA

Springer
Berlin
Heidelberg
New York
Hong Kong
London
Milan
Paris
Tokyo

Jan Awrejcewicz Vadim A. Krys'ko

Nonclassical Thermoelastic Problems in Nonlinear Dynamics of Shells

Applications of the Bubnov-Galerkin and Finite Difference Numerical Methods

With 222 Figures

Springer

Professor Jan Awrejcewicz
Technical University of Łodz
Department of Automatics
and Biomechanics
1/15 Stefanowskiego St.
90-924 Łodz, Poland
E-mail: awrejcew@ck-sg.p.lodz.pl

Professor Vadim A. Krys'ko
Saratov State Technical University
Department of Mathematics
77 Polyteshnycheskaya St.
41005 Saratov, Russia

ISSN 1434-8322
ISBN 3-540-43880-7 Springer-Verlag Berlin Heidelberg New York

Library of Congress Cataloging-in-Publication Data

Awrejcewicz, J. (Jan)
Nonclassical thermoelastic problems in nonlinear dynamics of shells: applications of the Bubnov–Galerkin and finite difference numerical methods/Jan Awrejcewicz, Vadim A. Krys'ko.
p.cm. – (Scientific computation, ISSN 1434-8322)
Includes bibliographical references and index.
ISBN 3-540-43880-7 (acid-free paper)
1. Elastic plates and shells. 2. Galerkin methods. 3. Finite differences. I. Krys'ko, V. A. (Vadim Anatol'evich), 1937– . II. Title. III. Series. QA935.A97 2003 624.1'776–dc21 2002030638

Springer-Verlag Berlin Heidelberg New York
a member of BertelsmannSpringer Science+Business Media GmbH

http://www.springer.de

Typesetting by the author
Final page layout: LE-TEX, Leipzig
Cover design: *design & production* GmbH, Heidelberg
Printed on acid-free paper SPIN: 10885444 55/3141/tr - 5 4 3 2 1 0

Preface

This monograph describes some approaches to the nonlinear theory of plates and shells. By nonclassical approaches we mean the desciption of problems with mathematical models of different sizes (two- and three-dimensional differential equations) and different types (differential equations of hyperbolic and parabolic type in the spatial coordinates). The nonlinearities investigated are also of various categories: geometrical, physical, elasto-plastic, and periodic. Creating such types of mathematical models and their detailed justification allows us to achieve the most accurate description of the real behaviour of shell-type structures. These models allow us to include interaction between the strain and temperature fields and coupling between the displacement field and the external influence of a transonic gas flow. The mathematical treatment of such models helps us greatly in obtaining reliable results by numerical computation.

It appears that the most dangerous situation for thin shallow shells is the conjunction of a static load with dynamic interactions. Such combined loads very often cause buckling of shell structures, and in many cases a series of bucklings, which can cause fracture. The failure of a structure usually needs a small amount of time. Therefore the lifetime of a shell structure depends strongly on nonelastic deflections and it is important to mathematically model shell structures as precisely as possible. This monograph is one of several devoted to this subject.

Now we shall briefly describe the contents of the book. Note that not all of the results presented here have been published in textbook format.

Chapter 1 of this monograph is devoted to an analysis of the current literature concerning the problems that the title of the book refers to. We emphasize, among other things, the lack of systematic presentation of these problems and their solutions in the existing literature.

In Chap. 2, a general statement of the coupled thermomechanical problems of flexible shells, with application of threedimensional heat transfer equations is presented, emphasizing a serious problem from the point of view of both the theoretical and the numerical approaches. However, the approach proposed here is one of the most powerful approaches, since it allows one to aviod any a priori hypotheses with respect to the temperature distribution through the shell thickness. In addition, a very important coupled problem of the interaction of shell-type structures with a transonic flow is addressed.

Note that in the case of a transonic gas flow, a special, singular flow behaviour can occur for certain profiles. In such flows, shock waves are born, which then move and vanish (they can be recognized as pressure jumps). In addition, the velocity of the excitations generated by the deformed profile is close to the velocities of the gas particles. These phenomena appear during the action of a flow on a profile and the simultaneous action of the profile on the flow.

In fact the problem of the dynamic interaction of a shell with a transonic flow considered here reduces to a two-dimensional problem of the interaction of a flow with an infinite-length cylindrical panel. A mathematical model of the latter problem is presented and a numerical method for its solution is formulated. The dynamic stability of such an infinite panel while interacting with a transonic ideal gas is analysed.

Research into the existence and uniqueness of solutions of linear coupled problems of the thermomechanics of plates and shells, within the framework of both the Kirchhoff–Love kinematic model and Timoshenko-like models with a three-dimensional parabolic equation of heat transfer, is described in Chap. 3. The problem is reduced to an abstract Cauchy problem of a coupled system of two differential equations in a Hilbert space, which generalizes the set of the above coupled problems of thermoelasticity. The solutions to the problems formulated are obtained by applying the Bubnov–Galerkin method in a Sobolev space. Many theorems on the existence, uniqueness, regularity and continuous dependence of solutions on the initial conditions are formulated and discussed.

For linear coupled problems of shallow shells that are rectangular in plan, prior estimates of the errors of the Bubnov–Galerkin method are obtained. The estimates are obtained for the general case of a shell of variable thickness, subjected to time-dependent mechanical and thermal loads. These estimates are of great theoretical importance, since they guarantee strong convergence of a series of approximate solutions to the exact solution with a speed not smaller than that described in this chapter, for a wide class of problems.

In Chap. 4, results of numerical analysis of the estimated errors of the Bubnov–Galerkin method for linear problems in the theory of plates are presented. These results show that the estimates provide high efficiency.

In Chap. 5, a mathematical model of coupled, geometrically nonlinear problems of the thermoelasticity of shallow shells, using a Timoshenko-type kinematic theory and taking into account the rotational inertia of the shell and a three-dimensional thermal field (as in previous chapters), is formulated. The problem of the existence and uniqueness of a solution, as well as the convergence speed of the Bubnov–Galerkin method, is analysed; this generalizes results presented in Chap. 3.

In Chap. 6, a more exact mathematical model of the theory of shells is introduced and analysed. In addition to the hypotheses and assumptions of Chap. 5, nonlinearity of the physical properties of the shell material is included. The kinematic Kirchhoff–Love model is used. The mathematical model is obtained by the Biot variation principle.

Chapter 7 contains numerical methods for solution of the problems introduced in the previous chapter a finite-difference method of order $O(h^2)$ and the method of "elastic" solutions are applied in the numerical computations.

An analysis of the convergence of the solutions and an investigation of the dynamic stability loss of shells for different values of the parameters governing the geometric, physical and thermal properties are carried out.

The Bubnov–Galerkin method is applied to problems concerning the dynamic stability of shells using Timoshenko-type or Kirchhoff–Love models of coupled thermoelasticity with respect to displacements. In the Sects. 7.7–7.11, parametric vibrations of flexible shells are investigated. Solutions are approximated using finite-difference methods of order $O(h^2)$ and $O(h^4)$ with respect to the spatial coordinates.

Symmetric and nonsymmetric vibrations are analysed. Plates are treated as systems with an infinite number of degrees of freedom. Soliton phenomena in elastic plates in the presence of parametric vibrations are detected and discussed, among other topics. It is shown that both symmetric and nonsymmetric vibrations are included in the mathematical model. This part of the monograph was prepared in collaboration with Anton V. Krys'ko.

Chapter 8 presents an even more complicated mathematical model than that applied in Chap. 6. Here, besides geometric nonlinearities, the periodicity of material properties during vibration is taken into account. Numerical algorithms for the solution of such problems are presented. Analysis of the dynamic stability and vibrations of plates and shells for different physical and geometric parameters is performed, and the results are presented in a series of graphs.

In the last chapter, some mathematical problems which have been briefly mentioned in this monograph but which have not yet been solved are addressed.

Both authors greatly appreciate the help of Dr. Jacek Nowakowski and Mr. Marek Kaźmierczak. Their revisions and many suggestions have certainly improved the quality of the book.

The first author (J.A.) wishes to acknowledge the financial support by the Polish National Scientific Research Committe Grant No. 8 T11F 014 16.

Finally, the help of Professor Wolf Beiglboeck, Senior Physics Editor of Springer, during the whole process leading to the appearance of the book, is highly appreciated.

Łódź, Saratov
August 2002

J. Awrejcewicz
V.A. Krys'ko

Contents

1 Introduction

In this chapter, we give a brief discussion of the literature on the nonlinear theory of plates and shells, paying attention particularly to Eastern references, where many interesting results have been obtained and which are (unfortunately) still not well distributed among the worldwide scientific community.

Interest in coupled thermoelastic problems has developed during last 30–35 years. This was caused by the need for mathematical calculations of structures subjected to the action of a nonuniform thermal load (for example in aerospace and nuclear engineering).

During deformation caused by high-speed mechanical and thermal loads, the "coupling effect" appears, which depends on the interaction of the temperature and deformation fields. The first suggestion of this coupling effects was presented by Duhamel [70] in 1837. He introduced into the thermal-conductivity equations a term representing the change of volume of a body. Investigations during the next 100 years were performed mainly in the framework of the classical theory of elasticity and thermal conductivity. The theory of the thermal stresses was based on a simplified assumption that there is no influence of the deformation of an elastic body on the heat transfer process. Various aspects of these theories were presented in monographs by Karsloy and Eger [111], Lykov [150] and Kutateladze [138].

A comparison of methods suitable for solving linear coupled problems of thermoelasticity up to 1970 was presented by Nowacki [176–180].

Kozlov [128, 129] has considered the problem of thermoelastic oscillations of an isotropic rectangular plate subjected to a thermal shock. With some generalizing assumptions, a solution to this problem was obtained analytically using a Fourier series approach. The investigations indicated that damping of the oscillations of the plate occurred as a result of the coupling of the deformation and temperature fields. No additional simplifying hypotheses regarding the distribution of the temperature field through the thickness of the plate were used.

The majority of researchers have assumed a polynomial distribution of the temperature through the thickness of the plate or a shell to decrease the order of the thermoelasticity equations. Malkin [153] first introduced an infinite-series representation for the temperature. Marguerre [155] obtained equations governing the stationary thermoconductivity of a plate by averaging the temperature through the thickness of the plate. This work made the assumption that the temperature is linearly distributed through the thickness of the plate. Podstrigach [187] has extended this method to the case of a nonstationary thermal load for thin shells, without taking their curvature into account. The generalized case of a polynomial temperature distribution through the thickness of a shell was investigated by Borisyuk and Motovilovitz [53]. A linear temperature distribution through the thickness of the shell was also assumed by Shvetz and Lun' [202] in deriving the equations of coupled thermoelasticity of orthotropic shells in the framework of Timoshenko's theory. Shvetz [203] and other authors used the same assumption to obtain generalized variational principles of dynamic coupled problems for isotropic and anisotropic shells. Shvetz and Flachok [200] presented theorems about uniqueness, mutuality and variational principles. These authors derived those theorems for coupled linear equations for an anisotropic shell with a cubic distribution of the temperature through the thickness. The same authors presented solutions extended to problems of mechanical thermodiffusion, also with a cubic distribution of the temperature through the thickness of the shell. Different variants of a polynomial distribution of the temperature through the thickness of the shell were presented in the monograph [168] by Motovilovitz and Kozlov. One of the most promising methods is undoubtedly the operator method [188], which is extremely well applicable to a certain class of linear problems.

In the above-mentioned references, a complete, detailed description of the physical principles of the coupled linear problems of thermoelasticity for thin shell structures is given. Methods of solution and detailed bibliographies are also presented.

Vakhlaeva and Krys'ko [223] have presented solutions of a certain class of problems on the static loss of stability of a shallow shell in a stationary temperature field. Some problems in the theory of plates with boundary conditions not suitable for two-dimensional cases were solved. An article by Kirichenko et al. [116] presented results of calculations characterizing the influence of the coupling of the temperature and deformation fields on the value of the critical dynamic load for a clamped shallow shell.

In all works listed above concerning coupled thermoelasticity problems, all thermal transfer equations are treated as parabolic equations. This treatment was derived from an assumption of immediate heat transfer inside the body. Lord and Shulman [149] presented a new mathematical model of thermomechanics, including the velocity of heat transfer. In this new model, terms with thermal inertia were introduced into the heat transfer equations. This changed the type of these equations to hyperbolic. The thermomechanical theory presented in that work was called a generalized theory, in contrast to the classical theory, which did not take into account thermal inertia.

In the monograph by Podstrigach and Kolyano [186] there was presented a theory of generalized thermomechanics, and also some theoretical results and practical examples concerning the effects of a finite speed of heat transfer. In particular, during motion of a plane harmonic wave in an infinite thermoelastic body, an important influence of the heat transfer speed on the increase of the phase velocity was observed. Also, a detailed bibliography was presented. In paper by Shvetz and Lopat'ev [201], some characteristics of dynamic processes at higher frequencies based on linear, three-dimensional equations of a generalized thermodynamics were presented. In particular, an investigation of a distribution corresponding to a plane harmonic wave showed that, for an oscillation frequency lower than a certain frequency characteristic of the material, parabolic-type equations can be applied. Kolyano and Shter [120] applied the principle of conservation of energy to obtain coupled equations of generalized thermomechanics for anisotropic bodies. Kil'chinskaya [115] presented applications of the small-excitation Gauss method to cases of generalized thermomechanics.

Many theorems related to the existence of solutions in the nonlinear theory of shells have been proved by Destuynder [68], whereas in the case of shallow shells a serious mathematical treatment has been carried out by Bernadou and Oden [46]. We are not going to cite all references concerning rigorous mathematical approaches to problems of plates and shells, but the reader is encouraged to follow up the results included in [54, 60–62, 101, 144, 217, 227]. The general conclusions derived from the investigations of coupled problems of thermoelasticity are the following:

1. The quantitative effect of coupling depends on the type of material. For metals this effect is less important; for polymers the effect is relatively large.
2. Consideration of the coupling effect leads to qualitatively new properties in the distribution of shock waves (damping and dispersion).

Kupradze et al. [137], in their monograph, presented the mathematical principles of the theory of elasticity and thermoelasticity. They discussed in detail the problem of the existence and uniqueness of classical solutions (i.e. solutions that are continuously differentiable the necessary number of times) of linear problems of elasticity and of coupled thermoelasticity problems for three-dimensional bodies. They applied potential methods and the theory of multidimensional, singular integral equations. Fischer [77], in his papers, applied Laplace transformation to the generalized theory of strongly nonlinear elliptic systems and obtained results concerning the uniqueness of the solution of a linear coupled thermoelasticity problem, for the Sobolev class and also for continuously differentiable functions. Mikhaylovskaya and Novik [157] presented theorems concerning the solvability of the Cauchy problem and the stability of its periodic approximation for a system of differential–integral equations. As a special case, a system of linear coupled thermoelasticity equations was investigated. Some mathematical aspects of various problems of elasticity were also presented in papers by Vasil'kovskiy [226], Parton and Perlin [182].

Simultaneously with the development and investigation of mathematical models, the creation of approximate methods of solution for problems in the theory of plates and shells has been of great importance. Obtaining exact solutions requires great effort – and does not always give an adequate return on the investment. Exact solutions are not always possible in technological applications [185]. During the last 20–25 years, numerical computation has become one of the most important methods of solution of problems in the theory of plates and shells. In this review we include only a few of the publications concerning the principles and numerical realizations of the Bubnov–Galerkin method and the Ritz method, which is a special case of the Bubnov–Galerkin method.

It is known [141, 142] that investigations of any mechanical system may be made on the basis of the Hamilton–Ostrogradski stationary principle. In other words, for every mechanical system, a functional of a certain structure can be created in such a way that every point of the existence of the functional maps to a certain state of the system. Also, stationary points of the functional correspond to real, practical situations (states). This leads to the conclusion that it is possible to obtain the state of the system in two ways: either indirectly, by applying the method of variational calculation, searching for stationary points of the functional, or by applying methods of solution of partial differential equations, to search for stationary points as solutions of Euler's equations for the given functional. Numerical realizations of the first method correspond to projection methods, and those of the second one correspond to difference methods. Note that, for thermoelastic problems, different options for using the variational principle are presented by Nickell and Sackman [175] and by Kaczkowski [108] for three-dimensional bodies, and in the references mentioned earlier [200, 203] for thin-walled structures. More recent results can be found in the monographs [91, 92, 95, 102, 103, 107, 109, 118, 145, 154, 195, 211, 212, 220, 221, 225, 228].

The Ritz method and Bubnov–Galerkin method belong to the class of projection methods, where a unique functional, obtainable by the Ritz method, possesses extrema in the form of stationary points. The principle of the methods presented here can be described in the following form. The space of solutions of the investigated system is approximated by subspaces, and a search for a stationary point of functional is performed in these subspaces.

This is usually realized by a transformation of the original system of partial differential equations into a system of algebraic equations (in the case of static problems) or ordinary differential equations (for dynamic problems). Note that the Bubnov–Galerkin method is a projection method and can be applied for solving differential equations which are not Euler's equations for any functional (or if the corresponding functional is unknown). In this case the method relies on the projection of the original equations into a certain subspace of the functional space in which there exists an exact solution of the problem. An approximate solution can be obtained from the projected equations as an element of this subspace.

The first application of this method was presented by Ritz [190]. Further development of the Ritz method and the Bubnov–Galerkin method and of the application of those methods in mechanics was presented by Galerkin [80], Galimov [81] and others. Keldysh [113] presented a proof of the convergence of the Bubnov–Galerkin method for linear problems of ordinary differential equations of even order and for partial differential equations of elliptic type with lower-order elements. Mikhlin [161, 162] extended the results obtained by Keldysh to a generalized case of a linear operator equation containing a nonadjoint operator, which describes a wide class of boundary problems for elliptic equations. Krasnosel'skiy [130] presented general theorems about the convergence of the Bubnov–Galerkin method for nonlinear operator-type equations and about error estimation for this method. For this purpose, the author used the error of the best approximation to a sought element by a combination of coordinate elements.

In papers by Petryshyn [184] and Shalov [198], the construction of a squared functional, analogous to the functional of the Ritz method for adjoint problems, was presented for linear equations with a nonadjoint operator, as well as the convergence of the corresponding variational method was discussed. Nashed [173] extended the above-mentioned results to the case of a nonlinear equation with a nonpotential operator.

Unlike the Ritz method, the Bubnov–Galerkin method can be applied effectively to solving not only stationary problems (boundary-type problems for elliptic equations) but also to evolution-type problems (initial–boundary problems for parabolic and hyperbolic equations). It is generally agreed, that the first investigation of the use of the Bubnov–Galerkin method for nonstationary problems was performed by Faedo [75] in 1949; for this reason, the Bubnov–Galerkin method is also known as the Faedo–Galerkin method. In 1931 Krylov and Bogolyubov [131] applied what was effectively the Bubnov–Galerkin method to an investigation of the Cauchy–Dirichlet problem for hyperbolic-type equations of the second order. The way in which that method was applied was quite similar to the currently used method.

A review and comparison of the development of the Ritz method and Bubnov–Galerkin method and their application to nonlinear problems in the theory of plates and shells was presented in a monograph by Volmir [231] in 1956, where methods using the first- and second-order approximation were presented. Further development of computational methods allowed the use of higher-order approximations of the Bubnov–Galerkin method. Results of the application of the Ritz method and Bubnov–Galerkin method to stability problems in the theory of shells were presented in a monograph [94] by Grigolyuk and Kabanov. These authors pointed out that it is necessary to describe the stability loss mode with great nonhomogeneity, either by taking into account a large number of terms in a series or by choosing an appropriate number of basis functions after making a prior estimate.

A detailed description of the use of the Bubnov–Galerkin method in dynamical, geometrically nonlinear problems in the theory of plates and shells was presented in the monograph by Volmir [232]. This monograph also contains a large bibliography presenting the history of the solutions up to 1972. In [234] there were presented some results of work on the dynamic stability of structures interacting with fluids and gases.

Kirichenko et al. [116] applied the Bubnov–Galerkin method to the solution of the problem of the dynamic stability of a shallow shell clamped on the boundary, both with and without taking into account the coupling of the temperature and deformation fields. Various aspects of the numerical application of the Bubnov–Galerkin method were presented in papers by the following authors: Svirskiy [215], Bacinov [40], Mukhopandhayay [170], Chen and Hwang [57], and Gelos et al. [84]. In a monograph by Krys'ko and Kutsemako [135], the application of the Bubnov–Galerkin method to the problem of the stability and oscillations of nonhomogeneous shallow shells was presented.

The Bubnov–Galerkin method initially developed as a numerical method, is currently widely used for solving a large class of problems in mathematical physics; stationary and evolutionary as well as linear and nonlinear. The development of powerful mathematical schemes for proving the existence of solutions obtained by the Bubnov–Galerkin method, based on fundamental theorems of functional analysis, became possible after Sobolev's [209] research on the generalized differential rational function of the Lebesgue class and the formulation of principles for obtaining generalized solutions of problems of mathematical physics. Research on the convergence of the Ritz method and Bubnov–Galerkin method for linear problems, including the theory of plates and shells, and proof of the solvability of particular problems in a Sobolev space was published by Ladyzhenskaya [140]. The mathematical principles of various approximation methods, including the Bubnov–Galerkin method, were summarized in a monograph by Krasnosel'skiy et al. [151] and by Lions [148]. A detailed review of research on the convergence of Bubnov–Galerkin method for nonlinear problems was presented in a monograph by Lions [147], where a large bibliography is presented. Research on the solvability of nonlinear problems in the theory of plates and shells was presented by Vorovich [235]. Fundamental results in this area were obtained by Morozov [164, 165] in publications concerning the solvability of nonlinear problems in the theory of thin plates. Morozov stated (applying Sobolev's theorems of inclusion) the existence and uniqueness of the solution. Some aspects of the solvability of various problems in the mechanics of thin-walled structures by means of the Bubnov–Galerkin method were presented by Vorovich and Lebedev [236], Skrypnik [204] and others. The detailed background of the application of the Bubnov–Galerkin method and Ritz method to static problems in the nonlinear theory of shallow shells was presented by Vorovich [235].

Theorems about the existence, uniqueness, regularity and continuous dependence of the solution on the data have been formulated for solutions obtained by the Bubnov–Galerkin method. Chrzeszczyk [58, 59] presented conclusions about the uniqueness and smoothness of the solution to a problem considered also in [117]. Kowalski and Piskorek [125], Gawinecki [82,83], and Kowalski et al. [124,125], by the application of the Bubnov–Galerkin method presented theorems about the existence, uniqueness and regularity of the solutions to problems in the linear, three-dimensional theory of thermal stresses and coupled thermoelasticity for both isotropic and anisotropic bodies, with hyperbolic and parabolic types of heat transfer equation.

In the majority of the papers listed above, the convergence of the Ritz method and Bubnov–Galerkin method appears as an extra result of proving the existence of a solution of the problem under investigation. The standard solution (for an arbitrarily selected basis) appears to show convergence in the energy norm for static problems, and weak convergence in the corresponding "energy" spaces for dynamic problems. For properly selected basis systems the results concerning convergence can be stronger and, besides, effective estimates of the convergence speed can be determined. Important results concerning this problem were presented by Mikhlin [163], in which convergence operators were introduced. Here, the deviation of the Ritz method converged to zero if a basis system of eigenelements of an auxiliary operator was introduced that converged to the equation under investigation. In [159] a more exact result was presented, where the main and auxiliary operators satisfied an extra condition of acute-angle form. Fundamental results concerning converging operators satisfying an acute-angle inequality were presented by Sobolevskiy [210] and Ladyzhenskaya [139]. On the basis of these results Mikhlin [159] presented for a series of examples special bases that ensured convergence to null of the deviation of the Ritz method. Bogaryan [49] extended Mikhlin's results concerning the Ritz method to the Bubnov–Galerkin method. Dzhishkaryani [71] obtained prior estimates of the error of the Ritz method in an energy norm, where eigenvalues of an auxiliary operator converging to the original equation were applied. A similar problem was investigated by Vaynikko [222] and Dzhishkaryani [72]. These authors obtained similar results for the Bubnov–Galerkin method. Vaynikko obtained error estimates for the Bubnov–Galerkin method for stationary problems in cases where the spectrum of the auxiliary operator did not appear to be exact. Dzhishkaryani extended the error estimates for the Ritz method to a case of a quasi-linear equation containing a nonlinear operator, as lower terms of a continuous, potential operator having a positive differential. Zarubin [243] presented quite generalized estimates of the convergence speed of the Galerkin–Petrov method for linear and quasi-linear stationary problems. From these results, error estimates can be derived for the Bubnov–Galerkin method, taking as a basis eigenelements of a self-adjoint operator that converges to an operator of the main part of the original system and projects onto it at an acute angle. A detailed review of work on the theory of the errors of numerical methods, including the Bubnov–Galerkin method for stationary problems, is presented in the monograph by Mikhlin [160].

The main publication that dealt with the convergence of the Bubnov–Galerkin method for evolutionary problems was a work by Sobolevskiy [210]. This publication presented theorems about the strong convergence of approximate solutions to the exact solution and about the convergence of errors to zero with a quasi-linear parabolic equation for an arbitrarily selected basis. Zarubin and Tiunchik [245] applied the Bubnov–Galerkin method to a parabolic equation with a nonlinear operator of small-element type, corresponding to a dissipation condition introduced in that publication. As a basis, a system of eigenelements of an auxiliary operator satisfying the acute-angle inequality was used. In that work, results were presented that showed convergence to zero of the deviation of the Bubnov–Galerkin method and a strong convergence of the series approximating the exact solution. Currently there are no publications concerning investigations of the convergence speed of the Bubnov–Galerkin method for hyperbolic equations or for coupled systems similar to thermoelasticity equations.

Note there are a very large number of books, papers and commercial software pacages devoted to other numerical methods for the analysis of plates and shells (see, for instance, [99, 102, 216]).

Since, the finite-element method is one of the most popular of those methods especially in the field of engineering, a treatment of this approach is omitted in this book; descriptions can be found elsewhere ([208, 239]).

An analysis of the current research on the theory of plates and shells and of the convergence speed of the Bubnov–Galerkin method leads to the following conclusions:

1. The Bubnov–Galerkin method appears to be one of the most effective and widely applied numerical methods for solving static and dynamic problems in the theory of plates and shells. Current computational techniques and computer power allow us to apply the Bubnov–Galerkin method with a higher level of accuracy, i.e. to take into account a greater number of series elements in the approximate solution.
2. The main problem related to numerical realization of the Bubnov–Galerkin method for dynamic problems is the lack of a relatively simple estimate of the errors (currently, the reliability of numerical methods is verified experimentally) and, in many cases, the lack of a mathematical background for the proper choice of basis functions. These problems arise not only for dynamic problems in the theory of elasticity and coupled thermoelasticity, but also for a large class of evolutionary problems containing the second derivative with respect to time of a sought function.

The rapid development of computational methods and applied mathematics has allowed researchers to apply more reliable mathematical models. It has forced us to take into account geometric nonlinearities, elasto-plastic deformations and periodic loads when solving dynamic problems. The approaches presented by Il'yushin [104] and Birger [48] are widely applied.

The research on the dynamic stability of plates and shells is based on fundamental work by Bolotin [51, 52]. Various models for describing the behaviour of a structure have been applied to solve nonlinear problems in the dynamics of plates and shells. Historically, one of the first approaches was the application of a model of a rigid–plastic body [97]. For large plastic deformation, this model allows us to obtain sufficiently accurate solutions. When the plastic deformation is much smaller than the elastic deformation, however, this method is not suitable.

Various approaches can be applied to solving problems of cyclic deformation with a small number of cycles. One such approach is based on differential relations between stresses and strains, presented by Birger [48] and Krys'ko [133].

For solving certain problems, finite relations between stresses and strains are widely applied. Although the spectrum of solvable problems is limited to those with loads that are simple or close to simple, such an approach allows a numerical verification.

Results of research on circular plates and cylindrical shells subjected to dynamic loads, where the material elasto-plastic characteristics, were presented by Filippov [78].

Shallow shells were investigated theoretically and experimentally in [56]. The failure of a shell after elasto-plastic deformation was investigated. The boundary problem for a shell subjected to an arbitrary pressure load, with static nonhomogeneous boundary conditions of generalized form was solved by application of an orthogonal-projection Godunov method [86, 87].

The dynamic behaviour of an elasto-plastic plate clamped at an internal edge and not supported at the outside edge was investigated by Stepanov and Kovalenko [214]. An approximate solution for estimating the elasto-plastic deflection of a plate subjected to a short-time impact pressure was presented, with application of an iterative method. Results of theoretical and experimental investigations of the dynamic deflection of a round plate subjected to a uniformly distributed impact pressure load were also presented.

There is a large number of works concerning analysis of the elasto-plastic (physically nonlinear) strain of rectangular plates and of shallow shells that are rectangular in plan.

Krys'ko [133], in his monograph, presented a wide range of problems in the theory of nonhomogeneous shallow shells taking transverse shear into account. The convergence of the Bubnov, Vlasov–Kantorovich and finite-difference variational methods was numerically investigated for static and dynamic problems. The author analysed the influence of different types of transverse impact loads on the stability of shells and also the influence of the deflection on the critical load.

Bauer [41] investigated bending oscillations of simply supported and clamped plates subjected to impact loads of step-like and exponentially decreasing type. A possibility of separation of the spatial and time-dependent elements of the solution was presented.

The reaction of metal plates to an impact load modelling an underwater explosion was presented in [181]. The dynamic problem was solved in Lagrange coordinates by means of a finite-difference method for a two-dimensional axisymmetric state, with an elasto-plastic approximation. Different forms of pressure impulse, different impulse durations, different distributions of the pressure on the plate surface and variable boundary conditions were investigated, and their influence on the deformation process was analysed.

The loss of stability in bifurcation form of a rectangular plate under combined loads was considered in [74]. The problem was solved by representing the equations of the deformation theory in an incremental form, which allowed the authors to include the load history without investigating local loads. The finite-element method (FEM) was applied to cases of plates subjected to transverse and longitudinal loads.

For solving problems in the dynamics of plates and shells, it is necessary not only to know the material characteristics but also, in many cases, to take into account the deformation speed, acceleration and other parameters characterizing the dynamics of the process. These elements have been taken into account in numerical solutions of various problems in the theory of plates and of spherical and cylindrical shells [114, 143, 241, 242].

An analysis of the current research on the dynamic stability of geometrically and physically nonlinear plates and shells, under nonstationary loads, leads to the following conclusions:

1. Research on the dynamic elasto-plastic deformation of plates and shells with geometric nonlinearity has been presented in a large number of publications. However, the dynamic loss of stability of slender elasto-plastic plates and shells that are rectangular in plan, under impact loads of arbitrary character, is poorly represented.
2. The dynamic behaviour of elasto-plastic structures has mainly been investigated in axisymmetric cases . The solutions to such cases are very complex, i.e. the deformation process has to be represented as a function of time for all points of the structure; therefore an analysis of every point in the deformation chart must be performed.
3. In practically all publications, the following assumption is made: the collapse (failure) of a structure takes place when any characteristic point of the structure fails. Bi-linear charts with linear or incremental reinforcement were used in most of the calculations. There were no comparisons of different deformation charts.

Along with problems of coupled thermoelasticity, other cases, including other physical–chemical effects, magnetic fields and transonic gas flows, have been widely investigated.

In a series of works Day [66, 67] presented some qualitative and quantitative aspects of linear coupled thermoelasticity. The author investigated the behaviour of a beam under a thermal load and on one side investigated the conditions, for certain types of problems, under which an uncoupled thermoelasticity theory can be applied. The results presented the possibility of a negative temperature increase for coupled thermoelasticity under boundary conditions that excludedg such a phenomenon in classical thermodynamics.

Of the various methods applied for numerical solution of coupled problems, the finite-difference method (FDM) and the FEM were used extremely often.

The FEM was first applied to coupled problems by Nickell and Sackman [175]. The application of the FEM was analysed in details by Gribanov and Panichkin [93].

Analysis of the scientific literature shows that solutions to certain problems with coupling of the temperature and deformation fields had been obtained for linear models only. In a publication by Krys'ko and Mishnik [136], three-dimensional problems in the theory of plates with physical nonlinearities were presented. For the discretization of the spatial coordinates, a finite-difference method with an $O(h^2)$ approximation was applied, and for integration with respect to time, the fourth-order Runge–Kutta method was applied. Numerical results confirmed the effect of coupling on the stress–strain state in nonlinear plate problems.

An analysis of work on current problems in dynamic thermoelasticity with two types of nonlinearity and coupling of the temperature and deformation fields leads to the following conclusions:

1. Only a small number of coupled problems have been solved with nonlinear models. There are no publications concerning the analysis of the dynamic stability of slender shells with coupling of the temperature and deformation fields. The problems of the complex oscillations of such structures have not been solved.
2. Numerical solution of doubly nonlinear problems of coupled thermoelasticity with a three-dimensional heat transfer equation is very complicated, but it allows one to apply different hypotheses for the temperature distribution through the thickness of the shell.

Problems of the coupled interaction of a shell-type structure with a transonic flow are a special case and have a special complexity. When a gas flows with transonic velocity around a profile, some particular phenomena appear in the flow. In such a flow shock waves, which are pressure jumps, appear, move and disappear. Also, the velocity of movementof an excitation caused by a deformation of the profile is close to the speed of the gas particles. These phenomena appear when a gas flow interacts with a profile and, vice versa, when a profile interacts with a gas flow.

An important scientific goal is the application of mathematical approaches to transonic-flow problems. As stated by Godunov et al. [87], an aircraft can cover the greatest distance at a transonic speed. Also, an aircraft flying at a transonic speed has the best ability to manoeuvre while still preserving high speed, i.e. the relation of manoeuvrability to speed appears to be optimal.

Despite the theoretical and practical aspects, problems of the interaction of a shell and a transonic flow have not received proper attention in the scientific literature. An investigation of travelling pressure jumps, pulsing with a given frequency, was presented by Croker [64] in a publication about a vertical-take-off aircraft. An analysis of such shock waves acting on an aircraft, performed for linear cases, led to the conclusion that these shock waves appear to be the main reason for the relatively high increase of shell vibrations when an aircraft passes through the transonic speed zone.

Volmir [232] and Skurlatov [205,206] presented pictures from a high-speed camera showing the failure of a steel shell as a result of interaction with a transonic flow with heat transfer.

Analytic research on cylindrical shells in a wind tunnel was presented by Skurlatov and Solomenko [207]. These authors stated that during the interaction, the incoming flow changes from subsonic to supersonic. Two monographs [7, 133] presented a solution obtained by the Bubnov–Galerkin method to a nonlinear problem of forced vibrations of a shallow, simply supported cylindrical panel. The Kirchhoff–Love model was applied.

Investigations of the longitudinal flow of a transonic gas stream around rigid and elastic profiles of conical shape and around a cylindrical panel were performed. The model and the experimental results showed oscillating movements of pressure jumps, i.e. shock waves and their zone of motion was described. Experiments were performed for incoming flows of Mach number M_∞ varying from 0.8 to 1.2 in steps of 0.1. Shock waves were observed visually with a shadow device.

Some publications [232, 234, 247] have presented some properties of the interaction of a transonic flow with the profiles or shells obtained experimentally. These publications showed that during transonic flow, dynamic parameters reach high values and that pulsing shock waves lead to the appearance of grooves in structures. Measurements of the pressure on the surface of a rigid profile showed that the amplitude of oscillations rapidly increased when the Mach number of the incoming stream M_∞ reached 1. Further increase of M_∞ caused a strong decrease of the oscillation amplitude to a level characteristic of the noise of a turbulent boundary layer.

The number of problems investigated in which a shell passes through a transonic-flow region has increased rapidly. The pressure amplitudes and shell deflections reach very high values.

Some other problems of aeroelasticity of plates and shells have been considered in a monograph by Dowell [69].

Analysis of the current literature shows that at this moment there is no theory of the dynamic interaction of a shell with a transonic gas flow. Further development of solution methods for nonstationary transonic flows and their interaction with slender shells would allow better research on this problem.

The easiest problem of dynamic interaction of a shell with a transonic flow is the case of the two-dimensional interaction of a transonic flow with an infinitely long cylindrical panel. The solution for a shallow shell in this case allows us to investigate transonic flow in an asymptotic nonlinear theory when the flow is close to a homogeneous potential flow.

Section 2.2 of this monograph is devoted to current problems of aeroelasticity and presents a method for the numerical solution of the above problem, together with research on the dynamic stability of a cylindrical panel interacting with a transonic flow of an ideal gas. The characteristics of the interaction of a transonic flow with a shallow elastic shell are deduced.

Many problems of the vibration of plates and shells have been solved by the authors of this book in a series of monographs [5–9], and in papers [11–19, 132] and reviewed conference proceedings [20–36].

In this book we address problems related to the theory of plates and shells from the point of view of mathematical and numerical investigations only. This means that all problems related to the very efficient and effective asymptotic approaches are omitted here (perhaps with the exception of Sect. 2.2). However, the reader is encouraged to follow up the recent results in this field presented in the monographs [37–39] and in the review paper [2] and the references there in.

2 Coupled Thermoelasticity and Transonic Gas Flow

This chapter includes considerations of the coupled linear thermoelasticity of shallow shells and of a cylindrical panel within a transonic gas flow. First, the fundamental assumptions related to the stress–strain relation of the Timoshenko kinematic model are formulated, and then the differential equations are derived. Both the Timoshenko and the Kirchhoff–Love models are taken into account. Then the boundary and initial conditions are formulated. Next, an abstract Cauchy problem for a coupled system of two differential equations in a Hilbert space is considered. This includes the thermoelastic problems of shallow shells modelled by the Kirchhoff–Love and Timoshenko theories defined earlier. In Sect. 2.1.5, theorems related to the existence and uniqueness of a general, “classical” solution to the coupled abstract program are given, and then the corresponding theorems for coupled thermoelastic problems of shallow shells are formulated.

The second part of this chapter is focused on the dynamics of an elastic, infinite cylindrical panel within a transonic gas flow.

First, the equations of motion and boundary conditions of the panel are formulated. It is shown that the problem of the oscillations of a flexible, elastic, infinite, circular cylinder panel subjected to a transverse load can be reduced to an initial-boudary-value problem of mathematical physics governed by an integral–differential equation, which is derived here.

Further, the equations of the transonic motion of an ideal gas are formulated. Next, the equations and boundary conditions of the panel–flow interaction are discussed and illustrated. Free and forced panel oscillations are analysed using the finite-difference method. A new, modified Godunov method for calculation of a transonic flow is presented. An algorithm for calculation of the interaction between a flow and a panel are given, with results of calculations. A stable vibrating panel in a transonic flow including shock wave motion, and the interaction of a flow and a panel are discussed in some detail. The last subsection of this chapter is focused on the stability loss of a panel within a transonic flow. This part includes stability criteria, as well as an analysis of the stress–strain state of a panel and its plastic deformation.

2.1 Coupled Linear Thermoelasticity of Shallow Shells

2.1.1 Fundamental Assumptions

A shallow shell with variable thickness $h(x, y)$, covering a plane surface Ω_1 and occupying a three-dimensional spatial volume Ω_2, is considered. We assume the shell material to be isotropic, homogeneous and elastic. We introduce in the usual way the coordinate system x, y, z [232]. The coordinates x, y coincide with the directions of the principal curvatures of the middle surface, whereas z has a normal direction towards the centre of the shell curvature.

The components of the displacement vector of a point (x, y) of the middle surface of the shell at time t are denoted by $u(x, y, t)$, $v(x, y, t)$, $w(x, y, t)$. The angles of rotation of the normal to the middle surface in the xz and yz planes are denoted by $\Psi_x(x, y, t)$ and $\Psi_y(x, y, t)$, respectively. The initial shell curvatures related to the coordinates x and y are denoted by k_x and k_y, respectively.

In the analysis of the stress–strain state, we use the Timoshenko kinematic model including inertial rotation [7, 232]. In accordance with this model, the shell fibres that are normal to the middle surface before deformation remain linear but are no longer normal to the middle surface. The deformations related to the fibres normal to the average surface are defined by a plane stress state. For the shell under consideration, the following relations between deformations and displacements are satisfied on the mean shell surface [232]:

$$\varepsilon_{11}^z = \varepsilon_{11} + z æ_{11}, \quad \varepsilon_{22}^z = \varepsilon_{22} + z æ_{22}, \quad \varepsilon_{12}^z = \varepsilon_{12} + z æ_{12}, \tag{2.1}$$

$$\varepsilon_{13}^z = \varepsilon_{13}, \quad \varepsilon_{23}^z = \varepsilon_{23}, \tag{2.2}$$

where $|z| \leq \frac{1}{2} h(x, y)$; $\varepsilon_{ij} (i, j = 1, 2)$ are the tangential deformations of the middle surface; ε_{13}, ε_{23} are the shear deformations; and $æ_{ij} (i, j = 1, 2)$ are the bending deformations. The following relations hold:

$$\varepsilon_{11} = \frac{\partial u}{\partial x} - k_x w, \quad \varepsilon_{22} = \frac{\partial v}{\partial y} - k_y w, \quad \varepsilon_{12} = \frac{\partial u}{\partial y} + \frac{\partial v}{\partial y}, \tag{2.3}$$

$$æ_{11} = \frac{\partial \Psi_x}{\partial x}, \quad æ_{22} = \frac{\partial \Psi_y}{\partial y}, \quad æ_{12} = \frac{\partial \Psi_x}{\partial y} + \frac{\partial \Psi_y}{\partial x}, \tag{2.4}$$

$$\varepsilon_{13} = \Psi_x + \frac{\partial w}{\partial x}, \quad \varepsilon_{23} = \Psi_y + \frac{\partial w}{\partial y}. \tag{2.5}$$

It should be noted that the deformations ε_{11}, ε_{22}, ε_{12}, ε_{13}, ε_{23} are small in comparison with one, and that the products of deformations can be neglected.

Suppose that in the nondeformed and nonstressed state the shell has a temperature T_0. The temperature of the shell starts to change owing to the surface load and inertial forces, as well as owing to internal heat sources and heat exchange with the surrounding medium. By $\theta(x, y, z, t)$, we denote the temperature increment at the point (x, y, z) at a time instant t. We assume that $|\theta/T_0| \ll 1$, which leads to the conclusion that the temperature change does not vary the elastic and thermodynamic properties of the shell material. The following isothermal constants are used in what follows: E, Young's modulus; ν, Poisson's ratio; α_T, coefficient of linear thermal expansion; ϱ, density; λ_q, coefficient of thermal conductivity; and c_ε, specific thermal capacity for a constant deformation tensor . Assuming that the shell is in a local quasi-equilibrium condition [230], using the relations of noninvertible dynamic processes and taking into account the small values of the deformations and temperature increase, the following Duhamel–Neumann relations are obtained:

$$\varepsilon_{11}^z = \frac{1}{E}\sigma_{13} - \frac{\nu}{E}\sigma_{22} + \alpha_T\theta, \quad \varepsilon_{22}^z = \frac{1}{E}\sigma_{22} - \frac{\nu}{E}\sigma_{11} + \alpha_T\theta, \tag{2.6}$$

$$\varepsilon_{33}^z = -\frac{\nu}{E}(\sigma_{11} + \sigma_{22}) + \alpha_T\theta, \tag{2.7}$$

$$\varepsilon_{12}^z = \frac{2(1+\nu)}{E}\sigma_{12}, \quad \varepsilon_{13}^z = \frac{2(1+\nu)}{E}\sigma_{13}, \quad \varepsilon_{23}^z = \frac{2(1+\nu)}{E}\sigma_{23}. \tag{2.8}$$

These describe the relations between the components of the stress tensor σ_{ij} and the deformation tensor ε_{ij}^z in a plane stress state ($\sigma_{33} = 0$). Solving the system of equations (2.6)–(2.8) for σ_{ij}, we obtain:

$$\sigma_{11} = \frac{E}{1-\nu^2}(\varepsilon_{11}^z + \nu\varepsilon_{22}^z) - \frac{E}{1-\nu}\alpha_T\theta, \qquad (1 \leftrightarrow 2), \tag{2.9}$$

$$\sigma_{12} = \frac{E}{2(1+\nu)}\varepsilon_{12}^z, \quad \sigma_{13} = \frac{E}{2(1+\nu)}\varepsilon_{13}^z, \quad \sigma_{23} = \frac{E}{2(1+\nu)}\varepsilon_{23}^z. \tag{2.10}$$

Integrating the stresses (2.9), (2.10) with respect to z and taking into account (2.1)–(2.5), we find the corresponding normal and transverse forces in the middle plane:

$$N_x = \int_{-\frac{h}{2}}^{\frac{h}{2}} \sigma_{11}\,dz = \frac{Eh}{1-\nu^2}\left[\frac{\partial u}{\partial x} - k_x w + \nu\left(\frac{\partial v}{\partial y} - k_y w\right)\right] - \frac{E\alpha_T}{1-\nu}\int_{-\frac{h}{2}}^{\frac{h}{2}} \theta\,dz$$
$$(x \leftrightarrow y, \quad u \leftrightarrow v, \quad 1 \leftrightarrow 2), \tag{2.11}$$

$$S = \int_{-\frac{1}{2}}^{\frac{1}{2}} \sigma_{12}\,dz = \frac{Eh}{2(1+\nu)}\left(\frac{\partial u}{\partial y} + \frac{\partial v}{\partial x}\right), \tag{2.12}$$

$$Q_x = k^2\int_{-\frac{1}{2}}^{\frac{1}{2}} \sigma_{32}\,dz = \frac{k^2 Eh}{2(1+\nu)}\left(\Psi_x + \frac{\partial v}{\partial x}\right), \qquad (x \leftrightarrow y, \quad 1 \leftrightarrow 2), \tag{2.13}$$

where

$$\frac{1}{k^2} = \frac{1}{h}\int\limits_{-\frac{h}{2}}^{\frac{h}{2}} f^2(\frac{z}{h})\,dz = \int\limits_{-\frac{h}{2}}^{\frac{h}{2}} f^2(\xi)\,d\xi,$$

and $f(\frac{z}{h})$ is a function characterizing the distribution of tangential stresses through the thickness of the shell [232]. The moments are obtained after integration of the stresses σ_{11}, σ_{22}, σ_{12}, taking account of (2.1)–(2.4), over the shell thickness:

$$M_x = \int\limits_{-\frac{1}{2}}^{\frac{1}{2}} \sigma_{11} z\,dz = \frac{Eh^3}{12(1-\nu^2)}\left(\frac{\partial \Psi_x}{\partial x} + \nu\frac{\partial \Psi_y}{\partial y}\right) - \frac{E\alpha_T}{1-\nu}\int\limits_{-\frac{1}{2}}^{\frac{1}{2}} \theta z\,dz, \tag{2.14}$$

$$(x \leftrightarrow y, \quad 1 \leftrightarrow 2),$$

$$H = \int\limits_{-\frac{1}{2}}^{\frac{1}{2}} \sigma_{12} z\,dz = \frac{Eh^3}{24(1+\nu)}\left(\frac{\partial \Psi_x}{\partial y} + \nu\frac{\partial \Psi_y}{\partial x}\right). \tag{2.15}$$

2.1.2 Differential Equations

Taking into account the entropy balance and the Fourier relation between the components of the heat flow vector and the thermodynamic forces [179], the following heat transfer equation is obtained:

$$c_\varepsilon \frac{\partial \theta}{\partial t} - \lambda_q \nabla^2 \theta = -\underline{\frac{E\alpha_T T_0}{1-2\nu}\frac{\partial}{\partial t}(\varepsilon_{11}^z + \varepsilon_{22}^z + \varepsilon_{33}^z)} + q_2. \tag{2.16}$$

The underlined term links the temperature increase to the speed of the volume change. In this equation, $q_2(x, y, z, t)$ denotes the specific power of the heat sources located within the volume occupied by a shell.

In order to obtain an equation governing shell vibrations in terms of displacements, the following system of equations is used:

$$\frac{\partial N_x}{\partial x} + \frac{\partial S}{\partial y} + p_1 - \rho h\frac{\partial^2 u}{\partial t^2} = 0 \quad (x \leftrightarrow y, u \leftrightarrow v, 1 \leftrightarrow 2), \tag{2.17}$$

$$\frac{\partial Q_x}{\partial x} + \frac{\partial Q_y}{\partial y} + k_x N_x + k_y N_y + q_1 - \rho h\frac{\partial^2 w}{\partial t^2} = 0, \tag{2.18}$$

$$\frac{\partial M_x}{\partial x} + \frac{\partial H}{\partial y} - Q_x - \underline{\rho\frac{h^3}{12}\frac{\partial^2 \Psi_x}{\partial t^2}} = 0 \quad (x \leftrightarrow y). \tag{2.19}$$

This describes the motion of a shell element, including inertial effects associated with motion along the x, y, z axes as well as the inertia of its rotation about x, y [52] (p_1, p_2 and q_1 correspond to the intensity of the external load along the axes x, y and z, respectively).

Substituting (2.11)–(2.15) into (2.17)–(2.19) and attaching to the system of equations obtained the heat transfer equation (2.16), transformed by use of (2.1), (2.3), (2.4), (2.7) and (2.9), the following full system of equations governing the thermoelastic thermoelasticity behaviour of a shell in terms of to displacements is obtained:

$$\rho h \frac{\partial^2 u}{\partial t^2} - \frac{E}{1-\nu^2}\frac{\partial}{\partial x}\left\{ h\left[\frac{\partial u}{\partial x} - k_x w + \nu\left(\frac{\partial v}{\partial y} - k_y w\right)\right]\right\}$$
$$- \frac{E}{2(1+\nu)}\frac{\partial}{\partial y}\left[h\left(\frac{\partial u}{\partial y} + \frac{\partial v}{\partial x}\right)\right] + \frac{E\alpha_T}{1-\nu}\frac{\partial}{\partial x}\int_{-\frac{h}{2}}^{\frac{h}{2}} \theta\, dz = p_1 \tag{2.20}$$

$$(x \leftrightarrow y, \quad u \leftrightarrow v, \quad p_1 \leftrightarrow p_2),$$

$$\rho h \frac{\partial^2 w}{\partial t^2} - \frac{k^2 E}{2(1+\nu)}\left\{\frac{\partial}{\partial x}\left[h\left(\Psi_x + \frac{\partial w}{\partial x}\right)\right] + \frac{\partial}{\partial y}\left[h\left(\Psi_y + \frac{\partial w}{\partial y}\right)\right]\right\}$$
$$- \frac{E}{1-\nu^2}\left[(k_x + \nu k_y)\left(\frac{\partial u}{\partial x} - k_x w\right) + (k_y + \nu k_x)\left(\frac{\partial v}{\partial y} - k_y w\right)\right] \tag{2.21}$$
$$+ \frac{E\alpha_T}{1-\nu}\frac{\partial}{\partial x}(k_x + k_y)\int_{-\frac{h}{2}}^{\frac{h}{2}} \theta\, dz = q_1,$$

$$\rho\frac{h^3}{12}\frac{\partial^2 \Psi_x}{\partial t^2} - \frac{\partial}{\partial x}\left[D\left(\frac{\partial \Psi_x}{\partial x} + \nu\frac{\partial \Psi_y}{\partial y}\right)\right]$$
$$- \frac{1-\nu}{2}\frac{\partial}{\partial y}\left[D\left(\frac{\partial \Psi_x}{\partial y} + \frac{\partial \Psi_y}{\partial x}\right)\right] - \frac{k^2 E h}{2(1+\nu)}\left(\Psi_x + \frac{\partial w}{\partial x}\right) \tag{2.22}$$
$$+ \frac{E\alpha_T}{1-\nu}\frac{\partial}{\partial x}\int_{-\frac{h}{2}}^{\frac{h}{2}} \theta\, dz = 0 \qquad (x \leftrightarrow y),$$

$$c_\varepsilon(1+\varepsilon)\frac{\partial \theta}{\partial t} - \lambda_q \nabla^2 \theta + \frac{E\alpha_T T_0}{1-\nu}\frac{\partial}{\partial t}\left[\frac{\partial u}{\partial x} + \frac{\partial v}{\partial y}\right.$$
$$\left. - (k_x + k_y)w + z\left(\frac{\partial \Psi_x}{\partial x} + \frac{\partial \Psi_y}{\partial y}\right)\right] = q_2, \tag{2.23}$$

$$D = \frac{Eh^3}{12(1-\nu^2)}, \qquad \varepsilon = \frac{E\alpha_T^2 T_0(1+\nu)}{c_\varepsilon(1-\nu)(1-2\nu)}. \tag{2.24}$$

The system of differential equations (2.20)–(2.23) is referred to as the full system of equations of the coupled dynamic problem of linear thermoelasticity. It should be emphasized that we have not introduced any limitations on the temperature distribution through the shell thickness, which could lead to a system of equations with different dimensions. The function θ in the parabolic-type heat transfer equation depends on the three spatial variables x, y, z and the time t. The functions u, v, w, Ψ_x, Ψ_y appearing in the hyperbolic-type equation governing motion of a shell element motion depend on the two spatial coordinates x, y and the time t.

The system of equations obtained above governing the thermoelastic behaviour of a shallow shell in the framework of the Timoshenko model can be transformed to a system of equations in the framework of the Kirchhoff–Love model. To this end, we neglect the effect of transverse shear deformations. The rotation angles of the normal to the middle surface have the following form [232]:

$$\Psi_x = -\frac{\partial w}{\partial x}, \qquad \Psi_y = -\frac{\partial w}{\partial y}. \tag{2.25}$$

In addition, rotational inertia is neglected, which leads to omission of the underlined term in (2.19). This is introduced simplification leads to the following equations:

$$M_x = -D\left(\frac{\partial^2 w}{\partial x^2} + \nu\frac{\partial^2 w}{\partial y^2}\right) - \frac{E\alpha_T}{1-\nu}\int\limits_{-\frac{h}{2}}^{\frac{h}{2}} \theta z\, dz = 0 \qquad (x \leftrightarrow y), \tag{2.26}$$

$$H = -D(1-\nu)\frac{\partial^2 w}{\partial x\, \partial y}, \tag{2.27}$$

$$\theta_x = \frac{\partial M_x}{\partial x} + \frac{\partial H}{\partial y} \qquad (x \leftrightarrow y). \tag{2.28}$$

Substituting (2.28) into (2.18) and attaching (2.17), the following equations governing the motion of a shell element are obtained:

$$\begin{gathered}\frac{\partial N_x}{\partial x} + \frac{\partial S}{\partial y} + p_1 - \rho h\frac{\partial^2 u}{\partial t^2} = 0 \quad (x \leftrightarrow y,\ u \leftrightarrow v,\ p_1 \leftrightarrow p_2),\\ \frac{\partial^2 M_x}{\partial x^2} + 2\frac{\partial^2 H}{\partial x\, \partial y} + \frac{\partial^2 M_y}{\partial y^2} + k_x N_x + k_y N_y + q_1 - \rho h\frac{\partial^2 w}{\partial t^2} = 0.\end{gathered} \tag{2.29}$$

Next, the relations (2.11), (2.12), (2.26) and (2.27) are substituted into (2.29) and, additionally, we attach the general heat transfer equation (2.23) modified by (2.25). As a result, we obtain

$$\begin{gathered}\rho h\frac{\partial^2 u}{\partial t^2} - \frac{E}{1-\nu^2}\frac{\partial}{\partial x}\left\{h\left[\frac{\partial u}{\partial x} - k_x w + \nu\left(\frac{\partial v}{\partial y} - k_y w\right)\right]\right\}\\ -\frac{E}{2(1+\nu)}\frac{\partial}{\partial y}\left[h\left(\frac{\partial u}{\partial y} + \frac{\partial v}{\partial x}\right)\right] + \frac{E\alpha_T}{1-\nu}\frac{\partial}{\partial x}\int\limits_{-\frac{h}{2}}^{\frac{h}{2}} \theta\, dz = p_1\\ (x \leftrightarrow y, \quad u \leftrightarrow v, \quad p_1 \leftrightarrow p_2),\end{gathered} \tag{2.30}$$

$$\rho h \frac{\partial^2 w}{\partial t^2} + \frac{\partial^2}{\partial x^2}\left[D\left(\frac{\partial^2 w}{\partial x^2} + \nu \frac{\partial^2 w}{\partial y^2}\right)\right] + 2(1-\nu)\frac{\partial^2}{\partial x \partial y}(D)\left(\frac{\partial^2 w}{\partial x \partial y}\right)$$
$$+\frac{\partial^2}{\partial y^2}\left[D\left(\frac{\partial^2 w}{\partial y^2} + \nu \frac{\partial^2 w}{\partial x^2}\right)\right] - \frac{E}{1-\nu^2}\left[(k_x + \nu k_y)\left(\frac{\partial u}{\partial x} - k_x w\right)\right.$$
$$\left. +(k_y + \nu k_x)\left(\frac{\partial v}{\partial y} - k_y w\right)\right] + \frac{E\alpha_T}{1-\nu}\left[\nabla^2 \int_{-\frac{h}{2}}^{\frac{h}{2}} \theta\, z dz + (k_x + k_y)\int_{-\frac{h}{2}}^{\frac{h}{2}} \theta\, dz\right] = q_1, \tag{2.31}$$

$$c_\varepsilon(1+\varepsilon)\frac{\partial \theta}{\partial t} - \lambda_q \nabla^2 \theta$$
$$+\frac{E\alpha_T T_0}{1-\nu}\frac{\partial}{\partial t}\left[\frac{\partial u}{\partial x} + \frac{\partial v}{\partial y} - (k_x + k_y)w - z\nabla^2 w\right] = q_2. \tag{2.32}$$

The system of differential equations (2.30)–(2.32) governs linear thermoelastic dynamics within the Kirchhoff–Love model (without the inertial effect of shell element rotation). It is clear that in this case we have a system of equations of different types (hyperbolic and parabolic) with different dimensions.

2.1.3 Boundary and Initial Conditions

In order to describe in full a coupled dynamical problem of thermoelasticity, the boundary and initial conditions need to be attached to the system of equations obtained above. This means that we need to define the initial state of the shell and to formulate the conditions that describe the mechanical and thermal interaction with the surrounding medium on both the edge and the surface of the shell. The boundary and initial conditions given below are obtained from a stationary condition of the action of the corresponding Hamilton functional.

As the initial conditions attached to the system (2.30)–(2.32), we take the following distributions (for the time instant $t = 0$) inside the shell of the displacements u, v, w, their velocities $\frac{\partial u}{\partial t}, \frac{\partial v}{\partial t}, \frac{\partial w}{\partial t}$, and the temperature increase θ (the last requirement is equivalent to the introduction of a temperature field at the initial time instant):

$$u\big|_{t=0} = u_0(x,y), \quad v\big|_{t=0} = v_0(x,y), \quad w\big|_{t=0} = w_0(x,y), \tag{2.33}$$

$$\left.\frac{\partial u}{\partial t}\right|_{t=0} = u_1(x,y), \quad \left.\frac{\partial v}{\partial t}\right|_{t=0} = v_1(x,y), \quad \left.\frac{\partial w}{\partial t}\right|_{t=0} = w_1(x,y), \tag{2.34}$$

$$\theta\big|_{t=0} = \theta_0(x,y,z), \tag{2.35}$$

where $(x,y) \in \Omega_1, (x,y,z) \in \Omega_2$, and $u_0, v_0, w_0, u_1, v_1, w_1, \theta_0$ are given functions.

In considering the equations (2.20)–(2.23), we need to add to the initial conditions (2.33)-(2.35) the distributions (for $t = 0$) of the angles of rotation Ψ_x, Ψ_y of the normal to the middle surface and their derivatives $\dfrac{\partial \Psi_x}{\partial t}, \dfrac{\partial \Psi_y}{\partial t}$. These distributions have the form

$$\Psi_x\big|_{t=0} = \Psi_{x0}(x,y), \quad \Psi_y\big|_{t=0} = \Psi_{y0}(x,y), \tag{2.36}$$

$$\left.\frac{\partial \Psi_x}{\partial t}\right|_{t=0} = \Psi_{x1}(x,y), \quad \left.\frac{\partial \Psi_y}{\partial t}\right|_{t=0} = \Psi_{y1}(x,y), \tag{2.37}$$

where $(x,y) \in \Omega_1$ and $\Psi_{x0}, \Psi_{y0}, \Psi_{x1}, \Psi_{y1}$ are given functions.

At each point of the shell surface $\partial\Omega_2$, one of the following thermal boundary conditions must be applied [122]:

– the temperature of the shell surface,

$$\theta = \theta^0(x,y,z,t); \tag{2.38}$$

– the density of the heat flow through the shell surface,

$$\frac{\partial \theta}{\partial n} = \theta_n^0(x,y,z,t); \tag{2.39}$$

– a condition of free heat exchange with the surrounding medium,

$$\frac{\partial \theta}{\partial n} + \alpha\theta = \theta_n^\alpha(x,y,z,t), \quad \alpha = const > 0. \tag{2.40}$$

Here $(x,y,z) \in \partial\Omega_2$, $t \in [0,T]$, and $\theta^0, \theta_n^0, \theta_n^\alpha$ are given functions, and T is the total time of observation of the shell.

In the following investigation of rectangular shells, on the sides (for $x = const$ and $y = const$) only a homogeneous condition of type (2.38) will be considered, while on the shell surfaces $\left(z = \pm\dfrac{h}{2}\right)$, homogeneous conditions of type (2.38) or (2.39) (condition of thermal isolation) will be considered.

The mechanical conditions on the shell edge $\partial\Omega_1$ for the equations (2.30)–(2.32) have the following form (at each point of $\partial\Omega_1$, four boundary conditions are applied):

$$u = u^0(x,y,t) \quad \text{or} \quad N_x n_x + S n_y = A^0(x,y,t), \tag{2.41}$$

$$v = v^0(x,y,t) \quad \text{or} \quad N_y n_y + S n_x = B^0(x,y,t), \tag{2.42}$$

$$\begin{aligned} w = w^0(x,y,t) \quad \text{or} \quad & \left(\frac{\partial M_x}{\partial x} + \frac{\partial H}{\partial y}\right) n_x + \left(\frac{\partial M_y}{\partial y} + \frac{\partial H}{\partial x}\right) n_y \\ & + \frac{\partial}{\partial s}\left[(M_y n_y + H n_x) n_x - (M_x n_x + H n_y) n_y\right] = C^0(x,y,t), \end{aligned} \tag{2.43}$$

$$\frac{\partial w}{\partial n} = w_n^0(x,y,t) \quad \text{or} \quad M_x n_x^2 + 2H n_x n_y + M_y n_y^2 = D^0(x,y,t), \tag{2.44}$$

where $(x, y) \in \partial\Omega_2$, $t \in [0, T]$, and n_x, n_y are the components of the external normal to the edge $\partial\Omega_1$. The symbols $\frac{\partial}{\partial n}, \frac{\partial}{\partial s}$ denote differentiation in the direction of the external normal to the edge $\partial\Omega_1$ and in the direction of the edge $\partial\Omega_1$, respectively, and $u^0, v^0, w^0, w_n^0, A^0, B^0, C^0, D^0$ are given functions.

When the Timoshenko kinematic model is used ((2.20)–(2.23)), we need to formulate five mechanical conditions , i.e. the conditions (2.41), (2.42) and the following additional three:

$$w = w^0(x, y, t) \quad \text{or} \quad Q_x n_x + Q n_y = E^0(x, y, t), \tag{2.45}$$

$$\Psi_x = \Psi_x^0(x, y, t) \quad \text{or} \quad M_x n_x + H n_y = F^0(x, y, t), \tag{2.46}$$

$$\Psi_y = \Psi_y^0(x, y, t) \quad \text{or} \quad M_y n_y + H n_x = G^0(x, y, t), \tag{2.47}$$

where $(x, y) \in \partial\Omega_2$, $t \in [0, T]$, and $w^0, \Psi_x^0, \Psi_y^0, E^0, F^0, G^0$ are given functions.

In the each pair of alternative boundary conditions (2.41)–(2.47), the first is the main condition (this means that during the search for a stationary point of Hamilton's functional this condition is specified in advance), whereas the second is a physically realistic ("real") condition (it is derived from a stationary condition of Hamilton's functional). In the case of a rectangular shell, these boundary conditions have been formulated in a series of monographs (see, for instance [133, 232, 234]).

The following mechanical boundary conditions for different types of support of the shell edge will be considered (for shells that are rectangular in plan):

– simple support on an elastic rib (no compression or elongation in the tangential plane):

$$w = \frac{\partial u}{\partial x} = v = \frac{\partial \Psi_x}{\partial x} = \Psi_y = 0, \quad x = const, \tag{2.48}$$

$$w = \frac{\partial^2 w}{\partial x^2} = \frac{\partial u}{\partial x} = v = 0, \quad x = const; \tag{2.49}$$

– simply supported edge:

$$w = u = v = \frac{\partial \Psi_x}{\partial x} = \Psi_y = 0, \quad x = const, \tag{2.50}$$

$$w = \frac{\partial^2 w}{\partial x^2} = u = v = 0, \quad x = const; \tag{2.51}$$

– clamped edge:

$$w = u = v = \Psi_x = \Psi_y = 0, \quad x = const, \tag{2.52}$$

$$w = \frac{\partial w}{\partial x} = u = v = 0, \quad x = const, \tag{2.53}$$

$$x \leftrightarrow y, \quad u \leftrightarrow v.$$

The conditions (2.48), (2.50), (2.52) correspond to the Timoshenko-type model, whereas the conditions (2.49), (2.51), (2.53) correspond to the Kirchhoff–Love model. Each of the groups of boundary conditions (when a homogeneous boundary condition of type (2.38) is satisfied on the sides of the shell) refers to a particular case of the general conditions, either (2.41)–(2.44) (Kirchhoff–Love model) or (2.41), (2.42), (2.45)–(2.47) (Timoshenko model). For instance, in the group (2.48) the conditions $w = v = \Psi_y = 0$ are related to the homogeneous main conditions (2.45), (2.42), (2.47), whereas the conditions $\frac{\partial u}{\partial x} = \frac{\partial \Psi_x}{\partial x} = 0$ are equivalent to the homogeneous real conditions (2.41), (2.46), and so on.

2.1.4 An Abstract Coupled Problem

Here we consider an abstract Cauchy problem for a coupled system of two differential equations in a Hilbert space which generalizes a series of coupled thermoelastic problems. This also includes the thermoelastic problems of shallow shells modelled by the Kirchhoff–Love and Timoshenko theories defined earlier and is an important subject of this chapter.

First we introduce the necessary notation. Let H be a certain Hilbert space. By $L_p(0,T;H)$ we denote a space of measurable (in an interval $[0,T]$ of the time axis) functions, which have their values in H and are integrable $[0,T]$ (bounded for $p=\infty$), with a norm of order p ($p<\infty$)

$$\|u\|_{L_p(0,T;H)} = \left[\int_0^T \|u(t)\|_H^p \, dt\right]^{\frac{1}{p}},$$

which for $p=\infty$ takes the form

$$\|u\|_{L_\infty(0,T;H)} = \sup_{0\le t\le T} ess \|u(t)\|_H .$$

By $W_p^k(0,T;H)$ we denote a space of rational functions $u(t)$, which have values in H, possess in the interval $[0,T]$ generalized derivatives (in the Sobolev sense) $u^{(j)}(t)$ up to the k_Ω^{th} order (inclusive), and are integrable with a norm of power p. The space $W_p^k(0,T;H)$ for $p<\infty$ is defined by the norm

$$\|u\|_{W_p^k(0,T;H)} = \left[\int_0^T \sum_{j=0}^k \left\|u^{(j)}(t)\right\|_H^p dt\right]^{\frac{1}{p}},$$

whereas for $p=\infty$ it is defined by the norm

$$\|u\|_{W_\infty^k(0,T;H)} = \sup_{0\le t\le T} ess \sum_{j=0}^k \|u(t)\|_H .$$

For $1 \leq p \leq \infty$, each of the spaces $L_p(0,T;H)$, $W_p^k(0,T;H)$ $(k = 1,2,\dots)$ is a Banach space. The spaces $L_2(0,T;H)$, $W_2^k(0,T;H)$ are Hilbert spaces, and a scalar product is defined as follows:

$$(u,v)_{L_2(0,T;H)} = \int_0^T (u(t),v(t))_H \, dt,$$

$$(u,v)_{W_2^k(0,T;H)} = \int_0^T \sum_{j=0}^{k} (u^{(j)}(t),v^{(j)}(t))_H \, dt.$$

By $W_2^k(\Omega)$ we denote the Sobolev space of measurable functions $u(x)$, defined in the space Ω of n-dimensional arithmetic space R^n, which have all generalized partial derivatives $D^\alpha u(x)$ of order up to k inclusive, and are integrable with a norm of the second power in Ω. We define $\alpha = (\alpha_1 + \dots + \alpha_n)$ as the integer multi-index of differentiation, $|\alpha| = \alpha_1 + \dots + \alpha_n$, and $D^\alpha = D_1^{\alpha_1} \dots D_n^{\alpha_n}, D_i = \partial / \partial x_i$, denotes a general differentiation operation in the Sobolev sense. The space $W_k^2(\Omega)$ is a Hilbert space and a scalar product is defined in it by the relation

$$(u,v)_{W_2^k(\Omega)} = \int_\Omega \sum_{|\alpha| \leq k} D^\alpha u D^\alpha v \, d\Omega.$$

For $k = 0$, the space $W_k^2(\Omega)$ overlaps with the Lebesgue space $L_2(\Omega)$.

Let $H_1, \dots, H_m$ be m samples of the Hilbert space. We denote by $H_1 \times \dots \times H_m$ their Cartesian product, i.e. the Hilbert space created by all possible ordered samples of elements $\overline{u} = \{u_i\}_{i=1}^m$, where $u_i \in H_i$ and where the scalar product has the form $(\overline{u},\overline{v})_{H_1 \times \dots \times H_m} = (u_1,v_1)_{H_1} + \dots + (u_m,v_m)_{H_m}$.

Let $u(x,t)$ be a rational function defined for $x \in \Omega \subset R^n$, $t \in [0,T]$, which has its values in R^m and possesses the following property: for all $t \in [0,T]$, the function $x \to u(x,t)$ belongs to a Hilbert space H of the type $W_2^{k_1}(\Omega) \times \dots \times W_2^{k_m}(\Omega)$. Each function defined in this way is related to a map $u(t)$, which for an arbitrary number $t \in [0,T]$ defines the function $x \to u(x,t)$ as an element of H. Therefore, we shall use the following notation: $\|u(t)\|_H$, $(u(t),v(t))_H$, $\|u\|_{W_p^k(0,T;H)}$, $(u,v)_{W_2^k(0,T;H)}$, where u,v are functions depending on x,t.

In the case of continuous and continuously differentiable functions, the following established notation is applied: $C(\overline{\Omega})$, $C^k(\overline{\Omega})$, $C([0,T];H)$.

Let A be a self-adjoint, positive definite and (possibly) unbounded operator acting in the Hilbert space H, which we attach to the space $D(A)$; the action of A creates a Hilbert space with the same number of elements and with a scalar product $(u,v)_A = (Au,Av)_H$. Therefore, the following notation is applied: $L_p(0,T;D(A))$, $W_p^k(0,T;D(A))$.

Let H_1, H_2 be two specimens of the Hilbert space. Let us formulate the Cauchy problem for the following system of two differential equations:

$$I_1 w''(t) + Lw(t) + M\theta(t) = g_1(t), \tag{2.54}$$

$$I_2\theta'(t) + K\theta(t) + Nw'(t) = g_2(t), \tag{2.55}$$

$$w(0) = w_0, \quad w'(0) = w_1, \quad \theta(0) = \theta_0, \tag{2.56}$$

where w, θ are the sought functions mapping the interval $[0, T]$ into H_1 and H_2, respectively. The given functions g_1, g_2 satisfy the conditions

$$g_1 \in L_2(0, T; H_1), \quad g_2 \in L_2(0, T; H_2); \tag{2.57}$$

I_1 and I_2 are bounded self-adjoint, positive definite operators acting in H_1 and H_2, respectively; and L and K are unbounded, self-adjoint, positive definite operators acting in H_1 and H_2, respectively, defined on dense sets $D(L) \subset H_1$, $D(K) \subset H_2$. The inverse operators L^{-1}, K^{-1} are continuous. M and N are linear, unbounded operators acting from H_2 to H_1 and from H_1 to H_2, respectively, linked by the relation $N \subset -M^*$, i.e. for arbitrary elements $w \in D(N)$, $\theta \in D(M)$, the following "linking condition" holds:

$$(M\theta, w)_{H_1} + (Nw, \theta)_{H_2} = 0. \tag{2.58}$$

Also $D(K) \subset D(M), D(L^{1/2}) \subset D(N)$. The bounded operators MK^{-1} and $NL^{-1/2}$ correspond to the operators $H_2 \to H_1$ and $H_1 \to H_2$. The operator $L^{-1/2}MK^{-1/2}$ can be extended to a bounded operator from H_2 to H_1. Finally, w_0, w_1, θ_0 are given elements satisfying the conditions

$$w_0 \in D(L^{1/2}), \quad w_1 \in H_1, \quad \theta_0 \in H_2. \tag{2.59}$$

Definition 2.1 *An ordered pair of functions $\{w, \theta\}$ is called a generalized solution to the Cauchy problem (2.54)–(2.56) if $w \in W^1_\infty(0, T; H_1) \cap L_\infty(0, T; D(L^{1/2}))$, $w(0) = w_0$, $\theta \in L_\infty(0, T; H_2) \cap L_2(0, T; D(K^{1/2}))$ and, for arbitrary functions ν, η satisfying the conditions $\nu \in W^1_1(0, T; H_1) \cap L_1(0, T; D(L^{1/2}))$, $\nu(T) = 0$, $\eta \in W^1_1(0, T; H_2) \cap L_2(0, T; D(K^{1/2}))$, $\eta(T) = 0$, the following assumptions are satisfied:*

$$\int_0^T \Big[-(I_1 w'(t), \nu'(t))_{H_1} + (L^{1/2}w(t), L^{1/2}\nu(t))_{H_1}$$
$$+ (L^{-1/2}M\theta(t), L^{1/2}\nu(t))_{H_1} \Big] dt - (I_1 w_1, \nu(0))_{H_1} = \int_0^T (g_1(t), \nu(t))_{H_1} dt, \tag{2.60}$$

$$\int_0^T \Big[-(I_1\theta(t), \eta'(t))_{H_2} + (K^{1/2}\theta(t), K^{1/2}\eta(t))_{H_2} - (Nw(t)\eta'(t))_{H_2}$$
$$-L^{1/2}\nu(t))_{H_1} \Big] dt - (I_1 w_1, \nu(0))_{H_1} = \int_0^T (g_1(t), \nu(t))_{H_1} dt. \tag{2.61}$$

Definition 2.2 *An ordered pair of functions $\{w, \theta\}$ is called the classical solution to the Cauchy problem (2.54)–(2.56) if*

$$w \in W^2_\infty(0,T;H_1) \cap W^1_\infty(0,T;D(L^{1/2})) \cap L_\infty(0,T;D(L)),$$
$$\theta \in W^1_\infty(0,T;H_2) \cap W^1_2(0,T;D(K^{1/2})) \cap L_\infty(0,T;D(K)),$$

and the functions w, θ satisfy the relations (2.54), (2.55) almost always on the interval $[0,T]$ and satisfy the relations (2.56).

Remark 2.1 *In Definitions 2.1 and 2.2, maximal smoothness of the functions w, θ is needed, which can be achieved using Theorems 2.1 and 2.2. The integral relations (2.60), (2.61) can be obtained by a general approach used in solving problems of mathematical physics, namely by a formal multiplication of the differential equations (2.54), (2.55) by the functions $\nu(t), \eta(t)$ and integration by parts, with consideration of the initial condition (2.56). In particular, this means that the "classical" solution to the problem (2.54)–(2.56) is its generalized solution.*

Let us check if the formally defined Cauchy problem (2.54)–(2.56) is a real generalization of the coupled thermoelasticity problem of a shallow shell defined in Sects. 2.1.2, 2.1.3 (in the case of homogeneous boundary conditions). To do this, we analyse the initial–boundary problem for the system of differential equations (2.20)–(2.23) with the initial conditions (2.33)–(2.37), the homogeneous mechanical boundary conditions (2.41), (2.42), (2.45)–(2.47) and the homogeneous thermal boundary conditions (2.38) (main condition) or (2.39) (real condition). Let $\Omega_1 \subset R^2$ denote a bounded space with a piecewise smooth boundary of C^1 class. With each member of the pair of alternative boundary conditions we associate either the part Ω_1, of the boundary (for mechanical conditions) or Ω_2 (for heating conditions); these parts do not intersect each other. It should be noted that the measure of the part of the boundary corresponding to the main condition is strictly positive, and on the whole of the side surfaces of the shell the main thermal condition is applied. This means that all mechanical boundary conditions may be described via the displacements u, v, w and rotation angles Ψ_x, Ψ_y only (the function θ does not appear in these conditions). Let us denote by $(\mathring{W}^2_2(\Omega_1))^5$ the space of vector functions $\overline{\omega} = (x,y)$ of the form

$$\overline{\omega} = \{u, v, w, \Psi_x, \Psi_y\}; \tag{2.62}$$

all components of this space belong to the space $W^2_2(\Omega_1)$ and satisfy the main and real mechanical boundary conditions of the problem under consideration. By $(\mathring{W}^1_2(\Omega_1))^5$ we denote the space of vector functions of the form (2.62), all components of which belong to $W^1_2(\Omega_1)$ and satisfy the main mechanical conditions of the problem. By $\mathring{W}^2_2(\Omega_1)$ we denote the space of functions $\theta \in W^2_2(\Omega_1)$ satisfying both the main and the real thermal boundary conditions, and by $\overset{\circ}{W}{}_2^1(\Omega_1)$ we denote the space of the functions $\theta \in W_2^1(\Omega_2)$ satisfying the main thermal condition.

The following conditions are applied to the functions and constants occurring in (2.20)–(2.23), (2.33)–(2.37):

$$h \in C^1(\overline{\Omega}_1), \quad h(x,y) > 0, \quad (x,y) \in \overline{\Omega}_1, \tag{2.63}$$

$$\rho, E, \alpha_T, c_\varepsilon, \lambda_q, T_0, T = const > 0, \quad k = const \neq 0,$$

$$k_x, k_y = const, \quad 0 < \nu = const < \frac{1}{2}, \tag{2.64}$$

$$p_1, p_2, q_1 \in L_2(\Omega_1 \times (0,T)), \quad q_2 \in L_2(\Omega_2 \times (0,T)), \tag{2.65}$$

$$\overline{\omega}_0 = \{u_0, v_0, w_0, \Psi_{x0}, \Psi_{y0}\} \in (\overset{\circ}{W}{}_2^1(\Omega_1))^5, \tag{2.66}$$

$$\overline{\omega}_1 = \{u_1, v_1, w_1, \Psi_{x1}, \Psi_{y1}\} \in (L_2(\Omega_1))^5, \tag{2.67}$$

$$\theta_0 \in L_2(\Omega_1). \tag{2.68}$$

The initial–boundary problem under consideration can be transformed to the form (2.54)–(2.56), if the following assumptions are made:

$$H_1 = (L_2(\Omega_1))^5, \quad H_2 = L_2(\Omega_1), \tag{2.69}$$

$$I_1\overline{\omega} = \left\{\rho h u, \rho h v, \rho h w, \frac{1}{12}\rho h^3 \Psi_x, \frac{1}{12}\rho h^3 \Psi_y\right\}, \tag{2.70}$$

$$I_2\theta = \frac{c_\varepsilon}{T_0}(1+\varepsilon)\theta, \tag{2.71}$$

$$\begin{aligned} L\overline{\omega} = \Bigg\{ & l_x\overline{\omega}, l_y\overline{\omega}, \\ & -\frac{k^2 E}{2(1+\nu)}\left\{\frac{\partial}{\partial x}\left[h\left(\Psi_x + \frac{\partial w}{\partial x}\right)\right] + \frac{\partial}{\partial y}\left[h\left(\Psi_y + \frac{\partial w}{\partial y}\right)\right]\right\} \\ & -\frac{Eh}{1-\nu^2}\left[(k_x + \nu k_y)\left(\frac{\partial u}{\partial x} - k_x w\right) + (k_x + \nu k_y)\left(\frac{\partial v}{\partial y} - k_y w\right)\right], \\ & \tilde{l}_x\overline{\omega}, \tilde{l}_y\overline{\omega}\Bigg\}, \end{aligned} \tag{2.72}$$

$$\begin{aligned} & D(L) = (\overset{\circ}{W}{}_2^2(\Omega_1))^5, \\ & l_x\overline{\omega} = -\frac{E}{1-\nu^2}\frac{\partial}{\partial x} h\left[\frac{\partial u}{\partial x} - k_x w + \nu\left(\frac{\partial v}{\partial y} - k_y w\right)\right] \\ & -\frac{E}{2(1+\nu)}\frac{\partial}{\partial y}\left[h\left(\frac{\partial u}{\partial y} + \frac{\partial v}{\partial x}\right)\right] \quad (x \leftrightarrow y, \quad u \leftrightarrow v), \end{aligned} \tag{2.73}$$

$$\tilde{l}_x\overline{\omega} = -\frac{\partial}{\partial x}\left[D\left(\frac{\partial \Psi_x}{\partial x} + \nu\frac{\partial \Psi_y}{\partial y}\right)\right] - \frac{1-\nu}{2}\frac{\partial}{\partial y}\left[D\left(\frac{\partial \Psi_x}{\partial y} + \frac{\partial \Psi_y}{\partial x}\right)\right]$$
$$+\frac{k^2 Eh}{2(1+\nu)}\left(\Psi_x + \frac{\partial w}{\partial x}\right) \quad (x \leftrightarrow y), \tag{2.74}$$

$$K\theta = -\frac{\lambda_q}{T_0}\nabla^2\theta, \quad D(K) = \overset{\circ}{W}{}_2^2(\Omega_2), \tag{2.75}$$

$$M\theta = \left\{ m_x\theta, m_y\theta, \frac{E\alpha_T}{1-\nu}(k_x+k_y)\int\limits_{-\frac{h}{2}}^{\frac{h}{2}}\theta\, dz, \tilde{m}_x\theta, \tilde{m}_y\theta \right\}, \tag{2.76}$$

$$D(M) = \overset{\circ}{W}{}_2^1(\Omega_2),$$

$$m_x\theta = \frac{E\alpha_T}{1-\nu}\frac{\partial}{\partial x}\int\limits_{-\frac{h}{2}}^{\frac{h}{2}}\theta\, dz, \quad \tilde{m}_x\theta = \frac{E\alpha_T}{1-\nu}\frac{\partial}{\partial x}\int\limits_{-\frac{h}{2}}^{\frac{h}{2}}\theta z\, dz, \quad (x \leftrightarrow y), \tag{2.77}$$

$$N\overline{\omega} = \frac{E\alpha_T}{1-\nu}\left[\frac{\partial u}{\partial x} + \frac{\partial v}{\partial y} - (k_x+k_y)w + z\left(\frac{\partial \Psi_x}{\partial x} + \frac{\partial \Psi_y}{\partial y}\right)\right], \tag{2.78}$$

$$D(N) = (\overset{\circ}{W}{}_2^1(\Omega_1))^5,$$

$$\overline{\omega}_0 = \{u_0, v_0, w_0, \Psi_{x0}, \Psi_{y0}\}, \quad \overline{\omega}_1 = \{u_1, v_1, w_1, \Psi_{x1}, \Psi_{y1}\}, \tag{2.79}$$

$$g_1 = \{p_1, p_2, q_1, 0, 0\}, \quad g_2 = \frac{1}{T_0}q_2. \tag{2.80}$$

The vector function $\overline{\omega} = \{u, v, w, \Psi_x, \Psi_y\}$ and the given vector functions, $\overline{\omega}_0, \overline{\omega}_1$ (2.79) play the roles of the sought function $w(t)$ and the initial data w_0, w_1. In addition, the following relations hold:

$$D(L^{1/2}) = (\overset{\circ}{W}{}_2^1(\Omega_1))^5, \quad D(K^{1/2}) = (\overset{\circ}{W}{}_2^1(\Omega_1))^5, \tag{2.81}$$

with the accuracy of the equivalent scalar product.

Let us consider further the initial–boundary value problem for the system of equations (2.30)–(2.32) with the initial conditions (2.33)–(2.35) and homogeneous boundary conditions of the form (2.41)–(2.44), (2.38) (or (2.39)). We retain the earlier assumptions about the smoothness of Ω_1 and about the partitioning of the boundary into Ω_1, Ω_2, corresponding to each member of the pair of alternative boundary conditions. We denote by $((\overset{\circ}{W}{}_2^2(\Omega_1))^2 \times \overset{\circ}{W}{}_2^4(\Omega_1)$ the space of vector functions $\overline{\omega}(x, y)$ of the form

$$\overline{\omega}_0 = \{u, v, w\}, \tag{2.82}$$

the components of which belong to $W_2^2(\Omega_1), W_2^2(\Omega_1), W_2^4(\Omega_1)$, respectively and satisfy the main mechanical boundary conditions of the problem. The conditions (2.64), (2.65), (2.68) are applied to the functions and constants occurring in (2.30)–(2.35), and in addition the following conditions are applied:

$$h \in C^2(\overline{\Omega}_1), \quad h(x,y) > 0, \quad (x,y) \in \overline{\Omega}_1, \tag{2.83}$$

$$\overline{\omega}_0 = \{u_0, v_0, w_0\} \in (\overset{\circ}{W}{}_2^1(\Omega_1))^2 \times \overset{\circ}{W}{}_2^2(\Omega_1), \tag{2.84}$$

$$\overline{\omega}_1 = \{u_1, v_1, w_1\} \in (L_2(\Omega_1))^3. \tag{2.85}$$

The initial–boundary value problem under consideration is reduced to that described by (2.54)–(2.56) if we make the assumptions

$$H_1 = (L_2(\Omega_1))^3, \quad H_2 = L_2(\Omega_2), \tag{2.86}$$

$$I_1\overline{\omega} = \{\rho h u, \rho h v, \rho h w\}, \quad I_2\theta = \frac{c_\varepsilon}{T_0}(1+\varepsilon)\theta, \tag{2.87}$$

$$\begin{aligned} L\overline{\omega} = \Bigg\{ & l_x\overline{\omega}, l_y\overline{\omega}, \frac{\partial^2}{\partial x^2}\left[D\left(\frac{\partial^2 w}{\partial x^2} + \nu\frac{\partial^2 w}{\partial y^2}\right)\right] \\ & +2(1-\nu)\frac{\partial^2}{\partial x \partial y}\left(D\frac{\partial^2 w}{\partial x \partial y}\right) + \frac{\partial^2}{\partial y^2}\left[D\left(\frac{\partial^2 w}{\partial y^2} + \nu\frac{\partial^2 w}{\partial x^2}\right)\right] \\ & -\frac{Eh}{1-\nu^2}\left[(k_x+\nu k_y)\left(\frac{\partial u}{\partial x} - k_x w\right) + (k_y + \nu k_x)\left(\frac{\partial v}{\partial y} - k_y w\right)\right]\Bigg\}, \end{aligned} \tag{2.88}$$

$$D(L) = (\overset{\circ}{W}{}_2^2(\Omega_1))^2 \times \overset{\circ}{W}{}_2^4(\Omega_1),$$

$$K\theta = -\frac{\lambda_q}{T_0}\nabla^2\theta, \quad D(K) = \overset{\circ}{W}{}_2^2(\Omega_2), \tag{2.89}$$

$$M\theta = \left\{ m_x\theta, m_y\theta, \frac{E\alpha_T}{1-\nu}\left[\nabla^2 \int_{-\frac{h}{2}}^{\frac{h}{2}} \theta\, dz + (k_x + k_y)\int_{-\frac{h}{2}}^{\frac{h}{2}} \theta\, dz\right]\right\}, \tag{2.90}$$

$$D(M) = \overset{\circ}{W}{}_2^2(\Omega_2),$$

$$N\overline{\omega} = \frac{E\alpha_T}{1-\nu}\left[\frac{\partial u}{\partial x} + \frac{\partial v}{\partial y} - (k_x + k_y)w + z\nabla^2 w\right], \tag{2.91}$$

$$D(N) = (\overset{\circ}{W}{}_2^1(\Omega_1))^2 \times \overset{\circ}{W}{}_2^2(\Omega_1),$$

$$\overline{\omega}_0 = \{u_0, v_0, w_0\}, \quad \overline{\omega}_1 = \{u_1, v_1, w_1\} \tag{2.92}$$

$$g_1 = \{p_1, p_2, q_1\}, \quad g_2 = \frac{1}{T_0}q_2. \tag{2.93}$$

The operators l_x, l_y, m_x, m_y appearing in (2.88), (2.90) are defined by the relations (2.73), (2.77). The sought function and initial data play a being sought vector-function $\overline{\omega} = \{u, v, w\}$ and the given vector functions $\overline{\omega}_0, \overline{\omega}_1$ (2.92) play the roles of the sought function $w(t)$ and the initial data w_0, w_1. It should be noted that following relations hold:

$$D(L^{1/2}) = (\overset{\circ}{W}{}_2^1(\Omega_1))^2 \times \overset{\circ}{W}{}_2^2(\Omega_1), \quad D(K^{1/2}) = \overset{\circ}{W}{}_2^1(\Omega_2). \tag{2.94}$$

Remark 2.2 *An action space of the operators L, K may be defined as a self-conjugating extension of the initial differential operators defined by (2.72), (2.75) or (2.88), (2.89) initially on sets of smooth enough functions satisfying the corresponding homogeneous boundary conditions. (This definition is due to Friedrichs.)*

Remark 2.3 *All of the general properties mentioned above of the operators I_1, I_2, L, K, M, N appearing in (2.54), (2.55) either are explicitly obvious (see (2.69)–(2.78) or (2.86)–(2.91)) or can be obtained from known general results. In particular, full continuity of the operators L^{-1}, K^{-1} can be established using theorems of equivalent normalization of the Sobolev spaces described in [139, 156] or using theorems of compact inclusion [139], satisfied for instance for spaces with a boundary of class C^1.*

Remark 2.4 *The class of Cauchy problems defined by (2.54)–(2.56) is not limited to the problems of the theory of shallow shells theory presented in the Sects. 2.1.2, 2.1.3. It is not difficult to check that a thermoelastic problem of a three-dimensional body can be reduced to a problem governed by (2.54)–(2.56) [179], as well as to various other problems of thermoelasticity.*

2.1.5 Existence and Uniqueness of Solutions of Thermoelasticity Problems

In this section, theorems related to the existence and uniqueness of the general "classical" solution to the coupled abstract problem (2.54)–(2.56) will be given, and then the corresponding theorems for coupled thermoelastic problems of shallow shells will be formulated.

In proving results related to the existence of solutions, the known scheme of the compactness method will be used [147], and special attention will be paid to prior estimates of exact and approximate solutions of the problem (2.54)–(2.56), which will then be used in the next chapter in obtaining estimates for the Bubnov–Galerkin method.

Let $\{\nu_i\}_{i=1}^{\infty}, \{\eta_j\}_{j=1}^{\infty}$ be linearly independent systems of elements from $D(L)$ and $D(K)$, respectively (fully defined in $D(L^{1/2})$ and $D(K^{1/2})$), and let $i(n), j(n)$ be two subseries of a natural series. Let H_1^n and H_2^n be linear vicinities of finite systems of elements $\{\nu_i\}_{i=1}^{i(n)}$ and $\{\eta_j\}_{j=1}^{j(n)}$. Let P_G denote the operator of orthogonal projection of the Hilbert space H ($= H_1$ or H_2).

We are going to solve the initial problem (2.54)–(2.56) using the Bubnov–Galerkin method. We consider the following problem:

$$\begin{cases} P_{H_1^n}(I_1 w_n''(t) + Lw_n(t) + M\theta_n(t)) = P_{H_1^n} g_1(t), \\ P_{H_2^n}(I_2\theta_n'(t) + K\theta_n(t) + Nw_n'(t)) = P_{H_2^n} g_2(t), \\ w_n(t) \in H_1^n, \quad \theta_n(t) \in H_2^n, \quad \forall t \in [0,T], \\ w_n(0) = w_{0n}, \quad w_n' = w_{1n}, \quad \theta_n(0) = \theta_{0n}, \end{cases} \tag{2.95}$$

where $\{w_{0n}\}_{n=1}^{\infty}$, $\{w_{1n}\}_{n=1}^{\infty}$, $\{\theta_{0n}\}_{n=1}^{\infty}$ are arbitrary subseries of elements taken from H_1^n, H_1^n, H_2^n, respectively, satisfying the conditions

$$w_{0n} \to w_0 \text{ in } D(L^{1/2}), \quad w_{1n} \to w_1 \text{ in } H_1, \quad \theta_{0n} \to \theta_0 \text{ in } H_2. \tag{2.96}$$

The exact solution $\{w_n, \theta_n\}$ to the problem (2.95) will be called the approximate solution to the problem (2.54)–(2.56) obtained using the Bubnov–Galerkin method. Because the problem (2.95) is equivalent to a Cauchy problem for a systems of $2i(n) + J(n)$ ordinary differential equations with $2i(n) + J(n)$ sought functions, this solution exists and is unique for an arbitrary natural number n.

Theorem 2.1 *A general solution $\{w, \theta\}$ to the problem (2.54)–(2.56) exists and is unique when the conditions (2.57)–(2.59) are satisfied.*

Proof. Let us begin with the existence of a general solution to the problem (2.54)–(2.56). Multiplying scalarly the first equation of (2.95) by $2w_n'(t)$ and the second equation of (2.95) by $2\theta_n(t)$, we take the sum of them and take into account the condition (2.58). Integrating again the equality obtained over the interval $[0,T]$ (where $t \le T$), we obtain:

$$\begin{aligned} &\left\|I_1^{1/2} w_n'(t)\right\|_{H_1}^2 + \left\|L^{1/2} w_n(t)\right\|_{H_1}^2 + \left\|I_2^{1/2}\theta_n(t)\right\|_{H_2}^2 \\ &+ 2\int_0^t \left\|K_1^{1/2}\theta_n(\tau)\right\|_{H_2}^2 d\tau = \left\|I_1^{1/2} w_{1n}\right\|_{H_2}^2 + \left\|L^{1/2} w_{0n}\right\|_{H_1}^1 \\ &+ \left\|I_2^{1/2}\theta_{0n}\right\|_{H_2}^2 + 2\int_0^t \left[(g_1(\tau), w_n'(\tau))_{H_1} + (g_2(\tau), \theta_n(\tau))_{H_2}\right] d\tau. \end{aligned}$$

Applying the Cauchy inequality to the right-hand side of the relation given above, we obtain

$$\begin{aligned} &\left\|I_1^{1/2} w_n'(t)\right\|_{H_1}^2 + \left\|L^{1/2} w_n(t)\right\|_{H_1}^2 + \left\|I_2^{1/2}\theta_n(t)\right\|_{H_2}^2 \\ &+ 2\int_0^t \left\|K^{1/2}\theta_n(\tau)\right\|_{H_2}^2 d\tau \le w + \frac{1}{\delta}\left\|I^{-1/2} g\right\|_{L_2(0,T;H_1\times H_2)}^2 \\ &+ \delta\int_0^t \left[\left\|I_2^{1/2} w_n'(\tau)\right\|_{H_1}^2 + \left\|I_2^{1/2}\theta_n(\tau)\right\|_{H_2}^2\right] d\tau, \end{aligned} \tag{2.97}$$

where $I^{-1/2}g = \{I_1^{-1/2}g_1, I_2^{-1/2}g_2\}$, δ is an arbitrary positive number, and

$$w = \sup_n \left[\left\| I_1^{1/2} w_{1n} \right\|_{H_1}^2 + \left\| L^{1/2} w_{0n} \right\|_{H_1}^2 + \left\| I_2^{1/2} \theta_{0n} \right\|_{H_2}^2 \right]. \tag{2.98}$$

From (2.96), one can conclude that $w < \infty$. We apply to the inequality (2.97) the Gronwill lemma,[1] and finally obtain

$$\left\| I_1^{1/2} w_n'(t) \right\|_{H_1}^2 + \left\| L^{1/2} w_n(t) \right\|_{H_1}^2 + \left\| I_2^{1/2} \theta_n(t) \right\|_{H_2}^2$$
$$+2 \int_0^t \left\| K^{1/2} \theta_n(\tau) \right\|_{H_2}^2 d\tau \le \left(w + \frac{1}{\delta} \left\| I^{-1/2} g \right\|_{L_2(0,T;H_1 \times H_2)}^2 \right) \exp(\delta T).$$

Minimizing the right-hand side of the last inequality and taking account of the fact that $\delta > 0$, we obtain the following prior estimate of an approximate solution to the problem (2.54)–(2.56):

$$\left\| I_1^{1/2} w_n'(t) \right\|_{H_1}^2 + \left\| L^{1/2} w_n(t) \right\|_{H_1}^2 + \left\| I_2^{1/2} \theta_n(t) \right\|_{H_2}^2$$
$$+ 2 \int_0^t \left\| K^{1/2} \theta_n(\tau) \right\|_{H_2}^2 d\tau \le m(g, w, T), \tag{2.99}$$

where

$$m(g,w,T) = \min_{\delta > 0} \left[\left(w + \frac{1}{\delta} \left\| I^{-1/2} g \right\|_{L_2(0,T;H_1\times H_2)}^2 \right) \exp(\delta T) \right]$$
$$\equiv \left\{ w + \frac{1}{2} \left[\left(\left\| I^{-1/2} g \right\|_{L_2(0,T;H_1\times H_2)}^4 \right) T^2 \right.\right.$$
$$\left.\left. + 4 \left(\left\| I^{-1/2} g \right\|_{L_2(0,T;H_1\times H_2)}^2 wT \right)^{1/2} + \left\| I^{-1/2} g \right\|_{L_2(0,T;H_1\times H_2)}^2 T \right] \right\}$$
$$\times \exp\left\{ 2 \left\| I^{-1/2} g \right\|_{L_2(0,T;H_1\times H_2)}^2 T \left[\left(\left\| I^{-1/2} g \right\|_{L_2(0,T;H_1\times H_2)}^4 \right) T^2 \right.\right.$$
$$\left.\left. + 4 \left(\left\| I^{-1/2} g \right\|_{L_2(0,T;H_1\times H_2)}^2 wT \right)^{1/2} + \left\| I^{-1/2} g \right\|_{L_2(0,T;H_1\times H_2)}^2 T \right]^{-1} \right\}. \tag{2.100}$$

---

[1] The different variants of the Gronwill lemma are given in a series of monographs (see, for instance, [98, 148]). Here and later the following version of this lemma is applied: if $f \in L([0,T])$ and almost always $(t \in [0,T])$, $0 \le f(t) \le C_1 + C_2 \int_0^t f(\tau)\, d\tau$, where $C_1, C_2 = const \ge 0$, then almost always $(t \in [0,T])$, $f(t) \le C_1 \exp(C_2 T)$.

The estimate (2.99) holds for all n, almost always ($t \in [0,T]$). The series of function pairs $\{w_n, \theta_n\}$ is bounded in the Hilbert space $[W_1^2(0,T;H_1) \cap L_2(0,T;D(L^{1/2}))] \times L_2(0,T;D(K^{1/2}))$ owing to (2.99) (from now on this space will be referred to as $\tilde{H}$). Because any bounded set in the Hilbert space is weakly compact, there exists a sequence of function pairs $\{w_\nu, \theta_\nu\}$ that converges weakly in $\tilde{H}$. Let us denote its weak limit by $\{w, \theta\}$, and let us derive a prior estimate of type (2.99) for the pair of functions $\{w, \theta\}$.

In order to simplify the notation, we write

$$< w, \theta; \nu, \eta; t > = (I_1^{1/2} w'(t), (I_1^{1/2} \nu'(t))_{H_1} + (L^{1/2} w(t), L^{1/2} \nu(t))_{H_1}$$

$$+ (I_2^{1/2} \theta(t), I_2^{1/2} \eta(t))_{H_2} + 2 \int_0^t (K^{1/2} \theta(\tau), K^{1/2} \eta(\tau))_{H_2} d\tau,$$

$$\|w, \theta; t\| = < w, \theta; w, \theta; t >^{1/2} .$$

Let C be an arbitrary number greater than $m(g, w, T)$. Let M be the set of such $t \in [0,T]$ for which the following inequality holds:

$$\|w, \theta; t\| \geq C^{1/2}. \tag{2.101}$$

Let us consider the linear bounded functional Φ over $\tilde{H}$ defined by the relation

$$\Phi(\{\nu, \eta\}) = \int_M \frac{< w, \theta; \nu, \eta; t >}{\|w, \theta; t\|} dt, \ \{\nu, \eta\} \in \tilde{H}.$$

From (2.99), we obtain

$$\Phi(\{\nu, \eta\}) = \lim_{\nu \to \infty} \Phi(\{w_\nu, \theta_\nu\})$$

$$= \lim_{\nu \to \infty} \int_M \frac{< w, \theta; w_\nu, \theta_\nu; t >}{\|w, \theta; t\|} dt \leq \overline{\lim_{\nu \to \infty}} \int_M \|w_\nu, \theta_\nu; t\| \, dt \leq$$

$$\leq (m(g, w, T))^{1/2} mes \ M, \tag{2.102}$$

and also we get

$$\Phi(\{w, \theta\}) = \int_M \|w, \theta; t\| \, dt \geq C^{1/2} mes \ M. \tag{2.103}$$

From (2.102), (2.103), owing to the inequality $C > m(g, w, T)$, one obtains $mes \ M = 0$. Therefore, for an arbitrary $C > m(g, w, T)$, the measure of the set of $t \in [0,T]$ for which the inequality (2.101) is satisfied equals zero. In what follows, almost always ($t \in [0,T]$), $\|w, \theta, t\| \leq (m(g, w, T))^{1/2}$, i.e. for the pair of functions $\{w, \theta\}$ a prior estimate of type (2.99) can be used:

$$\left\|I_1^{1/2}w_n'(t)\right\|_{H_1}^2 + \left\|L^{1/2}w_n(t)\right\|_{H_1}^2 + \left\|I_2^{1/2}\theta_n(t)\right\|_{H_2}^2$$

$$+2\int_0^t \left\|K^{1/2}\theta(\tau)\right\|_{H_2}^2 d\tau \le m(g,w,T), \tag{2.104}$$

(almost always ($t \in [0,T]$)).

We now show that $\{w,\theta\}$ is a general solution to the problem (2.54)–(2.56). From (2.104), we obtain

$$w \in W_\infty^2(0,T;H_1) \cap L_\infty(0,T;D(L^{1/2})),$$
$$\theta \in L_\infty(0,T;H_2) \cap L_2(0,T;D(K^{1/2})).$$

The relation $w(0) = w_0$ and the integral identities (2.60), (2.61) are obtained on the basis of (2.95) using a limiting transition similar to that described in [139]. To sum up, the pair of functions $\{w,\theta\}$ satisfies all requirements of Definition 1.1. Therefore, the existence of a generalized solution to the problem (2.54)–(2.56) has been proved.

Now we are going to prove the uniqueness of the generalized solution to the problem (2.54)–(2.56). For this purpose it is sufficient to show that the corresponding homogeneous problem possesses only a trivial solution. Let us make the following substitutions in the integral identities (2.60), (2.61), corresponding to the case of homogeneous data:

$$g_1(t) \equiv 0,\ g_2(t) \equiv 0,\ w_0 = w_1 = 0,\ \theta_0 = 0,\ \nu = 2\nu_\tau,\ \eta = 2\eta_\tau,$$

where $0 \le \tau \le T$, and

$$\nu_\tau(t) = \begin{cases} \int_t^\tau w(\xi)\,d\xi, & t \le \tau, \\ 0, & t > \tau, \end{cases}$$

$$\eta_\tau(t) = \begin{cases} \int_t^\tau \int_0^\xi \theta(\sigma)\,d\sigma\,d\xi, & t \le \tau, \\ 0, & t > \tau. \end{cases}$$

We add together the relations obtained, perform integration by parts and use the condition (2.58) again. As a result, we obtain

$$\left\|I_1^{1/2}w_n(\tau)\right\|_{H_1}^2 + \left\|L^{1/2}\int_0^\tau w(\xi)\,d\xi\right\|_{H_1}^2$$

$$+\left\|I_1^{1/2}\int_0^\tau \theta(\sigma)\,d\sigma\right\|_{H_2}^2 + 2\int_0^\tau \left\|K^{1/2}\int_0^t \theta(\sigma)\,d\sigma\right\|_{H_2}^2 dt = 0.$$

This implies that $w(\tau) \equiv 0$, $\theta(\tau) \equiv 0$, $0 \le \tau \le T$, i.e. a homogeneous problem of the form (2.54)–(2.56) possesses only a trivial solution. Therefore, the uniqueness of the generalized solution to the problem (2.54)–(2.56) has been established. Therefore, Theorem 2.1 has been fully proved.

Remark 2.5 *In Chap. 3, in deriving a prior estimate for the Bubnov–Galerkin method, the case of homogeneous initial conditions (2.56) will be considered:* $w(0) = w'(0) = 0$, $\theta(0) = 0$. *Therefore, we take* $w_{0n} = w_{1n} = 0$, $\theta_{0n} = 0$ *in (2.95) also. It should be noted that if* w *as defined by (2.98) is constant, then*

$$m(g, w, T) = eT \left\| I^{-1/2} g \right\|^2_{L_2(0,T;H_1 \times H_2)},$$

and the estimates (2.99), (2.104) are reduced to the form

$$\left\| I_1^{1/2} w_n'(t) \right\|^2_{H_1} + \left\| L^{1/2} w_n(t) \right\|^2_{H_1} + \left\| I_2^{1/2} \theta_n(t) \right\|^2_{H_2}$$
$$+ 2 \int_0^t \left\| K^{1/2} \theta_n(\tau) \right\|^2_{H_2} d\tau \le eT \left\| I^{-1/2} g \right\|^2_{L_2(0,T;H_1 \times H_2)}, \tag{2.105}$$

$$\left\| I_1^{1/2} w_n'(t) \right\|^2_{H_1} + \left\| L^{1/2} w(t) \right\|^2_{H_1} + \left\| I_2^{1/2} \theta(t) \right\|^2_{H_2}$$
$$+2 \int_0^t \left\| K^{1/2} \theta(\tau) \right\|^2_{H_2} d\tau \le eT \left\| I^{-1/2} g \right\|^2_{L_2(0,T;H_1 \times H_2)}, \tag{2.106}$$

almost always ($t \in [0, T]$).

In our further considerations, special systems composed of eigenelements of certain operators (convergent and making an acute angle with the operators L, K) will be used as the bases for the Bubnov–Galerkin method. Let us introduce the definitions needed.

Definition 2.3 [246]. *The self-adjoint positive definite operators* A *and* A_0 *acting in the same Hilbert space* H *are called convergent if* $D(A) = D(A_0)$.

It is known (see [246]) that if the operators A and A_0 are convergent, then $D(A^{1/2}) = D(A_0^{1/2})$, and all of the operators AA_0^{-1}, $A^{-1}A_0$, $A^{1/2}A_0^{-1/2}$, $A^{-1/2}A_0^{1/2}$ are bounded in H, together with their inverses.

Definition 2.4 [126]. *Let* A, A_0 *be self-adjoint positive definite operators acting in the same Hilbert space. The operators* A *and* A_0 *are said to make an acute angle if*

$$\gamma(A, A_0) \equiv \inf_{u \in D(A) \cap D(A_0)} \frac{(Au, A_0 u)_H}{\left\| A_0 u \right\|^2_H} > 0. \tag{2.107}$$

Let L_0, K_0 be self-adjoint positive definite operators that are convergent and make an acute angle with the operators L and K, respectively. Let $\{\nu_i\}_{i=1}^{\infty}$ be a basis in H_1 composed of eigenelements of the operator L_0, and let $\{\eta_j\}_{j=1}^{\infty}$ be a basis in H_2 composed of eigenelements of the operator K_0. The eigenelements of the operators L_0 and K_0 will be denoted by λ_i and $æ_j$, respectively: $L_0\nu_i = \lambda_i\nu_i$, $K_0\eta_j = æ_j\eta_j$, $i, j = 1, 2, \ldots$. It should be noted that because the operators L^{-1}, K^{-1} are fully continuous, the operators L_0^{-1}, K_0^{-1} are also fully continuous, which implies that

$$\lambda_i \to \infty, \quad æ_j \to \infty \quad (i, j \to \infty). \tag{2.108}$$

The systems of elements $\{\nu_i\}_{i=1}^{\infty}$, $\{\eta_j\}_{j=1}^{\infty}$ will be used in what follows as the basis for the equations of the Bubnov–Galerkin method (2.95). As the initial data for the problem (2.95), we use the projections of the initial data (2.56) onto the finitely measurable subspaces:

$$w_{0n} = P_{H_1^n} w_0, \quad w_{1n} = P_{H_1^n} w_1, \quad \theta_{0n} = P_{H_1^n}\theta_0. \tag{2.109}$$

The following stronger conditions are applied to the given functions g_1, g_2 and the elements w_0, w_1, θ_0 (instead of (2.57) and (2.59)):

$$g_1 \in W_2^1(0, T; H_1), \quad g_2 \in W_2^1(0, T; H_2), \tag{2.110}$$

$$w_0 \in D(L), \quad w_1 \in D(L^{1/2}), \quad \theta_0 \in D(K). \tag{2.111}$$

Theorem 2.2 *If the conditions (2.58), (2.110), (2.111) are satisfied then the "classical" solution to the problem (2.54)–(2.56) exists and is unique.*

Proof. Let us define the existence of a "classical" solution to the problem (2.54)–(2.56). We differentiate the first two relations of (2.95), perform scalar multiplication by $2w_n''(t)$ and $2\theta_n'(t)$, and sum the resulting expression with the use of (2.58). We integrate expression obtained over the interval $[0, T]$, where $t \le T$, and finally obtain

$$\begin{aligned}
&\left\|I_1^{1/2} w_n''(t)\right\|_{H_1}^2 + \left\|L^{1/2} w_n'(t)\right\|_{H_1}^2 + \left\|I_2^{1/2}\theta_n(t)\right\|_{H_2}^2 \\
&+ 2\int_0^t \left\|K^{1/2}\theta_n'(\tau)\right\|_{H_2}^2 d\tau = \left\|I_1^{1/2} w_n''(0)\right\|_{H_1}^2 + \left\|L^{1/2} w_{1n}\right\|_{H_1}^2 \\
&+ \left\|I_1^{1/2}\theta_n'(0)\right\|_{H_1}^2 + 2\int_0^t \left[(g_1'(\tau), w_n''(\tau))_{H_1} + (g_2'(\tau), \theta_n''(\tau))_{H_2}\right] d\tau.
\end{aligned}$$

After applying the Cauchy inequality to the right-hand side of the equation obtained, we obtain

$$\left\|I_1^{1/2}w_n''(t)\right\|_{H_1}^2 + \left\|L^{1/2}w_n'(t)\right\|_{H_1}^2 + \left\|I_2^{1/2}\theta_n'(t)\right\|_{H_2}^2$$
$$+2\int_0^t \left\|K^{1/2}\theta_n'(\tau)\right\|_{H_2}^2 d\tau \le \chi + \frac{1}{\delta}\left\|I^{-1/2}q'\right\|_{L_2(0,T;H_1\times H_2)}^2 \qquad (2.112)$$
$$+\delta\int_0^t \left[\left\|I_1^{1/2}w_n''(\tau)\right\|_{H_1}^2 + \left\|I_1^{1/2}\theta_n'(\tau)\right\|_{H_2}^2\right] d\tau,$$

where $I^{1/2}g' = \{I_1^{1/2}g_1',\ I_2^{1/2}g_n'\}$, δ is an arbitrary positive number, and

$$\chi = \sup_n \left[\left\|I_1^{1/2}w_n''(0)\right\|_{H_1}^2 + \left\|L^{1/2}w_{1n}\right\|_{H_1}^2 + \left\|I_1^{1/2}\theta_n'(0)\right\|_{H_2}^2\right]. \qquad (2.113)$$

From the first two relations of (2.95), taking into account the initial data $w_n(0) = w_{0n}$, $w_n'(0) = w_{1n}$, $\theta_n(0) = \theta_{0n}$, we obtain the following inequalities:

$$\left\|I_1^{1/2}w_n''(0)\right\|_{H_1} \le \left\|I_1^{-1/2}(g_1(0) - Lw_{0n} - M\theta_{0n})\right\|_{H_1}, \qquad (2.114)$$
$$\left\|I_1^{1/2}\theta_n'(0)\right\|_{H_2} \le \left\|I_1^{-1/2}(g_2(0) - K\theta_{0n} - Nw_{1n})\right\|_{H_2}. \qquad (2.115)$$

Owing to the special choice of basis systems $\{\nu_i\}_{i=1}^\infty$, $\{\eta_j\}_{j=1}^\infty$, (2.109) and (2.111) imply that

$$Lw_{0n} \to Lw_0,\quad L^{1/2}w_{1n} \to L^{1/2}w_1 \text{ in } H_1,\quad K\theta_{0n} \to K\theta \text{ in } H_2. \qquad (2.116)$$

From (2.113)–(2.116) and the boundedness conditions on the operators MK^{-1}, $NL^{-1/2}$, the following estimate is obtained:

$$\chi \le \sup_n \left[\left\|I_1^{-1/2}(g_1(0) - Lw_{0n} - M\theta_{0n})\right\|_{H_1}^2 + \left\|L^{1/2}w_{1n}\right\|_{H_1}^2\right.$$
$$\left. + \left\|I_1^{-1/2}(g_2(0) - K\theta_{0n} - Nw_{1n})\right\|_{H_2}^2\right] < \infty. \qquad (2.117)$$

Applying the Gronwill lemma to the inequality (2.112) leads to the result

$$\left\|I_1^{1/2}w_n''(t)\right\|_{H_1}^2 + \left\|L^{1/2}w_n'(t)\right\|_{H_1}^2 + \left\|I_2^{1/2}\theta_n(t)\right\|_{H_2}^2$$
$$+2\int_0^t \left\|K^{1/2}\theta_n'(\tau)\right\|_{H_2}^2 d\tau \le \chi + \frac{1}{\delta}\left(\left\|I^{-1/2}q'\right\|_{L_2(0,T;H_1\times H_2)}^2\right)\exp(\delta T).$$

Minimizing the right-hand side of the last inequality and taking account of the fact that $\delta > 0$, we obtain the following prior estimate of the approximate solution $\{w_n, \theta_n\}$ to the problem (2.54)–(2.56):

$$\left\| I_1^{1/2} w_n''(t) \right\|_{H_1}^2 + \left\| L^{1/2} w_n'(t) \right\|_{H_1}^2 + \left\| I_2^{1/2} \theta_n'(t) \right\|_{H_2}^2 + 2 \int_0^t \left\| K^{1/2} \theta_n'(\tau) \right\|_{H_2}^2 d\tau \le m(g_1, \chi, T), \tag{2.118}$$

where $m(\cdot,\cdot,\cdot)$ is defined by (2.100). The estimate (2.118) holds for every n almost always ($t \in [0,T]$).

Furthermore, owing to the definition (2.107) of the quantities $\gamma(L, L_0)$, $\gamma(K, K_0)$ for the pairs of operators L, L_0 and K, K_0, the following relations are obtained from the first two expressions of (2.95):

$$\begin{aligned} \gamma(L, L_0) \left\| L_0 w_n(t) \right\|_{H_1}^2 &\le (L w_n(t), L_0 w_n(t))_{H_1} \\ &= (g_1(t) - I_1 w_n''(t) - M\theta_n(t), L_0 w_n(t))_{H_1} \\ &\le \Big[\left\| g_1(t) \right\|_{H_1} + \left\| I_1^{1/2} \right\|_{H_1 \to H_1} \left\| I_1^{1/2} w_n''(t) \right\|_{H_1} \\ &+ \left\| M K^{-1} \right\|_{H_2 \to H_1} \left\| K\theta_n(t) \right\|_{H_2} \Big] \left\| L_0 w_n(t) \right\|_{H_1}, \end{aligned} \tag{2.119}$$

$$\begin{aligned} &\gamma(K, K_0) \left\| K_0 \theta_n(t) \right\|_{H_2}^2 \le (K\theta_n(t), K_0\theta_n(t))_{H_2} \\ &= (g_2(t) - I_2 \theta_n'(t) - N w_n'(t), K_0 \theta_n(t))_{H_2} \le \Big[\left\| g_2(t) \right\|_{H_2} \\ &+ b_1 \left(\left\| I_2^{1/2} \theta_n'(t) \right\|_{H_2}^2 + \left\| L^{1/2} w_n'(t) \right\|_{H_1}^2 \right)^{1/2} \Big] \left\| K_0 \theta_n(t) \right\|_{H_2}, \end{aligned} \tag{2.120}$$

where

$$B_1 = \left(\left\| I_2^{1/2} \right\|_{H_2 \to H_2}^2 + \left\| N L^{-1/2} \right\|_{H_1 \to H_2}^2 \right)^{1/2}. \tag{2.121}$$

Applying the inequality of an acute angle (2.107) to the operator pairs L, L_0 and K, K_0 and applying the boundedness condition on the operators $L L_0^{-1}$, $K K_0^{-1}$, the following prior estimate of the components of the approximate solution to the problem (2.54)–(2.56) are obtained from (2.119) and (2.120):

$$\begin{aligned} \left\| L w_n(t) \right\|_{H_1} &\le c_L \Big[\left\| g_1(t) \right\|_{H_1} + \left\| M K^{-1} \right\|_{H_2 \to H_1} c_K \left\| g_2(t) \right\|_{H_2} \\ &+ b_2 \left(\left\| I_2^{1/2} w_n''(t) \right\|_{H_1}^2 + \left\| L^{1/2} w_n'(t) \right\|_{H_1}^2 + \left\| I_2^{1/2} \theta_n'(t) \right\|_{H_2}^2 \right)^{1/2} \Big], \end{aligned} \tag{2.122}$$

$$\left\| K\theta_n(t) \right\|_{H_2} \le c_K \left\| g_2(t) \right\|_{H_2} + b_1 \left(\left\| I_2^{1/2} \theta_n'(t) \right\|_{H_2}^2 + \left\| L^{1/2} w_n'(t) \right\|_{H_1}^2 \right)^{1/2}, \tag{2.123}$$

where

$$c_L = \frac{\|LL_0^{-1}\|_{H_1 \to H_1}}{\gamma(LL_0)}, \quad c_K = \frac{\|KK_0^{-1}\|_{H_2 \to H_2}}{\gamma(KK_0)}, \tag{2.124}$$

$$b_2 = \left(\|MK^{-1}\|^2_{H_2 \to H_1} c_k^2 b_1^2 + \|I^{1/2}\|^2_{H_1 \to H_1}\right)^{1/2}. \tag{2.125}$$

The estimates (2.122), (2.123) are satisfied for all n, almost always ($t \in [0, T]$).

Let $\tilde{\tilde{H}}$ be the Hilbert space $[W_2^2(0,T; H_1) \cap W_2^1(0,T; D(L^{1/2})) \cap L_2(0,T; D(L))] \times [W_2^1(0,T; D(K^{1/2})) \cap L_2(0,T; D(K))]$. According to the prior estimates (2.118), (2.122), (2.123), the sequence of approximate solutions $\{w_n, \theta_n\}$ is bounded in $\tilde{\tilde{H}}$. Because any bounded set in a Hilbert space is weakly compact, a special subsequence of approximate solutions $\{w_\nu, \theta_\nu\}$ exists which is weakly convergent in $\tilde{\tilde{H}}$. Let us denote its weak limit by $\{w, \theta\}$. We shall show that the pair of functions $\{w, \theta\}$ is the "classical" solution to the problem (2.54)–(2.56).

In a manner analogous to that used in the proof of Theorem 1.1, an estimate of (w, θ) similar to (2.104) can be obtained from (2.99). Taking into account (2.118), (2.122) and (2.123), the following prior estimates are obtained:

$$\left\|I_1^{1/2} w_n''(t)\right\|^2_{H_1} + \left\|L^{1/2} w_n'(t)\right\|^2_{H_1} + \left\|I_2^{1/2} \theta_n'(t)\right\|^2_{H_2} + 2\int_0^t \left\|K^{1/2}\theta_n'(\tau)\right\|^2_{H_2} d\tau \le m(g', \chi, T), \tag{2.126}$$

$$\left\|Lw(t)\right\|_{H_1} \le \mathit{const} < \infty, \quad \left\|Kw(t)\right\|_{H_2} \le \mathit{const} < \infty, \tag{2.127}$$

valid almost always ($t \in [0, T]$) (the last two estimates were obtained using $g_1 \in L_\infty(0,T; H_1)$, $g_2 \in L_\infty(0,T,H_2)$, using the conditions (2.110)). The estimates (2.126), (2.127) imply

$$w \in W_\infty^2(0,T; H_1) \cap W_\infty^1(0,T; D(L^{1/2}) \cap L_\infty(0,T; D(L)),$$

$$\theta \in W_\infty^1(0,T; H_2) \cap W_2^1(0,T; D(K^{1/2}) \cap L_\infty(0,T; D(K)).$$

The uniqueness of the "classical" solution of the problem (2.54)–(2.56) follows from Theorem 2.1, because if the "classical" solution to the problem (2.54)–(2.56) exists then it is also the general solution. Thus, Theorem 2.2 has been proved.

Remark 2.6 *For later convenience, we evaluate separately the variants of the prior estimates (2.118), (2.126) which correspond to the case of homogeneous initial conditions (2.56). In this case we take* $w_{0n} = w_{1n} = 0$, $\theta_{0n} = 0$

in (2.95), and therefore the estimate (2.117) of the constant χ defined by (2.113) implies $x \leq \left\| I^{-1/2} g(0) \right\|^2_{H_1 \times H_2}$. Using the latter inequality, we make an estimate from above of the right-hand sides of the inequalities (2.118), (2.126), and we obtain

$$\left\| I_1^{1/2} w_n''(t) \right\|^2_{H_1} + \left\| L^{1/2} w_n'(t) \right\|^2_{H_1} + \left\| I_2^{1/2} \theta_n'(t) \right\|^2_{H_2} + 2 \int_0^t \left\| K^{1/2} \theta_n'(\tau) \right\|^2_{H_2} d\tau \leq M_g(T), \tag{2.128}$$

$$\left\| I_1^{1/2} w''(t) \right\|^2_{H_1} + \left\| L^{1/2} w'(t) \right\|^2_{H_1} + \left\| I_2^{1/2} \theta'(t) \right\|^2_{H_2} + 2 \int_0^t \left\| K^{1/2} \theta'(\tau) \right\|^2_{H_2} d\tau \leq M_g(T), \tag{2.129}$$

for $t \in [0, T]$, where

$$\begin{aligned} M_g(T) &= m(g', \left\| I^{-1/2} g(0) \right\|^2_{H_1 \times H_2}, T) \\ &\equiv \Bigg\{ \left\| I^{-1/2} g(0) \right\|^2_{H_1 \times H_2} + \frac{1}{2} \Bigg[\Bigg(\left\| I^{-1/2} g' \right\|^4_{L_2(0,T;H_1 \times H_2)} T^2 \\ &+ 4 \left\| I^{-1/2} g(0) \right\|^2_{H_1 \times H_2} \left\| I^{-1/2} g' \right\|^2_{L_2(0,T;H_1 \times H_2)} T \Bigg)^{1/2} \\ &+ \left\| I^{-1/2} g' \right\|^2_{L_2(0,T;H_1 \times H_2)} T \Bigg] \Bigg\} \exp \Bigg\{ 2 \left\| I^{-1/2} g' \right\|^2_{L_2(0,T;H_1 \times H_2)} \\ &\times T \Bigg[\Bigg(\left\| I^{-1/2} g' \right\|^4_{L_2(0,T;H_1 \times H_2)} T^2 + 4 \left\| I^{-1/2} g' \right\|^2_{L_2(0,T;H_1 \times H_2)} \\ &\times \left\| I^{-1/2} g(0) \right\|^2_{H_1 \times H_2} T \Bigg)^{1/2} + \left\| I^{-1/2} g' \right\|^2_{L_2(0,T;H_1 \times H_2)} T \Bigg]^{-1} \Bigg\}. \end{aligned} \tag{2.130}$$

Finally, a theorem about the existence and uniqueness of solutions to coupled thermoelastic problems of shallow shells derived from Theorems 2.1 and 2.2 will be formulated.

Theorem 2.3 *If the conditions (2.63)–(2.68) are satisfied, then the initial–boundary value problem for the system of differential equations (2.20)–(2.23) with the initial conditions (2.33)–(2.37) and homogeneous boundary conditions has a unique (general) solution $\{u, v, w, \Psi_x, \Psi_y, \theta\}$ satisfying the conditions*

$$\{u, v, w, \Psi_x, \Psi_y\} \in W^1_\infty(0, T; (L_2(\Omega_1))^5) \cap L_\infty(0, T; (\mathring{W}^1_2(\Omega_1))^5),$$
$$\theta \in L_\infty(0, T; L_2(\Omega_2)) \cap L_2(0, T; \mathring{W}^1_2(\Omega_2)).$$

Theorem 2.4 *If the conditions (2.63), (2.64) and*

$$p_1, p_2, q_1 \in W_2^1(0,T; L_2(\Omega_1)),\ \ q_2 \in W_2^1(0,T; L_2(\Omega_2)),$$
$$\overline{\omega}_0 = \{u_0, v_0, w_0, \Psi_{x0}, \Psi_{y0}\} \in (\overset{\circ}{W}{}_2^2(\Omega_1))^5,$$
$$\overline{\omega}_1 = \{u_1, v_1, w_1, \Psi_{x1}, \Psi_{y1}\} \in (\overset{\circ}{W}{}_2^1(\Omega_1))^5,$$
$$\theta \in \overset{\circ}{W}{}_2^2(\Omega_2),$$

are satisfied, then the initial–boundary value problem for the system of differential equations (2.20)–(2.23) with the initial conditions (2.33)–(2.37) and homogeneous boundary conditions has a unique ("classical") solution $\{u, v, w, \Psi_x, \Psi_y, \theta\}$ *satisfying the conditions*

$$\{u, v, w, \Psi_x, \Psi_y\} \in W_\infty^2(0,T; (L_2(\Omega_1))^5)$$
$$\cap\ W_\infty^1(0,T; (\overset{\circ}{W}{}_2^1(\Omega_1))^5) \cap L_\infty(0,T; (\overset{\circ}{W}{}_2^2(\Omega_1))^5),$$
$$\theta \in W_\infty^1(0,T; L_2(\Omega_2)) \cap W_2^1(0,T; \overset{\circ}{W}{}_2^1(\Omega_2))$$
$$\cap\ L_\infty(0,T; \overset{\circ}{W}{}_2^2(\Omega_2)).$$

Theorem 2.5 *If the conditions (2.64), (2.65), (2.68) and (2.83)–(2.85) are satisfied, then the initial–boundary value problem for the system of differential equations (2.30)–(2.32) with the initial conditions (2.33)–(2.35) and homogeneous boundary conditions has a unique (general) solution* $\{u, v, w, \theta\}$ *satisfying the conditions*

$$\{u, v, w\} \in W_\infty^1(0,T; (L_2(\Omega_1))^3) \cap L_\infty(0,T; (\overset{\circ}{W}{}_2^1(\Omega_2))^2 \times \overset{\circ}{W}{}_2^2(\Omega_1)),$$
$$\theta \in L_\infty(0,T; L_2(\Omega_2)) \cap L_2(0,T; \overset{\circ}{W}{}_2^1(\Omega_2)).$$

Theorem 2.6 *If the conditions (2.64), (2.83) and*

$$p_1, p_2, q_1 \in W_2^1(0,T; L_2(\Omega_1)),\ \ q_2 \in W_2^1(0,T; L_2(\Omega_2)),$$
$$\overline{\omega}_0 = \{u_0, v_0, w_0\} \in (\overset{\circ}{W}{}_2^2(\Omega_1))^2 \times \overset{\circ}{W}{}_2^4(\Omega_1),$$
$$\overline{\omega}_1 = \{u_1, v_1, w_1\} \in (\overset{\circ}{W}{}_2^1(\Omega_1))^2 \times \overset{\circ}{W}{}_2^2(\Omega_1),$$
$$\theta_0 \in \overset{\circ}{W}{}_2^2(\Omega_2),$$

are satisfied, then the initial–boundary value problem for the system of differential equations (2.30)–(2.32) with the initial conditions (2.33)–(2.35) and homogeneous boundary conditions has a unique ("classical") solution $\{u, v, w, \theta\}$ *satisfying the conditions*

$$\{u,v,w\} \in W^2_\infty(0,T;(L_2(\Omega_1))^3)$$
$$\cap\ W^1_\infty(0,T;(\mathring{W}^1_2(\Omega_1))^2 \times \mathring{W}^2_2(\Omega_1))$$
$$\cap\ L_\infty(0,T;(\mathring{W}^2_2(\Omega_1))^2 \times \mathring{W}^4_2(\Omega_1)),$$

$$\theta \in W^1_\infty(0,T;L_2(\Omega_2)) \cap W^1_2(0,T;\mathring{W}^1_2(\Omega_2)) \cap L_\infty(0,T;L_2\mathring{W}^2_2(\Omega_1)).$$

These results can be obtained directly from Theorems 2.1 and 2.2 when either the relations (2.69)–(2.81) (for (2.20)–(2.23) with the initial conditions (2.33)–(2.37)) or the relations (2.86)–(2.94) (for (2.30)–(2.32) with the initial conditions (2.33)–(2.35)) hold.

2.2 Cylindrical Panel Within Transonic Gas Flow

2.2.1 Statement and Solution of the Problem

2.2.1.1 Equations of Motion and Boundary Conditions of a Panel. We start with the fundamental relations governing the dynamics of an elastic, infinite, cylindrical panel. We assume that the shell has a constant thickness, is shallow and slender, and is made of an isotropic material for which Hooke's law holds. We assume also that the deformations are small and that the Kirchhoff–Love hypothesis hold [232]. In addition, we neglect the stresses normal to the middle surface. We assume that the properties of the shell material do not depend on temperature, and we consider dynamic processes to occur in the shell without propagation of longitudinal elastic waves over a finite distance (i.e. we assume that the velocity of those waves is infinitely high). A right-handed Cartesian coordinate system x, y, z is used. The z axis coincides with the normal to the middle surface of the shell, with its direction pointing towards the initial centre of curvature. The cylindrical panel, infinitely long, is part of a shallow cylinder. The panel has rectangular sides along generators of the cylinder. It is assumed that all physical properties in all perpendicular cross-sections, i.e. cross-sections perpendicular to the generators of the infinite cylindrical shell, are the same. Therefore, all of the quantities are independent of coordinate x. In this case we take

$$u = 0, \quad P_x = 0.$$

The equations governing the motion of an elastic, infinite, cylindrical panel have been cited in many publications. However, those nonlinear equations were first derived by Mushtari and Galimov [172]. Following [232], we briefly sketch the derivation of the equations for such a panel starting from a variational principle.

The Hamilton–Ostrogradskij principle has the form

$$\delta \int_{t'_0}^{t'_1} (K - \Pi)\, dt' + \delta' \int_{t'_0}^{t'_1} E\, dt' = 0. \tag{2.131}$$

The geometrical relations for the deformations can be described through the displacements of the middle surface of the infinite cylindrical panel in the following form:

$$\begin{aligned}
\gamma_{11}^z = 0, \quad \gamma_{12}^z = 0, \quad \gamma_{23}^z = 0, \\
\gamma_{23}^z = -\frac{h^2 \partial^3 w'}{6(1-\nu)\partial y^3}, \\
\gamma_{22}^z = \frac{\partial v}{\partial y} - \frac{w'}{R} + \frac{1}{2}\left(\frac{\partial w'}{\partial y}\right)^2 - z\left(\frac{\partial^2 w'}{\partial y^2}\right), \\
\gamma_{33}^z = -\frac{\nu}{1-\nu}\left(\frac{\partial v}{\partial y} - \frac{w'}{R} + \frac{1}{2}\left(\frac{\partial w'}{\partial y}\right)^2 - z\frac{\partial^2 w'}{\partial y^2}\right).
\end{aligned} \tag{2.132}$$

By applying Hooke's law to the stresses, the following formulas are obtained:

$$\begin{aligned}
P_{11}^z = \frac{E'\nu}{1-\nu^2}\left(\frac{\partial v}{\partial y} - \frac{w'}{R} + \frac{1}{2}\left(\frac{\partial w'}{\partial y}\right)^2 - z\frac{\partial^2 w'}{\partial y^2}\right), \\
P_{12}^z = 0, \quad P_{13}^z = 0, \quad P_{23}^z = -\frac{E'h^2}{12(1-\nu^2)}\frac{\partial^3 w'}{\partial y^3}, \quad P_{33}^z = 0, \\
P_{22}^z = \frac{E'}{1-\nu^2}\left(\frac{\partial v}{\partial y} - \frac{w'}{R} + \frac{1}{2}\left(\frac{\partial w'}{\partial y}\right)^2 - z\frac{\partial^2 w'}{\partial y^2}\right).
\end{aligned} \tag{2.133}$$

We now define the internal forces and stresses averaged over a transverse cross-section, parallel to the y axis, of the infinite cylindrical shell. Substituting the expressions in (2.133) into the formulas for the internal forces and moments, we obtain

$$\begin{aligned}
N_x &= \frac{E'h\nu}{1-\nu^2}\left(\frac{\partial v}{\partial y} - \frac{w'}{R} + \frac{1}{2}\left(\frac{\partial w'}{\partial y}\right)^2\right), \\
N_y &= \frac{E'h}{1-\nu^2}\left(\frac{\partial v}{\partial y} - \frac{w'}{R} + \frac{1}{2}\left(\frac{\partial w'}{\partial y}\right)^2\right), \\
Q_x &= 0, \\
Q_y &= -\frac{E'h^3}{12(1-\nu^2)}\frac{\partial^3 w'}{\partial y^3}, \\
T &= 0, \\
M_x &= -\frac{E'h^3\nu}{12(1-\nu^2)}\frac{\partial^2 w'}{\partial y^2}, \\
M_y &= -\frac{E'h^3}{12(1-\nu^2)}\frac{\partial^2 w'}{\partial y^2}, \\
M_k &= 0.
\end{aligned} \tag{2.134}$$

From (2.131), the following variational equation for the infinite circular cylindrical shell in the time interval from t'_0 to t'_1 is obtained:

$$\int_{t'_0}^{t'_1}\int_{0}^{L}\int_{-\infty}^{+\infty}\left\{\left(\frac{\partial N_y}{\partial y}+P_y\right)\delta v+\left(\frac{N_y}{R}+P'_z-\rho'_0 h\frac{\partial^2 w'}{\partial t'^2}+\frac{\partial}{\partial y}\left(N_y\frac{\partial w'}{\partial y}\right)\right.\right.$$

$$\left.\left.+\frac{\partial Q_y}{\partial y}\right)\delta w'-\left(\frac{\partial M_y}{\partial y}-Q_y\right)\delta\left(\frac{\partial w'}{\partial y}\right)\right\}dx\,dy\,dt'$$

$$-\int_{t'_0}^{t'_1}\int_{-\infty}^{+\infty}\left\{N_y\delta v+\left(N_y\frac{\partial w'}{\partial y}+Q_y\right)\delta w'-M_y\delta\left(\frac{\partial w'}{\partial y}\right)\right\}\Bigg|_{y=0}^{y=L}dx\,dt' \quad (2.135)$$

$$+\rho'_0 h\int_{0}^{L}\int_{-\infty}^{+\infty}\left(\frac{\partial v}{\partial t'}\delta v+\frac{\partial w'}{\partial t'}\delta w'\right)\Bigg|_{t'=t'_0}^{t'=t'_1}dx\,dy=0.$$

In this equation the force N_y, the moment M_y and the perpendicular force Q_y are defined via the displacements of the middle surface through the formulas (2.134). In (2.135), all variations δv, $\delta w'$, $\delta\left(\frac{\partial w'}{\partial y}\right)$ are considered as functions of the time t'. Therefore, from the first integral of (2.135) we obtain the following system of three differential equations governing the motion of the infinite circular cylindrical panel:

$$\frac{\partial N_y}{\partial y}+P_y=0, \quad (2.136)$$

$$\frac{\partial Q_y}{\partial y}+\frac{N_y}{R}+\frac{\partial}{\partial y}\left(N_y\frac{\partial w'}{\partial y}\right)+P'_z-\rho'_0 h\frac{\partial^2 w'}{\partial t'^2}=0, \quad (2.137)$$

$$Q_y-\frac{\partial M_y}{\partial y}=0. \quad (2.138)$$

From the last equation (2.138) and the relations (2.134), one can observe that (2.138) is identically satisfied.

From the second integral of the variational equation (2.135), we deduce that the following three boundary conditions on the boundaries $y=0$, $y=L$ of the panel must be satisfied:

1. The value of the external load acting parallel to the y axis or the values of the displacements of the edge points of the middle surface in the direction of y axis are given:

$$N_y=N_y^0(t') \quad \text{or} \quad v=v^0(t'). \quad (2.139)$$

2. The value of the external transverse load or of the normal displacement of the edge points is given:

$$Q_y + \left(N_y \frac{\partial w'}{\partial y} \right) = Q_y^0(t') \quad \text{or} \quad w' = w^0(t'). \tag{2.140}$$

3. The external bending moment or the angle of rotation of the normal to an element of the shell around an axis are given:

$$M_y = M_y^0(t') \quad \text{or} \quad \frac{\partial w'}{\partial y} = \alpha_0(t'), \tag{2.141}$$

where $N_y^0(t')$, $v_y^0(t')$, $Q_y^0(t')$, $M_y^0(t')$ and $\alpha_y^0(t')$ are time-dependent given functions.

The third integral of the variational equation (2.135) requires initial and boundary conditions, i.e. conditions for either $t' = t_0'$ or $t' = t_1'$:

1. The displacements of each point of the panel along the y axis or their velocities are given:

$$v_y = v_0(y) \quad \text{or} \quad \frac{\partial v}{\partial t'} = v_0'(y). \tag{2.142}$$

2. The normal (to the middle surface) displacements of each point of the panel or their velocities are given:

$$w' = w_0'(y) \quad \text{or} \quad \frac{\partial w'}{\partial t'} = w_0''(y), \tag{2.143}$$

where $v_0(y)$, $v_0'(y)$, $w_0'(y)$ and $w_0''(y)$ are given functions dependent on y.

We assume that the final conditions are applied at an infinitely large time $t_1' \to +\infty$. Therefore, the final conditions do not influence the solution and can be neglected.

If a simple support (on balls) along the linear boundaries is applied, such that only a rotation of the panel around a line parallel to the x axis is possible, then the boundary conditions (2.139), (2.140), (2.141) read

$$v = 0, \; w' = 0, \; M_y = 0 \qquad (y = 0, \; y = L).$$

Using the formula for M_y from (2.134), the above condition can be rewritten in the following way:

$$v = 0, \; w' = 0, \; \frac{\partial^2 w'}{\partial y^2} = 0 \qquad (y = 0, \; y = L). \tag{2.144}$$

When a rigid panel support (clamp) is applied along the linear edges, the boundary conditions (2.139), (2.140), (2.141) have the form

$$v = 0, \; w' = 0, \; \frac{\partial w'}{\partial y} = 0 \qquad (y = 0, \; y = L). \tag{2.145}$$

Both of the boundary conditions (2.144) and (2.145) will be considered later in a problem of transonic aeroelasticity. Integrating (2.136) along the y variable, we obtain

$$N_y(y, t') = -\int_0^y P_y(y, t')\, dy + N_y^0(t'), \tag{2.146}$$

where $N_y^0(t')$ is an arbitrary function of t', and $N_y(0, t') = N_y^0(t')$.

Substituting the expressions (2.134) and (2.136) into the equation of motion (2.137), we obtain the following differential equation:

$$-\frac{E'h^3}{12(1-\nu^2)}\frac{\partial^4 w'}{\partial y^4} + \frac{N_y}{R} - P_y\frac{\partial w'}{\partial y} + N_y\frac{\partial^2 w'}{\partial y^2} + P_z' - \rho_0' h\frac{\partial^2 w'}{\partial t'^2} = 0. \tag{2.147}$$

Now the expressions for N_y from (2.134) are substituted into (2.136) and (2.137), and we obtain the following system:

$$\frac{\partial^2 v}{\partial y^2} + \frac{\partial w'}{\partial y}\frac{\partial^2 w'}{\partial y^2} - \frac{1}{R}\frac{\partial w'}{\partial y} = \frac{1-\nu^2}{E'h}P_y, \tag{2.148}$$

$$\begin{aligned} &\frac{E'h^3}{12(1-\nu^2)}\frac{\partial^4 w'}{\partial y^4} - \frac{E'h}{1-\nu^2}\left(\frac{1}{R} + \frac{\partial^2 w'}{\partial y^2}\right)\left(\frac{\partial v}{\partial y} - \frac{w'}{R} + \left(\frac{\partial w'}{\partial y}\right)^2\right) \\ &\qquad + P_y\frac{\partial w'}{\partial y} + \rho_0' h\frac{\partial^2 w'}{\partial t^2} = P_z'. \end{aligned} \tag{2.149}$$

This governs the motion of the infinite circular cylindrical panel under consideration with a continuously distributed load P_y in the direction of the y axis. Here we consider the interaction of the panel with a flow of an ideal gas. Neglecting tangential damping forces in the gas flow, we obtain

$$P_y = 0. \tag{2.150}$$

Therefore, in accordance with (2.146), the normal force N_y in an arbitrary cross-section $y = const$ is constant at any given time instant t', i.e. the normal internal force N_y depends only on the time t' and is equal to the compressional longitudinal force on the boundary $y = 0$ or to the modulus of the load on the boundary $y = L$ acting in the opposite direction.

Therefore, the assumptions about the infinite length of the cylindrical panel, the stress state discussed above and the infinite velocity of elastic waves along the shell lead to the result that the normal average force N_y at a fixed time moment t' in an arbitrary cross-section $y = const$ is constant. From the second relation of (2.134), by integration from 0 to y, we obtain

$$v = v|_{y=0} + \frac{1-\nu^2}{E'h}\int\limits_0^y N_y\,dy + \int\limits_0^y \frac{w'}{R}\,dy - \frac{1}{2}\int\limits_0^y \left(\frac{\partial w'}{\partial y}\right)^2 dy. \tag{2.151}$$

According to the earlier discussion, we use only the boundary conditions (2.144) and (2.145). Extracting from them the value of v for $y = 0$ and $y = L$ and taking into account the fact that N_y depends only on t', the relation (2.151) can be rewritten in the following way:

$$N_y = \frac{E'h}{(1-\nu^2)L}\left(\frac{1}{2}\int\limits_0^L \left(\frac{\partial w'}{\partial y}\right)^2 dy - \frac{1}{R}\int\limits_0^L w'\,dy\right). \tag{2.152}$$

The expression (2.151) yields

$$v = v|_{y=0} + \frac{1-\nu^2}{E'h} yN_y + \int\limits_0^y \frac{w'}{R}\,dy - \frac{1}{2}\int\limits_0^y \left(\frac{\partial w'}{\partial y}\right)^2 dy. \tag{2.153}$$

Therefore, for the model under consideration of an infinite cylindrical panel in the case of a purely transverse load (i.e. when (2.150) holds), the equation of motion (2.136) and also the equation (2.148) are transformed to identities.

The condition $v = 0$ from the boundary conditions (2.144) and (2.145) is already satisfied, whereas (2.137), i.e. (2.149) is transformed into the following integral–differential equation:

$$\begin{aligned} &\frac{E'h^3}{12(1-\nu^2)}\frac{\partial^4 w'}{\partial y^4} - \frac{E'h}{L(1-\nu^2)}\left(\frac{1}{R} + \frac{\partial^2 w'}{\partial y^2}\right) \\ &\times \left(\frac{1}{2}\int\limits_0^L \left(\frac{\partial w'}{\partial y}\right)^2 dy - \int\limits_0^l \frac{w'}{R}\,dy\right) + \rho_0' h\frac{\partial^2 w'}{\partial t'^2} = P_z'. \end{aligned} \tag{2.154}$$

This integral–differential equation includes only the normal deflection w' of the middle surface and governs the dynamics of the infinite circular panel in the case under consideration. The following boundary conditions from the relations (2.144) and (2.145) are applied to (2.154):

simple support (on balls):

$$w' = 0, \quad \frac{\partial^2 w'}{\partial y^2} = 0 \quad (y = 0,\ y = L); \tag{2.155}$$

stiff support (clamp):

$$w' = 0, \quad \frac{\partial w'}{\partial y} = 0 \quad (y = 0,\ y = L). \tag{2.156}$$

This is because the equality $v = 0$ from the conditions (2.144) and (2.145) is already satisfied.

Finally, we need to apply to (2.154) the initial conditions (2.142) and (2.143). We need to transform these conditions in order to obtain only one transverse deflection w'.

Substituting N_y from (2.152) into (2.153), we express the displacement v in terms of the transverse displacement w':

$$\begin{aligned} v = v\big|_{y=0} + \frac{y}{L}\left(\frac{1}{2}\int_0^L \left(\frac{\partial w'}{\partial y}\right)^2 dy - \int_0^L \frac{w'}{R}\,dy\right) \\ + \int_0^y \frac{w'}{R}\,dy - \frac{1}{2}\int_0^y \left(\frac{\partial w'}{\partial y}\right)^2 dy. \end{aligned} \tag{2.157}$$

This means that the first equalities in the initial conditions (2.142) and (2.143) include functions $v_0(y)$ and $w_0'(y)$ that are not completely arbitrary. If $w_0'(y)$ is given, then (2.157) uniquely defines $v_0(y)$ and vice versa. We now differentiate the equality (2.157) with respect to the time t', which results in the following conclusion. Setting the functions $w_0'(y)$ and $w_0''(y)$ is same as setting the following pairs of functions: $w_0'(y)$ and $v_0'(y)$; $v_0'(y)$ and $w_0''(y)$; $v_0(y)$ and $w_0''(y)$. This means that the following functions can be taken as the initial functions:

$$w' = w_0'(y), \quad \frac{\partial w'}{\partial t} = w_0''(y) \quad (t' = t_0'). \tag{2.158}$$

Therefore, the problem of the oscillations of a flexible, elastic, infinite, circular cylindrical panel subjected to a transverse load is reduced to an initial–boundary value problem of mathematical physics governed by the integral–differential equation (2.154) with the boundary conditions (2.155) (or (2.156)) and the initial conditions (2.158). Note that the continuum equation, in the case of the infinite cylindrical panel considered here, is satisfied identically.

It turns out that one of the equations obtained possesses only one unknown variable, w', and therefore we can express the other quantities via w'. Differentiating (2.151) with respect to y, we obtain

$$\frac{\partial v}{\partial y} = \frac{1-\nu^2}{E'h}N_y + \frac{w'}{R} - \frac{1}{2}\left(\frac{\partial w'}{\partial y}\right)^2.$$

Substituting N_y from (2.152) into the above equation, we obtain

$$\frac{\partial v}{\partial y} = \frac{1}{L}\left(\frac{1}{2}\int_0^L \left(\frac{\partial w'}{\partial y}\right)^2 \partial y - \int_0^L \frac{w'}{R}\,\partial y\right) + \frac{w'}{R} - \frac{1}{2}\left(\frac{\partial w'}{\partial y}\right)^2.$$

Substituting the expression for $\frac{\partial v}{\partial y}$ from the last equation into (2.132), we obtain

$$\begin{aligned}
\gamma_{11}^z &= 0,\\
\gamma_{22}^z &= \frac{1}{L}\left(\frac{1}{2}\int_0^L\left(\frac{\partial w'}{\partial y}\right)^2 \partial y - \int_0^L \frac{w'}{R}\,\partial y\right) - z\frac{\partial^2 w'}{\partial y^2},\\
\gamma_{12}^z &= 0,\\
\gamma_{13}^z &= 0,\\
\gamma_{23}^z &= -\frac{h^2}{6(1-\nu)}\frac{\partial^3 w'}{\partial y^3},\\
\gamma_{33}^z &= -\frac{\nu}{1-\nu}\left(\frac{1}{L}\left(\frac{1}{2}\int_0^L\left(\frac{\partial w'}{\partial y}\right)^2 dy - \int_0^L \frac{w'}{R}\,dy - z\frac{\partial^2 w'}{\partial y^2}\right)\right).
\end{aligned} \tag{2.159}$$

Therefore, the following expressions for the stress tensor components described via the transverse deflection w' can be obtained from the relation (2.133):

$$\begin{aligned}
P_{11}^z &= \frac{E'\nu}{1-\nu^2}\left(\frac{1}{L}\left(\frac{1}{2}\int_0^L\left(\frac{\partial w'}{\partial y}\right)^2 dy - \int_0^L \frac{w'}{R}\,dy\right) - z\frac{\partial^2 w'}{\partial y^2}\right),\\
P_{22}^z &= \frac{E'}{1-\nu^2}\left(\frac{1}{L}\left(\frac{1}{2}\int_0^L\left(\frac{\partial w'}{\partial y}\right)^2 dy - \int_0^L \frac{w'}{R}\,dy\right) - z\frac{\partial^2 w'}{\partial y^2}\right),\\
P_{12}^z &= 0,\\
P_{13}^z &= 0,\\
P_{23}^z &= -\frac{E'h^2}{12(1-\nu^2)}\frac{\partial^3 w'}{\partial y^3},\\
P_{33}^z &= 0.
\end{aligned} \tag{2.160}$$

These results will be used later.

2.2.1.2 Equations of Transonic Ideal-Gas Motion. Now we investigate the motion of a transonic flow of an ideal gas. Let the gas move in a two-dimensional space with coordinates x_1, x_2. It is assumed that there are no external forces (including volume forces), and that in the volume occupied by the gas there are no sources of gas mass and no input of energy. We consider a gas without internal friction and without heat transfer, i.e. the coefficients of viscosity and thermal conductivity are equal to zero. It is known that gas particles can change their entropy at discontinuities [119, 191, 196]. However, in a first approximation, it can be assumed that transonic flow with discontinuities is isentropic and vortex free and preserves those properties with time [196]. Therefore, such a gas flow remains approximately in a quasi-equilibrium state everywhere.

Derivations of the asymptotic equations of motion of a transonic (and close to homogeneous) flow of an ideal gas have been given before [146, 193, 197]. Our approach slightly differs from the previous ones, because we do not use any asymptotic series with respect to a small perturbation parameter. Instead we use Taylor series of certain dependences in the neighbourhood of critical values of some quantities and use integral estimates. We are going now to present briefly the derivation of the equations.

The integral law of mass conservation of the gas, for the space Ω bounded by the closed contour $\partial\Omega$ on the plane x_1, x_2, in the case under consideration has the following form [191, 196]:

$$\frac{\partial}{\partial t'}\int\limits_{w} \rho'\,dw + \int\limits_{\partial w} \rho'(\overline{q}\,\overline{n})\,d(\partial w) = 0. \tag{2.161}$$

The absence of a vortex in the velocity field implies that there exists a potential $\Phi(x_1, x_2, t')$ for the velocity vector, i.e. [119, 192, 196]

$$\overline{q} = grad\ \Phi. \tag{2.162}$$

As is known, one can derive the following Lagrange–Cauchy integral from the equations of motion [196]:

$$\frac{\partial \Phi}{\partial t'} + \frac{q^2}{2} + \int \frac{dP'}{\rho'(P')} = f_a(t'), \tag{2.163}$$

where $f_a(t')$ is a function of the time t'.

It is assumed that both an ideal and a real gas are governed by the following equation of state equation [119, 191, 196]:

$$P'\rho'^{-\chi} = const. \tag{2.164}$$

We consider the flow of the gas stream to be close to a homogeneous supersonic flow, in the direction of the axis of the profile of the shell, the x_1 axis. It is assumed that the homogeneously flowing gas stream is close to a critical condition determined by the speed of sound. The equation of state (2.164) can be rewritten in the form

$$P'\rho'^{-\chi} = P_\infty \rho_\infty^{-\chi}, \tag{2.165}$$

and the term $P'\rho'$ can be expressed by use of the integral in (2.163). We obtain

$$\int \frac{dP'}{\rho'(P')} = \frac{\chi}{\chi - 1}\frac{P_\infty}{\rho_\infty}\left(\frac{\rho'}{\rho_\infty}\right)^{\chi-1}. \tag{2.166}$$

According to the known definitions (see, for instance, [119,191,196]), the sound velocity is given by

$$a^2 = \frac{dP'}{d\rho'}, \tag{2.167}$$

for a constant entropy. Using the equation of state of the gas (2.165), the velocity a of sound in the gas stream (2.167) can be represented in the following form:

$$a^2 = \chi \frac{P_\infty}{\rho_\infty} \left(\frac{\rho'}{\rho_\infty} \right)^{\chi - 1}.$$

This formula serves to define the sound velocity in a homogeneously flowing gas stream; i.e. infinitely far from the profile, we have

$$a_\infty^2 = \chi \frac{P_\infty}{\rho_\infty}. \tag{2.168}$$

Using (2.166) and (2.168), the Lagrange–Cauchy integral can be represented in the following form:

$$\frac{\partial \Phi}{\partial t'} + \frac{q^2}{2} + \frac{a_\infty^2}{\chi - 1} \left(\frac{\rho'}{\rho_\infty} \right)^{\chi - 1} = f_1(t'). \tag{2.169}$$

Because on homogeneously flowing stream does not change as it flows, at an infinite distance from the profile one obtains

$$\left. \frac{\partial \Phi}{\partial t} \right|_\infty = 0.$$

Thus, the Lagrange–Cauchy integral (2.169) for the flowing stream possesses the following form:

$$\frac{q_\infty^2}{2} + \frac{a_\infty^2}{\chi - 1} = f_a(t').$$

In the above equation we have a function $f_a(t')$; we substitute this into (2.169), and obtain

$$\frac{\partial \Phi}{\partial t'} + \frac{q^2}{2} + \frac{a_\infty^2}{\chi - 1} \left(\frac{\rho'}{\rho_\infty} \right)^{\chi - 1} = \frac{q_\infty^2}{2} + \frac{a_\infty^2}{\chi - 1}. \tag{2.170}$$

Multiplying (2.170) by $(\chi - 1)/a_\infty^2$, we obtain

$$\frac{\chi - 1}{a_\infty^2} \frac{\partial \Phi}{\partial t'} + \frac{\chi - 1}{2} \frac{q^2}{a_\infty^2} + \left(\frac{\rho'}{\rho_\infty} \right)^{\chi - 1} = 1 + \frac{\chi - 1}{2} M_\infty^2, \tag{2.171}$$

where the Mach number for the incoming stream has been introduced. Using the equation of state of the gas (2.165), the Lagrange–Cauchy integral can be represented in the form

$$\frac{\chi-1}{a_\infty^2}\frac{\partial\Phi}{\partial t'}+\frac{\chi-1}{2}\frac{q^2}{a_\infty^2}+\left(\frac{\rho'}{\rho_\infty}\right)^{\frac{\chi-1}{\chi}}=1+\frac{\chi-1}{2}M_\infty^2. \tag{2.172}$$

We introduce the following nondimensional variables and nondimensional time:

$$\xi=\frac{x_1}{L},\quad \eta=\frac{x_2}{L},\quad t=t'\frac{q_\infty}{L}\mu. \tag{2.173}$$

We also introduce a nondimensional perturbation of the velocity vector with components $U(\xi,\eta,t)$ and $V(\xi,\eta,t)$:

$$q_1=q_\infty(1+V),\quad q_2=q_\infty V. \tag{2.174}$$

The relation of the perturbed stream to the homogeneous stream, i.e. to a uniform stream moving at the velocity of sound, is expressed by

$$|U|\ll 1,\quad |V|\ll 1.$$

We introduce the nondimensional density and pressure within the stream in the form

$$\rho=\frac{\rho'}{\rho_\infty},\quad P=\frac{P'}{P_\infty}. \tag{2.175}$$

Finally, the following potential for the perturbations to the velocity vector $\varphi(\varepsilon,\eta,t)$ is introduced:

$$\Phi(x_1,x_2,t')=q_\infty L(\varepsilon+\varphi(\varepsilon,\eta,t)). \tag{2.176}$$

The relations (2.173) and (2.176) yield

$$\frac{\partial\Phi}{\partial t'}=\mu q_\infty^2\frac{\partial\varphi}{\partial t},$$

where the number μ characterizes the velocity of mechanical changes in the gas stream (note that in the case of harmonic oscillations it corresponds to the frequency).

Therefore, using the last equation and (2.174) and (2.175), the Lagrange–Cauchy integral (2.171) can be expressed in the following nondimensional form:

$$(\chi-1)\mu M_\infty^2\frac{\partial\varphi}{\partial t}+\frac{\chi-1}{2}M_\infty^2\left((1+U)^2+V^2\right)+\rho^{\chi-1}$$
$$=1+\frac{\chi-1}{2}M_\infty^2. \tag{2.177}$$

Equation (2.172) takes the form

$$(\chi - 1)\mu M_\infty^2 \frac{\partial \varphi}{\partial t} + \frac{\chi - 1}{2} M_\infty^2 \left((1+U)^2 + V^2\right) + \rho^{\frac{\chi-1}{\chi}} = 1 + \frac{\chi - 1}{2} M_\infty^2. \tag{2.178}$$

The nondimensional density ρ can be obtained from (2.177), and we obtain

$$\rho = \left(1 - \frac{\chi-1}{2} M_\infty^2 \left(2U + U^2 + V^2 + 2\mu \frac{\partial \varphi}{\partial t}\right)\right)^{\frac{1}{\chi-1}}. \tag{2.179}$$

Analogously, for the nondimensional pressure, we obtain the following from (2.178):

$$P = \left(1 - \frac{\chi-1}{2} M_\infty^2 \left(2U + U^2 + V^2 + 2\mu \frac{\partial \varphi}{\partial t}\right)\right)^{\frac{\chi}{\chi-1}}. \tag{2.180}$$

With the definition

$$G = M_\infty^2 \left(2U + U^2 + V^2 + 2\mu \frac{\partial \varphi}{\partial t}\right) \tag{2.181}$$

and taking into account the relation (2.181), the Lagrange–Cauchy integral (2.179) has the form

$$\rho = \left(1 - \frac{\chi-1}{2} G\right)^{\frac{1}{\chi-1}}. \tag{2.182}$$

In an analogous way, we obtain the following relation from (2.180):

$$P = \left(1 - \frac{\chi-1}{2} G\right)^{\frac{\chi}{\chi-1}}. \tag{2.183}$$

Let us find the critical value of the derivative $\left(\frac{\partial \varphi}{\partial t}\right)_*$ from the Lagrange–Cauchy integral (2.177) for the critical values of the relevant quantities:

$$\mu M_\infty^2 \left(\frac{\partial \varphi}{\partial t}\right)_* = 1 + \frac{\chi-1}{2} M_\infty^2 - \rho_*^{\chi-1} - \frac{\chi-1}{2} M_\infty^2 \left((1+U_*)^2 + V_*^2\right). \tag{2.184}$$

If we assume, as has already been mentioned, that the incoming stream infinitely far from the profile is close to a critical condition, i.e. $M_\infty \approx 1$, $\rho_* \approx 1$, $U_* \approx 0$, $V_* \approx 0$, then we obtain approximately from (2.184)

$$\left(\frac{\partial \varphi}{\partial t}\right)_* = 0. \tag{2.185}$$

Therefore, in this case, when the gas particles reach the critical sound velocity, we obtain from (2.181) and (2.185) the result

$$G_* = 0.$$

The relation (2.182) between the density ρ and G for a transonic stream can then be approximated by the following Taylor series in the neighbourhood of $G = 0$:

$$\rho = 1 - \frac{1}{2}G + \frac{2-\chi}{8}G^2 + R_\rho, \tag{2.186}$$

where R_ρ is the rest of the series, which is of order G^3. In a similar way, for a transonic flow, the dependence of P on G (2.183) can be approximated by a series of the form

$$P = 1 - \frac{\chi}{2}G + \frac{\chi}{8}G^2 + R_p, \tag{2.187}$$

where R_p is the rest of the series, of order G^3. Multiplying (2.186) by $(\overline{q}, \overline{n})$ the following equation is obtained:

$$\rho\frac{(\overline{q}\,\overline{n})}{q_\infty} = \left(1 - \frac{G}{2} + \frac{2-\chi}{2}G^2 + R_p\right)\left((1+U)\cos(n,\xi) + V\cos(n,\eta)\right). \tag{2.188}$$

Therefore, from the above considerations, one can conclude that the values of U, V and $\dfrac{\partial\varphi}{\partial t}$ are much smaller than one, i.e. there exist positive numbers ε_1, ε_2 and ε_3 such that everywhere in the stream we have

$$|U| < \varepsilon_1 \ll 1, \quad |V| < \varepsilon_2 \ll 1, \quad \left|\frac{\partial\varphi}{\partial t}\right| < \varepsilon_3 \ll 1. \tag{2.189}$$

In these inequalities ($\varepsilon_i \ll 1$), we understand that each of the ε_i is about ten times smaller than 1. Also, the expression "an order smaller" is understood as meaning approximately ten times smaller. We do not know anything about the comparitive of the values of ε_1, ε_2 and ε_3, and for a derivation of the transonic asymptotic equations such a comparison is not needed. Of course, any multiplication of those values results in a value order lower than the quantities multiplied. For instance,

$$|UV| \ll |V|, \quad |UV| \ll |U|, \quad \left|U\frac{\partial\varphi}{\partial t}\right| \ll \left|\frac{\partial\varphi}{\partial t}\right|,$$

because the following inequalities hold:

$$|UV| \le \varepsilon_1\varepsilon_2 \ll \varepsilon_2, \quad |UV| \le \varepsilon_1\varepsilon_2 \ll \varepsilon_1, \quad \left|U\frac{\partial\varphi}{\partial t}\right| \le \varepsilon_1\varepsilon_3 \ll \varepsilon_3.$$

It is assumed that the orders of the quantities and their derivatives in relation to the nondimensional time are similar:

$$\left|\frac{\partial U}{\partial t}\right| \le \varepsilon_1, \quad \left|\frac{\partial U}{\partial t}\right| \le \varepsilon_2, \quad \left|\frac{\partial^2 U}{\partial t^2}\right| \le \varepsilon_3, \tag{2.190}$$

i.e. the order of a derivative with respect to the nondimensional time is μ times larger than the order of the quantity which is differentiated. Differentiating (2.186) with respect to t', we obtain

$$\frac{d\rho}{dt'} = \frac{q_\infty \mu}{L}\left(-\frac{1}{2} + \frac{2-\chi}{4}G + \frac{dR_\rho}{dG}\right)\frac{dG}{dt}. \tag{2.191}$$

We assume the existence of the derivatives and a unique convergence of the series and its derivatives. Taking into account these assumptions, we can conclude that the left-hand side of (2.191) and the right-hand side of (2.167) are equal. Substituting the necessary relations from (2.178) and (2.191) into the law of mass conservation (2.161) we obtain

$$\int_\Omega \mu\frac{dG}{dt}\left(-\frac{1}{2} + \frac{2-\chi}{4}G + \frac{dR_\rho}{dG}\right)d\Omega + \int_{\partial\Omega}\left[\left(1 - \frac{G}{2} + \frac{2-\chi}{8}G^2 + R_\rho\right)\right.$$
$$\left. + (1+V)\cos(n,\xi) + V\cos(n,\eta)\right] d(\partial\Omega) = 0, \tag{2.192}$$

where both the space Ω and its contour $\partial\Omega$ are now expressed in nondimensional form in new nondimensional coordinates ξ and η. If a function g under an integral is bounded, $|g| \le \varepsilon$, then for a limited space Ω the following inequality holds:

$$\int_w g\,d\Omega \le \varepsilon \int_w d\Omega. \tag{2.193}$$

An analogous inequality can be written for the integral along the closed contour $\partial\Omega$. Integral estimates of type (2.193) will be used for estimation of the terms occurring in (2.192). Terms which are an order smaller in comparison with other terms will be omitted.

If we substitute the expression for G from (2.181) into the relation (2.188) and neglect small terms, we obtain

$$\rho(\overline{q}\,\overline{n}) = q_\infty\left\{\left[1 - \mu M_\infty^2\frac{\partial\varphi}{\partial t} + (1 - M_\infty^2)U\right.\right.$$
$$\left.\left. + \frac{1}{2}M_\infty^2(2M_\infty^2 - \chi M_\infty^2 - 3)U^2\right]\cos(n,\xi) + V\cos(n,\eta)\right\}. \tag{2.194}$$

In (2.194), the term containing U^2 remains, because for $M_\infty = 1$ the coefficient of U is equal to zero. Because the relative orders of the terms occurring in (2.194) and of the value of the coefficient of U are unknown, we need to consider the nonlinear term U^2. Of course, depending on the value of the frequency μ, the relation (2.191) can be simplified.

Substituting G from (2.181) into (2.191) and taking into account the inequalities (2.189) and (2.190), we can identify several separate intervals of positive values of the frequency μ for which the relation (2.191) has various approximate forms:

1. $\mu = 0$ (stationary flow):
$$\frac{d\rho}{dt'} = 0.$$

2. $0 < \mu \leq \varepsilon_5 \ll 1$, $\varepsilon_5 \ll \min\left(\frac{\varepsilon_1}{\varepsilon_3}, \frac{\varepsilon_2^2}{\varepsilon_3}\right)$:
$$\frac{d\varphi}{dt'} = -\frac{M_\infty^2 q_\infty \mu}{L}\left(\frac{\partial U}{\partial t} + V\frac{\partial V}{\partial t}\right).$$
This case is sometimes referred to as the low-frequency case and is denoted by $\mu \to 0$.

3. $\varepsilon_5 < \mu \leq \varepsilon_6$, $\varepsilon_6 \gg \max\left(\frac{\varepsilon_1}{\varepsilon_3}, \frac{\varepsilon_2^2}{\varepsilon_3}\right)$:
$$\frac{d\rho}{dt'} = -\frac{M_\infty^2 q_\infty \mu}{L}\left(\mu\frac{\partial^2 \varphi}{\partial t^2} + \frac{\partial U}{\partial t} + V\frac{\partial V}{\partial t}\right).$$
This case is referred to as the medium-frequency case.

4. $\varepsilon_6 \leq \mu < \infty$:
$$\frac{d\rho}{dt'} = -\frac{M_\infty^2 q_\infty \mu}{L}\frac{\partial^2 \varphi}{\partial t^2}. \tag{2.195}$$
This case is referred to as the high-frequency case.

It is clear that, depending on the relations between the constants ε_1, ε_2 and ε_3, one can also identify other intervals of the frequency μ. However, we have considered the case where the constants ε_1, ε_2 and ε_3 are arbitrary and much smaller than 1, i.e. they are small and the relations between them are not known. In other words, the frequency intervals have been separated for arbitrarily small values ($\varepsilon_i \ll 1$) of the positive constants ε_1, ε_2 and ε_3.

Our aim is to create an asymptotic equation governing the dynamics of a transonic ideal-gas stream which is valid for all intervals of the frequency μ. The most general formula is that governed by (2.195), which holds for medium frequencies. The cases of low and high frequencies are included in the expression (2.195).

Using (2.194) and (2.195), the equation (2.192) for a transonic stream reads

$$\begin{aligned}-\mu M_\infty^2 \int\limits_w \left(\frac{\partial U}{\partial t} + \mu \frac{\partial^2 \varphi}{\partial t^2} + V \frac{\partial V}{\partial t}\right) d\Omega + \int\limits_{\partial \Omega} \Big\{ (1 - \mu M_\infty^2 \frac{\partial \varphi}{\partial t} \\ + (1 - M_\infty^2) V + \frac{M_\infty^2}{2} (2M_\infty^2 - \chi M_\infty^2 - 3) U^2) \cos(n, \xi) \\ + V \cos(n, \eta) \Big\} d(\partial \Omega) = 0. \end{aligned} \tag{2.196}$$

We compare the terms in (2.196) occurring in the first and second integrals (we compared before only the terms in the same integrals), which yields

$$\int\limits_\Omega \left| \mu M_\infty^2 V \frac{\partial V}{\partial t} \right| d\Omega \ll \int\limits_{\partial \Omega} |V \cos(n, \eta)| \, d(\partial \Omega). \tag{2.197}$$

Because of the property (2.193) and the condition (2.189), we obtain

$$\int\limits_\Omega \left| \mu M_\infty^2 V \frac{\partial V}{\partial t} \right| d\Omega \ll \mu M_\infty^2 \varepsilon_2^2 \int\limits_\Omega d\Omega,$$

$$\int\limits_\Omega |V \cos(n, \eta)| \, d(\partial w) \leq \varepsilon_2 \int\limits_{\partial \Omega} d(\partial w).$$

Hence, the inequality (2.197) is true if the following relation holds:

$$\mu \int\limits_\Omega d\Omega \ll \frac{1}{\varepsilon_2 M_\infty^2} \int\limits_{\partial \Omega} d(\partial \Omega). \tag{2.198}$$

It can be seen that the ratio between the integrals appearing in the inequality (2.198) can give arbitrary positive real number. This property can be understood in the following manner. The larger Ω is, the smaller the ratio of the integrals is, and vice versa. However, this property is not always true, because a bounded space with a given area can have a boundary with an arbitrary large length. But the smallest length of the boundary of a space with a given area is the circumference of a circle, in this case the space corresponds to a circle. Hence, the asymptotic transonic equation can be considered in a space Ω that is a circle, a square or other simple geometrical figure. Hence, taking into account the relation (2.197), the equality (2.196) can be rewritten in the form

$$\begin{aligned} &- \mu M_\infty^2 \int\limits_\Omega \left(\frac{\partial U}{\partial t} + \mu \frac{\partial^2 \varphi}{\partial t^2}\right) d\Omega + \int\limits_{\partial \Omega} \Big\{ \Big[(1 - M_\infty^2) U - \mu M_\infty^2 \frac{\partial \varphi}{\partial t} \\ &+ \frac{M_\infty^2}{2} (2M_\infty^2 - \chi M_\infty^2 - 3) U^2 \Big] \cos(n, \xi) + V \cos(n, \eta) \Big\} d(\partial \Omega) = 0. \end{aligned} \tag{2.199}$$

We now use a property of a contour integral of the form $\int\limits_{\partial\Omega} \cos(n,\varepsilon)\, d(\partial\Omega) = 0$. If we consider a low-frequency supersonic flow, then $\mu \to 0$, and from (2.199) one of the equations of motion can be represented in the form

$$
\begin{aligned}
&-\mu M_\infty^2 \int\limits_{\Omega} \left(\frac{\partial U}{\partial t}\right) dw + \int\limits_{\partial\Omega} \left\{ \left[(1 - M_\infty^2 U - \mu M_\infty^2 \frac{\partial \varphi}{\partial t} \right.\right. \\
&\left.\left. + \frac{M_\infty^2}{2}(2M_\infty^2 - \chi M_\infty^2 - 3)U^2 \right] \cos(n,\xi) + V\cos(n,\eta) \right\} d(\partial w) = 0.
\end{aligned}
\tag{2.200}
$$

Applying the Osrogradskij–Gauss formula, assuming that U, V, $\frac{\partial \varphi}{\partial t}$ are continuous and differentiable, and transforming from the contour integral to a surface integral in (2.199), and then taking into account the arbitrariness of integration over the space Ω, we obtain the following transonic-stream equation:

$$
\begin{aligned}
&-\mu M_\infty^2 \frac{\partial^2 \varphi}{\partial t^2} - 2\mu M_\infty^2 \frac{\partial U}{\partial t} + (1 - M_\infty^2)\frac{\partial U}{\partial \xi} \\
&+ M_\infty^2 (2M_\infty^2 - \chi M_\infty^2 - 3) U \frac{\partial U}{\partial \xi} + \frac{\partial V}{\partial \eta} = 0,
\end{aligned}
\tag{2.201}
$$

which holds for low, medium and high frequencies.

For low frequencies, when $\mu \to 0$, the following differential equation for a transonic stream is obtained from (2.200):

$$
-2\mu M_\infty^2 \frac{\partial U}{\partial t} + (1 - M_\infty^2 + M_\infty^2)(2M_\infty^2 - \chi M_\infty^2 - 3)U\frac{\partial U}{\partial \xi} + \frac{\partial V}{\partial \eta} = 0.
\tag{2.202}
$$

This equation is almost the same as the Lin–Reissner–Tsien (LRT) equation for a nonstationary, transonic, vortex-free (close to homogeneous) flow of an ideal gas [146]. If we take $M_\infty = 1$ in the expression $(2M_\infty^2 - \chi M_\infty^2 - 3)$ in (2.202), obtain the LRT equation, because the following relation holds:

$$
2M_\infty^2 - \chi M_\infty^2 - 3 = -(\chi + 1).
$$

We now add the vortex-free equation (2.162) to (2.201) (or to (2.202) for low frequencies). Equation (2.162), using (2.174), can be represented in the following form:

$$
\frac{\partial U}{\partial \eta} - \frac{\partial V}{\partial \xi} = 0.
\tag{2.203}
$$

A simultaneous solution to (2.201) and (2.203) defines the velocity field of the gas particles in the transonic flow far from the profile, for various frequencies of flow vibrations. In the neighbourhood of the profile, we need to take into account the boundary layer and the influence of viscosity on the flow. The vortex-free equation (2.203) can be represented in the following integral form:

$$\int_{\partial\Omega} (U\cos(n,\eta) - V\cos(n,\xi))\, d(\partial\Omega) = 0. \tag{2.204}$$

Indeed, by applying the Osrogradskij–Gauss formula to the (2.203), one also obtains (2.204). However, that the class of functions $U(\xi,\eta,t)$ and $V(\xi,\eta,t)$ which satisfy (2.204) is wider than the class of the functions satisfying (2.203). The first class includes (unlike the second) discontinuous functions.

The components of the velocity vector satisfy the integral equation (2.204) for an arbitrary closed contour $\partial\Omega$. We call this flow a "generalized" vortex-free flow.

For the pressure P after truncation of the Taylor series (2.187), we obtain the following in the transonic approximation

$$P = 1 - \chi M_\infty^2 U - \chi\mu M_\infty^2 \frac{\partial\varphi}{\partial t} - \frac{\chi}{2} M_\infty^2 V^2. \tag{2.205}$$

In the low-frequency range, this reads

$$P = 1 - \chi M_\infty^2 U - \frac{\chi}{2} M_\infty^2 V^2.$$

Analysis of (2.202) leads to the conclusion that the quantity V^2 is an order smaller than U. Therefore the last relation can be rewritten in the following form:

$$P = 1 - \chi M_\infty^2 U. \tag{2.206}$$

For the dynamic coefficient of pressure [196] in the transonic approximation, the following relation holds:

$$C_p = -2U - 2\mu \frac{\partial\varphi}{\partial t}. \tag{2.207}$$

In the case of the stationary and low-frequency states, this has the form

$$C_p = -2U. \tag{2.208}$$

For a transonic flow around an arbitrary profile of infinite length, the nonpenetration condition for an ideal gas can be written [196]

$$\overline{q}_{ng} = \overline{q}_{nP}, \tag{2.209}$$

where $\overline{q}_{ng}$ is the normal component of the velocity of the gas particles close to the profile surface, and $\overline{q}_{nP}$ is the normal component of the velocity of the points of the profile surface. If the profile surface has the form $x_2 = H(x_1, t')$, then [119]

$$q_{nP} = \frac{\partial H}{\partial t'} \left(1 + \left(\frac{\partial H}{\partial x_1}\right)^2\right)^{-\frac{1}{2}} . \tag{2.210}$$

From the scalar product of the flow velocity vector and the unit vector normal to the profile surface, we obtain

$$q_{ng} = q_\infty \left(V - \frac{\partial H}{\partial x_1}(1+U)\right) \left(1 + \left(\frac{\partial H}{\partial x_1}\right)^2\right)^{-\frac{1}{2}} . \tag{2.211}$$

Substituting (2.210) and (2.211) into the nonpenetration condition, we obtain

$$q_\infty \left(V - \frac{\partial H}{\partial x_1}(1+U)\right) = \frac{\partial H}{\partial t'}.$$

The last expression is then transformed to nondimensional form by use of (2.173), and the condition of nonpenetration of the profile reads:

$$V = (1+U)\frac{\partial f}{\partial \xi} + \frac{\partial f}{\partial t} \quad \text{or} \quad V = \frac{\partial f}{\partial \xi} + \frac{\partial f}{\partial t}. \tag{2.212}$$

The conditions infinitely far from the profile read

$$U = 0, \quad V = 0 \quad (t \geq 0,\ \xi^2 + \eta^2 \to +\infty), \tag{2.213}$$

which express the fact that the transonic flow at infinity is nonexcited and homogeneous.

We assume that on a surface (ξ, η), a shock wave is governed by an equation $\eta = \eta(\xi, t)$. Expanding the contour ∂w associated with this shock wave, the integral equations (2.199) and (2.204) give the following relations:

$$2M_\infty^2 \frac{\partial \xi}{\partial t} + \left(\frac{V_- - V_+}{U_- - U_+}\right)^2 = M_\infty^2 - 1 + \frac{1}{2}M_\infty^2(\chi+1)(U_+ + U_-), \tag{2.214}$$

$$V_- - V_+ = \frac{\partial \xi}{\partial \eta}(U_+ - U_-), \tag{2.215}$$

which are the Reikin–Giugonio conditions in the transonic low-frequency approximation.

2.2.1.3 Equations and Boundary Conditions of Panel–Flow Interaction. We now consider the interaction between a transonic flow and an elastic panel. We assume that between two impenetrable, absolutely rigid, and immoveable semi-surfaces AB and CD an infinitely long circular cylindrical panel with straight, parallel boundaries B and C is fastened. Let us assume that the half-spaces AB and CD are located on the same level, i.e. if they were extended they would belong to one surface. Let us assume also that the panel has either free (ball) or stiff supports along sides B and C (other boundary conditions can also be taken into consideration). Assume that a homogeneous transonic gas flow runs over this panel in a direction transverse to the axis of the infinite circular cylinder, part of which is occupied by the panel under consideration. The direction of the flow is parallel to the half-planes AB and CD. It may be said that the panel, which can be called a profile panel (because it is an object of interest in terms of the theory of elasticity as well as the theory of gas dynamics), has zero attack angle. The interaction of the flow and the panel considered here is completely equivalent to a symmetric flow around a symmetric profile panel, where the plane ABCD becomes the plane of symmetry. This panel serves as a model for an aircraft wing or any part of the surface of any flying aircraft.

Now, let us assume that the motion of the panel is governed by the theory described in Sect. 2.2.1.1. Let us assume also that the motion of the flow is governed by the asymptotic theory of transonic gas flow described in Sect. 2.2.1.2.

In Fig. 2.1, the interaction described above of a transonic flow with an elastic, infinitely long, circular cylindrical panel is illustrated. According to the assumptions that we have made, the whole flow picture is flat.

In Fig. 2.2, an interaction related to the transverse cross-section of the panel is shown. The dashed curve shows the boundary of the supersonic zone of the flow, with a local Mach number $M > 1$. This zone ends, on moving along the direction of the flow, in an impact wave, which is marked by a double dashed line. Looking in the direction of the flow movement, the pressure

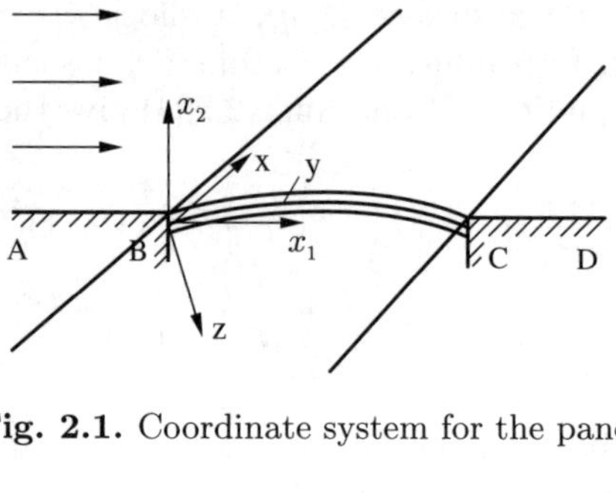

Fig. 2.1. Coordinate system for the panel

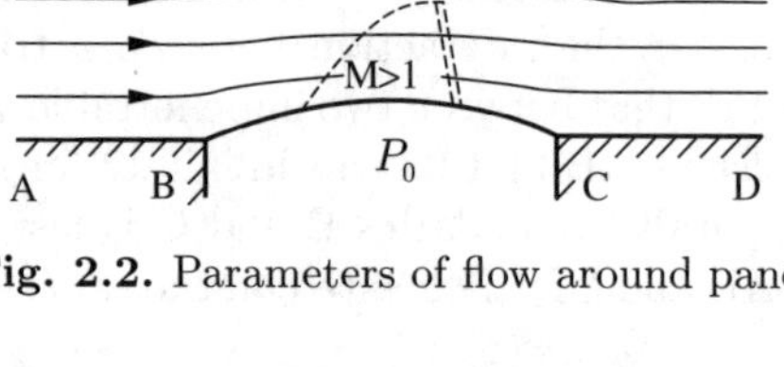

Fig. 2.2. Parameters of flow around panel

before the impact wave is smaller than behind the wave, i.e. the impact wave plays the role of a separatrix between regions of different pressure. If the impact wave is in contact with the surface of the profile panel, then the panel is attacked seriously by a nonuniformly distributed load. Hence, the most interesting case of interaction between a flow and a panel seems to be that in which an impact wave moves over the middle part of the panel. Such a motion of an impact wave is possible during flow around a panel in the case of transonic flow, when there are local supersonic zones. This means that for the panel considered here, shock wave motion happens during a flow corresponding to a homogeneous stream with a Mach number M_∞ lying in the interval from 0.75 to 1. Such interactions are analysed in this section.

Nondimensional Quantities. Although the panel motion is described here in the Lagrange coordinates x, y, z (given in Sect. 2.2.1.1 and shown in Fig. 2.1) and the motion of the gas flow is described in the Euler coordinates x_1, x_2 (defined in Sect. 2.2.1.2 and shown in Fig. 2.1), the following equalities hold approximately:

$$x_1 = y, \qquad x_2 = -z. \tag{2.216}$$

This is true if we take into account the fact that the panel is shallow and that large displacements do not occur in it. Therefore, the same for rectangular system of coordinates (x_1, x_2), shown in Fig. 2.1, has been chosen for both the flow and the panel.

The problem of interaction between a panel and a transonic flow considered here is governed by the following dimensional and nondimensional parameters:

$$x_1, x_2, t', P', P_\infty, P_0', P_z', \rho_\infty, \rho_0', L, H_*, h, w, q_1, q_2, q_\infty, E', \nu, \chi. \tag{2.217}$$

The radius R has not been included in the system, because there exists the geometrical relation

$$R^2 = \left(\frac{L}{2}\right)^2 + (R - H_*)^2, \tag{2.218}$$

given by Pythagoras' theorem. Of the above system of parameters, only three have independent dimensions. Hence, using Pythagoras' theorem and the theory of dimension, we can choose a system of 16 nondimensional control parameters of the following form [196]:

$$\begin{gathered}
\xi = \frac{x_1}{L}, \ \eta = \frac{x_2}{L}, \ t = t'\frac{q_\infty}{L}, \ P = \frac{P'}{P_\infty}, \ P_0 = \frac{P_0'}{P_\infty}, \ \nu, \\
P_z = \frac{P_z'}{P_\infty}, \ \rho_0 = \frac{\rho_0'}{\rho_\infty}, \ E = \frac{E'}{P_\infty}, \ \delta = \frac{h}{L}, \ w = \frac{w'}{h}, \ \chi, \\
K = \frac{8H_*}{h(1 + 4H_*^2/L^2)}, \ V = \frac{q_1}{q_\infty} - 1, \ V = \frac{q_2}{q_\infty}, \ M_\infty = q_\infty \sqrt{\frac{\rho_\infty}{\chi P_\infty}}.
\end{gathered} \tag{2.219}$$

These 16 parameters define the mathematical model of the interaction under consideration. However, in the analysis we shall introduce some other parameters, which will also be presented in a nondimensional form.

To illustrate our analysis, we introduce the parameter $f_* = H_*/L$, which can be interpreted as an initial relative deflection. It is geometrically linked to the relative thickness δ and to the nondimensional geometrical shell parameter K according to (2.219) via the relation

$$\delta f_* = K\delta(1 + 4f_*^2). \tag{2.220}$$

Hence, the equation of the middle surface has the form $\eta = f(\xi, t)$ instead of $x_2 = H(x_1, t')$. Finally, let us introduce a nondimensional parameter for the membrane stress P_{22} in the middle surface,

$$\sigma_M = \frac{1-\nu^2}{\delta^2}\frac{P_{22}}{E'}. \tag{2.221}$$

Mathematical Background. Returning to the nondimensional quantities (2.173) and (2.174) describing the transonic gas flow, it can be seen that not all of them are included in the system of nondimensional parameters (2.219). Hence, the equations governing the motion of the gas flow in nondimensional form (2.200), (2.204), (2.206) can be used in the statement of the problem of transonic aeroelasticity considered here. In nondimensional form (with the help of the nondimensional quantities (2.219) in the equation of motion of the panel (2.184), and the initial conditions for the transverse deflection w (2.158)), the problem of the interaction between the panel and the transonic gas flow can be stated as the following initial–boundary problem:

$$\begin{aligned} &\frac{\partial^4 w}{\partial \xi^4} - 12\left(K + \frac{\partial^2 w}{\partial \xi^2}\right)\sigma_M + \frac{12(1-\nu^2)}{E\delta^2}\rho_0 \chi M_\infty^2 \frac{\partial^2 w}{\partial t^2} \\ &= \frac{12(1-\nu^2)}{E\delta^4} P_z \quad (t > 0,\ 0 < \xi < 1,\ \eta = 0), \end{aligned} \tag{2.222}$$

$$\sigma_M = \frac{1}{2}\int_0^1 \left(\frac{\partial w}{\partial \xi}\right)^2 d\xi - K\int_0^1 w\, d\xi \quad (t \geq 0,\ \eta = 0), \theta \tag{2.223}$$

$$P_Z = P - P_0 \quad (t \geq 0,\ 0 \leq \xi \leq 1,\ \eta = 0), \tag{2.224}$$

$$P = 1 - \chi M_\infty^2 U \quad (t \geq 0,\ 0 < \xi < 1,\ \eta = 0), \tag{2.225}$$

$$\begin{aligned} &\int_\Omega (1 - M_\infty^2 - \frac{1}{2}M_\infty^2(\chi+1)U)U\, dt\, d\eta - \int_\Omega 2M_\infty^2 U\, d\xi\, d\eta \\ &+ \int_\Omega V\, dt\, d\xi = 0 \quad (t > 0,\ -\infty < \xi < +\infty,\ \eta > 0), \end{aligned} \tag{2.226}$$

$$\int_\Omega U\, dt\, d\xi - \int_\Omega V\, dt\, d\eta = 0 \quad (t > 0,\ -\infty < \xi < +\infty,\ \eta > 0), \tag{2.227}$$

$$U = 0,\ V = 0 \quad (t \geq 0,\ \xi^2 + \eta^2 \to +\infty), \tag{2.228}$$

$$U = U_0(\xi, \eta),\ V = V_0(\xi, \eta) \quad (t > 0,\ -\infty < \xi < +\infty,\ \eta > 0), \tag{2.229}$$

$$V = 0 \quad (t \geq 0,\ \xi < 0,\ \xi > 1,\ \eta = 0), \tag{2.230}$$

$$V = \frac{\partial f}{\partial \xi}(1 + U) + \frac{\partial f}{\partial t},$$

$$f(\xi, t) = f_0(\xi) - \delta w(\xi, t) \quad (t > 0,\ 0 < \xi < 1,\ \eta = 0), \tag{2.231}$$

$$w = 0,\ \frac{\partial w}{\partial t} = 0 \quad (t = 0,\ 0 \leq \xi \leq 1,\ \eta = 0), \tag{2.232}$$

$$w = 0,\ \frac{\partial^2 w}{\partial t^2} = 0 \quad (t \geq 0,\ \xi = 0,\ \xi = 1, \eta = 0). \tag{2.233}$$

Equation (2.222) governs the motion of the panel. Equation (2.223) defines the membrane stress σ_M, which appears in (2.222) through the transverse deflection w ((2.223) was obtained using (2.221) and (2.160)). The relation (2.224) defines the intensity of the transverse load P_Z through both the pressure of the transonic flow P and the internal shell pressure P_0. The formula (2.206) is transformed into (2.225). The equation of motion of a low-frequency transonic (close to homogeneous) flow of the an ideal gas (2.226) agrees with the equation given earlier for this case (2.200). The vortex-free equation (2.204) in integral form can be rewritten as the integral equation (2.227) using the Ostrogradskij–Gauss theorem. The condition (2.228) holds for the gas flow infinitely far from the panel. The initial flow is defined by the relations (2.229). As specified in the statement of the problem, the front and rear of the panel are fastened to absolutely rigid half-planes AB and CD, and the nonpenetration condition is expressed by (2.230). The nonpenetration condition (2.192) for a deformable panel has been transformed to the condition (2.231), where $\eta = f_0(\xi)$ governs the initial form of the panel. The initial state of the panel is governed by the condition (2.232), which has been obtained from the initial conditions (2.158) by taking $w_0(\xi) = 0$ and $w_0'(\xi) = 0$. The condition (2.233) describes free support (on balls) of the front and rear of the panel, which in nondimensional form is equivalent to the condition (2.155). Of course, instead of a free support along the panel sides we can consider a clamped support; in this case boundary conditions (2.156) have the following nondimensional form:

$$w = 0,\ \frac{\partial w}{\partial \xi} = 0 \quad (t = 0,\ 0 \leq \xi \leq 1,\ \eta = 0). \tag{2.234}$$

We now consider further the problem of the interaction of an elastic, infinite, cylindrical panel with a transonic flow of an ideal gas. The consideration starts from the case where the panel is absolutely rigid, and the flowing stream is stationary. We need to know the initial distribution of the velocity field, i.e. the functions $U_0(\xi, \eta)$ and $V_0(\xi, \eta)$. These functions are found from solutions to the equations (2.222)–(2.233), when $U_0(\xi,\eta) = 0$, $V_0(\xi,\eta) = 0$ and the given function $\eta = f(\xi,t)$. The last function is chosen in such a way that for $t = 0$ it describes a plate, whereas at a finite time it describes the given circular panel. Here we propose the following function:

$$f(\xi,t) = \begin{cases} 0, & t = 0, \\ f_0(\xi), & t = 0.02. \end{cases}$$

Therefore, this problem corresponds to the problem of a flow around a panel caused by a nonstationary transonic gas flow, in which a plate becomes a circular shell as a result of bending.

2.2.1.4 Analysis of Panel Oscillations. Now we analyse free and forced panel oscillations. We begin by considering the numerical solution of the problem. First we consider calculations for the panel. In order to check the reliability and correctness of the panel calculations, one first must compare the results of the calculations with known solutions. In this section we consider a method of performing calculations for a panel using the finite-difference method along the spatial coordinates and the Runge–Kutta method.

The equations of motion (2.154) of the panel with the specified initial and boundary conditions can be solved by different methods. The Bubnov–Galerkin method, as applied to this type of equations, is described in [94]. In [233], free vibrations of a panel were analysed using the finite-difference method.

Computational Method. This should satisfy the following requirements:

1. The computations must capture high frequencies in the distribution of the transverse deflection w of the transverse cross-section of the infinite cylindrical panel, because the relatively large and sudden nonuniformity in the distributed load caused by a shock wave can lead to a large local curvature of an elastic panel.
2. The algorithm must be sensitive to quick changes with time of the load, and consequently the deflection w in time, because a transonic flow can quickly change the pressure on the surface of the profile panel.
3. The computational method should be stable and must enable further tracing, after stability loss, of the shell dynamics.
4. The method must be economical from the point of view of computational time.

It turns out, that the Bubnov–Galerkin method is not suitable for solving the problem efficiently, because it requires higher-order approximations, which is in conflict to point 4.

The finite-difference method, if it is applied to both the spatial and the time variables, does not posses high stability [224], and therefore does not satisfy point 3.

However, if in the panel calculation one applies the finite-difference technique to the spatial coordinates, and applies one of the methods for solving Cauchy problems of ordinary differential equations (ODEs) with respect to the time variable, one of then [224] the stability of the calculation process increases and one can follow the behaviour of the panel over a long time.

Now we formulate the problem in its general form. In dimensional form, as has been mentioned in Sect. 2.2.2.1, the problem reads

$$\frac{E'h^3}{12(1-\nu^2)}\frac{\partial^4 w'}{\partial y^4} - \frac{E'h}{L(1-\nu^2)}\left(\frac{1}{R}+\frac{\partial^2 w'}{\partial y^2}\right) \times\left(\frac{1}{2}\int_0^L\left(\frac{\partial w'}{\partial y}\right)^2 dy - \int_0^L \frac{w'}{R}\,dy\right) + \rho_0' h\frac{\partial^2 w'}{\partial t'^2} = P_z', \tag{2.235}$$

$$w' = w_0'(y), \quad \frac{\partial w'}{\partial t'} = w_0''(y) \quad (t'=0,\ 0\le y\le L), \tag{2.236}$$

$$w' = 0, \quad \frac{\partial^2 w'}{\partial y^2} = 0 \quad (t'\ge 0,\ y=0,\ y=L). \tag{2.237}$$

Note that (2.235) is the same as (2.154). The initial conditions (2.236) are the same as the conditions (2.158). The conditions (2.237) are the simple support conditions (2.155). The panel and the system of coordinates are represented in Fig. 2.1. We introduce the following nondimensional quantities:

$$\xi = \frac{y}{L}, \quad t = At', \quad K = \frac{8H_*}{h(1+4h_*^2/L^2)}, \quad \delta = \frac{h}{L}, \quad w = \frac{w'}{h}, \quad P_Z = \frac{P_Z'}{E'}, \tag{2.238}$$

and transform the problem (2.235), (2.236) and (2.237) to the following nondimensional form:

$$\frac{\partial^4 w}{\partial \xi^4} - 12\left(K+\frac{\partial^2 w}{\partial \xi^2}\right)\left(\frac{1}{2}\int_0^1\left(\frac{\partial w}{\partial \xi}\right)^2 d\xi - K\int_0^1 w\,d\xi\right) + 12\frac{(1-\nu^2)\rho_0' L^2 A^2}{E'\delta^4}\frac{\partial^2 w}{\partial t^2} = \frac{12(1-\nu^2)}{\delta^4}P_Z, \tag{2.239}$$

$$w = w_0(\xi), \quad \frac{\partial w}{\partial t} = w_0'(\xi) \quad (t=0,\ 0\le\xi\le 1), \tag{2.240}$$

$$w = 0, \quad \frac{\partial^2 w}{\partial \xi^2} = 0. \tag{2.241}$$

In the system (2.238), the dimensional quantity A appears, which has different values in different problems and characterizes time step. The functions $w_0(\xi)$

and $w'_0(\xi)$ in the initial conditions (2.240) describe the nondimensional form of the corresponding dimensional functions $w'_0(y)$ and $w''_0(y)$ in the initial conditions (2.236). In order to solve (2.239), (2.240) and (2.241) numerically, the central finite differences are used:

$$\begin{aligned}\frac{\partial w}{\partial \xi} &= \frac{1}{2\Delta\xi}(w(\xi+\Delta\xi) - w(\xi-\Delta\xi)),\\ \frac{\partial^2 w}{\partial \xi^2} &= \frac{1}{(2\Delta\xi)^2}(w(\xi+\Delta\xi) - 2w(\xi) + w(\xi-\Delta\xi)),\\ \frac{\partial^4 w}{\partial \xi^4} &= \frac{1}{(2\Delta\xi)^4}(w(\xi+2\Delta\xi) - 4w(\xi+\Delta\xi)\\ &\quad + 6w(\xi) - 4w(\xi-\Delta\xi) + w(\xi+2\Delta\xi)).\end{aligned}$$

These are substituted into (2.239) and into the boundary conditions (2.241). Hence, the problem (2.239), (2.240) and (2.241) reduces to a Cauchy problem for ODEs:

$$\begin{aligned}\frac{\partial^2 w}{\partial t^2}(\xi,t) &= A_1(w(\xi+2\Delta\xi,t) + w(\xi-2\Delta\xi,t)) + A_2(w(\xi-\Delta\xi,t)\\ &\quad + w(\xi-\Delta\xi,t)) + A_3 w(\xi,t) + A_4 + A_p P_Z(\xi,t),\end{aligned} \tag{2.242}$$

$$w = w_0(\xi), \quad \frac{\partial w}{\partial t} = w'_0(\xi) \quad (t=0,\ 0 \le \xi \le 1), \tag{2.243}$$

$$w(0,t) = 0, \quad w(1,t) = 0 \quad (t \ge 0), \tag{2.244}$$

$$\begin{aligned}w(0+\Delta\xi,t) &= -w(0-\Delta\xi,t),\\ w(1-\Delta\xi,t) &= -w(1+\Delta\xi,t) \quad (t \ge 0),\end{aligned} \tag{2.245}$$

where

$$A_1 = -\frac{E\delta^2 A^2}{12(1-\nu^2)\rho'_0 L^2(\Delta\xi)^4}, \tag{2.246}$$

$$A_2 = 4A_1(6(\Delta\xi)^2\sigma_M - 1), \tag{2.247}$$

$$A_3 = 6A_1(4(\Delta\xi)^2\sigma_M + 1), \tag{2.248}$$

$$A_4 = 12A_1\sigma_M, \tag{2.249}$$

$$A_p = -\frac{12(1-\nu^2)}{\delta^4}A_1, \tag{2.250}$$

and σ_M is defined by (2.223).

Here we refer to the equation system (2.242) because, for every internal point with coordinate ξ, equations of type (2.242) and the initial conditions (2.243) are valid.

To apply central differences to the boundary nodes, we use points outside the contour, i.e. points with $\xi = 0 - \Delta\xi$, $\xi = 1 + \Delta\xi$. The values of the transverse deflection w at these points are derived from the boundary conditions

(2.245) and the values of the deflection w on the boundary from (2.244). If values of the transverse deflection w for boundary and off-boundary points are present on the right-hand side of the equation system (2.242), then these values are determined from (2.244) and (2.245).

The dimensionless tensile stress σ_M, following (2.224), after introducing the finite differences, has the following form:

$$\sigma_M = \frac{1}{2}\int_0^1 \frac{1}{(\Delta\xi)^2}\Big(w(\xi+\Delta\xi,t) - w(\xi-\Delta\xi,t)\Big)^2 d\xi - K\int_0^1 w(\xi,t)\,d\xi.$$

The integrals in the last equation were calculated by Simpson's rule with a step $\Delta\xi$ using the standard Fortran IV procedure QSF.

The Cauchy problem (2.243) and (2.242) was solved by a Runge–Kutta method of fourth order, using the standard procedure RGS in Fortran IV. Before applying the Runge–Kutta method, the system of ordinary differential equations of the second order was converted into a system of differential equations of the first order by introducing the derivative $\frac{\partial w}{\partial t}$ as a sought function.

The problem of free vibrations of as infinite, round, cylindrical elastic panel with simply supported edges was investigated in [234]. Following this work, we have assumed

$$K = 10, \quad \delta = 0.0091, \quad P'_z = 0, \quad A^2 = -\frac{\pi^2 E' h^2}{48(1-\nu^2)\rho'_0 L^4}\left(1 + 96\frac{K}{\pi^6}\right),$$

$$w_0(\xi) = 0.1\sin(\pi\xi), \quad w'_0(\xi) = 0.$$

Figure 2.3 presents, as solid lines, the results of calculations obtained with the method described above, and the circles show the results presented by Volmir [234].

The problem of forced vibrations of an elastic cylindrical panel of finite length conforming the Kirchhoff–Love model, described by the full nonlinear

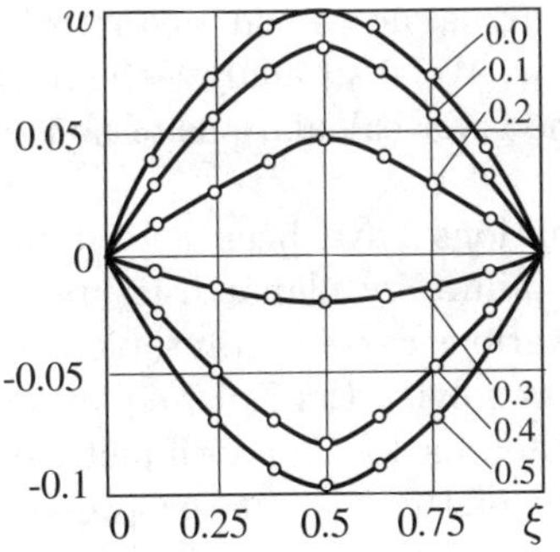

Fig. 2.3. Deflection of panel obtained by different methods, details in text

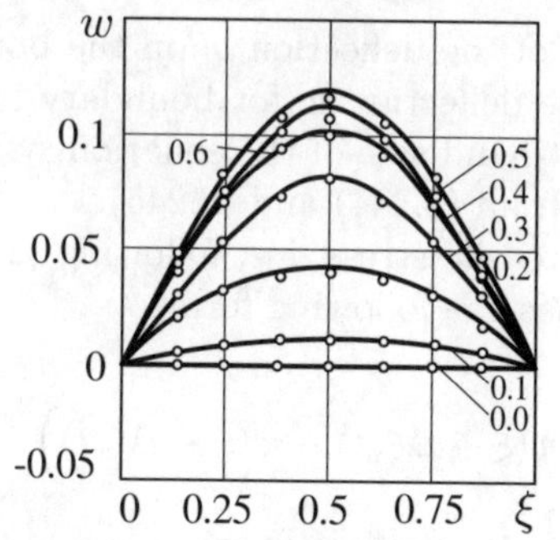

Fig. 2.4. Deflection of panel obtained by different methods; details in text

equation of motion for the displacements, was numerically solved using the finite-difference approach and the Runge–Kutta method for the parameters

$$K = 10, \quad \delta = 0.0091, \quad w_0(\xi) = 0, \quad w_0'(\xi) = 0,$$

$$A^2 = -\frac{\pi^2 E' h^2}{48(1-\nu^2)\rho_0' L^4}\left(1 + 96\frac{K}{\pi^6}\right), \quad P_z' = -\frac{\pi^2 \delta^4 E'}{24}\left(1 + 96\frac{K^2}{\pi^6}\right),$$

and simple support of the edges. The results were obtained by Varygin, the Ph.D student of V. A. Krys'ko.

Figure 2.4 shows by circles some results for points of the central section of a cylindrical panel with edges in the ratio 1:10 obtained by Varygin, and by solid lines, results obtained by the method described in this book. The results of calculations by the method described above agree reasonably with analogous data obtained by other authors.

2.2.1.5 Method of Solution of the Problem. Let us analyse the problem of the stationary and nonstationary flow around an ideally rigid profile. In this subsection, a new method of calculations of a transonic flow is presented. This is the modified Godunov method.

Recently, numerical methods have been applied to calculations of a nonstationary transonic flow of an ideal gas. These methods are based on the Euler equations for gas dynamics or on asymptotic equations of motion of a transonic potential flow. We shall analyse the main trends in the development of numerical methods for calculations of nonstationary transonic ideal-gas flows.

Euler's integral equations have been solved by the Godunov method, based on the theory of failure by planar fracture [86–88]. This method has been widely applied to various cases of transonic flow around absolutely rigid bodies [105,106,111]. For solving the Euler equations in the transonic region of flow velocities, a two-step Lacks–Wendorff method [152] was applied, which was used for calculation of the flow in the presence of sinusoidal longitudinal oscillations. The calculations showed that in that case a shock wave excited oscillations while moving along a surface. Calculations by Magnus and

Yoshihara [152] showed that an increase of the frequency of the oscillations decreases the amplitude of shock wave motion. Effective numerical solutions of the Euler equations for low-frequency cases using uncommon methods were presented by Beam and Warning [42, 43].

Various schemes for the full and low-frequency equations of the theory of small excitations of a transonic, close to homogeneous, nonstationary flow are presented in [233]. These schemes become the mixed Murman–Cole system [171] in the stationary case.

By extracting nonstationary addends from the solution (by a harmonic approach, applying a relaxation method with finite differences), Ehlers [73] and Weatherill [240], with co-workers, obtained numerical solutions of the full, transonic asymptotic equation. An analogous approach for solving not the full but the low-frequency asymptotic equations was presented by Traci et. al. [219]. As stated in reference [50], the disadvantage of these methods is the limitations caused by the assumption that the flow is harmonic.

Calculations of nonstationary transonic flows of an ideal gas have been presented in many works, not listed here, where analytical or other solutions were given. For numerical solution, we chose a commonly known and relatively simple method.

As stated by Gorshkov [90], the Godunov method can be used effectively for solving problems of the interaction of flexible bodies with gas flows. To simplify the numerical calculations of the transonic flow, we shall apply idea of the Godunov method to an approximate theory of transonic flow [79].

Let us investigate the flow around an elongated symmetric profile in a two-dimensional nonstationary, low-frequency transonic flow of an ideal gas. The profile is oriented to have zero attack angle relative to the incoming uniform flow. Figure 2.2 presents a map of the flow. The asymptotic transonic equations were derived in the Sect. 2.2.1.2. Hence, in dimensionless form, we obtain the following initial–boundary conditions:

$$\begin{aligned}\int\limits_{\Omega}((1-M_\infty^2)U-\frac{1}{2}M_\infty^2(\chi+1)U^2)\,d\eta\,dt-2\int\limits_{\Omega}M_\infty^2U\,d\xi\,d\eta\\+\int\limits_{\Omega}V\,d\xi\,dt=0\quad(t>0,\ -\infty<\xi<+\infty,\ \eta>0),\end{aligned}\tag{2.251}$$

$$\int\limits_{\Omega}U\,d\xi\,dt-\int\limits_{\Omega}V\,d\eta\,dt=0\quad(t>0,\ -\infty<\xi<+\infty,\ \eta>0),\tag{2.252}$$

$$V=\frac{\partial f}{\partial\xi}(1+U)+\frac{\partial f}{\partial t}\quad(t\geq0,\ 0\leq\xi\leq1,\ \eta=0),\tag{2.253}$$

$$U=0,\ V=0\quad(t\geq0,\ \xi<0,\ \xi>1,\ \eta=0),\tag{2.254}$$

$$U=U_0(\xi,\eta),\ V=V_0(\xi,\eta)\quad(t=0,\ -\infty<\xi<+\infty,\ \eta\geq0),\tag{2.255}$$

$$V=0\quad(t\geq0,\ \xi<0,\ \xi>1,\ \eta=0).\tag{2.256}$$

The boundary conditions of no intersection (2.253) on the surface of the elongated profile $\eta = f(\xi, t)$ are projected onto the chord $\eta = 0$. At an infinite distance from the profile, the flow is treated as homogeneous and nonexcited – the condition (2.254). The initial velocity field of the gas flow in a general case is random – the condition (2.255).

Following the assumption about the symmetry of the problem, only the upper half-plane $\eta \geq 0$ will be investigated. On the symmetry axis of the flow the condition (2.256) should be satisfied – no mass flow over the symmetry axis. All recalculated boundary and initial conditions should match.

Method of Calculation of the Transonic Flow. We assume that the surface is a side of an orthogonal parallelepiped, of which the edges are the axes of an orthogonal coordinate system ξ, η, t. We assume that the values of U and V are constant on the sides of the parallelepiped. The values of U are indicated in Fig. 2.5; similar indices are used for V. It is assumed also that these elements are constant in the internal volume and are described by $U_{j-1/2,k-1/2}$ and $V_{j-1/2,k-1/2}$, respectively. The nondeformed parallelepiped will be investigated.

Upper indices of similar form correspond to values at the time instant $t + \Delta t$. The integral equations of motion of the transonic gas flow (2.251) and (2.252) have the following form:

$$U^{j-\frac{1}{2},k-\frac{1}{2}} = U_{j-\frac{1}{2},k-\frac{1}{2}} + \frac{\Delta t}{2M_\infty^2 \Delta\xi}\Bigg(U_{j,k-\frac{1}{2}}\Big(1 - M_\infty^2$$
$$-\frac{1}{2}M_\infty^2(\chi+1)U_{j,k-\frac{1}{2}}\Big) - U_{j-\frac{1}{2},k-\frac{1}{2}}\Big(1 - M_\infty^2$$
$$-\frac{1}{2}M_\infty^2(\chi+1)U_{j-\frac{1}{2},k-\frac{1}{2}}\Big)\Bigg) + \frac{\Delta t}{2M_\infty^2 \Delta\eta}(V_{j-\frac{1}{2},k} - V_{j-\frac{1}{2},k-1}), \tag{2.257}$$

$$V_{j,k-\frac{1}{2}} = V_{j-\frac{1}{2},k-\frac{1}{2}} + \frac{\Delta\xi}{\Delta\eta}(U_{j-\frac{1}{2},k} - U_{j-\frac{1}{2},k-\frac{1}{2}}). \tag{2.258}$$

If U and V, which appear in the right-hand side of (2.257), are known at the time instant t, it is possible to calculate the value of $U^{j-\frac{1}{2},k-\frac{1}{2}}$ at the

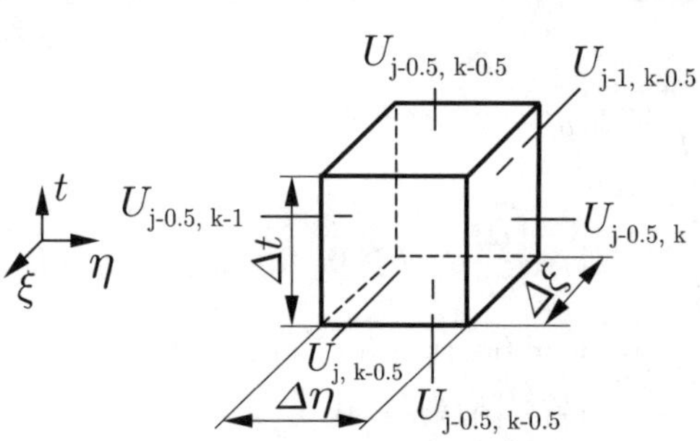

Fig. 2.5. Description of a panel by a parallelepiped

time instant $t + \Delta t$. At this time interval $t + \Delta t$, it is necessary to evaluate all variables in (2.257) and (2.258) before performing further calculations. It is possible to do this using (2.258) and some extra equations which can be found using the idea of the Godunov method for the analysis of a fracture, treating any of the sides of the parallelepiped, shown in Fig. 2.5 as a fracture surface. Here, differently from the Godunov method, the asymptotic problem of a transonic fracture is considered using the full set of the equations of ideal-gas dynamic. The case of a transonic, uniform fracture which is oriented transversly to the flow, treated on the basis of the asymptotic equations is called the Cauchy case. We have

$$\int\limits_{\Omega} \left((1 - M_\infty^2)U - \frac{1}{2}M_\infty^2(\chi + 1)U^2\right) dt - 2\int\limits_{\Omega} M_\infty^2 U\, d\xi = 0, \tag{2.259}$$

$$U(\xi, t_0) = \begin{cases} U_-, \ \xi < \xi_0, \\ U_+, \ \xi > \xi_0, \end{cases} \tag{2.260}$$

where U_-, U_+, t_0, ξ_0 are constants and Ω is a closed contour on the surface (ξ, t).

In differential form, the integral equations (2.259) have the form

$$-2M_\infty^2 \frac{\partial U}{\partial t} + \left(1 - M_\infty^2 - M_\infty^2(\chi + 1)U\right)\frac{\partial U}{\partial \xi} = 0, \tag{2.261}$$

which, after we make the substitutions

$$\begin{gathered} \overline{\xi} = -\xi, \quad 2M_\infty^2 \overline{t} = t, \\ \overline{U}(\overline{\xi}(\xi), \overline{t}(t)) = 1 - M_\infty^2 - M_\infty^2(\chi + 1)U(\xi, t), \end{gathered} \tag{2.262}$$

leads to the known equation [191]

$$\frac{\partial \overline{U}}{\partial \overline{t}} + \overline{U}\frac{\partial \overline{U}}{\partial \overline{\xi}} = 0. \tag{2.263}$$

Solving the characteristic equation, we obtain the equations of the characteristic curves:

$$\begin{aligned} &\xi = \xi_0 - \frac{t - t_0}{2M_\infty^2}(1 - M_\infty^2 - M_\infty^2(\chi + 1)U), \\ &U = U_0. \end{aligned} \tag{2.264}$$

These characteristics, in the coordinate plane (ξ, t), as can be seen from (2.264), have the form of straight semi-lines, starting from the line $t = t_0$ on which the value of the function is constant and equal to the initial value of U_0. Different cases of the characteristics for the initial condition (2.260) are presented in Fig. 2.6. Figures 2.6a, d show the case $U_- < U_+$. Figure 2.6b shows the case $U_- > U_+$ and Fig. 2.6c shows the case $U_- = U_+$. The case $U_- < U_+$ appears in two parts of the figure, Fig. 2.6a, d. This shows that the solution in this case is not unique.

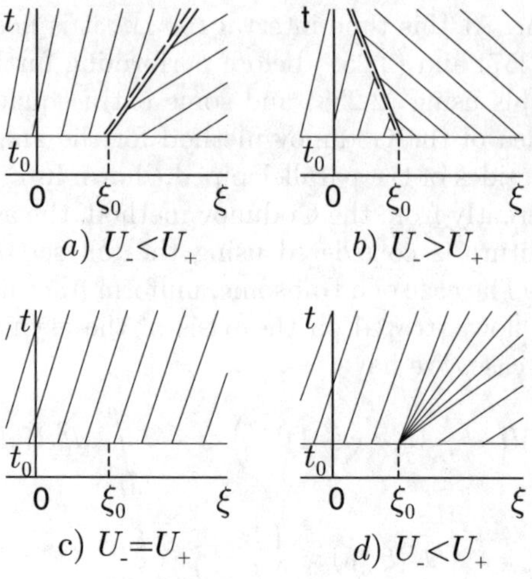

Fig. 2.6a–d. Characteristics for different cases of the initial condition (2.260)

A case similar to (2.259) and (2.260) with discontinuous initial conditions was studied by Riemann, who found a solution containing an arbitrary functional, called a simple Riemann wave [65]. The nonautonomous solution to the differential equation (2.261) containing an arbitrary functional has the following form [45]:

$$\begin{aligned} g(C) &= 1 - M_\infty^2 - M_\infty^2(\chi+1)U, \\ C &= \xi + \frac{t}{2M_\infty^2}\big(1 - M_\infty^2 - M_\infty^2(\chi+1)U\big), \end{aligned} \tag{2.265}$$

where $g(C)$ is any differentiable function of the single variable C. If we differentiate (2.265) with respect to ξ and t separately, we obtain

$$\frac{\partial U}{\partial \xi} = -\frac{2\dfrac{dg}{dC}}{(\chi+1)\left(2M_\infty^2 - t\dfrac{dg}{dC}\right)}, \tag{2.266}$$

$$\frac{\partial U}{\partial t} = \frac{\dfrac{dg}{dC}\big(1 - M_\infty^2 - M_\infty^2(\chi+1)U\big)}{(\chi+1)M_\infty^2\left(2M_\infty^2 - t\dfrac{dg}{dC}\right)}. \tag{2.267}$$

To prove that we can really obtain a solution of (2.259) from (2.266) and (2.267), we observe that (2.265) is always a solution of (2.261), excluding the case when

$$2M_\infty^2 - t\frac{dg}{dC} = 0,$$

and for air $\chi = 1.4$ and $\chi + 1 \neq 0$. The function $g(C)$ can be expressed through the principal values of the function $U(\xi, t)$. For $t = t_0$, we obtain from (2.265)

$$g(C) = 1 - M_\infty^2 - M_\infty^2(\chi + 1)U(\xi, t_0).$$

For the problem (2.259), (2.260) under investigation, the solution of type (2.265), in the case $U_- > U_+$, has the form

$$U(\xi, t) = \begin{cases} U_-, \; 2M_\infty^2(\xi - \xi_0) < A(t - t_0), \\ U_+, \; 2M_\infty^2(\xi - \xi_0) > A(t - t_0), \end{cases} \tag{2.268}$$

where

$$A = M_\infty^2 - 1 + \frac{1}{2}M_\infty^2(\chi + 1)(U_- + U_+).$$

The solution of (2.268) is presented in Fig. 2.6b. If $U_- < U_+$, then (2.265) gives the following solution for the problem (2.259), (2.260):

$$U(\xi, t) = \begin{cases} U_-, \; 2M_\infty^2(\xi - \xi_0) > A(t - t_0), \\ U_+, \; 2M_\infty^2(\xi - \xi_0) < A(t - t_0), \end{cases} \tag{2.269}$$

where

$$A = M_\infty^2 - 1 + \frac{1}{2}M_\infty^2(\chi + 1)(U_- + U_+).$$

This case is represented in Fig. 2.6a. If $U_- = U_+$ then (2.265) gives a trivial solution,

$$U(\xi, t) = U_- = U_+, \tag{2.270}$$

which is represented in Fig. 2.6c.

It can be seen that the solution of (2.259) and (2.260) is not a single one. Besides the solution of type (2.265), i.e. the solutions (2.258), (2.259) and (2.270), the following self-modelling solution exists:

$$\begin{aligned} U(\xi, t) &= (1 - M_\infty^2)/(M_\infty^2(\chi + 1)) + 2(\xi - \xi_0)/(t - t_0)(\chi + 1), \\ &M_\infty^2 - 1 + M_\infty^2(\chi + 1)U_- < 2M_\infty^2(\xi - \xi_0)/(t - t_0) \\ &< M_\infty^2 - 1 + M_\infty^2(\chi + 1)U_+. \end{aligned} \tag{2.271}$$

Let us prove it. The self-modelling solution (2.271) corresponds to a simple symmetric expanding wave. This is represented in Fig. 2.6d in the form of a family of straight semi-lines, starting from an initial point of the explosion at $\xi = \xi_0$. Besides this solution (2.271), the solution (2.269) is also shown outside the family of lines.

It is necessary to explain the physical interpretation for a solution of the transverse fracture (2.259), (2.260) in the case of the problem under consideration. In the currently theory of quasi-linear hyperbolic equations, the following criteria for the existence of generalized solutions have been established: criteria based on entropy, smoothness, and conformity of the physical model with a more accurate one [65, 191]. In general, those different criteria are equivalent [65].

The criterion assumed most often is that of conformity of the model of an ideal gas with the more accurate model of a viscous gas. The asymptotic, dimensionless differential equation of a low-frequency, transonic uniform flow of a viscous gas has the form [193]

$$-2M_\infty^2 \frac{\partial U_\alpha}{\partial t} + (1 - M_\infty^2 - M_\infty^2(\chi + 1)U_\alpha)\frac{\partial U_\alpha}{\partial \xi} = \alpha \frac{\partial^2 U_\alpha}{\partial \xi^2}. \tag{2.272}$$

Using the transformation (2.262), this equation can be converted into Burger's equation [65, 191]:

$$\frac{\partial \overline{U}_\alpha}{\partial \overline{t}} + \overline{U}_\alpha \frac{\partial \overline{U}_\alpha}{\partial \overline{\xi}} = \alpha \frac{\partial^2 \overline{U}_\alpha}{\partial \overline{\xi}^2}. \tag{2.273}$$

If we analyse the Cauchy problem for (2.272) with the initial conditions (2.260), then we obtain a solution of the equations (2.259), (2.260), which can be obtained from the solution U_α by the following transformation:

$$U(\xi, t) = \lim_{\alpha \to 0} U_\alpha(\xi, t). \tag{2.274}$$

The Cauchy problem governed by the equations (2.260), (2.270) with the generalized initial conditions $U(\xi, t_0) = U_0(\xi)$ was presented by Cole [63]. Germain and Bader [85] proved the unique of the solution of (2.259) and (2.260), if the condition $U_- > U_+$ is satisfied at the points of discontinuity of the the function $U(\xi, t)$; this yields the existence of a solution from the entropy criterion. Hopf [100] evaluated the solution of (2.272), (2.260) and formulated a strong limiting transformation (2.274). In this way, he proved the existence and unique of the generalized solution of the Cauchy problem (2.259) and (2.260) for bounded and rational initial functions $U_0(\xi)$. Hopf showed, that the generalized solution of (2.259), (2.260) obtained with the transformation (2.274) satisfies (2.261), which is described by the integral equation

$$\iint\limits_{t>0} \left(U \frac{\partial g}{\partial t} + \frac{U^2}{2} \frac{\partial g}{\partial \xi} \right) d\xi \, dt + \int\limits_{-\infty}^{+\infty} g(\xi, 0) U_0(\xi) \, d\xi = 0,$$

for any smooth finite function $g(\xi, t)$. Detailed bibliographies concerning the cases discussed in this section have been published in the literature [65, 191]. In the monograph [191], it is proved that the generalized solution of the Cauchy problem for (2.263), obtained by means of the limiting transformation (2.274) from the smooth solution of the Cauchy problem for (2.273), satisfies (2.263) in the sense of the integral equality

$$\oint_{\partial\Omega} \overline{U}(\bar{\xi}, \bar{t})\, d\bar{\xi} \; - \; \frac{1}{2}\overline{U}^2(\bar{\xi}, \bar{t})\, d\bar{t} = 0, \tag{2.275}$$

for any boundary consisting of a closed contour $\partial\Omega$ on which the measure of the set of points of discontinuity of $U(\xi, t)$ equals zero. The initial condition is satisfied in the sense of a weak convergence:

$$\lim_{\bar{t}\to\bar{t}_0} \int_0^{\bar{\xi}} \overline{U}(\xi, \bar{t})\, d\xi \rightrightarrows \int_0^{\bar{\xi}} U_0(\xi)\, d\xi. \tag{2.276}$$

Taking into account this result, the equations (2.262) can be substituted into (2.275):

$$\oint_{\partial\Omega} \left(M_\infty^2 - 1 + M_\infty^2(\chi + 1)U(\xi, t)\right) d\xi$$
$$- \frac{1}{4M_\infty^2}\left(1 - M_\infty^2 - M_\infty^2(\chi + 1)U(\xi, t)\right)^2 dt = 0.$$

Therefore, keeping in mind that the integral of a constant over a closed contour equals 0, one obtains the integral equation (2.259). Analogously, it can be proved from(2.276), using (2.262), that the initial conditions (2.260) are satisfied in the sense of the following limit:

$$\lim_{\bar{t}\to\bar{t}_0} \int_0^{\xi} U(\xi, t)\, d\xi = \int_0^{\xi} U_0(\xi, t_0)\, d\xi.$$

In that way it was proved that a single, generalized solution of the Cauchy problem (2.259), (2.260) exists. This solution is the limit (2.274) of the solution of the physically real Cauchy problem, with viscosity taken into account, for (2.268). This eigenvalue solution has the following form:

– if $U_- > U_+$, then

$$U(\xi, t) = \begin{cases} U_-, & 2M_\infty^2(\xi - \xi_0) < A(t - t_0), \\ U_+, & 2M_\infty^2(\xi - \xi_0) > A(t - t_0), \end{cases}$$

– if $U_- < U_+$, then

$$U(\xi,t)=\begin{cases} U_-, & 2M_\infty^2(\xi-\xi_0)<(t-t_0)B_-,\\ U_+, & 2M_\infty^2(\xi-\xi_0)>(t-t_0)B_+,\\ \dfrac{1-M_\infty^2}{M_\infty^2(\chi+1)}+\dfrac{2(\xi-\xi_0)}{(\chi+1)(t-t_0)}, & B_-<2M_\infty^2\dfrac{\xi-\xi_0}{t-t_0}<B_+,\end{cases}$$

– if $U_- = U_+$, then

$$U(\xi,t)=U_-=U_+. \tag{2.277}$$

Here,

$$A=M_\infty^2-1+\frac{1}{2}M_\infty^2(\chi+1)(U_-+U_+),$$
$$B_\mp=M_\infty^2-1+M_\infty^2(\chi+1)U_\mp).$$

In this case, the solution (2.269), which is represented in Fig. 2.6a, has no physical interpretation.The solution in the form of a fracture, represented in Fig. 2.6b, is said to be stable and represents a shock wave. Fracture occurs if the velocity components on the left and right sides of an initiating fracture satisfy the condition $U_- > U_+$. The shock wave moves along the flow. Its speed can be calculated from the Reikin–Gugiono condition for a shock wave, which, in the asymptotic transonic mode, has been included in (2.259) and has the form

$$2M_\infty^2\frac{\partial\xi}{\partial t}+1-M_\infty^2=\frac{1}{2}M_\infty^2(\chi+1)(U_-+U_+),$$

where $\xi=\xi(t)$ describes the shock wave front.

If the velocity components on the left and right sides of the initial fracture satisfy the condition $U_- < U_+$, then the central straight wave vanishes (solution (2.271), shown in Fig. 2.6d), and the initial fracture disappears.

Following the solution (2.277) of the problem of failure by fracture (2.259), (2.260) and assuming that the initiating side of the fracture is the side of the parallelepiped with indices j and $k-\dfrac{1}{2}$, we obtain the following formulas for use in calculations:

$$U_{j,k-\frac12}=\begin{cases} U_{j-\frac12,k-\frac12}, & (D_{j,k-\frac12}>0\wedge A_{j,k-\frac12}\ge 0)\\ & \vee(D_{j,k-\frac12}<0\wedge B_{j-\frac12,k-\frac12}\ge 0)\\ & \vee D_{j,k-\frac12}=0,\\ U_{j+\frac12,k-\frac12}, & (D_{j,k-\frac12}>0\wedge A_{j,k-\frac12}<0)\\ & \vee(D_{j,k-\frac12}<0\wedge B_{j+\frac12,k-\frac12}\le 0),\\ \dfrac{1-M_\infty^2}{M_\infty^2(\chi+1)}, & D_{j,k-\frac12}<0\wedge B_{j-\frac12,k-\frac12}<0<B_{j+\frac12,k-\frac12},\end{cases} \tag{2.278}$$

where

$$D_{j,k-\frac{1}{2}} = U_{j-\frac{1}{2},k-\frac{1}{2}} - U_{j+\frac{1}{2},k-\frac{1}{2}},$$

$$A_{j,k-\frac{1}{2}} = M_\infty^2 - 1 + \frac{1}{2} M_\infty^2 (\chi + 1)(U_{j+\frac{1}{2},k-\frac{1}{2}} + U_{j-\frac{1}{2},k-\frac{1}{2}}),$$

$$B_{j-\frac{1}{2},k-\frac{1}{2}} = M_\infty^2 - 1 + M_\infty^2 (\chi + 1) U_{j-\frac{1}{2},k-\frac{1}{2}}.$$

In this equation, for $D_{j,k-\frac{1}{2}} > 0$ and $A_{j,k-\frac{1}{2}} > 0$, the value of $U_{j,k-\frac{1}{2}}$ has to be defined. Also, for logical correctness, it is necessary to mention that for $D_{j,k-\frac{1}{2}} < 0$ the condition $B_{j+\frac{1}{2},k-\frac{1}{2}} > B_{j-\frac{1}{2},k-\frac{1}{2}}$ is always satisfied.

Following the Godunov method [86–89] for the component V tangential to the transverse fracture under consideration (remembering that the component U is an excitation of the longitudinal component of the velocity vector of the gas particles in the transonic flow), we obtain

$$V_{j,k-\frac{1}{2}} = V_{j-\frac{1}{2},k-\frac{1}{2}}. \tag{2.279}$$

The following Cauchy problem describes the destruction of a longitudinal failure by fracture in the asymptotic transonic flow investigated here:

$$\oint V \, dt = 0,$$

$$V(\eta, t_0) = \begin{cases} V_-, \, \eta < \eta_0, \\ V_+, \, \eta > \eta_0. \end{cases} \tag{2.280}$$

The solution

$$V(\eta, t) = \begin{cases} V_-, \, \eta - \eta_0 < t - t_0, \\ V_+, \, \eta - \eta_0 > t - t_0, \end{cases}$$

for $\eta = \eta_0$ gives the following calculation formula: if the initial fracture appears on the parallelepiped face with indices $j - \dfrac{1}{2}$ and k,

$$V_{j-\frac{1}{2},k} = V_{j-\frac{1}{2},k-\frac{1}{2}}. \tag{2.281}$$

In general, it is necessary to mention that the solution corresponding to the case (2.280) may be any of the integrable functions $f_1(t)$ of the variable t which satisfy the following condition:

$$\lim_{t \to t_0} f_1(t) = \begin{cases} V_-, \, \eta < \eta_0, \\ V_+, \, \eta > \eta_0. \end{cases}$$

The nonuniqueness of the solution to the problem (2.280) results in a formula for estimating the components $V_{j-\frac{1}{2},k}$. To obtain a unique solution, additional equations and limitations are necessary. For example, we can replace (2.281) with the formula

$$V_{j-\frac{1}{2},k} = \frac{1}{2}(V_{j-\frac{1}{2},k-\frac{1}{2}} + V_{j-\frac{1}{2},k+\frac{1}{2}}),$$

which corresponds to an analogous expression known from the theory of fracture in unidimensional acoustics [194]. Following the Godunov method, the component U tangential to a longitudinal fracture on the face of the parallelepiped with indices $j - \dfrac{1}{2}$ and k can be represented in the form

$$U_{j-\frac{1}{2},k} = \begin{cases} U_{j-\frac{1}{2},k-\frac{1}{2}}, \; V_{j-\frac{1}{2},k} \geq 0, \\ U_{j-\frac{1}{2},k+\frac{1}{2}}, \; V_{j-\frac{1}{2},k} \geq 0. \end{cases} \tag{2.282}$$

In this way, the equations (2.257), (2.259), (2.278), (2.279), (2.281), (2.282) give a set of algebraic equations from which it is possible to evaluate all values necessary for solving (2.251)–(2.256).

It can be seen from these equations that the way presented here of solving a transonic flow around a profile appears to give a real and unique for all regions of the flow, including the vicinity of the fracture.

It should be emphasized that in our case the flow is calculated from the same equations both in the vicinity of the fracture and far from it, because the Reikin–Giugonio conditions for the shock wave (2.214) and (2.215) and the equations (2.257), (2.258) are obtained from the same integral equations of motion of a transonic flow (2.251), (2.252).

The values of the components U and V on a boundary can be treated as values on the corresponding sides of bounding parallelepipeds. The values of those elements on the profile surface are not set, because the relationship between them is known (2.253). To calculate the value of the component U on the profile surface and to take into account the variation of the distribution of U, the Newton extrapolation method is applied, here. For two intervals, one of lenght $\Delta\eta$ and the other of lenght $\Delta\eta/2$, the following computational formula holds:

$$U_{j-\frac{1}{2},1} = 1.875U_{j-\frac{1}{2},1+\frac{1}{2}} - 1.25U_{j-\frac{1}{2},2+\frac{1}{2}} + 0.375U_{j-\frac{1}{2},3+\frac{1}{2}}. \tag{2.283}$$

Therefore the component $V_{j-\frac{1}{2},k}$ can be calculated from the boundary condition of nonpenetration of the profile (2.283). Of course, it is possible to satisfy the boundary conditions in other ways. It is necessary to mention that calculations according to (2.283) result in no oscillations in the vicinity of the shock wave, but calculations performed taking into account other boundary conditions on the profile, for example in the form

$$U_{j-\frac{1}{2},1} = U_{j-\frac{1}{2},1+\frac{1}{2}},$$

show oscillations.

In the vicinity of the profile, the mesh used in the calculations must be more dense. This means that the parallelepipeds used in the calculations are smaller.

Referring to the characteristic equations (2.264) for the case of uniform transonic destruction of a transverse fracture flow (2.259), (2.260) and assuming that excitations separating two neighbouring fractures cannot meet in the time interval Δt, we obtain a condition for stability of the numerical calculations of the transonic gas flow:

$$\Delta t \leq \min_{U} \frac{2M_\infty^2}{\left|M_\infty^2 - 1 + M_\infty^2(\chi + 1)U\right|} \Delta\xi,$$

which corresponds to the relationship between Δt and $\Delta\xi$.

Cases of Unstable and Stable Flow Around an Absolutely Rigid Profile. The following two problems will be considered. The first, is that of an unstable flow around a rigid profile as a result of a sudden start of the motion of the profile at a transonic speed or, equivalently, a sudden start of the gas flow around the profile. This problem corresponds to the following initial velocity field:

$$U_0(\xi,\eta) = 0, \;\; V_0(\xi,\eta) = 0. \tag{2.284}$$

The second problem is that of the stationary motion of a transonic gas flow around a rigid profile, which can be obtained from any unstable motion (this means that the stationary motion not dependent on the functions $U_0(\xi,\eta)$ and $V_0(\xi,\eta)$), by application of the method of stabilizing in time as $t \to \infty$. The initial conditions (2.255) and boundary condition (2.253) (when the equations (2.284) are satisfied) can coincide at $t = 0$ for points on the profile surface only when the profile is located in a plane oriented at zero angle of attack. This is valid when the equations (2.284) are satisfied. In other words, we can say that the first problem, concerning a sudden start of the motion of the profile, is equivalent to the problem of high-speed buckling of a plate into the given profile in the gas flow.

Let us investigate the flow around a symmetric profile, modelled as a long arc, oriented at zero attack angle to the incoming flow, with sudden acceleration. For the calculations, we have chosen the following values of the computational parameters: number of parallelepipeds for calculation on the surface of the profile, 30; time interval, $\Delta t = 0.001$. The values of the pressure factors c_p at different time instants for profile with a shape coefficient $f_* = 0.05$ are presented in Fig. 2.7. This profile is placed in flow with a Mach number $M_\infty = 0.785$. In Fig. 2.7, a shock wave is seen as a zone of very high gradient. The shock wave decreases in intensity on moving from the back to the front of the profile. The solid lines shows the results of calculations of the transonic flow obtained with the method described above. The dots show the results given in [50]. Comparison of these results shows reasonable agreement between them.

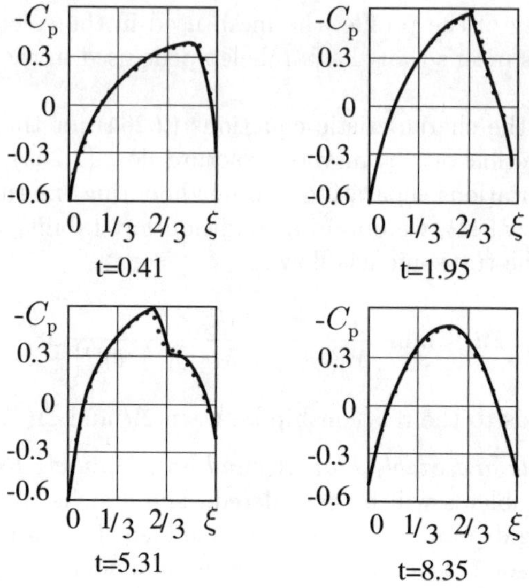

Fig. 2.7. Distribution of the pressure factor c_p at different time instants for $f_* = 0.05$, $M_\infty = 0.785$

To obtain better control of the calculation method, another flow around a long, symmetric profile, modelled as a long arc, oriented at zero attack angle to the incoming flow, was considered. This incoming flow was characterized by a Mach number $M_\infty = 0.864$, and the corresponding profile factor was $f_* = 0.03$. Calculations were performed applying the asymptotic method for the time.

In Fig. 2.8 the solid line represents the distribution of the pressure factor c_p on the profile surface, obtained by the calculation method described

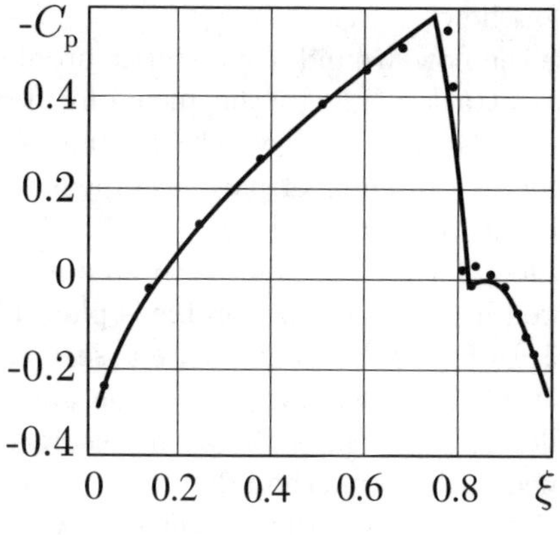

Fig. 2.8. Distribution of the pressure factor c_p for $f_* = 0.03$, $M_\infty = 0.864$

above. The dots represent analogous results taken from [213]. As can be seen from the Fig. 2.8 calculation results show good equivalency of these methods.

We conclude that the numerical method for calculation of transonic flows presented in this section allows us to calculate stable and unstable flows and gives good results.

2.2.1.6 Interaction Between a Flow and a Panel. In this section we shall describe an algorithm for calculation of the interaction between a flow and a panel. The numerical solution of the interaction of a transonic flow and a shallow panel (2.222)–(2.233) consists of two separate problems, of the motion of the panel and of the flow. In solving problems of the interaction of a shell with a flowing medium, it is very important to perform the calculations for the shell and the gas together.

Let us assume that at a time instant t, the values of w, $\partial w/\partial t$, U, V, P_z, f are known. Knowing the values of w, $\partial w/\partial t$ and P_z, from (2.222) and (2.223) (for the time instant t), taking into account the conditions (2.232) and (2.233), it is possible to calculate the values of w and $\partial w/\partial t$ at the next time instant $t + \Delta t$. From the known values of U, V and f obtained from (2.226) and (2.227) at the time instant t, with the conditions (2.228), (2.229), (2.230) and (2.231), the values of U and V at the time instant $t + \Delta t$ can be evaluated. The load at the time instant $t + \Delta t$ is calculated from (2.224) and (2.225). Knowing the value of w at the time instant $t + \Delta t$, the shape of the panel is computed from the second equation of (2.231); this means the value of the function f, at that time instant $t + \Delta t$. Iterating the time step, the calculations described above are performed repeatedly.

At every time step, the results of the calculation of the transonic flow or of the shallow shell must be verified. This is performed by repeating the calculations with a smaller time interval and a greater number of smaller iterations. The results of these calculations are then compared. If that comparison shows that the results agree with the required accuracy, then the calculations of the solution are continued. If the comparison gives a difference between the results, it is necessary to change the time integration interval Δt. For the optimal setting of the time interval Δt, it is possible to perform only one calculation with that value of the time interval for the flow and for the shell. Computations showed that, for the required accuracy, the calculations of the panel were less sensitive and allowed a higher value of the time interval Δt than for the calculation of the transonic flow. The calculations showed that for the chosen rounding of the results, the best values of the time interval Δt were between 0.001 and 0.003. The influence of the value of the time interval Δt will be discussed for the following numbers of elements in the mesh: on the shell, 30 equally spaced nodes were created; in the flow over the panel in the direction of the ξ axis, 30 equally spaced nodes were also created; in the direction of the ξ axis there were approximately 60 nodes in total in the flow, with the distance between them increasing with the distance from the panel. In the direction of the η axis approximately 30 nodes were created, with the distance between them increasing with the distance from the panel.

The stability of the numerical calculations of the problem was checked by comparing the results for different values of the time interval Δt. If the results agreed within the allowed limits then the calculations were treated as convergent.

The calculations of the transonic flow are described in Sect. 2.2.1.5. Here we mention only that for estimating the initial velocity field of the particles of the gas flow, an initial calculation of the flow around a rigid profile was applied.

Calculations of the dynamic motion of the shallow shell were performed according to the method presented in the Sect. 2.2.1.4. It is necessary to estimate the values of some dimensionless parameters. Comparing the sets of dimensionless equations (2.219) and (2.238), one can obtain the following difference in transverse load:

$$A = \frac{q_\infty}{L}.$$

In the calculations, the interval of dimensionless time spent in stabilizing the flow around a rigid initial profile was set to 2.4 in all cases considered. As can be seen from the numerical calculations of a transonic flow around a rigid panel presented in the Sect. 2.2.1.5, sufficiently good stability of the flow can be expected after this time interval.

For performing calculations, it is necessary to set the values of the main characteristic parameters. The pressure in the flow, at an infinite distance from the profile was taken as equal to atmospheric pressure, $P_\infty = 101\,330$ Pa. The density of air was taken as $\rho_\infty = 1.2928$ kg/m^3, and the adiabatic Poisson coefficient as æ $= 1.4$. It was also assumed that the panel was made from steel with density $\rho_0' = 7953$ kg/m^3, Young's modulus $E' = 20 \times 10^{10}$ MPa and Poisson's ratio $\nu = 0.3$ [1]. Transforming these parameters into dimensionless values, we obtain the following values: $\rho_0 = 6163.8$, $E = 1\,973\,749$. These parameters were not changed during the calculations. The Mach number for the incoming flow varied from 0.75 to 1.0. The relative thickness of the panel δ, the pressure P_0 on the side opposite to the transonic flow, and the geometric dimensionless parameter K of the circular cylindrical panel had varying values. In detail, the relative thickness of the panel δ varied from 0.0005 to 0.01. The initial maximum deflection of the infinite circular cylindrical panel f_* was set to two values, 0.03 and 0.05. For each value of f_*, values of δ were assumed such that the panel started to buckle in the estimated time of calculation. This time interval, during which the interaction between the panel and the transonic flow was investigated, was approximately 6–10 dimensionless time units.

2.2.2 Stable Vibrating Panel Within a Transonic Flow

2.2.2.1 Shock Wave Motion. Let us discuss the characteristics of a transonic flow. The complexity of the problems of a transonic flow around a profile is caused by specific characteristics of transonic flows in comparison with subsonic and supersonic flows. The occurrence of shock waves, which can expand and move along the profile (shell) and disappear in the transonic flow, is characteristic of this type of flow. A shock wave is a jump of the values of certain parameters, including the pressure. It is necessary to consider the description of the motion of shock waves.

Classification of the Motion of a Shock Waves. To classify the motion of a shock wave during the interaction of a panel and a transonic flow it is necessary first to itemize some simple cases of shock wave motions:

U - motion upstream the flow only,
D - motion downstream the flow only,
S - steady state of a shock wave,
N - no shock wave in a flow.

More complicated forms of shock waves can be created from those simple forms of motion. For example (UD) means that a shock wave moves up and then down, does not disappear and does not stand still (the parentheses mean that this motion repeats many times); (UDN) means that a shock wave moves up, moves down and disappears. This repeats many times. It is sometimes important to put in parentheses (after each letter describing a simple shock wave mode) the coordinates of the time interval and flow space in which such a motion of a shock wave takes place.

Creation, Existence and Destruction of a Shock Wave. A shock wave can appear in a transonic flow when a local transonic zone appears. This means that the Mach number must have a value greater than 1 in a certain sub-zone of a flow with $M < 1$. The case of existing local supersonic waves without a shock wave cannot be taken into account. By applying a known estimate of the local Mach number M in a transonic zone [197], we can obtain an approximate condition for the existence of a shock wave in a flow in the following form:

$$M^2 = M_\infty^2 \left(1 + (\chi + 1)U + (\chi - 1)\frac{\partial \varphi}{\partial t}\right) > 1. \tag{2.285}$$

If at a certain point of the flow the velocity has locally reached the sound speed, then $M = 1$ at this point, and, following (2.285), the longitudinal velocity excitation of the gas particles reaches some critical value U_*. This critical value can be calculated from (2.285) after setting $M = 1$:

$$U_* = \frac{1 - M_\infty^2 - (\chi - 1)\dfrac{\partial \varphi}{\partial t}}{(\chi + 1)M_\infty^2}. \tag{2.286}$$

For a low-frequency motion $\left(\frac{\partial\varphi}{\partial t}=0\right)$, the critical sound speed depends on the Mach number M_∞ of the incoming flow:

$$U_* = \frac{1 - M_\infty^2}{(\chi+1)M_\infty^2}. \tag{2.287}$$

The critical values of U_*, calculated from (2.287) for the values of M_∞ used in the computations described in this book, are the following: $U_*(0.785) = 0.2595$; $U_*(0.825) = 0.1955$; $U_*(0.864) = 0.1415$; $U_*(0.9) = 0.098$; $U_*(1) = 0$. Of course, as can be seen from (2.286), the values of the critical parameter U_* in a nonstationary flow differ slightly from the values given above. A shock wave which satisfies the condition for the existence of shock waves (2.285) disappears when this condition is disturbed; this means when, at all points of the flow, the local Mach number satisfies the inequality $M < 1$. In a subsonic flow, a moving shock wave may exist, but the shock wave first appears in a supersonic flow.

Direction of Movement of a Shock Wave. From the Reikin–Giugono condition, for a shock wave in the transonic low-frequency approximation (2.215), we obtain a solution for the velocity of motion of a shock wave:

$$\frac{\partial\xi}{\partial t} = \frac{1}{2M_\infty^2}\left(M_\infty^2 - 1 + \frac{1}{2}M_\infty^2(\chi+1)(U_- + U_+) - \left(\frac{V_- - V_+}{U_+ - U_-}\right)^2\right). \tag{2.288}$$

A shock wave can move up the flow if the following condition is satisfied:

$$M_\infty^2 - 1 + \frac{1}{2}M_\infty^2(\chi+1)(U_- + U_+) - \left(\frac{V_- - V_+}{U_+ - U_-}\right)^2 < 0. \tag{2.289}$$

This is because the velocity of a shock wave moving up the flow is negative. The inequality (2.289) contains values of parameters on both sides of the shock wave front, and the Mach number of the incoming uniform flow M_∞. Analogously, a shock wave moves down the flow if the following condition is satisfied:

$$M_\infty^2 - 1 + \frac{1}{2}M_\infty^2(\chi+1)(U_- + U_+) - \left(\frac{V_- - V_+}{U_+ - U_-}\right)^2 > 0. \tag{2.290}$$

It is necessary to mention that the inequalities (2.289) and (2.290) must be considered at a particular instant of time and have a local character. As is known, the velocity component tangential to a shock wave front is continuous on passing through the shock wave. Therefore, treating the shock wave as approximately almost vertical, it can be assumed that

$$V_+ = V_-. \tag{2.291}$$

Taking into account the last equation (2.291), the conditions for the direction of motion of a shock wave (2.289) and (2.290) can be converted into the following form:

up:

$$U_- + U_+ < 2U_*; \tag{2.292}$$

down:

$$U_- + U_+ > 2U_*. \tag{2.293}$$

For an existing shock wave, another condition can be derived from (2.285) and (2.287):

$$U_- > U_+. \tag{2.294}$$

From the values of U_* presented above one can conclude that with an increase of the Mach number of the incoming flow M_∞ in the interval under consideration, the relative influence of U_* decreases. This means that for a nonstationary flow around the same profile, motion of a shock wave down the flow may occur earlier for flows with a higher value of the Mach number of the incoming stream M_∞ than for flows with lower values of the Mach number. The value of U_* is the limit of the sum of components U on opposite sides of the shock wave front, $U_- + U_+$, for a certain value of the Mach number of the incoming flow. This provides an estimate of the direction of motion of the shock wave at any given time t. Of course, the value of U_* given here is approximate. As can be seen from (2.287), the value of U_* depends only on the Mach number of the incoming uniform flow M_∞ and appears on the right-hand side of (2.292) and (2.293). The left-hand side of these equations, which means the sum $U_+ + U_-$, depends on many coefficients, including the form of the flow profile.

In a stabilizing stream, the location of a shock wave should also stabilize, and the inequalities (2.289), (2.290), (2.292), (2.293) should become equalities. In such a stabilizing flow, the sum $U_+ + U_-$ can be approximated as $2U_*$. In this case the condition for a shock wave to be stationary is equivalent to the equation

$$U_+ + U_- = 2U_* \tag{2.295}$$

A shock wave, as can be seen from (2.294), can be stationary only in a supersonic flow; this means that the velocity U_- exceeds the local sound speed.

It is obvious that a change of the form of the profile changes the flow around it and the location of any shock waves. The pressure on the panel surface changes corresopndingly. The flow and pressure on the panel surface can vary for small changes of the profile. It should be emphasized that for certain transonic flows, very small variations of the profile shape may cause relatively strong changes in the flow and pressure on the panel surface.

The Ability of a Shock Wave to Move up the Stream is Not Obvious. Such a motion of shock waves can be periodic–one shock wave behind another. Let us investigate slightly supersonic flows; this means flows for which the local Mach number M in some bounded sub-regions of the flow is slightly greater than one.

We shall prove that slightly supersonic, almost uniform gas flows exist for which shock waves appear at the rear, move up and disappear. This scheme is repeated. As can be seen from (2.293), the values of the dimensionless longitudinal components U of the velocity excitation vector are close to zero. This means that there exists a positive number ε_U, much less than one, such that $|U_-| < \varepsilon_U$ and $|U_+| < \varepsilon_U$. Thus, for a slightly overcritical transonic flow, we obtain the following from the triangle inequality:

$$|U_- + U_+| \leq |U_-| + |U_+| < 2\varepsilon_U .$$

From this expression, the following inequality can be derived, which is also satisfied:

$$U_+ + U_- < 2\varepsilon_U . \tag{2.296}$$

This yields the result that the inequality (2.292) is also satisfied if

$$\varepsilon_U \leq U_* . \tag{2.297}$$

Let us assume that for any positive constant number ε_U (as mentioned earlier, close to zero), such a transonic flow exists for which the inequalities (2.296) and (2.297) are satisfied. It can be proved that for every U_*, such a flow can be found for which the condition (2.292) is satisfied.

This means that a flow in which a shock wave moves upstream is slightly overcritical and nonstationary. Such a flow appears during flow around a slightly deformed profile panel of a weakly overcritical gas stream with local supersonic zones. In the stabilized state of an overcritical stream, there are no shock waves. Thus, when small deformations of the profile occur, a shock wave is created as in a nonstationary forced flow. A similar situation appears in an explosion of a profile at transonic speed. For a nonstationary flow, the component $\left(\frac{\partial \varphi}{\partial t}\right)$ in (2.285) cannot be neglected. This gives the possibility for the local Mach number to reach a value of 1 earlier than in a stationary flow. This explains the appearance of a shock wave in a nonstationary flow in cases when, for the same parameters, there is no shock wave in a stationary flow. In our investigations, the low-frequency approximation has been investigated, but, as can be seen in Sect. 6.1, $\left(\frac{\partial \varphi}{\partial t}\right)$ is partially taken into account in the equations of motion of a transonic flow element, and it is possible to find shock waves in nonstationary flows when numerical calculations are performed. In that case a shock wave moves upstream and disappears, losing its intensity, while the flow is stabilizing.

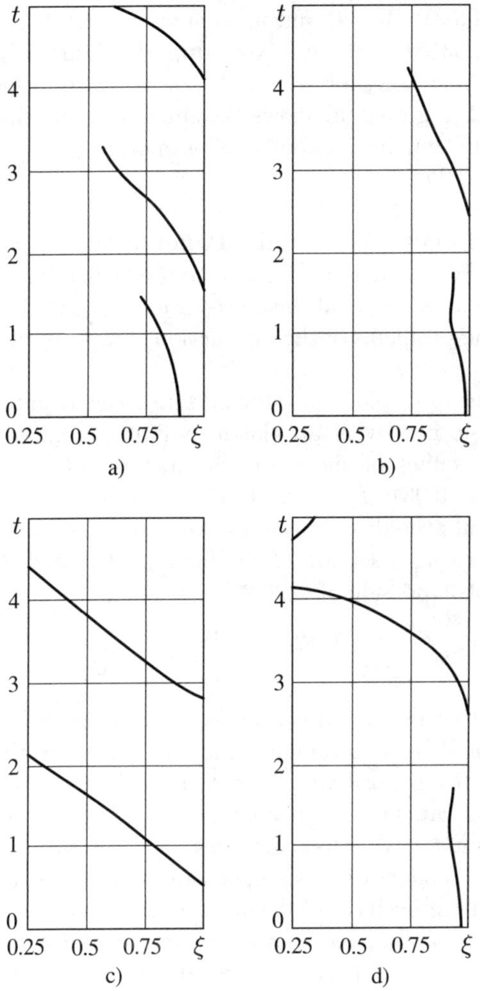

Fig. 2.9a–d. Distribution of shock wave motion for different parameters–details in text

Two reasons for shock wave motion are possible: (a) self-excited motion of a shock wave in a forced flow, and (b) motion caused by a change of the geometry of the profile. In slightly overcritical nonstationary flows, the motion of shock waves can be self-excited.

Let us consider some examples of the motion of shock waves. Figure 2.9 shows diagrams of the distribution of shock waves on the surface of a profile panel at certain instants of time t, for the following values of the parameters main flow and panel parameters: (a) $M_\infty = 0.864$, $K = 34.16$, $\delta = 0.007$ ($f_* = 0.03$), $P_0 = 0$ and fixed support; (b) $M_\infty = 0.9$, $K = 34.16$, $\delta = 0.007$ ($f_* = 0.03$), $P_0 = 0$ and fixed support; (c) $M_\infty = 0.785$, $K = 36.8$,

$\delta = 0.005$ ($f_* = 0.03$), $P_0 = 0$ and fixed support; (d) $M_\infty = 0.9$, $K = 36.8$, $\delta = 0.065$ ($f_* = 0.03$), $P_0 = 0$ and fixed support. Figure 2.9c shows a slightly overcritical transonic flow, for which the shock wave motion can be represented as type D2. Figure 2.9d shows the shock wave motion in a case where the panel loses its dynamic stability. This motion of a shock wave can be described as UDNUDN ...

2.2.2.2 Interaction of a Flow and a Panel. In this section we shall discuss some aspects of the interaction of a flow and a panel. The results presented here illustrate the application of the method described in the previous sections method for solving problems of the interaction of a transonic flow and a shock wave.

As one of the examples, we shall discuss the results for the interaction of a transonic flow with an elastic, infinite, circular cylindrical panel for the following values of the main parameters: $M_\infty = 0.9$, $K = 34.16$, $\delta = 0.007$ ($f_* = 0.03$), $P_0 = 0$. Both edges of the panel are clamped. Figure 2.10 presents the distribution at different time instants of the dimensionless pressure along a section of the panel, calculated from the following formula (the right-hand side of (2.222)):

$$\Delta p = \frac{12(1-\nu^2)}{E'\delta^4}(P' - P_0'). \tag{2.298}$$

This figure shows also the dimensionless transverse deflection w. The pressure Δp represented by a continuous line, and the deflection w by a dashed line. A zone of strong positive gradient on a diagram of the pressure Δp a shock wave. As can be seen from Fig. 2.10, a shock wave exists at the initial time instant $t = 0$. After the start of deformation of the panel, the shock wave moves a short distance upstream and moves down to the rear of the panel, loses its intensity and disappears. A shock wave appears again at $t = 2.8$ at the rear of the panel and starts to move forwards. Such a motion of a shock wave is repeated many times. In the classification of shock wave motion presented in the Sect. 2.2.2.1, shock wave has a motion of type (UDN). It can be seen that at the beginning, the deflection w increases over the whole panel. The rear half of the panel has negative deflections; this means that the panel buckles in that region. At the same time, in the front part of the panel, the positive value of the deflection w increases. After the time instant $t = 2.8$, the deflection w decreases in the front part and increases in the rear part.

For better representation of the changes of the transverse deflection w and pressure Δp, diagrams of the variations in time of w and Δp at different points of the panel in the case investigated here are presented in Figs. 2.11 and 2.12. Figure 2.13 presents the variation of the location of the shock wave with time. From Fig. 2.11 it can be seen that the absolute value of the deflection w does not exceed 1, and its value varies from -0.7 to 1. The deflection w at the point $\xi = 0.75$ has a negative value; at the points $\xi = 0.25$ and $\xi = 0.5$ the deflection

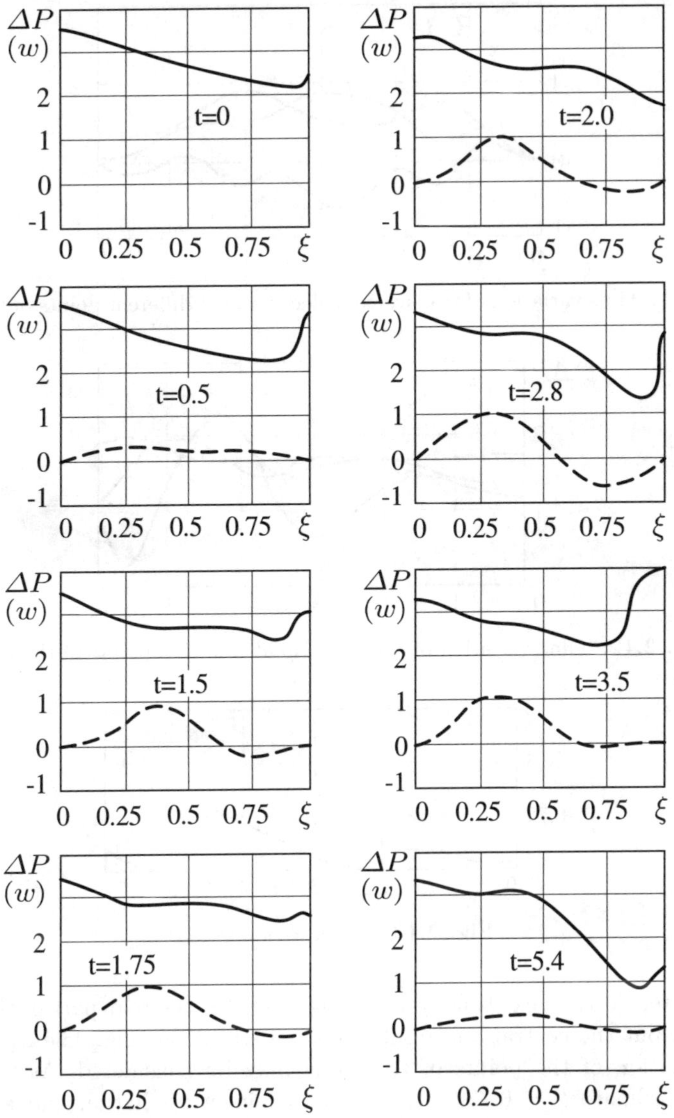

Fig. 2.10. Distribution of dimensionless pressure Δp (*solid line*) and transverse deflection w (*dashed line*) along a section of the panel at different time instants

w is positive. This means that the rear part of the panel buckles. This buckling is caused by the fact that, directly before a shock wave located near the rear of the panel, the pressure in the flow acting on the panel is significantly lower than that on the front part of the panel. From the diagrams presented in Fig. 2.11, it can be seen that in the time interval $(0 - -0.8)$ from the start of the interaction, the whole panel deflects under the pressure caused by the

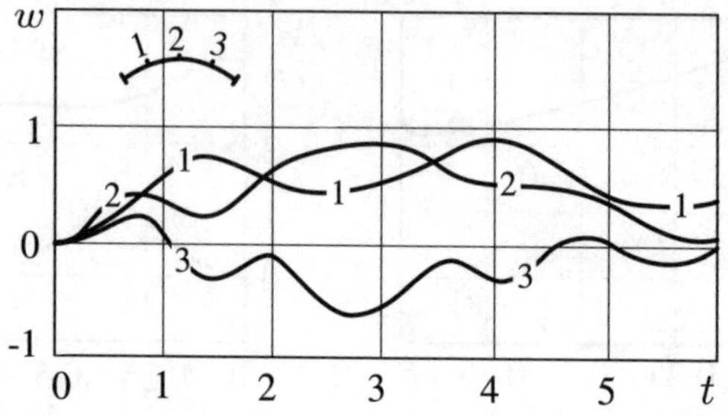

Fig. 2.11. Time variation of transverse deflection w at different points of the panel

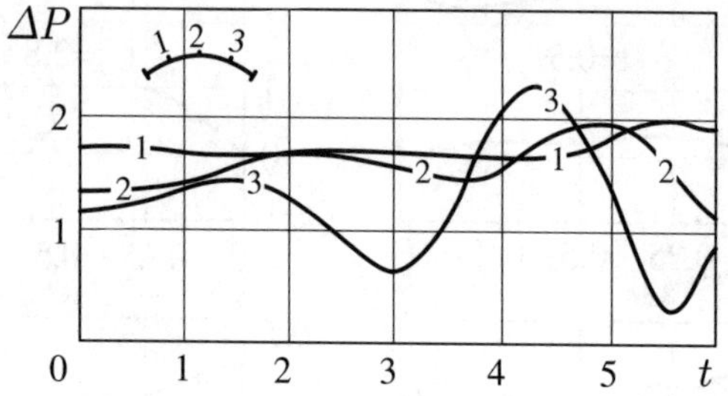

Fig. 2.12. Time variation of pressure Δp at different points of the panel

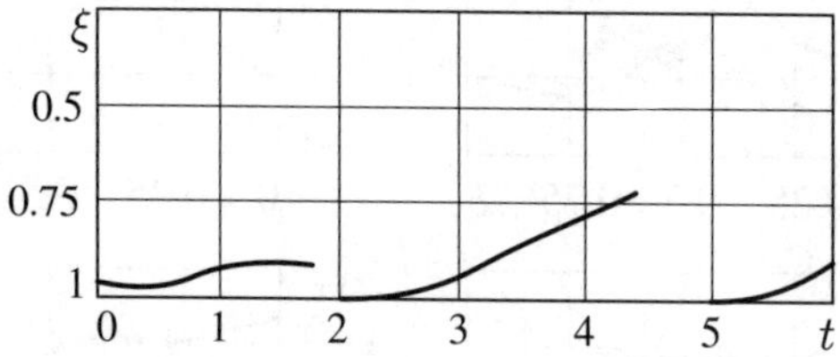

Fig. 2.13. Shock wave motion

flow. Later, in the next time interval (0.8 – –5.0), the rear part of the panel buckles, but the central and front parts are still deflecting. The frequencies of oscillation of the points of the panel may be considered. At the point with coordinate $\xi = 0.75$, a main low-frequency oscillation and a higher-frequency oscillation with lower amplitude can be observed. Similarly, low and high frequencies, as can be seen in Fig. 2.11, appear at the points with coordinates $\xi = 0.25$ and $\xi = 0.5$. The low frequencies of the points with $\xi = 0.25$ and $\xi = 0.75$ are almost equal, but the oscillations have opposite phases. The high frequencies of the oscillations of the deflection w at different points of the panel are approximately equal.

It can be seen from the diagrams presented in Fig. 2.12 that the lowest amplitude of the oscillations of the pressure Δp occurs at the point $\xi = 0.25$, which is located close to the front of the panel, and the highest amplitude of the pressure oscillations occurs at the point $\xi = 0.75$, in the rear part of the profile, in the zone of shock wave motion. From these diagrams and Fig. 2.13, it can be seen that the oscillation frequencies of Δp at different points of the panel are approximately the same and match the oscillations of the shock wave motion. This shows that motion of a shock wave in the rear part of panel causes changes of pressure Δp not only in the rear part of the profile, close to the shock wave, but also in other parts of the panel. It is obvious that at greater distances pressure changes are smaller.

Let us consider a second case of the interaction of this panel with a transonic flow. The distributions of the dimensionless pressure Δp and transverse deflection w along a section of the panel at different time instants are presented in Fig. 2.14 for the following parameters: $M_\infty = 0.864$, $K = 34.16$, $\delta = 0.007$ ($f_* = 0.03$), $P_0 = 0$.

The changes of the transverse deflections w with time at various points of the panel are shown in Fig. 2.15. Figure 2.16 presents diagrams of pressure Δp versus time for the same points as those for which deflections are shown in Fig. 2.15. A diagram of the shock wave motion in this case is presented in Fig. 2.17. The case presented in Figs. 2.14–2.17 differs from that presented in Figs. 2.10–2.13 only in the Mach number M_∞ of the incoming stream. In the first case $M_\infty = 0.9$, and in the second case $M_\infty = 0.864$. In general, the characteristic curves in Fig. 2.14, the transverse deflections w in Fig. 2.15 and the pressure Δp in Fig. 2.16 match the corresponding curves in Figs. 2.10–2.12. Analysis of the deflection diagrams in Figs. 2.14 and 2.15 leads to the following conlusions: the panel, after first buckling in the rear part buckles also in the front part, and the buckling in the rear part changes into a deflection inside the panel. The high frequencies of the oscillations of the transverse deflection w at the points with coordinates $\xi = 0.25$ and $\xi = 0.75$ approximately match each other. The low frequencies of the oscillations at these points, symmetrically placed around the centre of the panel, also almost match, but the oscillations have opposite phases. The panel buckles successively in the front and rear parts, as was observed earlier in the first case of this interaction.

Dependence on Main Parameters. Comparison of the diagrams in Figs. 2.13 and 2.17 leads to the conclusion that the frequency at which shock waves are created is higher for the flow with the lower value of the Mach number for the incoming stream, $M_\infty = 0.864$, than for the flow with the higher value of the Mach number, $M_\infty = 0.9$. Also, for the lower value of the Mach number, shock waves move faster, and further from the rear to the front of the panel, than for the higher value of the Mach number. It is easy to see from a comparison of the shock wave motion presented in Figs. 2.13 and 2.17 and the pressure distribution Δp presented in Figs. 2.12 and 2.16 that the peaks of the pressure Δp at the points of the panel surface investigated correspond to shock waves. A high pressure appears after a low pressure, as at a shock wave front. The peak of the pressure appears at the top of waves of pressure oscillations. For a slightly overcritical gas flow and a small amplitude

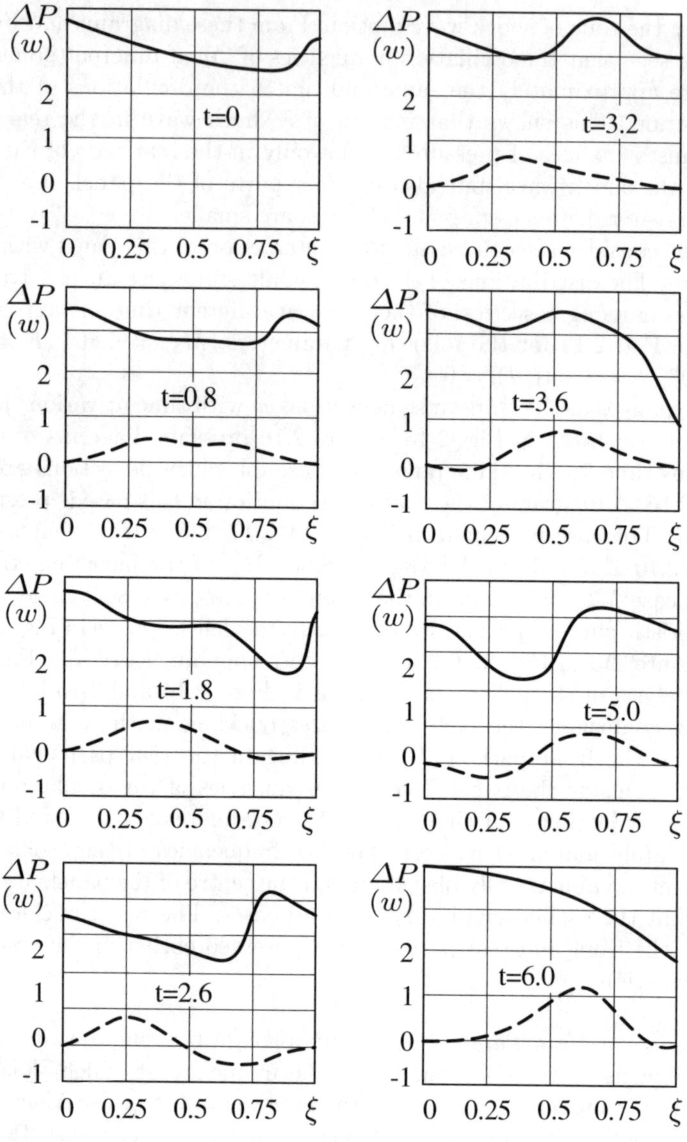

Fig. 2.14. Distribution of dimensionless pressure Δp (*solid line*) and transverse deflection w (*dashed line*) along a section of the panel at different time instants

of the oscillations of the transvers deflection w, moves along the panel surface almost independently of the changes of profile geometry, but with a strong interaction with the flow, which results in peaks of the pressure oscillations. The motion of shock waves described above for a slightly overcritical transonic flow around a profile appears to be the main reason for the strong increase of the amplitude of oscillations of the pressure Δp acting from the stream

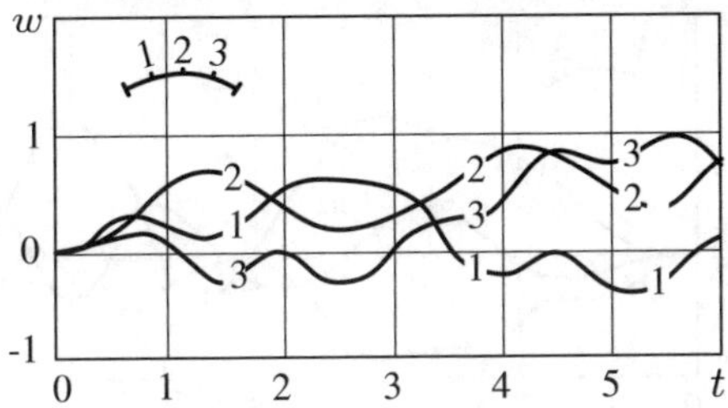

Fig. 2.15. Time variation of transverse deflection w at different points of the panel

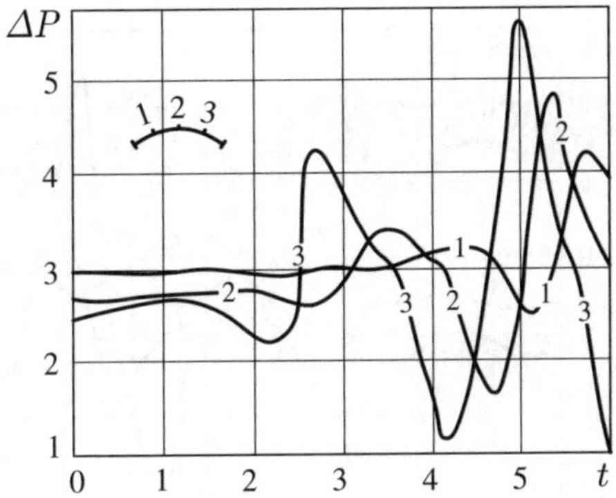

Fig. 2.16. Time vibration of pressure Δp at different points of the panel

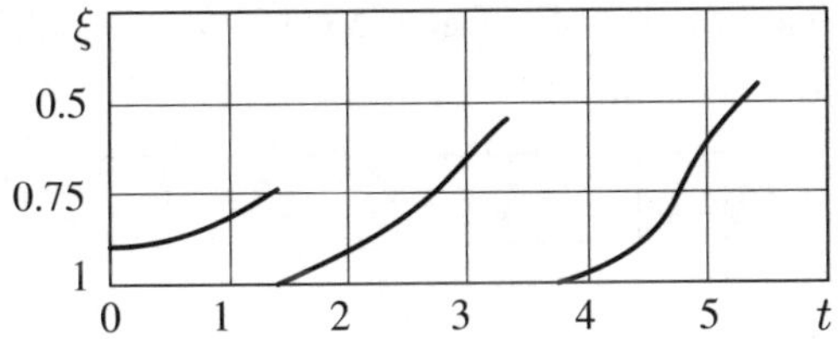

Fig. 2.17. Shock wave motion

on the panel. This also happens when the amplitude of oscillations of the transverse panel deflection w is relatively small and the panel does not lose its original dynamic stability.

Figure 2.18 shows the variation with time of the transverse deflection w for the following case: $M_\infty = 0.9$, $K = 47.8$, $\delta = 0.005$ ($f_* = 0.3$), $P_0 = 1$ and simple support on both edges of the panel. Figure 2.19 shows the variations of the pressure Δp at the same points of the panel as those for which the curves in Fig. 2.12 were plotted.

Figure 2.20 represents the motion of the shock wave in this case of interaction of a panel with a flow. This case is characterized by a simple support

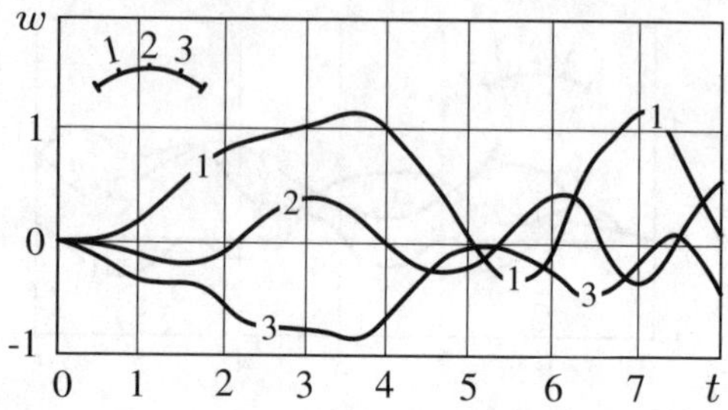

Fig. 2.18. Time variation of transverse deflection w at different points of the panel

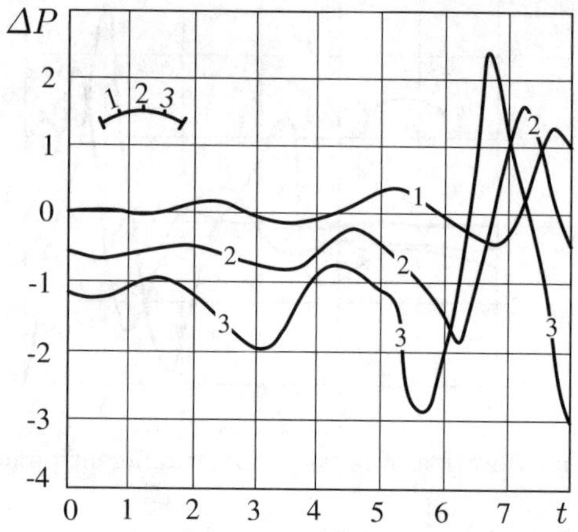

Fig. 2.19. Time variation of pressure Δp at different points of the panel

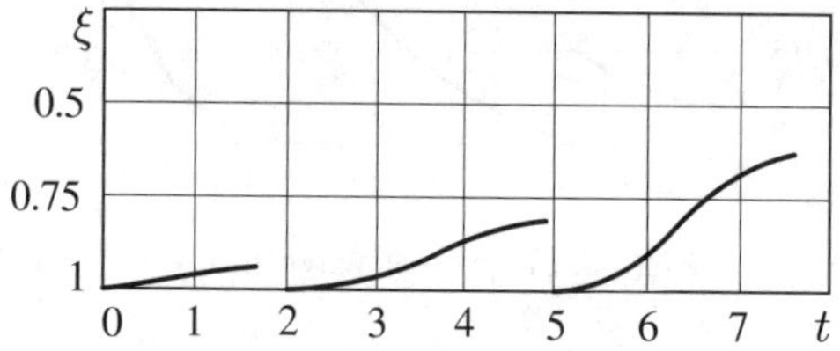

Fig. 2.20. Shock wave motion

and a dimensionless pressure under the panel $P_0 = 1$. From Fig. 2.19 it can be seen that Δp is negative at the points $\xi = 0.5$, $\xi = 0.75$, and for $\xi = 0.25$ it has both negative and positive values. That is why the deflections w shown in Fig. 2.18 for the points $\xi = 0.75$ and $\xi = 0.5$ are also sometimes negative. These values oscillate, covering both positive and negative values. The front part of the panel deflects only, the other parts of the panel can either slightly buckle or deflect.

In the initial time interval, from 0 to 5.5 dimensionless time units, the front ($\xi = 0.25$) and rear ($\xi = 0.75$) parts oscillate with a low principal frequency in opposite phase. After this time, the oscillations of these parts become more complicated. A high frequency of oscillations, as seen from Fig. 2.18, appears only for $\xi = 0.75$. At the initial time instant, this frequency is similar to the analogous high frequencies obtained in the cases of interaction of a flow and a panel described previously. In general, as can be seen in Fig. 2.18, the oscillations of the panel in the case $P_0 = 1$ are slightly more complicated than in the previous cases with $P_0 = 0$. It is necessary to mention that in the centre of the panel ($\xi = 0.5$), the oscillations of the deflection w and pressure Δp have similar frequencies, but have a small constant phase shift. Comparing Figs. 2.19 and 2.20, it can be seen that the peaks of pressure correspond to shock waves, which periodically appear and move upstream, losing intensity. Comparing Figs. 2.19 and 2.20 leads to the following conclusion: for the same Mach number of the incoming flow $M_\infty = 0.9$ and relative panel camber coefficient $f_* = 0.03$, in all cases of interaction between a flow and a panel investigated, differing mainly in the value of the pressure P_0 under the panel and the method of panel edge support, the principal features of the motion of shock waves are almost identical.

Therefore the motion of shock waves appears to be the reason for the large amplitudes of the oscillations of pressure at the surface points of the panel, and the frequency of these oscillations matches the frequency of shock wave motion. The motion of shock waves up the flow for slightly overcritical flows is repeated periodically. This excites a changing, periodic, nonuniformly distributed transverse load on the panel. A complicated deformation of the panel is caused by this load. Besides the main, low-frequency oscillations of the transverse deflection of the panel, high-frequency oscillations are also observed. Their amplitude is lower, but still comparable to the amplitude of the low-frequency oscillations. An oscillating buckling of the front and rear parts of the panel is characteristic of this case of interaction.

The Change of the Stress State of the Panel has a Complicated Oscillating Form. The changes in time of the dimensionless membrane stress on the middle surface of the panel, calculated according to (2.220), are presented in Fig. 2.21. These results were obtained for the following values of the main parameters: $M_\infty = 0.9$, $K = 36.16$, $\delta = 0.007$ ($f_* = 0.03$), $P_0 = 0$. Both edges of the panel were clamped. The figure shows that the membrane stress σ_M is negative. Comparing the oscillations of the transverse deflection w shown in Fig. 2.11 with the variation of the membrane stress shown in Fig. 2.21 leads to the conclusion that the oscillation frequency of the membrane stress is approximately equal to the high frequency of the oscillations of the transverse deflection. As explained earlier, motion of shock waves is the reason for the high-frequency oscillations of the deflection w, and this causes similar high-frequency oscillations of the membrane stress σ_M. Graphs versus time of the dimensionless bending moment M_u, which can be calculated from the

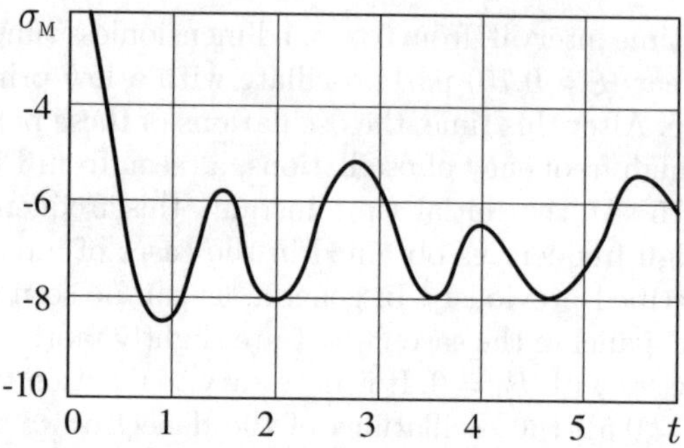

Fig. 2.21. Time variation of membrane stress on the middle surface for $M_\infty = 0.9$, $K = 36.16$, $\delta = 0.007$ ($f_* = 0.03$), $P_0 = 0$, clamped edges

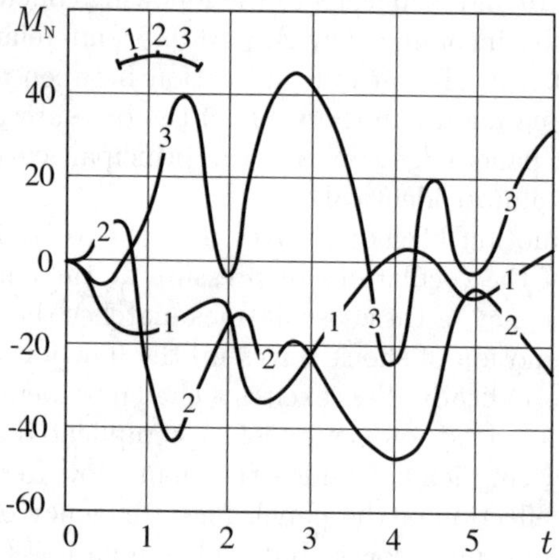

Fig. 2.22. Time variation of bending moment at different points of the panel

bending moment using

$$M_u = -\frac{12(1-\nu^2)}{E'h^3} M_y, \tag{2.299}$$

are presented in Fig. 2.22.

From this figure it can be seen that the values of the dimensionless bending moment are different at different points of the panel. As can be seen, even for stable oscillations of the panel the bending moment reaches relatively high values, which can lead to plastic deformation. The question of plastic deformation will be discussed in Sect. 2.2.3.2. The nonuniform distribution of the load applied to the panel may result in a large local curvature. As a result, high bending moments M_u may act inside the panel. In the present case, during stable motion of the panel, the bending moment has different signs at different points of the panel, as can be seen from the graphs in Fig. 2.22.

2.2.3 Stability Loss of Panel Within Transonic Flow

2.2.3.1 Stability Criteria. Let us consider the theory of the stability of an elastic panel. The criteria for the dynamic stability of an elastic shell interacting with a transonic flow must be obtained from the allowable load conditions of the panel, treated as a part of a structure. Also, the loss of stability of the structure is important, because after this happens, the structure is unable to satisfy the requirements on it.

We assume that a structure ceases to be functional when it undergoes sufficiently high deformation or when a change of material properties, for instance plastic deformation of an elastic material, occurs. We shall discuss two criteria.

The first criterion says that loss of dynamic stability of a panel happens when the transverse deflection w increases rapidly with small changes in the transverse load; this normally occurs in a nonstationary flow [51, 134].

The second criterion says that for an elastic structure, the appearance of plastic deformation is treated as a loss of dynamic stability [51, 134]. This criterion corresponds to loss of functionality of the structure.

Of course, in investigating the stability of a panel during interaction with a transonic flow, other criteria for dynamic stability can also be taken into account. We shall discuss only the two criteria mentioned above.

The first criterion means that the panel has reached large deflections and, in addition, the transition from one stable motion to another is characterized by a very high velocity. The limits on the deflection and speed of deflection, exceeding of which can be treated as a dynamic loss of stability, can be set arbitrarily.

In the case of interaction of an elastic panel and a transonic flow, investigated here, either a local or a global loss of stability may occur. Local loss of stability is understood as a rapid increase of the transverse deflection w in a small area of the panel. Global loss of stability takes place when the transverse deflection increases rapidly at all points of the panel. Of course, local and global loss of stability are related in that a local loss of stability may start a global loss of stability of the panel. Global loss of stability for a panel with nonzero curvature leads to a global buckling of the shell.

We shall start by investigating the dynamic loss of stability of an elastic cylindrical panel of infinite length during interaction with a transonic flow by applying the first criterion of dynamic stability loss. In Sect. 2.2.3.2 we shall analyse the stability of a panel starting from the second criterion, in which loss of stability occurs after plastic deformation of the panel.

Influence of Internal Pressure. Let us analyse the buckling of the panel for the following parameters: $M_\infty = 0.9$, $K = 36.8$, $\delta = 0.0065$ ($f_* = 0.3$), $P_0 = 0$. Both edges are clamped. Results are presented in Fig. 2.23. The graphs in this figure show, at different time instants, the distribution of the transverse deflection w and the pressure Δp in a transverse section of

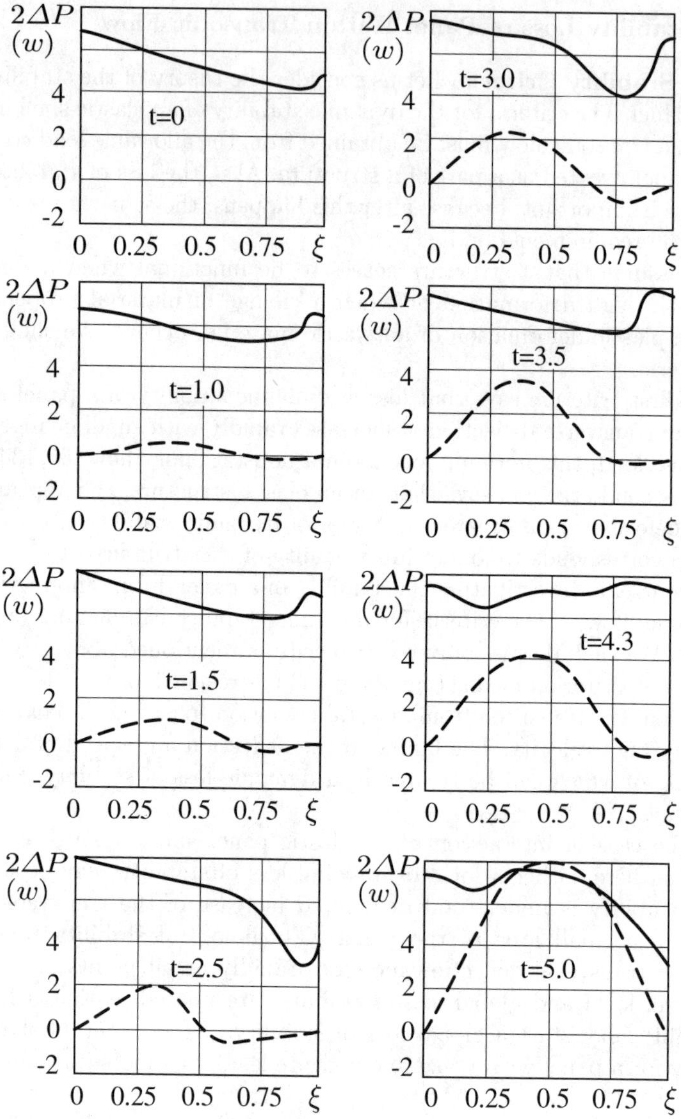

Fig. 2.23. Distribution of dimensionless pressure Δp (*solid line*) and transverse deflection w (*dashed line*) in a section of the panel at different time instants, for $M_\infty = 0.9$, $K = 36.8$, $\delta = 0.0065$ ($f_* = 0.3$), $P_0 = 0$, clamped edges

the infinite-length cylindrical shell. The values of Δp were calculated from (2.298), which characterizes the difference of pressure between the two sides of the shell. As mentioned earlier, a zone of large positive pressure gradient corresponds to a shock wave. A characteristic of the buckling of this panel is that the deflection w in the front part of the panel increases uniformly (see Fig. 2.23) and the deflection in the rear part is negative. This means that the panel buckles in the rear part, and the negative deflection oscillates, increases and decreases. The whole panel has positive deflections, and it buckles. In the case of no internal pressure ($P_0 = 0$) and a sufficiently thin panel, the main influence on buckling is the unidirectional pressure Δp, not the nonuniformly distributed transverse load nor moving shock waves. It is necessary to mention, however, the influence of travelling shock waves (initially moving, regardless of the form of the profile). As mentioned in the Sect. 2.2.2, shock waves undergo excited motion in addition to self-excited motion, depending on the changes of the geometrical form of the profile, which can be seen in Fig. 2.23 for high values of the time t. In other words, during buckling of the panel, the shock wave is located at the front part of the panel. The shock wave can undergo excited motion because of the change of form of the panel, but also self-excited motion if the conditions (2.292) and (2.293) are satisfied.

Let us now discuss a second case of dynamic loss of panel stability, for the following values of the main parameters: $M_\infty = 0.864$, $K = 198$, $\delta = 0.002$ ($f_* = 0.05$), $P_0 = 1$; both edges are clamped. Graphs of the transverse deflection w and pressure Δp at different time instants for this case are presented in Fig. 2.24.

The main difference between the two cases presented in Figs. 2.23 and 2.24 is in the internal pressure P_0. A comparison of the buckling graphs presented in Figs. 2.23 and 2.24 leads to the conclusion that the buckling graph for an internal pressure $P_0 = 1$ becomes complicated when different signs of the pressure Δp appear.

Figure 2.25 presents the variation of the transverse deflection w with time for different values of the pressure under the panel P_0, with the same values of the other parameters of the panel: $M_\infty = 0.785$, $K = 132$, $\delta = 0.003$ ($f_* = 0.05$). Both edges of the panel are clamped. In this figure characteristic signs of dynamic stability loss in the form of buckling are observed. For $P_0 = 0$, buckling happens earlier and faster than for $P_0 = 1$. For $P_0 = 0$, almost the whole panel deflects under a nearly uniformly distributed load, and the loss of stability has an almost symmetric form [133, 232], but is not exactly symmetric. For $P_0 = 1$, as can be seen in Fig. 2.25, during the transition to a new stable oscillating motion, the panel undergoes some transient oscillations that vary between different points of the panel. For $P_0 = 1$, the dynamic loss of stability has a nonsymmetric form [133, 232].

Influence of Geometric Parameters of the Panel. Graphs of the transverse deflection w at different points of the panel for $M_\infty = 0.785$, $K = 47.8$, $\delta = 0.005$ ($f_* = 0.03$), $P_0 = 0$ and clamping on both edges are presented in Fig. 2.26. In this diagram, the transition from one stable oscillating motion to another one is easily visible. This means that a dynamic loss of stability takes place. In the time interval from 0 to 2 dimensionless time units, oscillations with a small amplitude in the first stable motion are visible,

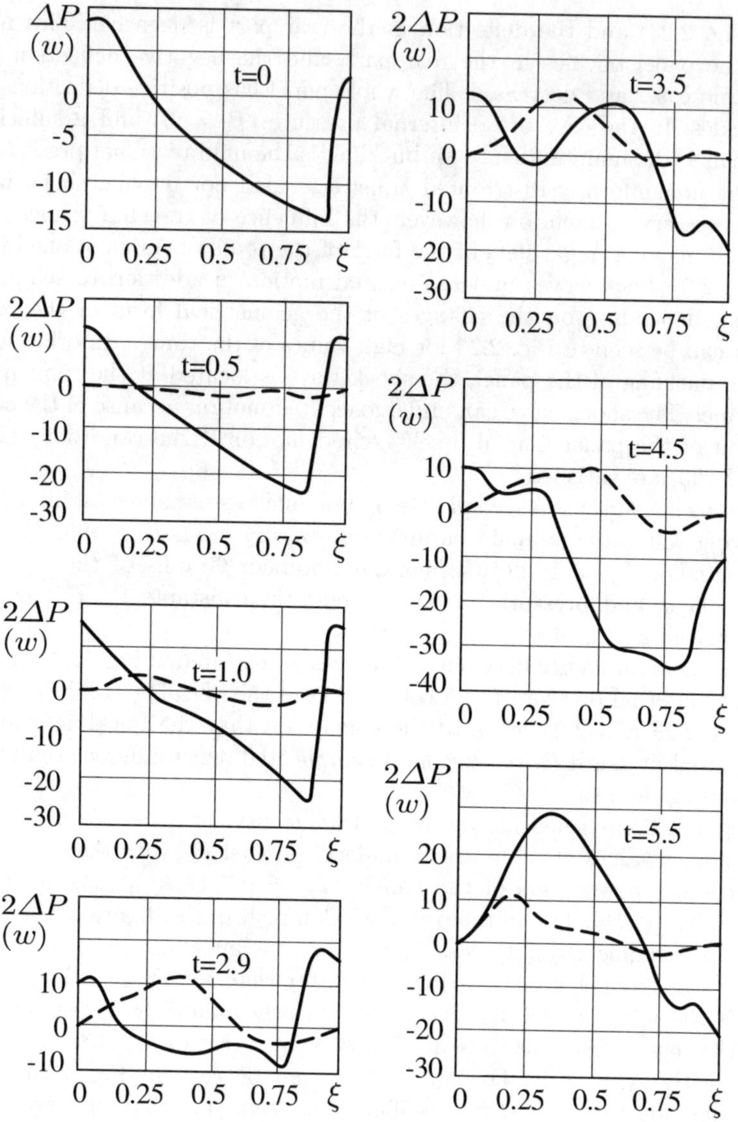

Fig. 2.24. Distribution of dimensionless pressure Δp (*solid line*) and transverse deflection w (*dashed line*) in a section of the panel at different time instants for $M_\infty = 0.864$, $K = 198$, $\delta = 0.002$ ($f_* = 0.05$), $P_0 = 1$, clamped edges

especially for the point with coordinate $\xi = 0.5$. In the time interval after 3.5 units, oscillations with a greater amplitude around a new central location can be observed. This is the new stable motion. The transition from one stable motion to the other in the time interval from 2 to 3.5 time units is characterized by high gradients of the deflection w. It is necessary to mention,

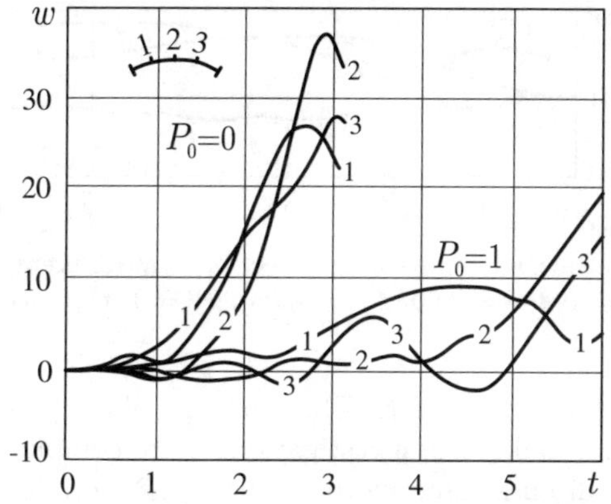

Fig. 2.25. Time variation of transverse deflection w at different points of the panel; $M_\infty = 0.785$, $K = 132$, $\delta = 0.003$ ($f_* = 0.05$), clamped edges

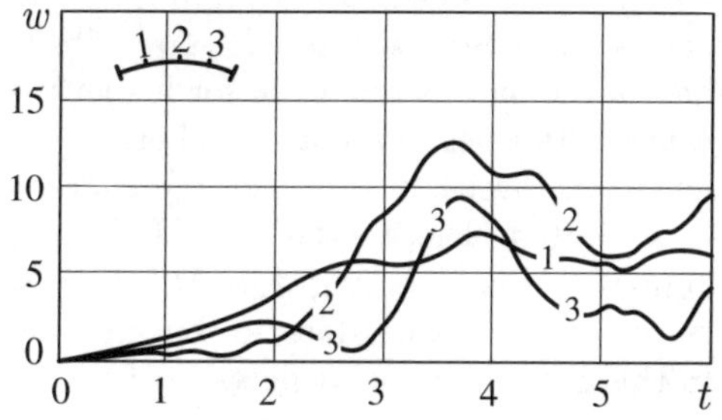

Fig. 2.26. Time variation of transverse deflection w at different points of the panel for $M_\infty = 0.785$, $K = 47.8$, $\delta = 0.005$ ($f_* = 0.03$), $P_0 = 0$, clamped edges

however, that not all points of the panel have a high value of the deflection gradient w at the same time. As can be seen in Fig. 2.26, the increase of deflection at the point $\xi = 0.25$ starts earlier than as the points $\xi = 0.5$ and $\xi = 0.75$. This means that buckling of the panel occurs in a complicated way: usually the front part of the panel deflects strongly enough to pull the rest of the panel. This effect is seen very well in the graphs of the deflection in Fig. 2.25.

Let us compare the changes of transverse deflection w for the two cases represented in Fig. 2.25 (for $P_0 = 0$) and Fig. 2.26. These two cases differ in the values of the geometric parameters K, δ (or f_*). From these two diagrams it can be seen that for high values of the dimensionless geometric parameter K, the panel buckles faster and the transverse deflections reach higher values. The infinite-length cylindrical panel investigated here is characterized, as stated in Sect. 2.2.1.3, by two independent geometric parameters: K, a dimensionless geometric parameter, and δ, the relative thickness of the panel. As was stated in Sect. 2.2.1.3, one of these calculated parameters can be

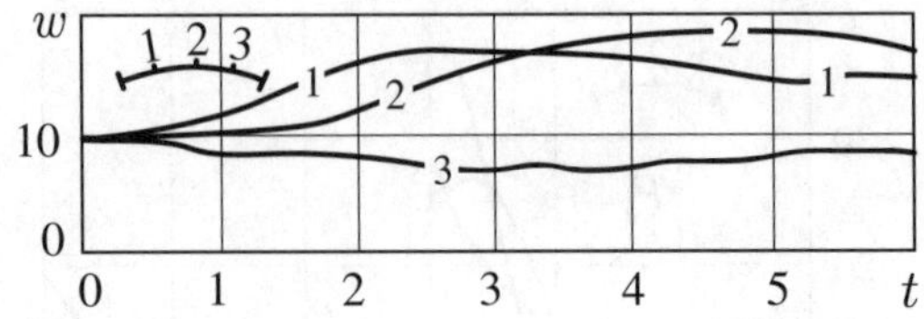

Fig. 2.27. Time variation of transverse deflection w at different points of the panel for $M_\infty = 0.9$, $K = 198$, $\delta = 0.002$ ($f_* = 0.05$), $P_0 = 1$, clamped edges

replaced by the relative camber f_*. The geometry of the panel, which depends on two parameters, makes the investigations more complicated.

We shall now discuss some more examples of the interaction of a panel and a transonic flow. Figure 2.27 presents the time variation of the transverse deflection w at different points of the panel for the following parameters: $M_\infty = 0.9$, $K = 198$, $\delta = 0.002$ ($f_* = 0.05$), $P_0 = 1$. Both edges of the panel are clamped. Figure 2.28 shows corresopndingly results for a panel with $K = 119.6$. By comparing these diagrams, it is possible again to conclude that for higher values of the parameter K (or for higher values of the camber factor f_*), the relative deflections w are also higher.

The cases presented in Figs. 2.27 and 2.28 are characterized by the fact that a shock wave is located close to the rear of the panel. This happens when the Mach number of the incoming flow M_∞ is close to one. In these diagrams it can be seen, that the panel does not buckle completely, and has an interesting form. In this case the panel deflects in the front part and buckles in the rear part. It stays for a relatively long period of time in that state and undergoes oscillations with small amplitude.

The motion of shock waves excites high-frequency, forced, oscillating deformation of the panel. In Figs. 2.27 and 2.28 the influence of the shock wave can be seen only for the point with coordinate $\xi = 0.75$ in the rear part of the panel. The graphs of the transverse deflection in Figs. 2.25 and 2.26 show a greater influence of travelling shock waves on the panel deflections than do the graphs in Figs. 2.27 and 2.28. In the case presented in Figs. 2.25 and 2.26, the deflection oscillations have a high frequency for

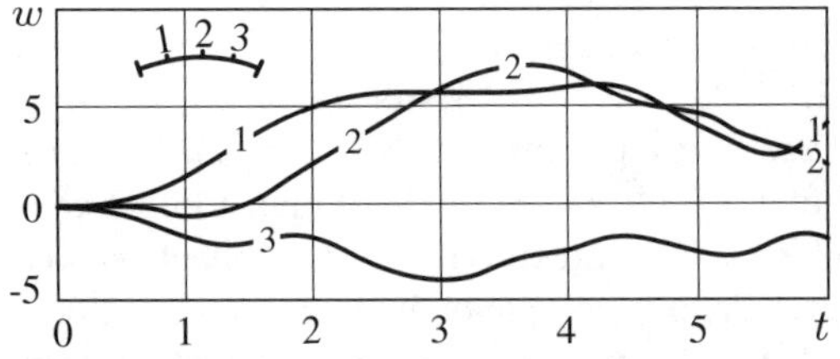

Fig. 2.28. Time variation of transverse deflection w at different points of the panel for $M_\infty = 0.9$, $K = 119.6$, $\delta = 0.002$ ($f_* = 0.3$), $P_0 = 1$

most of the points of the panel. For the case shown in Figs. 2.27 and 2.28, a shock wave moves in the central part of the panel; this happens in an overcritical flow, and takes place in small supersonic zones of the incoming stream for a Mach number $M_\infty = 0.785$. This confirms the high importance of the panel geometry in the interaction of a transonic flow and a panel.

Influence of Mach Number M_∞ of Incoming Flow. The influence of the Mach number of the incoming flow on the dynamic stability of a panel appears to be one of the most important influences, because the Mach number influences strongly the location and motion of shock waves. Figures 2.29 and 2.30 present time graphs of the transverse deflection w for three points on the profile for the following parameters: $M_\infty = 0.864$, $K = 119.6$, $\delta = 0.002$ ($f_* = 0.03$), $P_0 = 1$, clamping of edges; and $M_\infty = 0.785$, $K = 119.6$, $\delta = 0.002$ ($f_* = 0.03$), $P_0 = 1$, clamping of edges. Figures 2.28, 2.29 and 2.30 present cases which differ only in the Mach number M_∞ of the incoming flow. Comparing these three diagrams, it is possible to analyse the process of stability loss of the same panel for different velocities of the incoming flow, i.e. different values of the Mach number of the incoming stream. It can be seen that the oscillations of the same point of the panel differ in frequency, especially for the point with coordinate $\xi = 0.75$. This is caused by the

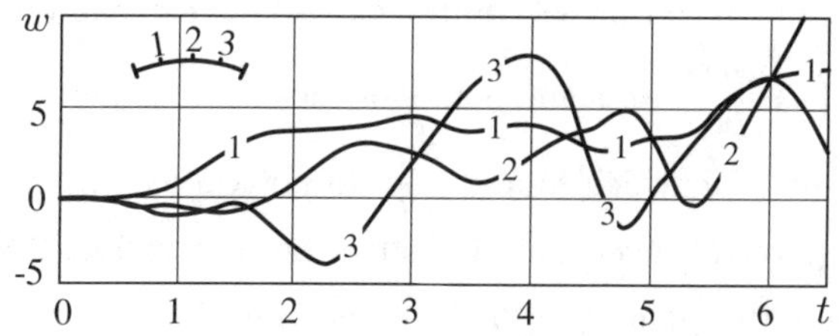

Fig. 2.29. Time variation of transverse deflection w at different points of the panel for $M_\infty = 0.864$, $K = 119.6$, $\delta = 0.002$ ($f_* = 0.03$), $P_0 = 1$, clamped edges

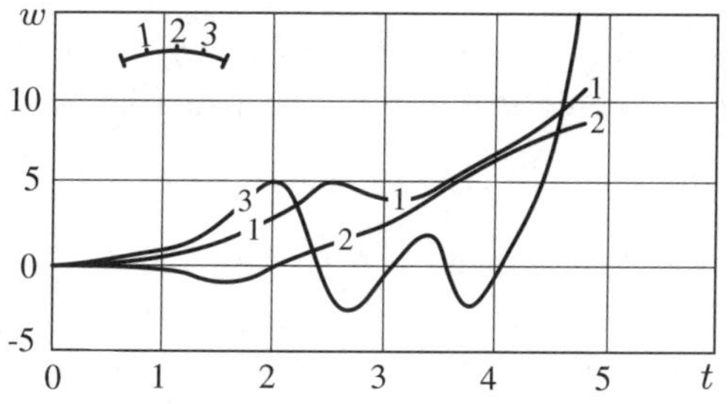

Fig. 2.30. Time variation of transverse deflection w at different points of the panel for $M_\infty = 0.785$, $K = 119.6$, $\delta = 0.002$ ($f_* = 0.03$), $P_0 = 1$, clamped edges

difference in the shock wave speed for different velocities of the incoming stream, as was described earlier.

A comparison of the graphs of transverse deflection w in Figs. 2.28, 2.29 and 2.30 leads to the conclusion that loss of panel stability happens earlier for lower values of the Mach number M_∞ of the incoming stream, i.e. for $M_\infty = 0.785$ (the case presented in Fig. 2.30). Of course, such rapid buckling in this case occurs during motion of a shock wave in the middle part of the panel, which happens for this value of M_∞. As was mentioned in Sect. 2.2.1, for greater values of the Mach number M_∞ of the incoming flow (values of M_∞ close to one), the zone of shock wave motion is located closer to the rear of the panel, which decreases the influence of the shock wave on buckling of the panel in the present case, for an internal pressure $P_0 = 1$.

In the cases presented in Figs. 2.29 and 2.30, the panel buckles after oscillation around a certain bent state. As can be seen, such oscillation around a bent state of the panel can be treated as stable in a limited sense in the time interval investigated. Following the method presented in [4], we can prove that such oscillations of an elastic panel around a deformed state can be treated as stable. Figure 2.31 presents diagrams of $\left(w, \dfrac{\partial w}{\partial t}\right)$ for different points of the panel, for the following parameters $M_\infty = 0.864$, $K = 119.6$, $\delta = 0.002$ ($f_* = 0.03$), $P_0 = 1$. Both sides are clamped. According to [4], the motion of a panel point is stable if the absolute values of the deflection w and deflection speed $\dfrac{\partial w}{\partial t}$ do not exceed preset values ε_w and $\varepsilon_{\dot{w}}$ for given excitations. On the graphs of $\left(w, \dfrac{\partial w}{\partial t}\right)$ for some points of the elastic panel, values of ε_w and $\varepsilon_{\dot{w}}$ can be chosen, for which w and $\dfrac{\partial w}{\partial t}$ do not exceed the set limits in the bounded time interval investigated. As can be seen in Fig. 2.31c, the graph of $\left(w, \dfrac{\partial w}{\partial t}\right)$ for the point $\xi = 0.25$ in the time interval investigated does not exceed the range $0 \le w < 7.5$, $-3.5 \le \dfrac{\partial w}{\partial t} < 9$. Figure 2.31b presents a graph of $\left(w, \dfrac{\partial w}{\partial t}\right)$ for the point $\xi = 0.5$. At the beginning this graph remains inside the limits $-2 \le w \le 5$, $-20 \le \dfrac{\partial w}{\partial t} \le 7.1$ but it then exceeds this range–loss of dynamic stability of the panel has occured around this point. Figure 2.31a presents a graph of $\left(w, \dfrac{\partial w}{\partial t}\right)$ for the point $\xi = 0.75$. This graph does not exceed the range $-5 \le w \le 4$, $-20 \le \dfrac{\partial w}{\partial t} \le 15$. This means that this point does not lose dynamic stability. Loss of dynamic stability begins around random point of the panel and then occurs in other points of the panel. The expanding loss of stability leads to buckling of the panel. After a transition process, which happens at the beginning of the interaction, the panel undergoes oscillations around a bent position.

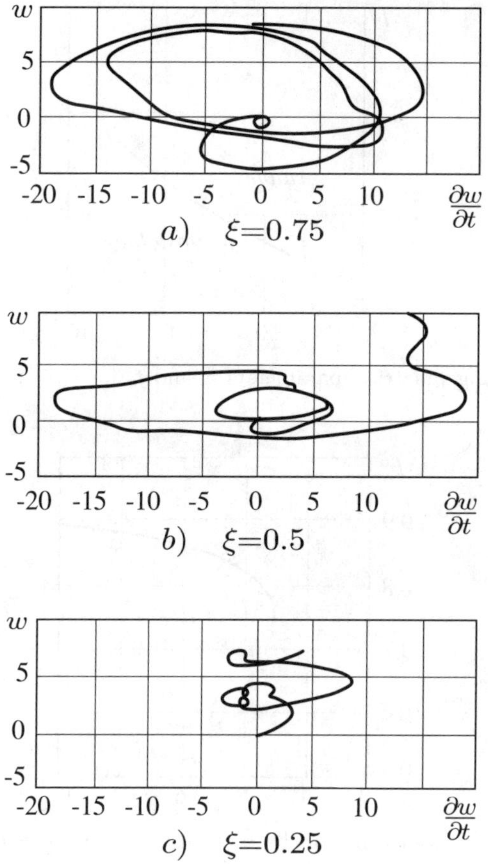

Fig. 2.31a–c. Diagrams of $\left(w, \dfrac{\partial w}{\partial t}\right)$ for different points of the panel; $M_\infty = 0.864$, $K = 119.6$, $\delta = 0.002$ ($f_* = 0.03$), $P_0 = 1$, clamped edges

Critical Values of Parameters. The critical values are those values of the parameters for which dynamic loss of stability takes place in the sense described above within the time interval analysed.

For an internal pressure $P_0 = 0$ and clamped edges of an elastic panel, numerical calculations performed according to the method presented above give the values of the critical geometrical parameters K and f_* which are presented in Fig. 2.32. A slight dependence of the results on the Mach number M_∞ of the incoming stream is observed in this case. As mentioned earlier, this results from the unidirectional load. The influence of the Mach number of the incoming flow and, consequently, of shock wave motion on dynamic loss of stability for the case $P_0 = 0$ depends on the geometrical form of the panel. Perhaps the importance of the Mach number in the stability loss of a panel for $P_0 = 0$ will be greater if the interac-

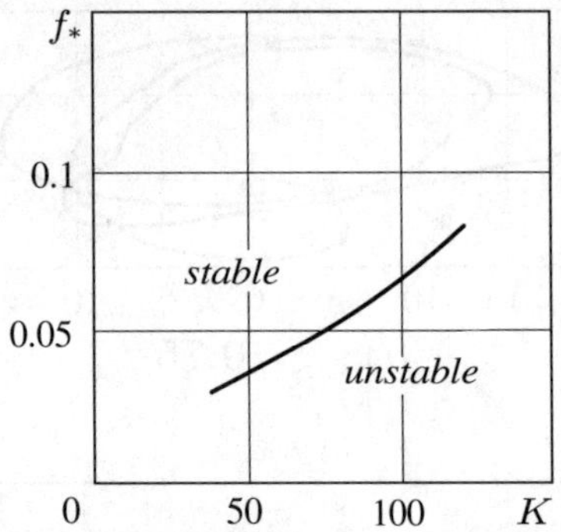

Fig. 2.32. Critical geometrical parameters K and f_* for $P_0 = 0$ and clamped edges

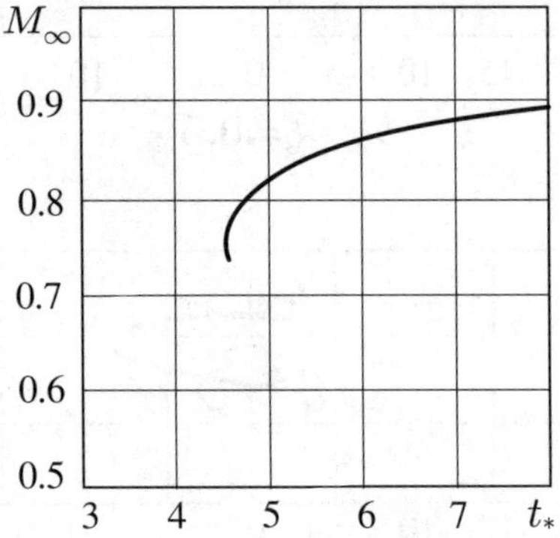

Fig. 2.33. Stability loss time for $P_0 = 1$, $K = 119.6$, $\delta = 0.002$ ($f_* = 0.03$), clamped edges

tion of a transonic flow and a panel is investigated over a longer period of time.

For the case $P_0 = 1$, an elastic panel can oscillate in a stable manner under excitations originating from the interaction of the panel with the flow (for example, in the case presented in Fig. 2.27).

In this case, during a small time interval, there is no loss of dynamic stability of the panel. As previously mentioned, for some values of the Mach number of the incoming flow M_∞ (see Figs. 2.29 and 2.30), dynamic loss of stability of the panel may happen. This loss of stability occurs at different time instants, depending on M_∞. Figure 2.33 presents a diagram of the stability loss time versus the Mach number M_∞ of the incoming flow for the following parameters of the elastic panel and transonic flow: $P_0 = 1$, $K = 119.6$, $\delta = 0.002$ ($f_* = 0.03$); both edges are clamped.

As has been stated in experimental studies [207, 232, 234], the following characteristic features of the interaction of a transonic flow with a shell can be examined: shock wave motion, and high amplitudes of oscillation of the pressure and transverse deflection. Although the case investigated here is different from those studied in [207, 232, 234], it is possible to talk about qualitative similarities of the behaviour of the panel and flow in the two cases.

2.2.3.2 Analysis of Stress–Strain State. This subsection gives an estimate of the stress–strain state of the panel and an analysis of the occurrence of plastic deformation during interaction of a panel and a flow.

The von Mises yield criterion has the following form:

$$\sigma = \frac{\sigma_S}{\sqrt{3}}, \tag{2.300}$$

where σ_S is the yield point of the shell material, and σ is the shear stress, which can be calculated from the following formula [165]:

$$\begin{aligned} \sigma = \frac{1}{\sqrt{6}}\Big(& (P^z_{11} - P^z_{22})^2 + (P^z_{22} - P^z_{33})^2 + (P^z_{33} - P^z_{11})^2 \\ & + 6({P^z_{12}}^2 - {P^z_{23}}^2 + {P^z_{13}}^2)\Big)^{1/2} . \end{aligned} \tag{2.301}$$

By use of the equations (2.160), the right-hand side of (2.301) can be converted to

$$\sigma = \frac{1}{\sqrt{3}}((1 + \nu^2 - \nu){P^z_{22}}^2 + 3{P^z_{23}}^2)^{\frac{1}{2}} . \tag{2.302}$$

Substituting the stress from (2.187) into (2.302), we obtain the following formula for calculating the value of the shear stress:

$$\begin{aligned} \sigma = \frac{E'}{\sqrt{3}(1-\nu^2)}(1+\nu^2-\nu)\left(\frac{1}{2L}\int_0^L \left(\frac{\partial w'}{\partial y}\right)^2 dy \right. \\ \left. -\frac{1}{RL}\int_0^L \left(w'\,dy - z\frac{\partial^2 w'}{\partial y^2}\right)^2 + \frac{h^4}{48}\left(\frac{\partial^3 w'}{\partial y^3}\right)^2\right)^{\frac{1}{2}} . \end{aligned} \tag{2.303}$$

Introducing the system of dimensionless parameters (2.216) and (2.219), we obtain the right-hand side of (2.303) in dimensionless form:

$$\begin{aligned} \frac{\sigma}{E'} = \frac{3^{-\frac{1}{2}}}{1-\nu^2}\left((1+\nu^2-\nu)\left(\frac{\delta^2}{2}\int_0^1 \left(\frac{\partial w}{\partial \xi}\right)^2 d\xi - \delta^2 K\int_0^1 w\,d\xi \right.\right. \\ \left.\left. + \eta\delta\frac{\partial^2 w}{\partial \xi^2}\right)^2 + \frac{\delta^4}{48}\left(\frac{\partial^3 w}{\partial \xi^3}\right)^2\right)^{\frac{1}{2}} . \end{aligned} \tag{2.304}$$

Here, the dimensionless coordinate η of the panel may take values in the interval $[-\frac{\delta}{2}, \frac{\delta}{2}]$ only. The von Mises yield condition then has the following form:

$$\frac{\sigma_S}{E'} = \frac{\delta^2}{1-\nu^2}\left(\frac{1+\nu^2-\nu}{4}\left(\int\limits_0^1\left(\frac{\partial w}{\partial \xi}\right)^2 d\xi - 2K\int\limits_0^1 w\, d\xi + 2\frac{\eta}{\delta}\frac{\partial^2 w}{\partial \xi^2}\right)^2 + \frac{\delta^4}{48}\left(\frac{\partial^3 w}{\partial \xi^3}\right)^2\right)^{\frac{1}{2}}. \tag{2.305}$$

If the right-hand side of (2.305) is greater than or equal to the left-hand side of this equation, then we assume that the shell yields at this point. If the right-hand side of (2.305) is less than the left-hand side, then elastic deformation occurs. The von Mises yield criterion (2.305) can also be presented in the following form:

$$\left(\frac{\sigma_S}{E'}\right)^2 \frac{4(1-\nu^2)^2}{(1+\nu^2-\nu)\delta^4} - \frac{\delta^2}{24(1+\nu^2-\nu)}\left(\frac{\partial^3 w}{\partial \xi^3}\right)^2 = \left(\int\limits_0^1\left(\frac{\partial w}{\partial \xi}\right)^2 d\xi - 2K\int\limits_0^1 w\, d\xi + 2\frac{\eta}{\delta}\frac{\partial^2 w}{\partial \xi^2}\right)^2. \tag{2.306}$$

After substituting in (2.306) the values of the parameters used in the studies described in this book, namely $E' = 20 \times 10^{10}$ N/m, $\sigma_S = 24 \times 10^7$ N/m^2, $\nu = 0.3$, the yield condition (2.306) has the form

$$\frac{1}{165\,623.4\,\delta^4} - 0.105\,48\,\delta^4\left(\frac{\partial^3 w}{\partial \xi^3}\right)^2 = \left(\int\limits_0^1\left(\frac{\partial w}{\partial \xi}\right)^2 d\xi - 2K\int\limits_0^1 w\, d\xi + 2\frac{\eta}{\delta}\frac{\partial^2 w}{\partial \xi^2}\right)^2. \tag{2.307}$$

If we assume, as stated in Sect. 2.2.1.6, that the relative thickness of the panel $\delta < 0.01$, the following inequality becomes true:

$$\frac{1}{165\,623.4\,\delta^4} > 603.78. \tag{2.308}$$

Assuming that

$$\left|\frac{\partial^3 w}{\partial \xi^3}\right| < 10^3, \tag{2.309}$$

which can be confirmed by numerical calculations, the von Mises yield criterion can be presented in the following simplified form:

$$165\,623.4\,\delta^4 \left(\int_0^1 \left(\frac{\partial w}{\partial \xi} \right)^2 d\xi - 2K \int_0^1 w \, d\xi + 2\frac{\eta}{\delta} \frac{\partial^2 w}{\partial \xi^2} \right)^2 = 1. \qquad (2.310)$$

In other words, if the conditions (2.308) and (2.309) are satisfied, as is true in the case investigated, the influence of the third derivative $\frac{\partial^3 w}{\partial \xi^3}$ on the appearance of yield zones in the panel is negligible. The second derivative $\frac{\partial^2 w}{\partial \xi^2}$ has, locally, the main influence, and it corresponds, following (2.161), to the bending moment M_y. After integration by parts of the first integral in (2.310) and taking into account the boundary condition (2.234) for simple support or the boundary condition (2.235) for clamping, the von Mises yield criterion can be presented in the following form:

$$165\,623.4\,\delta^4 \left(2\frac{\eta}{\delta} \frac{\partial^2 w}{\partial \xi^2} - \int_0^1 w \left(2K + \frac{\partial^2 w}{\partial \xi^2} \right) d\xi \right)^2 = 1. \qquad (2.311)$$

By solving (2.311) for η at a given time instant t, on the normal to the middle surface of the panel, values of the coordinate η can be found where are boundaries between yielded and elastic zones for the case $\frac{\partial^2 w}{\partial \xi^2} \neq 0$:

$$\begin{aligned} \eta_1 &= \frac{\delta}{2} \left(\frac{\partial^2 w}{\partial \xi^2} \right)^{-1} \left(\int_0^1 w \left(2K + \frac{\partial^2 w}{\partial \xi^2} \right) d\xi + 0.002\,46\,\delta^{-2} \right), \\ \eta_2 &= \frac{\delta}{2} \left(\frac{\partial^2 w}{\partial \xi^2} \right)^{-1} \left(\int_0^1 w \left(2K + \frac{\partial^2 w}{\partial \xi^2} \right) d\xi - 0.002\,46\,\delta^{-2} \right). \end{aligned} \qquad (2.312)$$

Of course, only those values of η which are contained in the range $\left(-\frac{\delta}{2}, \frac{\delta}{2} \right)$ are interesting. From (2.311) and (2.312), one can conclude that elastic deformation will appear for the following values of the variable η:

$$\min(\eta_1, \eta_2) < \eta < \max(\eta_1, \eta_2). \qquad (2.313)$$

We now present examples showing the occurrence of yield deformation in the cylindrical panel investigated here during interaction with a transonic flow of an ideal gas. Results of computations for the case $M_\infty = 0.9$, $K = 36.8$,

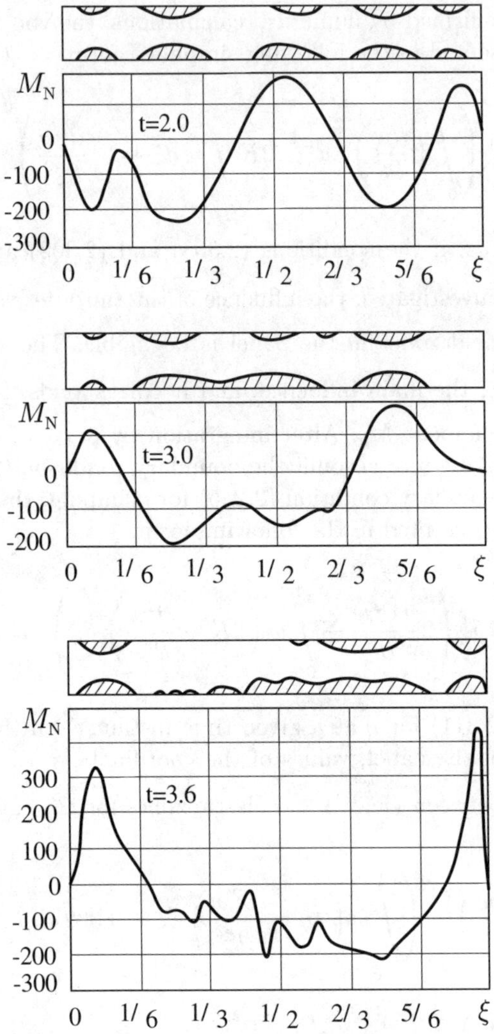

Fig. 2.34. Distribution of bending moment and yield deformation at different time instants t for $M_\infty = 0.9$, $K = 36.8$, $\delta = 0.0065$ $(f_* = 0.03)$, clamped edges

$\delta = 0.0065$ $(f_* = 0.03)$ and clamping of the edges are presented in Fig. 2.34. The graphs show the distribution of the bending moment M_u in transverse section of an infinite panel, computed at different time instants. The hatched areas in the diagrams above the graphs correspond to yield zones in sections of the panel. According to (2.161), M_u is proportional to the second derivative with respect to ξ of the transverse deflection, $\dfrac{\partial^2 w}{\partial \xi^2}$, and this, as can be seen from the von Mises criterion (2.311), has a strong influence on the occurrence

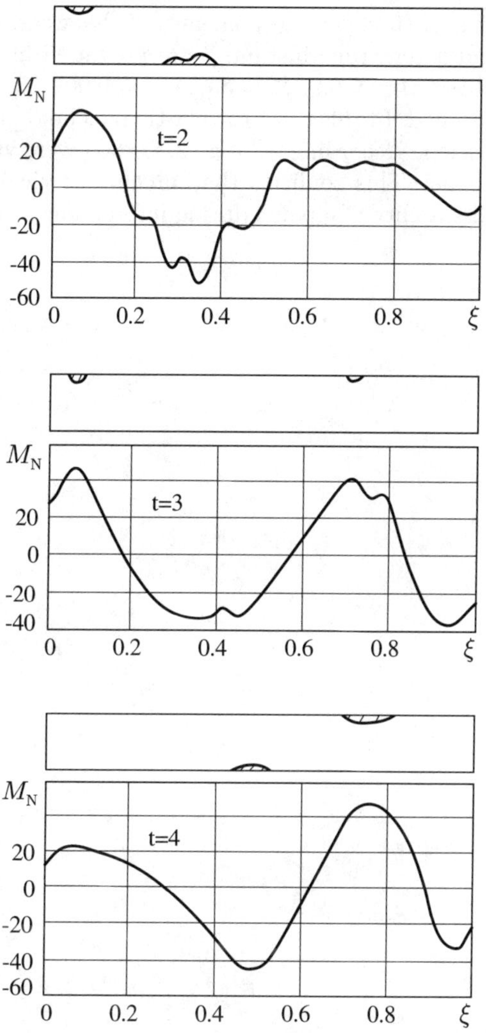

Fig. 2.35. Distribution of bending moment and yield deformation at different time instants t for $M_\infty = 0.9$, $K = 34.16$, $\delta = 0.007$ ($f_* = 0.03$), $P_0 = 0$, clamped edges

of yield deformations. To visualize the yield regions better, thickness of the panel for which results are presented in Fig. 2.34 was chosen to be abnormally high.

The boundaries of the yield regions were calculated from (2.312), taking into account the inequalities (2.313). In Fig. 2.34 it can be seen that yield deformation occurs, when the bending moment M_u has a sufficiently high value. The yield deformation appears initially on the surface of the panel. In Fig. 2.34 it can be seen that even in the initial stage of buckling of the panel at $t = 2$, relatively large yield regions appear.

The distribution of the bending moment and the areas of yield deformation (hatched) at different time instants t are presented in Fig. 2.35 for the following parameters: $M_\infty = 0.9$, $K = 34.16$, $\delta = 0.007$ ($f_* = 0.03$), $P_0 = 0$. Both edges are clamped. In this case, the elastic panel did not lose its initial, stable form of motion. According to Fig. 2.35, however, yield deformation appeared in the panel. This confirms the influence of yield deformation on the behaviour and stability of a panel during interaction with a transonic gas flow.

3 Estimation of the Errors of the Bubnov–Galerkin Method

In Sect. 3.1, an abstract coupled problem is considered and a few theorems related to the estimation of the accuracy of the Bubnov–Galerkin method are formulated and proved. The error estimates hold for a system of differential equations of a rather general form with homogeneous boundary conditions, which corresponds to coupled thermoelastic problems for plates and shallow shells with variable thickness. In addition, a particular case of this problem (with nonhomogeneous initial conditions), where a prior estimate of the errors of the Bubnov–Galerkin method is most effective, is illustrated and discussed. Finally, a prior estimate for the Bubnov–Galerkin method to a problem generalizing a class of dynamical problems of elasticity (without a heat transfer equation) for both three-dimensional and thin-walled elements of structures is given.

In Sect. 3.2, a coupled thermoelastic problem within the Kirchhoff–Love model is considered. We show how to find operators L_0, K_0 that are convergent and form an acute angle with the differential operators L, K, for which full systems of eigenfunctions and the corresponding eigenvalues are known. In Sect. 3.3, a coupled thermoelastic problem for a simply supported plate with constant thickness in the framework of the Kirchhoff–Love model is analysed. It is assumed that the mechanical and heat loads are constant in time and that the plate is square.

In the next section, a coupled problem of thermoelasticity of shallow shells, rectangular in plan with variable thickness is considered in the framework of a kinematic Timoshenko-type model . Some remarks related to estimation of the errors of the Bubnov–Galerkin method are made.

3.1 An Abstract Coupled Problem

We consider the abstract coupled Cauchy problem (2.54)–(2.56) and, in order to simplify our treatment we take $w = w_1 = 0$, $\theta_0 = 0$ (it is clear that an arbitrary problem can be reduced to one with homogeneous boundary conditions by a simple change of variables). Therefore, the problem analysed has the form

$$\begin{cases} I_1 w''(t) + Lw(t) + M\theta(t) = g_1(t), \\ I_2\theta'(t) + K\theta(t) + Nw'(t) = g_2(t), \\ w(0) = w'(0) = 0, \quad \theta(0) = 0. \end{cases} \tag{3.1}$$

The conditions imposed on the operator coefficients I_1, I_2, L, K, M, N are given in Sect. 2.1.4.

The Bubnov–Galerkin relations appropriate to the problem (3.1) have the form

$$\begin{cases} P_{H_1^n}(I_1 w_n''(t) + Lw_n(t) + M\theta_t(t)) = P_{H_1^n} g_1(t), \\ P_{H_2^n}(I_2\theta_n'(t) + K\theta_n(t) + Nw_n'(t)) = P_{H_2^n} g_2(t), \\ w_n(t) \in H_1^n, \quad \theta_n(t) \in H_2^n, \quad \forall t \in [0,T], \\ w_n(0) = w_n'(0) = 0, \quad \theta_n(0) = 0. \end{cases} \tag{3.2}$$

As the bases $\{\nu_i\}_{i=1}^{\infty}$, $\{\eta_j\}_{j=1}^{\infty}$, we take systems of eigenelements of certain self-adjoint, positive definite operators L_0 and K_0 that are convergent and form an acute angle with the operators L and K, respectively. The eigenvalues of the operators L_0 and K_0 are denoted by λ_i and æ$_i$: $L_0\nu_i = \lambda_i\nu_i$, $K_0\eta_j =$ æ$_j\eta_j$, $i,j = 1,2,\ldots$ Let us remind ourselves that the H_1^n, H_2^n occurring in (3.2) are composed of a linear vicinity and of finite systems of base elements $\{\nu_i\}_{i=1}^{i(n)}$, $\{\eta_j\}_{j=1}^{j(n)}$, $P_{H_1^n}$ is an orthoprojector of H_1 into H_1^n, and $P_{H_2^n}$ is an orthoprojector of H_2 into H_2^n.

Let us introduce the following notation:

$$\begin{aligned} \Delta w_n(t) &= w_n(t) - w(t), \quad \Delta\theta_n(t) = \theta_n(t) - \theta(t), \\ \delta_n^{(1)}(t) &= I_1 w_n''(t) + Lw_n(t) + M\theta_n(t) = g_1(t), \\ \delta_n^{(2)} &= I_2\theta_n'(t) + K\theta_n(t) + Nw_n'(t) = g_2(t), \end{aligned}$$

where $\{w,\theta\}$ is the exact solution to the problem (3.1), and $\{w_n,\theta_n\}$ is the approximated solution to the problem (3.1) constructed by the Bubnov–Galerkin method using (3.2).

The first theorem related to the convergence of the Bubnov–Galerkin relations (3.2) to the problem (3.1) is related to a case corresponding to differentiable free terms of (2.110).

Theorem 3.1 *If the conditions (2.58), (2.110) are satisfied then the accuracy of the Bubnov–Galerkin method based on (3.2) applied to the problem (3.1) can be estimated from the following inequality:*

$$\max_{0\le t\le T}\Bigg[\left\|I_1^{1/2}\Delta w_n'(t)\right\|_{H_1}^2 + \left\|L^{1/2}\Delta w_n(t)\right\|_{H_1}^2$$

$$+\left\|I_2^{1/2}\Delta\theta_n(t)\right\|_{H_2}^2 + \int\limits_0^t \left\|K^{1/2}\Delta\theta_n(\tau)\right\|_{H_2}^2 d\tau\Bigg]$$

$$\le c_1(M_g(T))^{1/2}\Big[\left\|g_1\right\|_{L_2(0,T;H_1)} + c_2\left\|g_2\right\|_{L_2(0,T;H_2)}$$

$$+ c_3 (M_g(T))^{1/2} \Big] \tilde{\lambda}_{i(n)+1}^{-1/2}$$

$$+ c_4 \Big[\Big\| g_2 \Big\|_{L_2(0,T;H_2)} + c_5 (M_g(T))^{1/2} \Big]^2 \tilde{æ}_{j(n)+1}^{-1}$$

$$\equiv 0(\max(\tilde{\lambda}_{i(n)+1}^{-1/2}, \tilde{æ}_{j(n)+1}^{-1})), \tag{3.3}$$

where

$$\tilde{\lambda}_k = \min_{i \geq k} \lambda_i, \quad \tilde{æ}_1 = \min_{j \geq l} æ_j, \tag{3.4}$$

$$\begin{aligned} c_1 &= 4 \Big\| L^{-1/2} L_0^{1/2} \Big\|_{H_1 \to H_1} T^{1/2}(1 + c_L), \quad c_2 = \Big\| M K^{-1} \Big\|_{H_2 \to H_1} c_K, \\ c_3 &= b_2 T^{1/2}, \quad c_4 = \Big\| K^{-1/2} K_0^{1/2} \Big\|_{H_2 \to H_2}^2 (1 + c_K)^2, \quad c_5 = b_1 T^{1/2}, \end{aligned} \tag{3.5}$$

and $M_g(T)$ *is the constant defined by the relation (2.130);* c_L, c_K *are the constants defined by (2.124); and* b_1, b_2 *are the constants defined by the relations* (2.121), (2.125).

Proof. Let us introduce the following notation:

$$R_{H_1^n} = E_1 - P_{H_1^n}, \quad R_{H_2^n} = E_2 - P_{H_2^n},$$

where E_i is the identity operator over H_i $(i = 1, 2)$. The relations (3.1), (3.2) imply

$$\begin{cases} I_1 \varDelta w_n''(t) + L \varDelta w_n(t) + M \varDelta \theta_n(t) = R_{H_1^n} \delta_n^{(1)}(t), \\ I_2 \varDelta \theta_n'(t) + K \varDelta \theta_n(t) + N \varDelta w_n'(t) = R_{H_2^n} \delta_n^{(2)}(t), \\ \varDelta w(0) = \varDelta w'(0) = 0, \quad \varDelta \theta(0) = 0. \end{cases} \tag{3.6}$$

We multiply first two equations of (3.6) scalarly by $2\,\varDelta w_n'(t)$ and by $2\,\varDelta \theta_n(t)$ and then sum them. Next we take into account the condition (2.58) and integrate the relation obtained over the interval $[0, T]$, where $t \leq T$. As the result, we obtain

$$\begin{aligned} &\Big\| I_1^{1/2} \varDelta w_n'(t) \Big\|_{H_1}^2 + \Big\| L^{1/2} \varDelta w_n(t) \Big\|_{H_1}^2 + \Big\| I_2^{1/2} \varDelta \theta_n(t) \Big\|_{H_2}^2 \\ &+ 2 \int_0^t \Big\| K^{1/2} \varDelta \theta_n'(\tau) \Big\|_{H_2}^2 d\tau = 2 \int_0^t (R_{H_1^n} \delta_n^{(1)}(\tau), \varDelta w_n'(\tau))_{H_1} \, d\tau \\ &+ 2 \int_0^t (R_{H_2^n} \delta_n^{(2)}(\tau), \varDelta \theta_n(\tau))_{H_2} \, d\tau. \end{aligned} \tag{3.7}$$

The first term of the right-hand side of (3.7) can be estimated in the following way:

$$2\int_0^t \left(R_{H_1^n}\delta_n^{(1)}(\tau), \Delta w_n'(\tau)\right)_{H_1} d\tau = 2\int_0^t \left(L^{-1/2}L_0^{1/2}L_0^{-1/2}R_{H_1^n}\delta_n^{(1)}(\tau),\right.$$
$$\left. L^{1/2}\Delta w_n'(\tau)\right)_{H_1} d\tau \le 2\left\|L^{-1/2}L_0^{1/2}\right\|_{H_1\to H_1}\left\|L_0^{1/2}R_{H_1^n}\right\|_{H_1\to H_1}$$
$$\times\left\|\delta_n^{(1)}\right\|_{L_2(0,T;H_1)}\left\|L^{1/2}\Delta w_n'\right\|_{L_2(0,T;H_1)}. \tag{3.8}$$

Owing to the special choice of the basis $\{\nu_i\}_{i=1}^{\infty}$, we obtain

$$\left\|L_0^{1/2}R_{H_1^n}\right\|_{H_1\to H_1} = \tilde{\lambda}_{i(n)+1}^{-1/2}. \tag{3.9}$$

Using the prior estimate (2.128), (2.129), we obtain

$$\left\|L^{1/2}\Delta w_n'\right\|_{L_2(0,T;H_1)} \le \left\|L^{1/2}w'\right\|_{L_2(0,T;H_1)} + \left\|L^{1/2}w_n'\right\|_{L_2(0,T;H_1)} \le 2T^{1/2}(M_g(T))^{1/2}. \tag{3.10}$$

From (3.8)–(3.10) we obtain

$$2\int_0^t \left(R_{H_1^n}\delta_n^{(1)}(\tau), \Delta w_n'(\tau)\right)_{H_1} d\tau \le 4\left\|L^{-1/2}L_0^{1/2}\right\|_{H_1\to H_1} T^{1/2}(M_g(T))^{1/2}\left\|\delta_n^{(1)}\right\|_{L_2(0,T;H_1)}\tilde{\lambda}_{i(n)+1}^{-1/2}. \tag{3.11}$$

The second term of the right-hand side of (3.7) can be estimated in the following way:

$$2\int_0^t \left(R_{H_2^n}\delta_n^{(2)}(\tau), \Delta w_n'(\tau)\right)_{H_2} d\tau$$
$$= 2\int_0^t \left(K^{-1/2}K_0^{1/2}K_0^{-1/2}R_{H_2^n}\delta_n^{(2)}(\tau), K^{1/2}\Delta\theta_n(\tau)\right)_{H_2} d\tau$$
$$\le 2\left\|K^{-1/2}K_0^{-1/2}\right\|_{H_2\to H_2}^2\left\|K_0^{1/2}R_{H_2^n}\right\|_{H_2\to H_2}^2 \times \left\|\delta_n^{(2)}\right\|_{K_2(0,T;H_1)}^2$$
$$+\int_0^t \left\|K^{1/2}\Delta\theta_n(\tau)\right\|_{H_2}^2 d\tau. \tag{3.12}$$

Thanks to the special choice of the basis $\{\eta_j\}_{j=1}^{\infty}$, we obtain

$$\left\|K_0^{1/2}R_{H_2^n}\right\|_{H_2\to H_2} = \tilde{æ}_{j(n+1)}^{-1/2},$$

and therefore from (3.12) we obtain

$$2\int_0^t \left(R_{H_2^n}\delta_n^{(2)}(\tau), \Delta\theta_n(\tau)\right)_{H_2} d\tau \le \left\|K^{-1/2}K_0^{1/2}\right\|^2_{H_2\to H_2}$$
$$\times\left\|\delta_n^{(2)}\right\|^2_{L_2(0,T;H_1)} \tilde{æ}^{-1}_{j(n+1)} + \int_0^t \left\|K^{1/2}\Delta\theta_n(\tau)\right\|^2_{H_2} d\tau. \tag{3.13}$$

Using the inequalities (3.11) and (3.13), we estimate the right-hand side of (3.7) form above as follow:

$$\left\|I_1^{1/2}\Delta w_n'(t)\right\|^2_{H_1} + \left\|L^{1/2}\Delta w_n(t)\right\|^2_{H_1} + \left\|I_2^{1/2}\Delta\theta_n(t)\right\|^2_{H_2}$$
$$+\int_0^t \left\|K^{1/2}\Delta\theta_n(\tau)\right\|^2_{H_2} d\tau \le 4\left\|L^{-1/2}L_0^{1/2}\right\|_{H_1\to H_1} T^{1/2}(M_g(T))^{1/2}$$
$$\times\left\|\delta_n^{(1)}\right\|_{L_2(0,T;H_1)} \tilde{\lambda}^{-1/2}_{i(n)+1} + \left\|K^{-1/2}K_0^{1/2}\right\|^2_{H_2\to H_2}\left\|\delta_n^{(2)}\right\|^2_{L_2(0,T;H_1)} \tilde{æ}^{-1}_{j(n+1)}. \tag{3.14}$$

Let us estimate from above the norms of the errors $\left\|\delta_n^{(i)}\right\|_{L_2(0,T;H_i)}$, $i = 1, 2$. For $\delta_n^{(1)}$, using the prior estimates (2.122), (2.123), (2.128), we obtain

$$\left\|\delta_n^{(1)}(t)\right\|_{H_1} \le \left\|g_1(t)\right\|_{H_1} + \left\|Lw_n(t)\right\|_{H_1} + \left\|MK^{-1}\right\|_{H_2\to H_1}\left\|K\theta_n(t)\right\|_{H_1}$$
$$+\left\|I_1^{1/2}(t)\right\|_{H_1\to H_1}\left\|I_1^{1/2}w_n''(t)\right\|_{H_1}$$
$$\le \left\|g_1(t)\right\|_{H_1} + c_L\left\|g_1(t)\right\|_{H_1} + \left\|MK^{-1}\right\|_{H_2\to H_1} c_K\left\|g_2(t)\right\|_{H_2}$$
$$+b_2\left(\left\|I_1^{1/2}w_n''(t)\right\|^2_{H_1} + \left\|L^{1/2}w_n'(t)\right\|^2_{H_1} + \left\|I_1^{1/2}\theta_n'(t)\right\|^2_{H_2}\right)^{1/2}$$
$$+\left\|MK^{-1}\right\|_{H_2\to H_1} c_K\Bigg[\left\|g_2(t)\right\|_{H_2}$$
$$+b_1\left(\left\|I_2^{1/2}\theta_n'(t)\right\|^2_{H_2} + \left\|L^{1/2}w_n'(t)\right\|^2_{H_1}\right)^{1/2}\Bigg] + \left\|I_1^{1/2}\right\|_{H_1\to H_1}\left\|I_1^{1/2}w_n''(t)\right\|_{H_1}$$
$$\le (1+c_L)\Bigg[\left\|g_1(t)\right\|_{H_1} + \left\|MK^{-1}\right\|_{H_2\to H_1} c_K\Bigg[\left\|g_2(t)\right\|_{H_2}$$
$$+b_2\left(\left\|I_1^{1/2}w_n''(t)\right\|^2_{H_1} + \left\|L^{1/2}w_n'(t)\right\|^2_{H_1}\right) + \left\|I_2^{1/2}\theta_n'(t)\right\|^2_{H_1}\Bigg]^{1/2}\Bigg]$$
$$\le (1+c_L)\left[\left\|g_1(t)\right\|_{H_1} + \left\|MK^{-1}\right\|_{H_2\to H_1} c_K\left\|g_2(t)\right\|_{H_2} + b_2(M_g(T))^{1/2}\right], \tag{3.15}$$

and then, by transition to the norm of the space $L_2(0,T;H_1)$, we obtain

$$\left\|\delta_n^{(1)}\right\|_{L_2(0,T;H_1)} \leq (1+c_L)\Big[\left\|g_1(t)\right\|_{L_2(0,T;H_1)} + \left\|MK^{-1}\right\|_{H_2\to H_1} \times c_K\left\|g_2(t)\right\|_{L_2(0,T;H_2)} + b_2(M_g(T))^{1/2}\Big]. \quad (3.16)$$

For the error $\delta_n^{(2)}$, using the prior estimates (2.123) and (2.128), we obtain

$$\begin{aligned}
&\left\|\delta_n^{(2)}(t)\right\|_{H_2} \\
&\leq \left\|g_2(t)\right\|_{H_2} + \left\|K\theta_n(t)\right\|_{H_2} + b_1\left(\left\|I_2^{1/2}\theta_n'(t)\right\|_{H_2}^2 + L^{1/2}w'(t)\right\|_{H_1}^2\right)^{1/2} \\
&\leq (1+c_k)\left[\left\|g_2(t)\right\|_{H_2} + b_1\left(\left\|I_2^{1/2}\theta_n'(t)\right\|_{H_2}^2 + L^{1/2}w'(t)\right\|_{H_1}^2\right)^{1/2}\right] \\
&\leq (1+c_k)\Big[\left\|g_2(t)\right\|_{H_2} + b_1(M_g(T))^{1/2}\Big],
\end{aligned} \quad (3.17)$$

and then, by transition to the norm of the space $L_2(0,T;H_2)$, we obtain

$$\left\|\delta_n^{(2)}\right\|_{L_2(0,T;H_2)} \leq (1+c_k)\Big[\left\|g_2(t)\right\|_{L_2(0,T;H_2)} + b_1T^{1/2}(M_g(T))^{1/2}\Big]. \quad (3.18)$$

Using (3.16) and (3.18), we estimate (from above) the right-hand side of the inequality (3.14), and we obtain

$$\begin{aligned}
&\left\|I_1^{1/2}\Delta w_n'(t)\right\|_{H_1}^2 + \left\|L^{1/2}\Delta w_n(t)\right\|_{H_1}^2 + \left\|I_2^{1/2}\Delta\theta_n(t)\right\|_{H_2}^2 \\
&+\int_0^t \left\|K^{1/2}\Delta\theta_n(\tau)\right\|_{H_2}^2 d\tau \leq 4\left\|L^{-1/2}L_0^{1/2}\right\|_{H_1\to H_1} T^{1/2}(M_g(T))^{1/2} \\
&\times(1+c_L)\Big[\left\|g_1(t)\right\|_{L_2(0,T;H_1)} + \left\|MK^{-1}\right\|_{H_2\to H_1} c_K\left\|g_2(t)\right\|_{L_2(0,T;H_2)} \\
&+b_2T^{1/2}(M_g(T))^{1/2}\Big]\tilde{\lambda}_{i(n)+1}^{-1/2} + \left\|K^{-1/2}K_0^{1/2}\right\|_{H_2\to H_2}^2 (1+c_K)^2 \\
&\times\Big[\left\|g_2(t)\right\|_{L_2(0,T;H_2)} + b_1T^{1/2}(M_g(T))^{1/2}\Big]^2 \tilde{æ}_{j(n)+1}^{-1},
\end{aligned}$$

where t is an arbitrary value from $[0,T]$. By introducing the notation (3.5), the last inequality can be transformed to that of (3.3). Therefore Theorem 2.1 has been proved.

Remark 3.1 *The estimate (3.3) is true owing to (2.108) in the sense that its right-hand side tends to zero when $n\to\infty$.*

Remark 3.2 *The estimate (3.3) has been obtained for a bounded operator MK^{-1}. For a series of real coupled thermoelasticity problems (for instance, for shallow shells in the framework of Timoshenko-type models), more restrictive requirements regarding the boundness of the operator $MK^{-1/2}$ are needed. In that case, for the problems (3.1), (3.3), the following prior estimate of the error of the Bubnov–Galerkin method based on (3.2) is valid:*

$$\max_{0\le t\le T}\Big[\Big\|I_1^{1/2}\Delta w_n'(t)\Big\|_{H_1}^2+\Big\|L^{1/2}\Delta w_n(t)\Big\|_{H_1}^2$$
$$+\Big\|I_2^{1/2}\Delta\theta_n(t)\Big\|_{H_2}^2+\int_0^t\Big\|K^{1/2}\Delta\theta_n(\tau)\Big\|_{H_2}^2 d\tau\Big]$$
$$\le c_1(M_g(T))^{1/2}\Big[\Big\|g_1\Big\|_{L_2(0,T;H_1)}+c_2^*\Big\|I^{-1/2}g\Big\|_{L_2(0,T;H_1\times H_2)}$$
$$+c_3^*(M_g(T))^{1/2}\Big]\tilde{\lambda}_{i(n)+1}^{-1/2}$$
$$+c_4\Big[\Big\|g_2\Big\|_{L_2(0,T;H_2)}+c_5(M_g(T))^{1/2}\Big]^2 æ_{j(n)+1}^{-1}, \tag{3.19}$$

where the constants c_1, c_4, c_5, $M_g(T)$ are the same as in (3.3), and c_2^*, c_3^* are constants defined by the relations

$$c_2^*=\frac{\sqrt{2}}{2}e^{1/2}T^{1/2}\Big\|MK^{-1/2}\Big\|_{H_2\to H_1},\quad c_3^*=T^{1/2}\Big\|I_1^{1/2}\Big\|_{H_1\to H_1}. \tag{3.20}$$

The estimate (3.19) is obtained via the following procedure. Instead of (2.122), we use an analogous but simpler estimate:

$$\Big\|Lw_n(t)\Big\|_{H_1}\le c_L\Big(\Big\|g_1(t)\Big\|_{H_1}+\Big\|I_1^{1/2}\Big\|_{H_1\to H_1}\Big\|I_1^{1/2}w_n''(t)\Big\|_{H_1}$$
$$+\Big\|MK^{-1/2}\Big\|_{H_2\to H_1}\Big\|K^{-1/2}\theta_n(t)\Big\|_{H_2}\Big),$$

and almost always ($t\in[0,T]$) the norm of the error $\Big\|\delta_n^{(1)}(t)\Big\|_{H_1}$ can be estimated as follows:

$$\Big\|\delta_n^{(1)}(t)\Big\|_{H_1}\le\Big\|g_1(t)\Big\|_{H_1}+\Big\|Lw_n(t)\Big\|_{H_1}+\Big\|I_1^{1/2}\Big\|_{H_1\to H_1}\Big\|I_1^{1/2}w_n''(t)\Big\|_{H_1}$$
$$+\Big\|MK^{-1/2}\Big\|_{H_2\to H_1}\Big\|K^{-1/2}\theta_n(t)\Big\|_{H_2}$$
$$\le(1+c_L)\Big[\Big\|g_1(t)\Big\|_{H_1}+\Big\|I_1^{1/2}\Big\|_{H_1\to H_1}\Big\|I_1^{1/2}w_n''(t)\Big\|_{H_1}$$
$$+\Big\|MK^{-1/2}\Big\|_{H_2\to H_1}\Big\|K^{-1/2}\theta_n(t)\Big\|_{H_2}\Big].$$

Then, using the prior estimates (2.105) and (2.128), we transform to the norm of the space $L_2(0,T;H_1)$:

$$\left\|\delta_n^{(1)}\right\|_{L_2(0,T;H_1)} \le (1+c_L)\Big[\left\|g_1(t)\right\|_{L_2(0,T;H_1)} + \frac{\sqrt{2}}{2} e^{1/2} T^{1/2} \left\|MK^{-1/2}\right\|_{H_2\to H_1} \times \left\|I^{-1/2} g\right\|_{L_2(0,T;H_1\times H_2)} + T^{1/2}\left\|I_1^{1/2}\right\|_{H_1\to H_1}(M_g(T))^{1/2}\Big].$$

Finally, using this newly obtained inequality (instead of (3.16)) and (3.18), we estimate from above the right hand-side of the inequality (3.14).

The estimates (3.3), (3.19) were obtained using an assumption about the nondifferentiability of the free forms in (2.110). Now we are going to consider a more general case, (2.57). Let f be one of the functions g_1, g_2, $I^{-1/2}g$. Let us denote by $\tilde{f}$ the following extension of the function f from the interval $[0,T]$ to the whole time axis:

$$\tilde{f}(t) = \begin{cases} 0, & |t| > T, \\ -f(-t), & -T \le t \le 0. \end{cases} \tag{3.21}$$

The integral modulus of continuity of the function $f \in L_2(0,T;H)$ is denoted by $w(f,\delta)$:

$$w(f,\delta) = \sup_{|\xi|\le\delta} \left[\int_0^T \left\|\tilde{f}(t+\xi) - \tilde{f}(t)\right\|_H^2 dt \right]^{1/2}, \tag{3.22}$$

where H is one of the spaces H_1, H_2, $H_1\times H_2$, and δ is an arbitrary positive number.

Theorem 3.2 *If the conditions (2.57), (2.58) are satisfied, then the error introduced by use of the equations of the Bubnov–Galerkin method (3.2) for the problem (3.1) allows one to obtain the following estimate:*

$$\begin{aligned} &\sup_{0\le t\le T} ess \Big[\left\|I_1^{1/2}\Delta w_n'(t)\right\|_{H_1}^2 + \left\|L^{1/2}\Delta w_n(t)\right\|_{H_1}^2 \\ &\quad + \left\|I_2^{1/2}\Delta\theta_n(t)\right\|_{H_2}^2 + \int_0^t \left\|K^{1/2}\Delta\theta_n(\tau)\right\|_{H_2}^2 d\tau\Big]^{1/2} \\ &\le d_1 w(I^{-1/2}g,\delta_n) + \Big[e_1(g,\delta_n) w(I^{-1/2}g,\delta_n)\tilde{\lambda}_{i(n)+1}^{-1/4} \\ &\quad + d_2(w(I^{-1/2}g,\delta_n))^2\Big] + c_4\Big[e_2(g,\delta_n)\text{æ}_{j(n+1)}^{-1/2} \\ &\quad + d_3\big(w(I^{-1/2}g,\delta_n)\big)^2\Big]^{1/2} \equiv 0\Big(\max\big(\delta_n, w(I^{-1/2}g,\delta_n)\big)\Big), \end{aligned} \tag{3.23}$$

where

$$\delta_n = \max\left(\tilde{\lambda}_{i(n)+1}^{-1/4}\, \tilde{æ}_{j(n+1)}^{-1/2}\right); \tag{3.24}$$

d_1, d_2, d_3 are constants defined by the relations (3.36); $e_1(g,\delta)$, $e_2(g,\delta)$ are quantities that are bounded for $\delta \to 0$, defined further by (3.37), (3.38); c_4 is a constant defined by the fourth relation of (3.5); and $\{\tilde{\lambda}_k\}_{k=1}^{\infty}$, $\{\tilde{æ}_l\}_{l=1}^{\infty}$ are the series defined by the relation (3.4).

Proof. Let $\varphi(t)$ be an arbitrary kernel of averaging of the class W_2^1 in the interval $[-1,1]$, i.e. let it be an even function of the Sobolev class $W_2^1([-1,1])$, having zero values for $t = \pm 1$ and satisfying the equation

$$\int_{-1}^{1} \varphi(t)\, dt = 1. \tag{3.25}$$

Let us introduce the following notation:

$$\varphi_\delta(t) = \frac{1}{\delta}\varphi\left(\frac{t}{\delta}\right) \quad (-\delta \le t \le \delta),$$

$$g_{j\delta}(t) = \int_{t-\delta}^{t+\delta} \tilde{g}_j(\tau)\varphi_\delta(t-\tau)\, d\tau \equiv \int_{-1}^{1} \tilde{g}_j(t+\delta\tau)\varphi(\tau)\, d\tau,$$

$$\Delta g_{j\delta}(t) = g_j(t) g_{j\delta}(t) \quad (0 \le t \le T),$$

where $\delta > 0$, $j = 1,2$, and $\tilde{g}_j$ are positive definite functions g_j of the form (3.21). In accordance with (3.25), we obtain

$$\Delta g_{j\delta}(t) = \int_{-1}^{1} \Big[g_j(t) - \tilde{g}_j(t+\delta\tau)\Big]\varphi(\tau)\, d\tau. \tag{3.26}$$

It is clear that, for arbitrary $\delta > 0$, $g_{j\delta} \in W_1^2(0,T;H_j)$ $(j = 1,2)$, and

$$g'_{j\delta}(t) = \int_{t-\delta}^{t+\delta} \tilde{g}_j(\tau)\varphi'_\delta(t-\tau)\, d\tau \equiv \frac{1}{\delta}\int_{-1}^{1} \tilde{g}_j(t+\delta\tau)\varphi'(\tau)\, d\tau.$$

The last expression, owing to the noneven kernel derivative $\varphi'(t)$, reads

$$g'_{j\delta}(t) = \frac{1}{\delta}\int_{-1}^{1} \Big[\tilde{g}_j(t+\delta\tau) - g_j(t)\Big]\varphi'(\tau)\, d\tau. \tag{3.27}$$

The relations (3.26), (3.27) imply the following inequalities:

$$\left\|\Delta g_{j\delta}\right\|_{L_2(0,T;H_j)} \le C_{0,\varphi} w(g_j,\delta), \quad j=1,2, \tag{3.28}$$

$$\left\|I^{-1/2}\Delta g_\delta\right\|_{L_2(0,T;H_1\times H_2)} \le C_{0,\varphi} w(I^{-1/2}g,\delta), \tag{3.29}$$

$$\left\|I^{-1/2}g'_\delta\right\|_{L_2(0,T;H_1\times H_2)} \le \frac{1}{\delta} C_{1,\varphi} w(I^{-1/2}g,\delta), \tag{3.30}$$

where

$$C_{0,\varphi} = \sqrt{2}\left[\int_{-1}^{1} (\varphi(\tau))^2\,d\tau\right]^{1/2}, \quad C_{1,\varphi} = \sqrt{2}\left[\int_{-1}^{1} (\varphi'(\tau))^2\,d\tau\right]^{1/2}$$

(for instance, for $\varphi(t) = 1-|t|$, $C_{0,\varphi} = 2\sqrt{3}/3$, $C_{1\varphi} = 2$).

Let us now consider a family of Cauchy-type problems (3.1) with differentiable free terms:

$$\begin{cases} I_1 w''_\delta(t) + L w_\delta(t) + M\theta_\delta(t) = g_{1\delta}(t), \\ I_2\theta'_\delta(t) + K\theta_\delta(t) + N w'_\delta(t) = g_{2\delta}(t), \\ w_\delta(0) = w'_\delta(0) = 0, \quad \theta_\delta(0) = 0, \end{cases} \tag{3.31}$$

where δ is a positive parameter. Let us denote by $\{w_{\delta,n}, \theta_{\delta,n}\}$ an approximate solution to the problem (3.31) constructed by the Bubnov–Galerkin method. Let us introduce the following notation:

$$\begin{aligned} \Delta w_\delta(t) &= w(t) - w_\delta(t), & \Delta\theta_\delta(t) &= \theta(t) - \theta_\delta(t), \\ \Delta w_{\delta,n}(t) &= w_n(t) - w_{\delta,n}(t), & \Delta\theta_{\delta,n}(t) &= \theta_n(t) - \theta_{\delta,n}(t), \\ \Delta w_{n,\delta}(t) &= w_{\delta,n}(t) - w_\delta(t), & \Delta\theta_{n,\delta}(t) &= \theta_{\delta,n}(t) - \theta_\delta(t). \end{aligned}$$

Let us consider the pair of functions $\{\Delta w_\delta, \Delta\theta_\delta\}$ as the exact solution to the Cauchy problem (3.1) with free terms $\Delta g_{1\delta}$, $\Delta g_{2\delta}$, and the pair of functions $\{\Delta w_{\delta,n}, \Delta\theta_{\delta,n}\}$ as an approximate solution to this problem constructed according to the Bubnov–Galerkin method. The following prior estimates of the form (2.105), (2.106) apply:

$$\begin{aligned} &\left[\left\|I_1^{1/2}\Delta w'_{\delta,n}(t)\right\|_{H_1}^2 + \left\|L^{1/2}\Delta w_{\delta,n}(t)\right\|_{H_1}^2 + \left\|I_2^{1/2}\Delta\theta_{\delta,n}(t)\right\|_{H_2}^2\right. \\ &\left. + 2\int_0^t \left\|K^{1/2}\Delta\theta_{\delta,n}(\tau)\right\|_{H_2}^2 d\tau\right]^{1/2} \le e^{1/2}T^{1/2}\left\|I^{-1/2}\Delta g_\delta\right\|_{L_2(0,T;H_1\times H_2)}^2, \end{aligned} \tag{3.32}$$

$$\begin{aligned} &\left[\left\|I_1^{1/2}\Delta w'_\delta(t)\right\|_{H_1}^2 + \left\|L^{1/2}\Delta w_\delta(t)\right\|_{H_1}^2 + \left\|I_2^{1/2}\Delta\theta_\delta(t)\right\|_{H_2}^2\right. \\ &\left. + 2\int_0^t \left\|K^{1/2}\Delta\theta_\delta(\tau)\right\|_{H_2}^2 d\tau\right]^{1/2} \le e^{1/2}T^{1/2}\left\|I^{-1/2}\Delta g_\delta\right\|_{L_2(0,T;H_1\times H_2)}^2 \end{aligned} \tag{3.33}$$

(almost everywhere ($t \in [0,T]$)).

By applying Theorem 2.1 to the problem (3.31), the following prior estimate of type (3.3) of the error introduced by the Bubnov–Galerkin method can be obtained.

$$\Bigg[\Big\|I_1^{1/2}\Delta w'_{n,\delta}(t)\Big\|^2_{H_1} + \Big\|L^{1/2}\Delta w_{n,\delta}(t)\Big\|^2_{H_1} + \Big\|I_2^{1/2}\Delta\theta_{n,\delta}(t)\Big\|^2_{H_2}$$

$$+\int_0^t \Big\|K^{1/2}\Delta\theta_{n,\delta}(\tau)\Big\|^2_{H_2} d\tau\Bigg]^{1/2} \le \Big\{c_1(M_{g\delta}(T))^{1/2}\Big[\big\|g_{1\delta}\big\|_{L_2(0,T;H_1)}$$

$$+ c_2\big\|g_{2\delta}\big\|_{L_2(0,T;H_2)} + c_3(M_{g\delta}(T))^{1/2}\Big]\tilde{\lambda}^{-1/2}_{i(n)+1}$$

$$+ c_4\Big[\big\|g_{2\delta}\big\|_{L_2(0,T;H_2)} + c_5(M_{g\delta}(T))^{1/2}\Big]^2 \tilde{æ}^{-1}_{j(n)+1}\Big\}^{1/2}, \tag{3.34}$$

where $0 \le t \le T$. Because the extension $\tilde{g}_j(0)$ is equal to 0 ($j = 1, 2, \delta > 0$), the following applies in (3.34):

$$M_{g\delta}(T) = eT\Big\|I^{-1/2}g'_\delta\Big\|^2_{L_2(0,T;H_1\times H_2)}. \tag{3.35}$$

The error of the Bubnov–Galerkin method for the initial problem (3.1) can be estimated in the following way:

$$\Bigg[\Big\|I_1^{1/2}\Delta w'_n(t)\Big\|^2_{H_1} + \Big\|L^{1/2}\Delta w_n(t)\Big\|^2_{H_1} + \Big\|I_2^{1/2}\Delta\theta_n(t)\Big\|^2_{H_2}$$

$$+\int_0^t \Big\|K^{1/2}\Delta\theta_n(\tau)\Big\|^2_{H_2} d\tau\Bigg]^{1/2} \le \Bigg[\Big\|I_1^{1/2}\Delta w'_{\delta,n}(t)\Big\|^2_{H_1} + \Big\|L^{1/2}\Delta w_{\delta,n}(t)\Big\|^2_{H_1}$$

$$+\Big\|I_2^{1/2}\Delta\theta_{\delta,n}(t)\Big\|^2_{H_2} + \int_0^t \Big\|K^{1/2}\Delta\theta_{\delta,n}(\tau)\Big\|^2_{H_2} d\tau\Bigg]^{1/2} + \Bigg[\Big\|I_1^{1/2}\Delta w'_\delta(t)\Big\|^2_{H_1}$$

$$+\Big\|L^{1/2}\Delta w_\delta(t)\Big\|^2_{H_1} + \Big\|I_2^{1/2}\Delta\theta_\delta(t)\Big\|^2_{H_2} + \int_0^t \Big\|K^{1/2}\Delta\theta_\delta(\tau)\Big\|^2_{H_2} d\tau\Bigg]^{1/2}$$

$$+\Bigg[\Big\|I_1^{1/2}\Delta w'_{n,\delta}(t)\Big\|^2_{H_1} + \Big\|L^{1/2}\Delta w_{n,\delta}(t)\Big\|^2_{H_1} + \Big\|I_2^{1/2}\Delta\theta_{n,\delta}(t)\Big\|^2_{H_2}$$

$$+\int_0^t \Big\|K^{1/2}\Delta\theta_{\delta,n}(\tau)\Big\|^2_{H_2} d\tau\Bigg]^{1/2}.$$

Using (3.32)–(3.35) and (3.28)–(3.30), we obtain

$$\left[\left\|I_1^{1/2}\Delta w_n'(t)\right\|_{H_1}^2 + \left\|L^{1/2}\Delta w_n(t)\right\|_{H_1}^2 + \left\|I_2^{1/2}\Delta\theta_n(t)\right\|_{H_2}^2 + \int_0^t \left\|K^{1/2}\Delta\theta_n(\tau)\right\|_{H_2}^2 d\tau\right]^{1/2} \le 2e^{1/2}T^{1/2}\left\|I^{-1/2}\Delta g_\delta\right\|_{L_2(0,T;H_1\times H_2)}$$

$$+\left\{c_1 e^{1/2}T^{1/2}\left\|I^{-1/2}g_\delta'\right\|_{L_2(0,T;H_1\times H_2)}\left[\left\|g_1\right\|_{L_2(0,T;H_1)} + c_2\left\|g_2\right\|_{L_2(0,T;H_2)} + \left\|\Delta g_{1\delta}\right\|_{L_2(0,T;H_1)} + c_2\left\|\Delta g_{2\delta}\right\|_{L_2(0,T;H_2)} + c_3 e^{1/2}T^{1/2}\left\|I^{-1/2}g_\delta'\right\|_{L_2(0,T;H_1\times H_2)}\right]\tilde{\lambda}_{i(n)+1}^{-1} + c_4\left[\left\|g_2\right\|_{L_2(0,T;H_2)} + \left\|\Delta g_{2\delta}\right\|_{L_2(0,T;H_2)} + c_5 e^{1/2}T^{1/2}\left\|I^{-1/2}g_\delta'\right\|_{L_2(0,T;H_1\times H_2)}\right]^2 \tilde{æ}_{j(n)+1}^{-1}\right\}^{1/2}$$

$$\le 2e^{1/2}T^{1/2}C_{0,\varphi}w(I^{-1/2}g,\delta)$$

$$+\left\{c_1 e^{1/2}T^{1/2}C_{1,\varphi}\frac{1}{\delta}w(I^{-1/2}g,\delta)\left[\left\|g_1\right\|_{L_2(0,T;H_1)} + c_2\left\|\Delta g_2\right\|_{L_2(0,T;H_2)} + C_{0,\varphi}w(g_1,\delta) + c_2 C_{0,\varphi}w(g_2,\delta) + c_3 e^{1/2}T^{1/2}C_{0,\varphi}\frac{1}{\delta}w(I^{-1/2}g,\delta)\right] \times \tilde{\lambda}_{i(n)+1}^{-1/2} + c_4\left[\left\|g_2\right\|_{L_2(0,T;H_1)} + C_{0,\varphi}w(g_2,\delta) + c_5 e^{1/2}T^{1/2}C_{1,\varphi}\times\frac{1}{\delta}w(I^{-1/2}g,\delta)\right]^2 \tilde{æ}_{j(n)+1}^{-1}\right\}^{1/2}.$$

In the last inequality, δ denotes an arbitrary positive number. Let us assume $\delta = \delta_n$, where δ_n is defined by the relation (3.24). The following inequality is then obtained:

$$\left[\left\|I_1^{1/2}\Delta w_n'(t)\right\|_{H_1}^2 + \left\|L^{1/2}\Delta w_n(t)\right\|_{H_1}^2 + \left\|I_2^{1/2}\Delta\theta_n(t)\right\|_{H_2}^2 + \int_0^t \left\|K^{1/2}\Delta\theta_n(\tau)\right\|_{H_2}^2 d\tau\right]^{1/2} \le 2e^{1/2}T^{1/2}C_{0,\varphi}\frac{1}{\delta}w(I^{-1/2}g,\delta_n)$$

$$+\left\{c_1 e^{1/2}T^{1/2}C_{1,\varphi}\frac{1}{\delta}w(I^{-1/2}g,\delta_n)\left[\left(\left\|g_1\right\|_{L_2(0,T;H_1)} + c_2\left\| g_2\right\|_{L_2(0,T;H_2)} + C_{0,\varphi}w(g_1,\delta_n) + c_2 C_{0,\varphi}w(g_2,\delta_n)\right)\tilde{\lambda}_{i(n)+1}^{-1/4} + c_3 e^{1/2}T^{1/2}C_{1,\varphi} \times w(I^{-1/2}g,\delta_n)\right] + c_4\left[\left(\|g_2\|_{L_2(0,T;H_2)} + C_{0,\varphi}w(g_2,\delta_n)\right)\tilde{æ}_{j(n)+1}^{-1/2} + c_5 e^{1/2}T^{1/2}C_{1,\varphi}\times w(I^{-1/2}g,\delta_n)\right]^2\right\}^{1/2}$$

(almost always everywhere ($t \in [0, T]$)). By introducing the notation

$$d_1 = 2e^{1/2}T^{1/2}C_{0,\varphi}, \quad d_2 = c_1c_3eTC_{1,\varphi}^2, \quad d_3 = c_5e^{1/2}T^{1/2}C_{1,\varphi}, \tag{3.36}$$

$$\begin{aligned} e_1(g,\delta) = c_1e^{1/2}T^{1/2}C_{1,\varphi}\Bigg(\Big\| g_1 \Big\|_{L_2(0,T;H_1)} + c_2 \Big\| g_2 \Big\|_{L_2(0,T;H_2)} \\ + C_{0,\varphi}w(g_1,\delta) + c_2C_{0,\varphi}w(g_2,\delta) \Bigg), \end{aligned} \tag{3.37}$$

$$e_2(g,\delta) = c_4 \Big\| g_2 \Big\|_{L_2(0,T;H_2)} + C_{0,\varphi}w(g_2,\delta), \tag{3.38}$$

the last inequality can be transformed to the form (3.23). Theorem 3.2 has been proved.

Remark 3.3 *The estimate (3.23) is true because of (2.108), in the sense that for $n \to \infty$ its right-hand side approaches zero.*

Remark 3.4 *The estimate (3.23) was obtained using an assumption about the bounds of the operator MK^{-1}. If the operator $MK^{-1/2}$ is also bounded, then (together with (3.23)) the following prior estimate of the error of the Bubnov–Galerkin method based on (3.2) for the problem (3.1) holds:*

$$\begin{aligned} \sup_{0 \le t \le T} ess \Bigg[\Big\| I_1^{1/2} \Delta w_n'(t) \Big\|_{H_1}^2 + \Big\| L^{1/2} \Delta w_n(t) \Big\|_{H_1}^2 \\ + \Big\| I_2^{1/2} \Delta\theta_n(t) \Big\|_{H_2}^2 + \int_0^t \Big\| K^{1/2} \Delta\theta_n(\tau) \Big\|_{H_2}^2 d\tau \Bigg]^{1/2} \\ \le d_1 w(I^{-1/2}g, \delta_n) + \Big[e_1^*(g,\delta_n) w(I^{-1/2}g,\delta_n) \tilde{\lambda}_{i(n)+1}^{-1/4} \\ + d_2^*(w(I^{-1/2}g,\delta_n))^2 + c_4\big(e_2(g,\delta_n) \tilde{æ}_{j(n)+1}^{-1/2} \\ + d_3 w(I^{-1/2}g,\delta_n)\big)^2 \Big]^{1/2}, \end{aligned} \tag{3.39}$$

where the constants d_1, d_3, d_4 and the quantity $e_2(g,\delta)$ are the same as those appearing in (3.23), d_2^* is a constant defined by the relation

$$d_2^* = c_1c_3^*eTC_{1,\varphi}^2, \tag{3.40}$$

and $e_1^*(g,\delta)$ is the bounded (for $\delta \to \infty$) quantity defined by the equation

$$e_1^*(g,\delta) = c_1e^{1/2}T^{1/2}C_{1,\varphi} \left(\Big\| g_1 \Big\|_{L_2(0,T;H_1)} + c_2^* \Big\| g_2 \Big\|_{L_2(0,T;H_2)} \right). \tag{3.41}$$

The scheme by which the estimate (3.39) is obtained is analogous to the scheme applied to prove Theorem 3.2. Here, instead of (3.34), a prior estimate of type (3.19) of the error introduced by the Bubnov–Galerkin method for the problem (3.31) is used.

All the error estimate for the Bubnov–Galerkin method given above have been obtained for a system of differential equations of a rather general form and with homogeneous boundary conditions, which in the case of its application to a thin-walled structure, corresponds to coupled thermoelastic problems for plates and shallow shells with variable thickness. The free terms g_1, g_2 (corresponding to the vector of mechanical load intensity and to the specific power of the internal heat sources) have been assumed to be time dependent. Let us consider now one of the particular cases of the problem (2.54)–(2.56) (with nonhomogeneous initial conditions), where a prior estimate of the error of the Bubnov–Galerkin method is most effective. This corresponds to a mechanical problem where the thickness of the shell or plate under consideration is constant, the intensity of the mechanical load and the power of the internal heat sources do not depend on time, and the full systems of eigenfunctions of the operators of the static elasticity theory of stationary heat transfer are known. These eigenfunctions are used as a basis. The problem corresponds to the following Cauchy problem:

$$\begin{cases} I_1 w''(t) + Lw(t) + M\theta(t) = g_1(t), \\ I_2\theta'(t) + K\theta(t) + Nw'(t) = g_2(t), \\ w(0) = w'(0) = 0, \quad \theta(0) = 0, \end{cases} \tag{3.42}$$

where $w(t)$, $\theta(t)$ are the sought functions; g_1, g_2 are given elements from the Hilbert spaces H_1, H_2; w_0, w_1, θ_0 are given elements satisfying the conditions (2.59); and I_1, I_2, L, K, M, N are operator coefficients satisfying all conditions given in Sect. 2.1.4 and the following conditions:

$$\begin{aligned} &I_1(H_1^n) = H_1^n, \quad L(H_1^n) = H_1^n, \quad M(H_2^n) \subset H_1^n, \\ &I_2(H_2^n) = H_2^n, \quad K(H_2^n) = H_2^n. \end{aligned} \tag{3.43}$$

Here H_1^n, H_2^n are the linear neighbourhoods of finite systems of basis elements $\{\nu_i\}_{i=1}^{i(n)}$, $\{\eta_j\}_{i=1}^{i(n)}$, and n is an arbitrary natural number. According to the conditions (3.43), for arbitrary elements $w_n \in H_1^n$, $\theta_n \in H_2^n$, the following relations are satisfied:

$$P_{H_1^n} I_1 w_n = I_1 w_n, \quad P_{H_2^n} I_2 \theta_n = I_2 \theta_n,$$

$$P_{H_1^n} L w_n = L w_n, \quad P_{H_2^n} K\theta_n = K\theta_n, \quad P_{H_1^n} M\theta_n = M\theta_n.$$

Taking these relations into account, (2.95), (2.109) can be written in the following form:

$$\begin{cases} I_1 w_n''(t) + Lw_n(t) + M\theta_t(t) = P_{H_1^n} g_1, \\ I_2\theta_n'(t) + K\theta_n(t) + P_{H_2^n} N w_t'(t) = P_{H_2^n} g_2, \\ w_n(t) \in H_1^n, \quad \theta_n(t) \in H_2^n, \quad \forall t \in [0,T], \\ w_n(0) = P_{H_1^n} w_0, \quad w_n'(0) = P_{H_1^n} w_1, \quad \theta_n(0) = P_{H_2^n} \theta_0. \end{cases} \tag{3.44}$$

According to the conditions formulated in Sect. 2.1.4 applied to the operators K, L, N, for arbitrary elements $w \in H_1^n$, $\theta \in D(K^{1/2})$, the following inequality is satisfied:

$$\left|(R_{H_2^n} Nw, \theta)_{H_2}\right| \leq s_n \left\| L^{1/2} w \right\|_{H_1} \left\| K^{1/2} w \right\|_{H_2}, \tag{3.45}$$

where

$$s_n = \left\| K^{-1/2} R_{H_2^n} \right\|_{H_2 \to H_2} \left\| N L^{-1/2} \right\|_{H_1 \to H_2} \to 0, \quad n \to \infty. \tag{3.46}$$

In the following treatment we mean by s_n an arbitrary numerical series (not necessarily defined by the relation (3.46)). The elements of this series do not depend on $w \in H_1^n$, $\theta \in D(K^{1/2})$, and then converge to zero as well as satisfy the inequality (3.45).

Theorem 3.3 *If the conditions (2.58), (2.59), (3.43), (3.45) are satisfied then the velocity of convergence of the Bubnov–Galerkin based on (3.44) (applied to the problem (3.42)) is characterized by the following inequality:*

$$\max_{0 \leq t \leq T} \left[\left\| I_1^{1/2} \Delta w_n'(t) \right\|_{H_1}^2 + \frac{1}{2} \left\| L^{1/2} \Delta w_n(t) \right\|_{H_1}^2 \right.$$
$$\left. + \left\| I_2^{1/2} \Delta \theta_n(t) \right\|_{H_2}^2 + \int_0^t \left\| K^{1/2} \Delta \theta_n(\tau) \right\|_{H_2}^2 d\tau \right]^{1/2}$$
$$\leq \left\| I_1^{1/2} R_{H_1^n} w_1 \right\|_{H_1}^2 + \left\| L^{1/2} R_{H_1^n} w_0 \right\|_{H_1}^2 + \left\| I_1^{1/2} R_{H_2^n} \theta_0 \right\|_{H_2}^2$$
$$+ 2 \left\| L^{-1/2} R_{H_1^n} g_1 \right\|_{H_1}^2 - 2(R_{H_1^n} g_1, R_{H_1^n} w_0)_{H_1}$$
$$+ \left(\left\| K^{-1/2} R_{H_2^n} g_2 \right\|_{H_2}^2 s_n c_n \right)^2 T, \tag{3.47}$$

$$\max_{0 \leq t \leq T} \left[\left\| I_1^{1/2} \Delta w_n'(t) \right\|_{H_1}^2 + \frac{1}{2} \left\| L^{1/2} \Delta w_n(t) \right\|_{H_1}^2 + \left\| I_2^{1/2} \Delta \theta_n(t) \right\|_{H_2}^2 \right]$$
$$\leq \left\| I_1^{1/2} R_{H_1^n} w_1 \right\|_{H_1}^2 + \left\| L^{1/2} R_{H_1^n} w_0 \right\|_{H_1}^2$$
$$+ \left\| I_1^{1/2} R_{H_2^n} \theta_0 \right\|_{H_2}^2 + 2 \left\| L^{-1/2} R_{H_1^n} g_1 \right\|_{H_1}^2 - 2(R_{H_1^n} g_1, R_{H_1^n} w_0)_{H_1} \tag{3.48}$$
$$+ \frac{1}{2} \left(\left\| K^{-1/2} R_{H_2^n} g_2 \right\|_{H_2}^2 s_n c_n \right)^2 T,$$

where

$$c_n = \left[\left\| I_1^{-1/2} (g_1 - L w_0 - M P_{H_2^n} \theta_0) \right\|_{H_1}^2 + \left\| L^{1/2} w_1 \right\|_{H_1}^2 \right.$$
$$\left. + \left\| I_2^{-1/2} (g_2 - K\theta_0 - N P_{H_1^n} w_1) \right\|_{H_2}^2 \right]^{1/2}. \tag{3.49}$$

Proof. We differentiate the first two relations of (3.42). The first is then scalarly multiplied by $2w_n''(t)$ and the second by $2\theta_n'(t)$. We then sum the relations, making use of the condition (2.58). As the result, we obtain

$$\frac{d}{dt}\left\|I_1^{1/2}w_n''(t)\right\|_{H_1}^2+\frac{d}{dt}\left\|L^{1/2}\Delta w_n'(t)\right\|_{H_1}^2+\frac{d}{dt}\left\|I_2^{1/2}\theta_n'(t)\right\|_{H_2}^2$$
$$+2\left\|K^{1/2}\theta_n'(t)\right\|_{H_2}^2=0.$$

Integrating this equation over the interval $[0,t]$, where $t\le T$, we obtain

$$\left\|I_1^{1/2}w_n''(t)\right\|_{H_1}^2+\frac{1}{2}\left\|L^{1/2}w_n'(t)\right\|_{H_1}^2+$$
$$\left\|I_2^{1/2}\theta_n'(t)\right\|_{H_2}^2+2\int\limits_0^t\left\|K^{1/2}\theta_n'(\tau)\right\|_{H_2}^2\,d\tau$$
$$=\left\|I_1^{1/2}w_n''(t)\right\|_{H_1}^2+\frac{1}{2}\left\|L^{1/2}w_n'(0)\right\|_{H_1}^2+\left\|I_2^{1/2}\theta_n'(0)\right\|_{H_2}^2$$
$$=\left\|I_1^{-1/2}(P_{H_1^n}g_1-Lw_n(0)-M\theta_n(0))\right\|_{H_1}^2$$
$$+\left\|L^{1/2}w_n'(0)\right\|_{H_1}^2+\left\|I_1^{-1/2}(P_{H_2^n}g_2-K\theta_n(0)-P_{H_2^n}Nw_n'(0))\right\|_{H_2}^2$$
$$=\left\|P_{H_1^n}I_1^{-1/2}(g_1-Lw_0-MP_{H_2^n}\theta_0)\right\|_{H_1}^2+\left\|P_{H_1^n}L^{1/2}w_1\right\|_{H_1}^2$$
$$+\left\|P_{H_2^n}I_2^{-1/2}(g_2-K\theta_0-NP_{H_1^n}w_1)\right\|_{H_2}^2\le c_n^2,$$

which means that the following inequality holds:

$$\left\|L^{1/2}w_n'(t)\right\|_{H_1}\le c_n. \tag{3.50}$$

From (3.45) and (3.50), the following inequality is obtained:

$$\left|\left(R_{H_2^n}Nw_n'(t),\theta\right)_{H_2}\right|\le c_ns_n\left\|K^{1/2}\theta\right\|_{H_2}, \tag{3.51}$$

where θ is an arbitrary element from $D(K^{1/2})$.

Taking into account (3.42) and (3.44), we then obtain

$$\begin{cases} I_1\Delta w_n''(t)+L\Delta w_n(t)+M\Delta\theta_n(t)=-R_{H_1^n}g_1,\\ I_2\Delta\theta_n'(t)+K\Delta\theta_n(t)+N\Delta w_n'(t)=-R_{H_2^n}g_2+R_{H_2^n}Nw_n'(t),\\ \Delta w_n(0)=-R_{H_1^n}w_0,\ \Delta w_n'(0)=-R_{H_1^n}w_1,\ \Delta\theta_n(0)=-R_{H_2^n}\theta_0. \end{cases} \tag{3.52}$$

We multiply scalarly the first two equations of (3.52) by $2\Delta w_n'(t)$ and $2\Delta\theta_n(t)$, respectively, and sum them. We then use the condition (2.58) and integrate the newly obtained expression over the interval $[0,t]$, where $t \le T$. As the result we obtain

$$\begin{aligned}
&\left\|I_1^{1/2}\Delta w_n'(t)\right\|_{H_1}^2 + \left\|L^{1/2}\Delta w_n(t)\right\|_{H_1}^2 + \left\|I_2^{1/2}\Delta\theta_n(t)\right\|_{H_2}^2 \\
&+2\int_0^t \left\|K^{1/2}\Delta\theta_n(\tau)\right\|_{H_2}^2 d\tau = \left\|I_1^{1/2}R_{H_1^n}w_1\right\|_{H_1}^2 + \left\|L^{1/2}R_{H_1^n}w_0'\right\|_{H_1}^2 \\
&+\left\|I_2^{1/2}R_{H_2^n}\theta_0\right\|_{H_2}^2 - 2(R_{H_1^n}g_1,\Delta w_n(t))_{H_1^n} - 2(R_{H_1^n}g_1,R_{H_1^n}w_0)_{H_1} \\
&\qquad +2\int_0^t \big(-R_{H_2^n}g_2 + R_{H_2^n}Nw_n'(\tau),\Delta\theta_n(\tau)\big)_{H_2} d\tau. \qquad (3.53)
\end{aligned}$$

The fourth term of the right-hand side of (3.53) allows us to obtain the estimate:

$$\begin{aligned}
&-2\big(R_{H_1^n}g_1,\Delta w_n(t)\big)_{H_1} \\
&\le \frac{1}{2}\left\|L^{1/2}\Delta w_n(t)\right\|_{H_1}^2 + 2\left\|L^{-1/2}R_{H_1^n}g_1(t)\right\|_{H_1}^2.
\end{aligned} \qquad (3.54)$$

The last term on the right-hand side of (3.53), owing to (3.51), leads to the estimate

$$\begin{aligned}
&2\int_0^t \Big(-R_{H_2^n}g_2 + R_{H_2^n}Nw_n'(\tau),\Delta\theta_n(\tau)\Big)_{H_2} d\tau \\
&\quad\le 2\left(\left\|K^{-1/2}R_{H_2^n}g_2\right\|_{H_2} + c_n s_n\right)\int_0^t \left\|K^{1/2}\Delta\theta_n(\tau)\right\|_{H_2} d\tau.
\end{aligned} \qquad (3.55)$$

Furthermore, by applying the Cauchy inequality, we obtain

$$\begin{aligned}
&2\left(\left\|K^{-1/2}R_{H_2^n}g_2\right\|_{H_2} + c_n s_n\right)\int_0^t \left\|K^{1/2}\Delta\theta_n(\tau)\right\|_{H_2} d\tau \\
&\le \left(\left\|K^{-1/2}R_{H_2^n}g_2\right\|_{H_2} + c_n s_n\right)^2 T + \int_0^t \left\|K^{1/2}\Delta\theta_n(\tau)\right\|_{H_2} d\tau.
\end{aligned} \qquad (3.56)$$

From (3.53)–(3.56) the estimate (3.47) can be derived.

The right-hand side of the inequality (3.55) can also be evaluated in the following way:

$$2\left(\left\|K^{-1/2}R_{H_2^n}g_2\right\|_{H_2}+c_n s_n\right)\int_0^t\left\|K^{1/2}\Delta\theta_n(\tau)\right\|_{H_2}d\tau$$
$$\leq\frac{1}{2}\left(\left\|K^{-1/2}R_{H_2^n}g_2\right\|_{H_2}+c_n s_n\right)^2T+2\int_0^t\left\|K^{1/2}\Delta\theta_n(\tau)\right\|_{H_2}^2d\tau. \quad (3.57)$$

From (3.53)–(3.56), we obtain the estimate (3.48). Theorem 3.3 has been proved.

Remark 3.5 *During the formulation of Theorem 3.3, two estimates of the accuracy of the Bubnov–Galerkin method based on (3.44) applied to the problem (3.42) have been used. They differ in the following way. Firstly, on the left hand side of the inequality (3.47), the following term appears:*

$$\int_0^t\left\|K^{1/2}\Delta\theta_n(\tau)\right\|_{H_2}^2d\tau, \quad (3.58)$$

which does not occur on the left-hand side of the inequality (3.48). Secondly, the last term of the right-hand side of the inequality (3.48) is half of the last term of the right-hand side of the inequality (3.47). To conclude, the estimate (3.47) should be used only when we need to estimate (3.58) (i.e. when we need to evaluate the convergence velocity $\theta_n \to \theta$ in the norm of the space $L_2(0,T;D(K^{1/2})))$. Otherwise, it is more convenient to apply the estimate (3.48) *with a smaller right-hand side.*

To end this section, we outline the prior estimates of the error of the Bubnov–Galerkin method applied to the Cauchy problem

$$I_1w''(t)+Lw(t)=g_1(t),\quad w(0)=w_0,\quad w'(0)=w_1, \quad (3.59)$$

which generalizes a class of dynamical problems of theory of the elasticity (without a heat transfer equation) for both three-dimensional and thin-walled elements of structures. The operators I_1 and L occurring in (3.59) satisfy all conditions formulated in Sect. 2.1.4.

If $w_0 = w_1 = 0$, $g_1 \in W_2^1(0,T;H_1)$ and a system of eigenelements ν_i of a self-adjoint, positive definite, convergent operator L_0 (forming an acute angle with the operator L) is used as a basis, then the error introduced by the Bubnov–Galerkin method for the problem (3.59) reads

$$\max_{0\le t\le T}\left[\left\|I_1^{1/2}\Delta w_n'(t)\right\|_{H_1}^2+\left\|L^{1/2}\Delta w_n(t)\right\|_{H_1}^2\right]$$
$$\le c_1(M_g(T))^{1/2}\left[\left\|g_1\right\|_{L_2(0,T;H_1)}+c_3(M_g(T))^{1/2}\right]\times\tilde{\lambda}_{i(n)+1}^{-1/2} \tag{3.60}$$
$$\equiv 0(\tilde{\lambda}_{i(n)+1}^{-1/2}),$$

where $\Delta w_n(t)=w_n(t)-w(t)$; w_n, w are the approximate and exact solutions to the problem (3.59); $\tilde{\lambda}_k=\min_{i\ge k}\lambda_i$; λ_i are the eigenvalues of the operator L_0; $i(n)$ is the set of basis functions ν_i used in constructiong the approximate solutions w_n; c_1 is a constant defined by the first relation of (3.5); c_3 is a constant defined by the relation

$$c_3=T^{1/2}\left\|I_1^{1/2}\right\|_{H_1\to H_1}; \tag{3.61}$$

and $M_g(T)$ is a constant defined by the relation (2.130), where we take $g_2(t)\equiv 0$. The estimate (3.60) is analogous to both of the estimates (3.3), (3.19), which hold for the approximate solutions of the corresponding coupled problems.

If $w_0=w_1=0$, $g_1\in L_2(0,T;H_1)$, then the error introduced by the Bubnov–Galerkin method for the problem (3.59) can be evaluated in the following way, using the basis defined above:

$$\sup_{0\le t\le T} ess\left[\left\|I_1^{1/2}\Delta w_n'(t)\right\|_{H_1}^2+\left\|L^{1/2}\Delta w_n(t)\right\|_{H_1}^2\right]^{1/2}$$
$$\le d_1 w(I^{-1/2}g,\tilde{\lambda}_{i(n)+1}^{-1/4})+\left[e_1(g_1,\tilde{\lambda}_{i(n)+1}^{-1/4})w(I^{-1/2}g_1,\tilde{\lambda}_{i(n)+1}^{-1/4})\tilde{\lambda}_{i(n)+1}^{-1/4}\right.$$
$$\left.+d_2(w(I_1^{-1/2}g_1,\tilde{\lambda}_{i(n)+1}^{-1/4}))^2\right]^{1/2}$$
$$\equiv 0\Big(\max\big(\tilde{\lambda}_{i(n)+1}^{-1/4},w(I_1^{-1/2}g_1,\tilde{\lambda}_{i(n)+1}^{-1/4})\big)\Big), \tag{3.62}$$

where the notations Δw_n, $\tilde{\lambda}_k$ and $i(n)$ have the same meaning as in (3.60); $w(f,\delta)$ is the integral modulus of continuity defined by the relation (3.22); d_1 is a constant defined by the first relation of (3.36); $e_1(g_1,\delta)$ is a bounded quantity for $\delta\to 0$ defined by the relation (3.37), where we take $g_2(t)\equiv 0$; d_2 is a constant defined by the second relation of (3.36), in which c_3 is a constant defined by the relation (3.61). The estimate (3.62) is analogous to both of the estimates (3.23), (3.39) satisfied for the corresponding coupled problems.

If the linear neighbourhood H_1^n of a finite system of basis elements $\{\nu_i\}_{i=1}^{i(n)}$ is the eigensubspace for both of the operators I_1 and L for an arbitrary n, $g_1(t)\equiv g_1=const$ ($g_1\in H_1$), $w_0\in D(L^{1/2})$, $w_1\in H_1$, then the error introduced by the Bubnov–Galerkin method for the problem (3.59) can be estimated as follows:

$$\max_{0\le t\le T}\left[\left\|I_1^{1/2}\Delta w_n'(t)\right\|_{H_1}^2+\frac{1}{2}\left\|L^{1/2}\Delta w_n(t)\right\|_{H_1}^2\right]$$
$$\le\left\|I_1^{1/2}R_{H_1^n}w_1\right\|_{H_1}^2+\left\|L^{1/2}R_{H_1^n}w_0\right\|_{H_1}^2+2\left\|L^{-1/2}R_{H_1^n}g_1\right\|_{H_1}^2 \tag{3.63}$$
$$-2(R_{H_1^n}g_1,R_{H_1^n}w_0)_{H_1},$$

where Δw_n possesses the same meaning as in (3.60), $R_{H_1^n}=E_1-P_{H_1^n}$, $P_{H_1^n}$ is an orthoprojector from H_1 to H_1^n, and E_1 is an identity operator over H_1. The estimates (3.63) are analogous to the estimates (3.47), (3.48) satisfied for the corresponding coupled problems.

The proofs of the estimates (3.60), (3.62), (3.63) are analogous to the proofs of Theorems 3.1–3.3. The estimates (3.60), (3.62), (3.63) can also be formally obtained using the corresponding estimates (3.3) (or (3.19)), (3.23) (or (3.39)), (3.47) (or (3.48)), by deleting all terms the existence of which leads to the occurrence of the second differential equation and the second sought function θ in (2.54)–(2.56).

3.2 Coupled Thermoelastic Problem Within the Kirchhoff–Love Model

A fundamental problem that arises in the application of the various error estimates for the Bubnov–Galerkin method to real thermoelastic problems is the need to find operators L_0, K_0 (convergent and forming an acute angle with the differential operators L, K) for which full systems of eigenfunctions and the corresponding eigenvalues are known. In this section, we show how to find operators L_0, K_0 with the required properties and the required precision of the estimates (3.3), (3.23) for coupled thermoelastic problems of shallow shells, rectangular in plan and with variable thickness, in the framework of the Kirchhoff–Love model. Assuming that the conditions (2.64), (2.65), (2.68), (2.83)–(2.85) are satisfied, we consider the system of differential equations (2.30)–(2.32) with the homogeneous initial conditions (2.33)–(2.35), where u, v, w, θ are the sought functions, defined for $(x,y)\in\Omega_1$, $|z|\le\frac{1}{2}h(x,y)$, $t\in[0,T]$. Let us assume that $\Omega_1=\left(-\frac{a}{2},\frac{a}{2}\right)\times\left(-\frac{b}{2},\frac{b}{2}\right)$, where a, b are positive constants, i.e. we are considering a shell that is rectangular in plan. Boundary conditions, corresponding either to simple support of the edges on a flexible rib that cannot be compressed or elongated in the tangential plane (2.49) or to clamping of the edges (2.51), are applied to the shell edge $d\Omega_1$. We apply the homogeneous boundary condition (2.38) ($\theta=0$) to the shell surface $d\Omega_2$.

The differential operator L for the problem under consideration is defined on $D(L) = (\overset{\circ}{W}{}_2^2(\Omega_1))^2 \times \overset{\circ}{W}{}_2^4(\Omega_1)$ by the relation (2.88). An operator L_0 with a known spectrum will be constructed in a way similar to that presented in [246]. Let us look for the operator L_0 in the form

$$L_0\overline{\omega} = \tilde{L}\overline{\omega} + P_L\overline{\omega}, \tag{3.64}$$

where $\overline{\omega} = \{u, v, w\}$, and $\tilde{L}$ is the self-conjugated extension of a differential operator in the Friedrichs sense, defined initially on $[(\overset{\circ}{W}{}_2^2(\Omega_1))^2 \times \overset{\circ}{W}{}_2^4(\Omega_1)] \cap [(C^2(\overline{\Omega_1}))^2 \times C^4(\overline{\Omega_1})]$ by the relation

$$\tilde{L}\overline{\omega} = \left\{ -\frac{\partial^2 u}{\partial x^2} - \frac{1-\nu}{2}\frac{\partial^2 u}{\partial y^2},\ -\frac{\partial^2 v}{\partial y^2} - \frac{1-\nu}{2}\frac{\partial^2 v}{\partial x^2},\ \nabla^2\nabla^2 w \right\},$$

from which P_L (see (3.64)) is going to be found.

We introduce the following notation:

$$w_k(x) = \begin{cases} \cos \pi k x, & k \text{ odd} \\ \sin \pi k x, & k \text{ even} \end{cases}, \tag{3.65}$$

$$w_k^*(x) = \begin{cases} \cos \pi k x, & k \text{ even} \\ \sin \pi k x, & k \text{ odd} \end{cases}, \tag{3.66}$$

$$w_{kl}^{(1,1)}(x,y) = w_k^*(x)w_l(y), \quad w_{kl}^{(1,2)}(x,y) = w_k(x)w_l^*(y), \tag{3.67}$$

$$w_{kl}^{(1,3)}(x,y) = w_k(x)w_l(y). \tag{3.68}$$

For the boundary conditions (2.49) the vector functions $\overline{v}_{kl}^{(j)}$ $(j = 1,2,3)$ of the operator L_0 have the form

$$\overline{v}_{kl}^{(1)}(x,y) = \left\{ w_{kl}^{(1,1)}\left(\frac{x}{a}, \frac{y}{b}\right), 0, 0 \right\}, \quad k = 0,1,\dots,\ l = 1,2,\dots, \tag{3.69}$$

$$\overline{v}_{kl}^{(2)}(x,y) = \left\{ 0, w_{kl}^{(1,2)}\left(\frac{x}{a}, \frac{y}{b}\right), 0 \right\}, \quad k = 1,2,\dots,\ l = 0,1,\dots, \tag{3.70}$$

$$\overline{v}_{kl}^{(3)}(x,y) = \left\{ 0, 0, w_{kl}^{(1,3)}\left(\frac{x}{a}, \frac{y}{b}\right) \right\}, \quad k,l = 1,2,\dots, \tag{3.71}$$

and the corresponding eigenvalues $\lambda_{kl}^{(j)}$ of the operator L_0 are defined by the equations

$$\lambda_{kl}^{(1)} = \pi^2 \left[\frac{k^2}{a^2} + \frac{(1-\nu)l^2}{2b^2} \right] + P_L, \quad \lambda_{kl}^{(2)} = \pi^2 \left[\frac{(1-\nu)l^2}{2a^2} + \frac{l^2}{b^2} \right] + P_L, \tag{3.72}$$

$$\lambda_{kl}^{(3)} = \pi^4 \left(\frac{k^2}{a^2} + \frac{l^2}{b^2} \right)^2 + P_L. \tag{3.73}$$

Using the boundary conditions (2.51), the eigenvector functions $\overline{v}_{kl}^{(j)}$ of the operator L_0 have the form

$$\overline{v}_{kl}^{(j)}(x,y) = \left\{ \delta_i^j w_{kl}^{(1,3)} \left(\frac{x}{a}, \frac{y}{b} \right) \right\}_{i=1}^{3}, \quad j = 1,2,3, \quad k,l = 1,2,\dots, \tag{3.74}$$

where δ_i^j is the Kronecker delta, i.e. $\delta_i^j = 1$, $\delta_i^j = 0$ $(i \neq j)$; the corresponding eigenvalues of the operator L_0 are again defined by (3.72), (3.73).

Let us check if, for an appropriate choice of the constant P_L, the operator L_0 is defined by the relation (2.88). Integrating by parts, with use of the homogeneous boundary conditions, leads to the following result:

$$\left\| L_0 \overline{\omega} \right\|^2_{(L_2(\Omega_1))^3} = \int_{\Omega_1} \left[\left(\frac{\partial^4 w}{\partial x^4} \right)^2 + \left(\frac{\partial^4 w}{\partial y^4} \right)^2 + 6 \left(\frac{\partial^4 w}{\partial x^2 \partial y^2} \right)^2 + 4 \left(\frac{\partial^4 w}{\partial x^3 \partial y} \right)^2 \right.$$
$$+4 \left(\frac{\partial^4 w}{\partial x \partial y^3} \right)^2 + \left(\frac{\partial^2 w}{\partial x^2} \right)^2 + \left(\frac{\partial^2 v}{\partial y^2} \right)^2 + (1-\nu) \left(\frac{\partial^2 w}{\partial x \partial y} \right)^2 + (1-\nu) \left(\frac{\partial^2 v}{\partial x \partial y} \right)^2$$
$$\left. + \frac{(1-\nu)^2}{4} \left(\frac{\partial^2 u}{\partial y^2} \right)^2 + \frac{(1-\nu)^2}{4} \left(\frac{\partial^2 v}{\partial x^2} \right)^2 \right] d\Omega_1 + 2P_L \int_{\Omega_1} \left[\left(\frac{\partial^2 w}{\partial x^2} \right)^2 + \left(\frac{\partial^2 w}{\partial y^2} \right)^2 \right.$$
$$\left. +2 \left(\frac{\partial^2 w}{\partial x \partial y} \right)^2 + \left(\frac{\partial u}{\partial x} \right)^2 + \left(\frac{\partial v}{\partial y} \right)^2 + \frac{1-\nu}{2} \left(\frac{\partial u}{\partial y} \right)^2 + \frac{1-\nu}{2} \left(\frac{\partial v}{\partial x} \right)^2 \right] d\Omega_1$$
$$+P_L \int_{\Omega_1} (u^2 + v^2 + w^2)\, d\Omega_1 .$$

From elementary estimates, we obtain the following inequality:

$$\left\| L \overline{\omega} \right\|_{(L_2(\Omega_1))^3} \leq \alpha_L \left\| L_0 \overline{\omega} \right\|_{(L_2(\Omega_1))^3}, \tag{3.75}$$

where $\overline{\omega} = \{u, v, w\}$ is an arbitrary vector function taken from $[(\mathring{W}_2^2(\Omega_1))^2 \times \mathring{W}_2^4(\Omega_1)] \cap [(C^2(\overline{\Omega_1}))^2 \times C^4(\overline{\Omega_1})]$, and α_L is a positive constant independent of $\overline{\omega}$.

The operator L defined by (2.88) can be transformed to the form

$$L\overline{\omega} = \left\{ -\frac{Eh}{1-\nu^2} + \left(\frac{\partial^2 u}{\partial x^2} + \frac{1-\nu}{2} \frac{\partial^2 u}{\partial y^2} + \frac{1-\nu}{2} \frac{\partial^2 v}{\partial x \partial y} \right) + l_1 \overline{\omega}, \right.$$
$$\frac{Eh}{(1-\nu)^2} \left(\frac{\partial^2 v}{\partial y^2} + \frac{1-\nu}{2} \frac{\partial^2 v}{\partial x^2} + \frac{1-\nu}{2} \frac{\partial^2 u}{\partial x \partial y} \right) + l_2 \overline{\omega}, \tag{3.76}$$
$$\left. D \bigtriangledown^2 \bigtriangledown^2 w + l_{31} w + l_{32} \overline{\omega} \right\},$$

where $l_1\overline{\omega}$, $l_2\overline{\omega}$ are linear differential operators that contain only first-order derivatives of the functions u, v, w; $l_{31}w$ is a linear operator that depends only on w and contains derivatives with respect to w of third order only; and $l_{32}\overline{\omega}$ is a linear differential operator that contains derivatives of the arbitrary functions u, v of first order only and derivatives of w of second order only. Using (3.76), any of the vector functions

$$\overline{\omega} = \{u, v, w\} \in \left[\left(\overset{\circ}{W}{}_2^2(\Omega_1)\right)^2 \times \overset{\circ}{W}{}_2^4(\Omega_1)\right] \cap \left[\left(C^2(\overline{\omega_1})\right)^2 \times C^4\left(\overline{\Omega_1}\right)\right]$$

can be represented in the following form:

$$(L\overline{\omega}, \tilde{L}\overline{\omega})_{(L_2(\Omega_1))^3} = \int_{\Omega_1} \left\{ D(\nabla^2\,\nabla^2\,w)^2 + \frac{Eh}{1-\nu^2}\left[\left(\frac{\partial^2 u}{\partial x^2} + \frac{1-\nu}{2}\frac{\partial^2 u}{\partial y^2}\right)^2\right.\right.$$

$$\left.\left. + \left(\frac{\partial^2 v}{\partial y^2} + \frac{1-\nu}{2}\frac{\partial^2 v}{\partial x^2}\right)^2\right]\right\} d\Omega_1 + \frac{1+\nu}{2}\int_{\Omega_1} \frac{Eh}{1-\nu^2}\left[\frac{\partial^2 v}{\partial x\partial y}\left(\frac{\partial^2 u}{\partial x^2}\right.\right.$$

$$\left.\left.\frac{1-\nu}{2}\frac{\partial^2 u}{\partial y^2}\right) + \frac{\partial^2 u}{\partial x\partial y}\left(\frac{\partial^2 v}{\partial y^2} + \frac{1-\nu}{2}\frac{\partial^2 v}{\partial x^2}\right)^2\right] d\Omega_1 + \int_{\Omega_1}\left[l_1\overline{\omega}\left(\frac{\partial^2 u}{\partial x^2}\right.\right.$$

$$\left.\left.+\frac{1-\nu}{2}\frac{\partial^2 u}{\partial y^2}\right) + l_2\overline{\omega}\left(\frac{\partial^2 v}{\partial y^2} + \frac{1-\nu}{2}\frac{\partial^2 v}{\partial x^2}\right) + (l_{31}w + l_{32}\overline{\omega})\,\nabla^2\,\nabla^2 w\right] d\Omega_1. \tag{3.77}$$

Using the Cauchy inequality, integrating by parts and using again the Cauchy inequality, the second term of the right-hand side of the relation (3.77) can be estimated in the following way:

$$\frac{1+\nu}{2}\left|\int_{\Omega_1} \frac{Eh}{1-\nu^2}\left[\frac{\partial^2 v}{\partial x\partial y}\left(\frac{\partial^2 u}{\partial x^2} + \frac{1-\nu}{2}\frac{\partial^2 u}{\partial y^2}\right) + \frac{\partial^2 u}{\partial x\partial y}\left(\frac{\partial^2 v}{\partial y^2} + \frac{1-\nu}{2}\right.\right.\right.$$

$$\left.\left.\left.\times\frac{\partial^2 v}{\partial x^2}\right)^2\right] d\Omega_1\right| \le \frac{(1+\nu)^2}{8}\int_{\Omega_1}\frac{Eh}{1-\nu^2}\left[\left(\frac{\partial^2 u}{\partial x\partial y}\right)^2 + \left(\frac{\partial^2 v}{\partial x\partial y}\right)^2\right] d\Omega_1$$

$$+\frac{1}{2}\int_{\Omega_1}\frac{Eh}{1-\nu^2}\left[\left(\frac{\partial^2 u}{\partial x^2} + \frac{1-\nu}{2}\frac{\partial^2 u}{\partial y^2}\right)^2 + \left(\frac{\partial^2 v}{\partial y^2} + \frac{1-\nu}{2}\frac{\partial^2 v}{\partial x^2}\right)^2\right] d\Omega_1$$

$$= \frac{(1+\nu)^2}{8}\int_{\Omega_1}\frac{Eh}{1-\nu^2}\left(\frac{\partial^2 u}{\partial x^2}\frac{\partial^2 u}{\partial y^2} + \frac{\partial^2 v}{\partial x^2}\frac{\partial^2 v}{\partial y^2}\right) d\Omega_1 + \frac{(1+\nu)^2}{8}$$

$$\times\int_{\Omega_1}\frac{Eh}{1-\nu^2}\left(\frac{\partial h}{\partial x} - \frac{\partial h}{\partial y}\right)\left(\frac{\partial u}{\partial x}\frac{\partial^2 u}{\partial x\partial y} - \frac{\partial v}{\partial y}\frac{\partial^2 v}{\partial x\partial y}\right) d\Omega_1 + \frac{1}{2}\int_{\Omega_1}\frac{Eh}{1-\nu^2}$$

$$\times\left[\left(\frac{\partial^2 u}{\partial x^2}+\frac{1-\nu}{2}\frac{\partial^2 u}{\partial y^2}\right)^2+\left(\frac{\partial^2 v}{\partial y^2}+\frac{1-\nu}{2}\frac{\partial^2 v}{\partial x^2}\right)^2\right]d\Omega_1 \le \frac{1}{2}\left[1+\frac{(1+\nu)^2}{8(1-\nu)}\right]$$

$$\times\int_{\Omega_1}\frac{Eh}{1-\nu^2}\left[\left(\frac{\partial^2 u}{\partial x^2}+\frac{1-\nu}{2}\frac{\partial^2 u}{\partial y^2}\right)^2+\left(\frac{\partial^2 v}{\partial y^2}+\frac{1-\nu}{2}\frac{\partial^2 v}{\partial x^2}\right)^2\right]d\Omega_1$$

$$+\frac{(1+\nu)^2}{8}\int_{\Omega_1}\frac{Eh}{1-\nu^2}\left(\frac{\partial h}{\partial x}-\frac{\partial h}{\partial y}\right)\left(\frac{\partial u}{\partial x}\frac{\partial^2 u}{\partial x\partial y}-\frac{\partial v}{\partial y}\frac{\partial^2 v}{\partial x\partial y}\right)d\Omega_1. \tag{3.78}$$

From (3.77), (3.78), we obtain

$$(L\overline{\omega},\tilde{L}\overline{\omega})_{(L_2(\Omega_1))^3}\le\int_{\Omega_1}\left\{D(\nabla^2\nabla^2 w)^2+\gamma(\nu)\frac{Eh}{1-\nu^2}\left[\left(\frac{\partial^2 u}{\partial x^2}\right.\right.\right.$$

$$\left.\left.\left.+\frac{1-\nu}{2}\frac{\partial^2 u}{\partial y^2}\right)^2+\left(\frac{\partial^2 v}{\partial y^2}+\frac{1-\nu}{2}\frac{\partial^2 v}{\partial x^2}\right)^2\right]\right\}d\Omega_1+I, \tag{3.79}$$

where $\gamma(\nu)=\frac{1}{2}\left[1-\frac{(1+\nu)^2}{8(1-\nu)}\right]$ (it is easy to check that for $0<\nu<\frac{1}{2}$, $\gamma(\nu)>0$), and

$$I=\int_{\Omega_1}\left[l_1\overline{\omega}\left(\frac{\partial^2 u}{\partial x^2}+\frac{1-\nu}{2}\frac{\partial^2 u}{\partial y^2}\right)+l_2\overline{\omega}\left(\frac{\partial^2 v}{\partial y^2}+\frac{1-\nu}{2}\frac{\partial^2 v}{\partial x^2}\right)+(l_{31}w\right.$$

$$\left.+l_{32}\overline{\omega})\nabla^2\nabla^2 w+\frac{(1+\nu)E}{8(1-\nu)}\left(\frac{\partial h}{\partial x}-\frac{\partial h}{\partial y}\right)\left(\frac{\partial u}{\partial x}\frac{\partial^2 u}{\partial x\partial y}-\frac{\partial v}{\partial y}\frac{\partial^2 v}{\partial x\partial y}\right)\right]d\Omega_1.$$

The inequality (3.79) implies

$$(L\overline{\omega},\tilde{L}\overline{\omega})_{(L_2(\Omega_1))^3}\ge m_1\left\|L\overline{\omega}\right\|_{(L_2(\Omega_1))^3}+I, \tag{3.80}$$

where

$$m_1=\min_{(x,y)\in\overline{\Omega_1}}\left\{D(x,y),\gamma(\nu)\frac{Eh(x,y)}{(1-\nu^2)}\right\}.$$

Furthermore, using the Cauchy inequality, we obtain

$$\begin{aligned}&\left|\int_{\Omega_1}l_1\overline{\omega}\left(\frac{\partial^2 u}{\partial x^2}+\frac{1-\nu}{2}\frac{\partial^2 u}{\partial y^2}\right)d\Omega_1\right|\le\frac{\varepsilon}{2}\int_{\Omega_1}\left(\frac{\partial^2 u}{\partial x^2}+\frac{1-\nu}{2}\frac{\partial^2 u}{\partial y^2}\right)d\Omega_1\\&+\frac{k_1}{\varepsilon}\left\|\tilde{L}^{1/2}\overline{\omega}\right\|^2_{(L_2(\Omega_1))^3}\qquad(x\leftrightarrow y,\quad u\leftrightarrow v,\quad l_1\leftrightarrow l_2,\quad k_1\leftrightarrow k_2),\end{aligned} \tag{3.81}$$

$$\left|\int_{\Omega_1}l_{32}w\,\nabla^2\nabla^2 w\,d\Omega_1\right|\le\frac{\varepsilon}{4}\int_{\Omega_1}(\nabla^2\nabla^2 w)^2\,d\Omega_1+\frac{1}{\varepsilon}\left\|l_{31}w\right\|^2_{L_2(\Omega_1)}, \tag{3.82}$$

$$\left\| l_{31} w \right\|^2_{L_2(\Omega_1)} \le \frac{\varepsilon^2}{4} \int\limits_{\Omega_1} (\nabla^2 \nabla^2 w)^2 \, d\Omega_1 + \frac{k_{31}}{\varepsilon^2} \left\| \tilde{L}^{1/2} \overline{\omega} \right\|^2_{(L_2(\Omega_1))^3}, \tag{3.83}$$

$$\left| \int\limits_{\Omega_1} l_{32} \overline{\omega} \, \nabla^2 \nabla^2 w \, d\Omega_1 \right| \le \frac{\varepsilon}{2} \int\limits_{\Omega_1} (\nabla^2 \nabla^2 w)^2 \, d\Omega_1 + \frac{k_{32}}{\varepsilon} \left\| \tilde{L}^{1/2} \overline{\omega} \right\|^2_{(L_2(\Omega_1))^3}, \tag{3.84}$$

$$\begin{aligned} &\frac{(1+\nu)E}{8(1-\nu)} \left| \int\limits_{\Omega_1} \left[\left(\frac{\partial h}{\partial x} - \frac{\partial h}{\partial y} \right) \left(\frac{\partial u}{\partial x} \frac{\partial^2 u}{\partial x \partial y} - \frac{\partial v}{\partial y} \frac{\partial^2 v}{\partial x \partial y} \right) \right] d\Omega_1 \right| \\ &\le \frac{\varepsilon}{2} \int\limits_{\Omega_1} \left[\left(\frac{\partial^2 u}{\partial x^2} + \frac{1-\nu}{2} \frac{\partial^2 u}{\partial y^2} \right)^2 + \left(\frac{\partial^2 v}{\partial y^2} + \frac{1-\nu}{2} \frac{\partial^2 v}{\partial x^2} \right)^2 \right] d\Omega_1 \\ &+ \frac{k_4}{\varepsilon} \left\| \tilde{L}^{1/2} \overline{\omega} \right\|^2_{(L_2(\Omega_1))^3}, \end{aligned} \tag{3.85}$$

where ε is an arbitrary positive number, and k_1, k_2, k_{31}, k_{32}, k_4 are positive constants independent of both ε and $\overline{\omega}$. From (3.82) and (3.83), we obtain

$$\left| \int\limits_{\Omega_1} l_{31} \overline{\omega} \, \nabla^2 \nabla^2 w \, d\Omega_1 \right| \le \frac{\varepsilon}{2} \int\limits_{\Omega_1} (\nabla^2 \nabla^2 w)^2 \, d\Omega_1 + \frac{k_{31}}{\varepsilon^3} \left\| \tilde{L}^{1/2} \overline{\omega} \right\|^2_{(L_2(\Omega_1))^3}. \tag{3.86}$$

The union of the estimates (3.81), (3.84)–(3.86) gives us

$$|I| \le \varepsilon \left\| \tilde{L} \overline{\omega} \right\|^2_{(L_2(\Omega_1))^3} + \left(\frac{m_2}{\varepsilon} + \frac{k_{31}}{\varepsilon^3} \right) \left\| \tilde{L}^{1/2} \overline{\omega} \right\|^2_{(L_2(\Omega_1))^3}, \tag{3.87}$$

where $m_2 = k_1 + k_2 + k_{32} + k_4$. Taking into account (3.80), (3.87), we obtain

$$\begin{aligned} &(L\overline{\omega}, \tilde{L}\overline{\omega})_{(L_2(\Omega_1))^3} \\ &\ge (m_1 - \varepsilon) \left\| \tilde{L} \overline{\omega} \right\|^2_{(L_2(\Omega_1))^3} - \left(\frac{m_2}{\varepsilon} + \frac{k_{31}}{\varepsilon^3} \right) \left\| \tilde{L}^{1/2} \overline{\omega} \right\|^2_{(L_2(\Omega_1))^3}, \end{aligned} \tag{3.88}$$

where $\overline{\omega}$ is an arbitrary vector function from $[(\overset{\circ}{W}{}^2_2(\Omega_1))^2 \times \overset{\circ}{W}{}^4_2(\Omega_1)] \cap [(C^2(\overline{\Omega_1}))^2 \times C^4(\overline{\Omega_1})]$, and ε is an arbitrary positive number.

With the help of the Korn inequalities [244], the following estimate can be obtained

$$\left\| L^{1/2} \overline{\omega} \right\|^2_{(L_2(\Omega_1))^3} \ge m_L \left\| \tilde{L}^{1/2} \overline{\omega} \right\|^2_{(L_2(\Omega_1))^3}, \tag{3.89}$$

where $\overline{\omega}$ is an arbitrary vector function from $(\overset{\circ}{W}{}^1_2(\Omega_1))^2 \times \overset{\circ}{W}{}^2_2(\Omega_1)$, and m_L is a positive constant independent of $\overline{\omega}$.

We put $\varepsilon = \varepsilon_0 = \dfrac{m_1}{2}$ into (3.88) and take

$$P_L = \frac{1}{m_L}\left(\frac{m_2}{\varepsilon_0} + \frac{k_{31}}{\varepsilon_0^3}\right); \tag{3.90}$$

then, owing to (3.64), (3.89), we obtain

$$\begin{aligned}
&(L\overline{\omega}, L_0\overline{\omega})_{(L_2(\Omega_1))^3} = (L\overline{\omega}, \tilde{L}\overline{\omega})_{(L_2(\Omega_1))^3} + P_L\left\|\tilde{L}^{1/2}\overline{\omega}\right\|_{(L_2(\Omega_1))^3} \\
&\geq (L\overline{\omega}, \tilde{L}\overline{\omega})_{(L_2(\Omega_1))^3} + m_L P_L\left\|\tilde{L}^{1/2}\overline{\omega}\right\|^2_{(L_2(\Omega_1))^3} \\
&\geq (m_1 - \varepsilon_0)\left\|\tilde{L}\overline{\omega}\right\|^2_{(L_2(\Omega_1))^3} + \left(m_L P_L - \frac{m_2}{\varepsilon_0} - \frac{k_{31}}{\varepsilon_0^3}\right)\left\|\tilde{L}^{1/2}\overline{\omega}\right\|^2_{(L_2(\Omega_1))^3} \\
&= \frac{m_1}{2}\left\|\tilde{L}\overline{\omega}\right\|^2_{(L_2(\Omega_1))^3}.
\end{aligned} \tag{3.91}$$

Because L_0 and $\tilde{L}$ are convergent, the following inequality is obtained from (3.91):

$$(L\overline{\omega}, L_0\overline{\omega})_{(L_2(\Omega_1))^3} \geq \beta_L\left\|\tilde{L}_0\overline{\omega}\right\|^2_{(L_2(\Omega_1))^3}, \tag{3.92}$$

where $\overline{\omega}$ is an arbitrary vector function from $[(\mathring{W}_2^2(\Omega_1))^2 \times \mathring{W}_2^4(\Omega_1)] \cap [(C^2(\overline{\Omega_1}))^2 \times C^4(\overline{\Omega_1})]$, and β_L is a positive constant independent of $\overline{\omega}$. The inequality (3.92) implies that when the constant P_L is chosen to have the value given in (3.90), the operators L and L_0 defined by (2.88), (3.64) form an acute angle, and the constant $\gamma(L, L_0)$ occurring in an inequality of type (2.107) can be estimated from

$$\gamma(L, L_0) \geq \beta_L. \tag{3.93}$$

Taking into account (3.92), we obtain

$$\left\|L_0\overline{\omega}\right\|_{(L_2(\Omega_1))^3} \leq \beta_L^{-1}\left\|L\overline{\omega}\right\|_{(L_2(\Omega_1))^3}, \tag{3.94}$$

where $\overline{\omega}$ is an arbitrary vector function from $[(\mathring{W}_2^2(\Omega_1))^2 \times \mathring{W}_2^4(\Omega_1)] \cap [(C^2(\overline{\Omega_1}))^2 \times C^4(\overline{\Omega_1})]$. It is clear, by taking into account (3.75) and (3.94), that the norms $\|L \cdot \|_{(L_2(\Omega_1))^3}$ and $\|L_0 \cdot \|_{(L_2(\Omega_1))^3}$ are equivalent. Therefore, the spaces of existence of the operators L and L_0 overlap because they have the same linear manifold $[(\mathring{W}_2^2(\Omega_1))^2 \times \mathring{W}_2^4(\Omega_1)] \cap [(C^2(\overline{\Omega_1}))^2 \times C^4(\overline{\Omega_1})]$ in equivalent norms, which means that the operators L and L_0 are convergent. To conclude, it has been shown, that when the constant P_L is chosen in the form (3.90), the operator L_0 defined by (3.64) is convergent and forms an acute angle with the operator L defined by the relation (2.88).

For the problem under consideration the space H_1 overlaps with $(L_2(\Omega_1))^3$, and therefore the inequality (3.75) can be represented in the form

$$\left\| LL_0^{-1} \right\|_{H_1 \to H_1} \le \alpha_L. \tag{3.95}$$

According to (3.93), (3.95), the constant C_L defined by the first relation of (2.124) can be estimated in the following manner:

$$C_L \le \alpha_L \beta_L^{-1}. \tag{3.96}$$

An auxiliary operator K_0 with a known spectrum can be constructed in the following way. Let $\overline{\Phi} = \{\Phi_1, \Phi_2, \Phi_3\}$ be a reciprocally unique mapping of the parallelepiped $\tilde{\Omega}_2 = \Omega_1 \times \left(-\frac{1}{2}, \frac{1}{2}\right)$ into the space Ω_2, defined by the relations

$$\Phi_1(\overline{x}) = x, \quad \Phi_2(\overline{x}) = y, \quad \Phi_3(\overline{x}) = zh(x,y), \quad \overline{x} = (x,y,z).$$

We now introduce the differential operator $\tilde{K}$, defined on $\mathring{W}_2^2(\Omega_2)$ by the relation

$$\tilde{K}\theta = -\nabla^2 \left(\theta \circ \overline{\Phi}\right) \circ \overline{\Phi}^{-1}, \tag{3.97}$$

and a new scalar product (equivalent to the usual scalar product) in the space $L_2(\Omega_2)$:

$$\begin{aligned} (\theta,\eta)_{L_2(\Omega_2)} &= \int\limits_{\tilde{\Omega}_2} (\theta \circ \overline{\Phi})(\eta \circ \overline{\Phi})\, d\tilde{\Omega}_2 \\ &\equiv \int\limits_{\Omega_2} \theta\eta \left| \det \frac{D(\overline{\Phi}^{-1})}{D\overline{x}} \right| d\Omega_2 \equiv \int\limits_{\Omega_2} \frac{1}{h} \theta\eta\, d\Omega_2. \end{aligned} \tag{3.98}$$

The operator $\tilde{K}$ defined by (3.97) is self-adjoint and positive definite in relation to the scalar product (3.98) (because the operator $-\nabla^2$ has the same properties in relation to the usual scalar product of a space $L_2(\tilde{\Omega}_2)$). We introduce the notation

$$w_{klm}^{(2)}(x,y,z) = w_k(x) w_l(y) w_m(z), \tag{3.99}$$

where $w_k(x)$, $w_l(y)$, $w_m(z)$ are the functions as defined by (3.65). Let us assume that

$$K_0\theta = \tilde{K}\theta + p_k\theta, \tag{3.100}$$

where p_k is a sought positive constant. The eigenfunctions η_{klm} of the operator (3.100), from $\mathring{W}_2^2(\Omega_2)$, are known and have the form

$$\eta_{klm}(x,y,z) = w_{klm}^{(2)}\left(\frac{x}{a}, \frac{y}{b}, \frac{z}{h(x,y)}\right), \quad k,l,m = 1,2,\ldots . \tag{3.101}$$

The corresponding eigenvalues $æ_{klm}$ of the operator (3.100) are defined by the relations

$$æ_{klm} = \pi^2 \left(\frac{k^2}{a^2} + \frac{l^2}{b^2} + m^2 \right) + p_k, \quad k, l, m = 1, 2, \ldots . \tag{3.102}$$

For the problem under consideration $H_2 = L_2(\Omega_2)$. We assume that the scalar product (3.98) is defined for the space $L_2(\Omega_2)$. It turns out that the operator K defined on $\mathring{W}_2^2(\Omega_2)$ by (2.89) is not self-adjoint in relation to the new scalar product. Let us multiply the two parts of the differential equation (2.32) by $T_0^{-1} h(x,y)$ and define the operator K again, i.e., instead of (2.89), we take

$$K\theta = -\frac{\lambda_q}{T_0} h \nabla^2 \theta, \qquad D(K) = \mathring{W}_2^2(\Omega_2). \tag{3.103}$$

Now the operator K is defined as self-adjoint and positive in relation to the scalar product (3.98). Using the method described in [246], the following estimate can be obtained:

$$\left\| K K_0^{-1} \right\|_{H_1 \to H_1} \le \alpha_K, \quad \gamma(K, K_0) \ge \beta_K, \tag{3.104}$$

for an appropriate choice of p_k (α_K, β_K are positive constants). This implies that the operators K_0 and K defined by the relations (3.100) and (3.103) are convergent and form an acute angle, and the constant c_K defined by the second relation of (2.124) can be estimated in the following way:

$$c_K \le \alpha_K \beta_K^{-1}. \tag{3.105}$$

The operators I_1, M for the problem under consideration are defined by the relations (2.87) and (2.90), correspondingly. Assuming that the scalar product (3.98) is determined for the space $H_2 = L_2(\Omega_2)$, the operators I_2, N can be defined using a differential equation obtained by multiplication of the two parts of (2.32) by $T_0^{-1} h(x,y)$:

$$I_2\theta = \frac{c_\varepsilon (1+\varepsilon) h}{T_0} \theta,$$

$$N\overline{\omega} = \frac{E\alpha_T h}{1-\nu} \left[\frac{\partial u}{\partial x} + \frac{\partial v}{\partial y} - (k_x + k_y) w - z \nabla^2 w \right],$$

$$D(N) = (\mathring{W}_2^1(\Omega_1))^2 \times \mathring{W}_2^2(\Omega_1).$$

It is clear that, for an arbitrary vector function $\overline{\omega} \in (L_2(\Omega_1))^3$, the following inequality holds:

$$(I_1 \overline{\omega}, \overline{\omega})_{(L_2(\Omega_1))^3} \le \rho H_0 \left\| \overline{\omega} \right\|^2_{(L_2(\Omega_1))^3}, \tag{3.106}$$

where $H_0 = \max_{(x,y)\in\overline{\Omega_1}} h(x,y)$. For an arbitrary function $\theta \in L_2(\Omega_2)$, the following inequality holds:

$$(I_2\theta,\theta)_{L_2(\Omega_1)} \leq c_\varepsilon(1+\varepsilon)T_0^{-1}H_0\left\|\theta\right\|^2_{L_2(\Omega_1)}, \tag{3.107}$$

where the constant H_0 in the inequalities (3.106) and (3.107) cannot be replaced by a smaller one. It follows that for the problem under consideration the norms of the operators $I_1^{1/2}$, $I_2^{1/2}$ are defined by the relations

$$\left\|I_1^{1/2}\right\|_{H_1\to H_1} = \rho^{1/2}H_0^{1/2}, \quad \left\|I_2^{1/2}\right\|_{H_2\to H_2} = c_\varepsilon^{1/2}(1+\varepsilon)^{1/2}T_0^{-1/2}H_0^{1/2}. \tag{3.108}$$

Using some methods due to Korn [158], the Poincaré–Friedrichs inequalities [139] the second fundamental inequality for elliptic operators [139] the and Cauchy–Buniakowski inequalities, the following estimates can be obtained:

$$\begin{aligned}
(L_0\overline{\omega},\overline{\omega})_{(L_2(\Omega_1))^3} &\leq A_L^2(L\overline{\omega},\overline{\omega})_{(L_2(\Omega_1))^3},\\
(K_0\theta,\theta)_{L_2(\Omega_2)} &\leq A_K^2(K\theta,\theta)_{L_2(\Omega_2)},\\
\left\|M\theta\right\|_{(L_2(\Omega_1))^3} &\leq B_1\left\|K\theta\right\|_{L_2(\Omega_2)},\\
\left\|N\overline{\omega}\right\|^2_{L_2(\Omega_2)} &\leq B_2^2(L\overline{\omega},\overline{\omega})_{(L_2(\Omega_1))^3},
\end{aligned}$$

where $\overline{\omega}$ is an arbitrary vector function from $D(L)$; θ is an arbitrary function from $D(K)$; A_L, B_2 are positive constants independent of $\overline{\omega}$; and A_K, B_1 are positive constants independent of θ. The estimates given above can be represented in the following way:

$$\left\|L_0^{1/2}L^{-1/2}\right\|_{H_1\to H_1} \leq A_L, \quad \left\|K_0^{1/2}K^{-1/2}\right\|_{H_2\to H_2} \leq A_K, \tag{3.109}$$

$$\left\|MK^{-1}\right\|_{H_2\to H_1} \leq B_1, \quad \left\|NL^{-1/2}\right\|_{H_1\to H_2} \leq B_2. \tag{3.110}$$

Because

$$\left\|L^{-1/2}L_0^{1/2}\right\|_{H_1\to H_1} = \left\|(L^{-1/2}L_0^{1/2})^*\right\|_{H_1\to H_1} = \left\|L_0^{1/2}L^{-1/2}\right\|_{H_1\to H_1},$$

and, analogously,

$$\left\|K^{-1/2}K_0^{1/2}\right\|_{H_2\to H_2} = \left\|K_0^{1/2}K^{-1/2}\right\|_{H_2\to H_2},$$

(3.109) implies the following inequalities:

$$\left\|L^{-1/2}L_0^{1/2}\right\|_{H_1\to H_1} \leq A_L, \quad \left\|K^{-1/2}K_0^{1/2}\right\|_{H_2\to H_2} \leq A_K. \tag{3.111}$$

The free term g_1 appearing in the first equation of (3.1) is (for the case under consideration) defined by the first relation of (2.93). The free term g_2 of the second equation of (3.1) can be obtained using a differential equation obtained by multiplication of the two sides of (2.32) by $T_0^{-1}h(x,y)$, i.e. we take $g_2 = T_0^{-1}hq_2$.

Using the relations (3.108) and the inequalities (3.96), (3.105), (3.110), (3.111), we can now estimate from above the constants c_i $(i = 1, \dots, 5)$ defined by the relations (3.5), the constants d_j $(j = 1, 2, 3)$ defined by the relations (3.36) and the varying quantities e_1, e_2 defined by (3.37), (3.38). By substituting these estimates into the right-hand sides of the inequalities (3.3), (3.23), we then obtain prior estimates of the errors of the Bubnov–Galerkin method for the system of differential equations (2.30)–(2.32) with homogeneous initial conditions (2.33)–(2.35) and boundary conditions. Let present, as an example, one of those estimates, obtained from (3.3):

$$\max_{0\le t\le T}\left\{\int\limits_{\Omega_1}\left[\rho h\left(\left(\frac{d}{dt}\Delta u_n\right)^2+\left(\frac{d}{dt}\Delta v_n\right)^2+\left(\frac{d}{dt}\Delta w_n\right)^2\right)\right.\right.$$
$$+\frac{Eh}{1+\nu}\left(\left(\frac{d}{dx}\Delta u_n-k_x\Delta w_n\right)^2+\left(\frac{d}{dy}\Delta v_n-k_y\Delta w_n\right)^2\right.$$
$$\left.+\frac{1}{2}\left(\frac{d}{dy}\Delta u_n+\frac{d}{dx}\Delta v_n\right)^2\right)+\frac{Eh\nu}{1-\nu^2}\left(\frac{d}{dx}\Delta u_n+\frac{d}{dy}\Delta v_n\right.$$
$$\left.-(k_x+k_y)\Delta w_n\right)^2+D(1-\nu)\left(\left(\frac{d^2}{dx^2}\Delta w_n\right)^2+\left(\frac{d^2}{dy^2}\Delta w_n\right)^2\right.$$
$$\left.\left.+2\left(\frac{d^2}{dxdy}\Delta w_n\right)^2\right)+D\nu(\nabla^2\Delta w_n)^2\right]d\Omega_1$$
$$\left.+\frac{c_\varepsilon(1+\varepsilon)}{T_0}\int\limits_{\Omega_1}(\Delta\theta_n)^2\,d\Omega_2+\frac{\lambda_q}{T_0}\int\limits_0^t\int\limits_{\Omega_1}\left|\nabla\Delta\theta_n\right|^2 d\Omega_2\,d\tau\right\}$$
$$\le 4A_LT^{1/2}(1+\alpha_L\beta_L^{-1})(M_g(T))^{1/2}\left\{\left[\int\limits_0^t\int\limits_{\Omega_1}(p_1^2+p_2^2+q_1^2)d\Omega_1\,d\tau\right]^{1/2}\right.$$
$$+B_1\alpha_k\beta_k^{-1}\left[\int\limits_0^T\int\limits_{\Omega_2}hq_2^2\,d\Omega_2\,d\tau\right]^{1/2}$$
$$+\left[B_1^2\alpha_k^2\beta_k^{-2}(c_\varepsilon(1+\varepsilon)H_0T_0^{-1}+B_2^2)+\rho H_0\right]^{1/2}$$

$$\left.\times T^{1/2}(M_g(T))^{1/2}\right\}\tilde{\lambda}_{i(n)+1}^{-1/2}+A_k^2(1+\alpha_k\beta_k^{-1})^2\left\{\left[\int\limits_0^T\int\limits_{\Omega_2} hq_2^2\,d\Omega_2\,d\tau\right]^{1/2}\right.$$

$$\left.+\left[c_\varepsilon(1+\varepsilon)H_0T_0^{-1}+B_2^2\right]^{1/2}T^{1/2}(M_g(T))^{1/2}\right\}^2\tilde{æ}_{j(n)+1}^{-1} \tag{3.112}$$

$$\equiv 0(\max(\tilde{\lambda}_{i(n)+1}^{-1/2},\tilde{æ}_{j(n)+1}^{-1})),$$

where

$$\Delta u_n=u_n-u,\ \Delta v_n=v_n-v,\ \Delta w_n=w_n-w,\ \Delta\theta_n=\theta_n-\theta,$$

$\{u,v,w,\theta\}$ is the exact ("classical") solution to the problem under consideration, and $\{u_n,v_n,w_n,\theta_n\}$ is its approximate solution found using the Bubnov–Galerkin method. The latter approximate solution is given by

$$\overline{\omega_n}(x,y,t)=\{u_n(x,y,t),v_n(x,y,t),w_n(x,y,t)\}$$
$$=\sum_{\overline{v}_{kl}^{(j)}\in H_1^n}\alpha_{kl}^{(j)}(t)\overline{v}_{kl}(x,y),$$
$$\theta(x,y,z,t)=\sum_{\eta_{klm}\in H_2^n}\beta_{klm}(t)\eta_{klm}(x,y,z),$$
$$\tilde{\lambda}_{i(n)+1}=\min\{\lambda_{kl}^{(j)}:\overline{v}_{kl}^{(j)}\overline{\in}H_1^n\},$$
$$\tilde{æ}_{j(n)+1}=\min\{æ_{klm}:\eta_{klm}\overline{\in}H_2^n\},$$

where H_1^n, H_2^n are finitely measured subspaces of $(L_2(\Omega_1))^3$ and $L_2(\Omega_2)$, spanned by $i(n)$ basis vector functions $\overline{v}_{kl}^{(j)}$ (defined either by the dependence of the mechanical boundary conditions, by the relations (3.69)–(3.71), or by the relations (3.74)) and on $j(n)$ basis functions η_{klm} defined by the relations (3.101); $\lambda_{kl}^{(j)}$, $æ_{klm}$ are the eigenvalues of the auxiliary operators L_0, K_0, defined by the relations (3.72), (3.73), (3.102); and

$$M_g(T)=M\left[\frac{1}{\rho}\int\limits_{\Omega_1}\frac{1}{h}\left(p_1^2\big|_{t=0}+p_2^2\big|_{t=0}+q_1^2\big|_{t=0}\right)d\Omega_1\right.$$

$$+\frac{T_0}{c_\varepsilon(1+\varepsilon)}\int\limits_{\Omega_2}q_2^2\big|_{t=0}\,d\Omega_2,\frac{1}{\rho}\int\limits_0^T\int\limits_{\Omega_1}\frac{1}{h}\left[\left(\frac{\partial p_1}{\partial t}\right)^2+\left(\frac{\partial p_2}{\partial t}\right)^2\right. \tag{3.113}$$

$$\left.\left.+\left(\frac{\partial q_1}{\partial t}\right)^2\right]d\Omega_1\,dt+\frac{T_0}{c_\varepsilon(1+\varepsilon)}\int\limits_0^T\int\limits_{\Omega_2}\left(\frac{\partial q_2}{\partial t}\right)^2d\Omega_2\,dt,T\right],$$

where

$$\begin{aligned} M[w,X,T] &= \min_{\delta>0}\left[\left(w+\frac{1}{\delta}X\right)\exp(\delta T)\right] \\ &\equiv \left\{w+\frac{1}{2}\left[(X^2T^2+4wXT)^{1/2}+XT\right]\right\} \\ &\times \exp\left\{2XT\times\right. \\ &\times \left.\left[(X^2T^2+4wXT)^{1/2}+XT\right]^{-1}\right\}. \end{aligned} \tag{3.114}$$

Let us remember, that owing to the known theorem about the boundedness of the operation $W_2^2(\Omega) \to C(\overline{\Omega})$, where Ω is a two-dimensional space with a sufficiently smooth boundary [209], an estimate of the convergence $w_n \to w$ in a norm uniform with respect to the variables x, y, t can be easily obtained from (3.112).

3.3 Case of a Simply Supported Plate Within the Kirchhoff Model

Let us consider a coupled thermoelastic problem for a simply supported plate ($k_x = k_y = 0$) with a constant thickness ($h = const$) in the framework of the Kirchhoff–Love kinematic model. This problem is related to our earlier general considerations of the problem described by (2.30)–(2.35), (2.38) and (2.51). In addition, we assume that (1) the mechanical and heat loads are constant in time, and (2) the plate under consideration is square ($a = b$). The last assumption is not essential and is made in order to simplify the prior estimate for the Bubnov–Galerkin method (the considerations below can be easily extended to the more general case of a rectangular plate).

The following nondimensional quantities are introduced:

$$\begin{aligned} &\overline{x} = \frac{x}{a}, \quad \overline{y} = \frac{y}{a}, \quad \overline{z} = \frac{z}{h}, \quad \overline{t} = \frac{\lambda_q}{c_\varepsilon(1+\varepsilon)h^2}t, \\ &\lambda = \frac{h}{a}, \quad \overline{w} = \frac{w}{h}, \quad \overline{\theta} = \frac{\alpha_T a^2}{h^2}\theta, \\ &\overline{q_1} = \frac{a^4}{Eh^4}q_1, \quad \overline{q_2} = \frac{\alpha_t a^2}{\lambda_q}q_2, \quad \overline{w_0} = \frac{w_0}{h}, \\ &\overline{w_1} = \frac{c_\varepsilon(1+\varepsilon)h}{\lambda_q}w_1, \quad \overline{\theta_0} = \frac{\alpha_T a^2}{h^2}\theta_0, \quad \overline{T} = \frac{\lambda_q}{c_\varepsilon(1+\varepsilon)h^2}T, \\ &c = \frac{\rho\lambda_q^2}{c_\varepsilon^2(1+\varepsilon)^2Eh^2}, \quad \beta = \frac{E\alpha_T^2T_0}{(1-\nu)c_\varepsilon(1+\varepsilon)}. \end{aligned} \tag{3.115}$$

The initial–boundary value problem considered here has the following form:

$$\frac{c}{\lambda^4}\frac{\partial^2 \overline{w}}{\partial \overline{t}^2} + \frac{1}{12(1-\nu^2)}\overline{\nabla}^2\overline{\nabla}^2\overline{w} + \frac{1}{1-\nu}\overline{\nabla}^2\int_{-\frac{1}{2}}^{\frac{1}{2}}\overline{\theta}\overline{z}\,d\overline{z} = \overline{q_1}(\overline{x},\overline{y}), \tag{3.116}$$

$$\frac{\partial\overline{\theta}}{\partial\overline{t}} - \left(\lambda^2\frac{\partial^2\overline{\theta}}{\partial\overline{x}^2} + \lambda^2\frac{\partial^2\overline{\theta}}{\partial\overline{y}^2} + \frac{\partial^2\overline{\theta}}{\partial\overline{z}^2}\right) - \beta\overline{z}\frac{\partial}{\partial\overline{t}}(\overline{\nabla}^2\overline{w}) = \overline{q_2}(\overline{x},\overline{y},\overline{z}), \tag{3.117}$$

$$\begin{gathered} \overline{w}|_{\overline{x}=\pm\frac{1}{2}} = \overline{w}|_{\overline{y}=\pm\frac{1}{2}} = \left.\frac{\partial^2\overline{w}}{\partial\overline{x}^2}\right|_{\overline{x}=\pm\frac{1}{2}} = \left.\frac{\partial^2\overline{w}}{\partial\overline{y}^2}\right|_{\overline{y}=\pm\frac{1}{2}} = 0, \\ \overline{\theta}|_{\overline{x}=\pm\frac{1}{2}} = \overline{\theta}|_{\overline{y}=\pm\frac{1}{2}} = \overline{\theta}|_{\overline{z}=\pm\frac{1}{2}} = 0, \quad 0 \le \overline{t} \le \overline{T}, \end{gathered} \tag{3.118}$$

$$\begin{gathered} \overline{w}|_{\overline{t}=0} = \overline{w}_0(\overline{x},\overline{y}), \quad \left.\frac{\partial\overline{w}}{\partial\overline{t}}\right|_{\overline{t}=0} = \overline{w}_1(\overline{x}\overline{y}), \\ \overline{\theta}|_{\overline{t}=0} = \overline{\theta}_0(\overline{x},\overline{y},\overline{z}), \quad -\frac{1}{2} \le \overline{x},\overline{y},\overline{z} \le \frac{1}{2}. \end{gathered} \tag{3.119}$$

In the differential equations (3.116), (3.117), the following notation has been used:

$$\overline{\nabla}^2 = \frac{\partial^2}{\partial\overline{x}^2} + \frac{\partial^2}{\partial\overline{y}^2}.$$

From now on, in this chapter and in Chap. 4, the bars above the nondimensional quantities will be omitted.

Remark 3.6 *The thermal-isolation condition* $\left.\frac{\partial\theta}{\partial z}\right|_{z=\pm\frac{1}{2}} = 0$ *can be used instead of the boundary condition* $\theta|_{z=\pm\frac{1}{2}} = 0$. *All of the analysis below can be applied in the case of this condition with insignificant changes.*

The following notation is introduced:

$$\Omega_1 = \left(-\frac{1}{2},\frac{1}{2}\right)^2, \quad \Omega_2 = \left(-\frac{1}{2},\frac{1}{2}\right)^3, \quad H_j = L_2(\Omega_j) \quad (j=1,2), \tag{3.120}$$

$$I_1 w = \frac{c}{\lambda^4}w, \quad I_2\theta = \frac{1}{\beta(1-\nu)}\theta, \tag{3.121}$$

$$\begin{gathered} Lw = \frac{1}{12(1-\nu^2)}\nabla^2\nabla^2 w, \\ K\theta = -\frac{1}{\beta(1-\nu)}\left(\lambda^2\frac{\partial^2\theta}{\partial x^2} + \lambda^2\frac{\partial^2\theta}{\partial y^2} + \frac{\partial^2\theta}{\partial z^2}\right), \end{gathered} \tag{3.122}$$

$$M\theta = \frac{1}{1-\nu}\nabla^2\int_{-\frac{1}{2}}^{\frac{1}{2}}\theta z\,dz, \quad Nw = -\frac{1}{1-\nu}z\nabla^2 w, \tag{3.123}$$

$$g_1 = q_1, \quad g_2 = \frac{1}{\beta(1-\nu)}q_2. \tag{3.124}$$

The space of the functions $w(x, y)$ from $W_2^2(\Omega_1)$ that are equal to zero on the boundary $\partial\Omega_1$ is denoted by $\overset{\circ}{W}{}_2^2(\Omega_1)$, and the space of the functions $w(x, y)$ from $W_2^4(\Omega_1)$ satisfying the mechanical boundary conditions (3.118) is denoted by $\overset{\circ}{W}{}_2^4(\Omega_1)$. The notations $\overset{\circ}{W}{}_2^1(\Omega_2)$, $\overset{\circ}{W}{}_2^2(\Omega_2)$ introduced in Sect. 2.1.4 are used for the manifolds of functions $\theta(x, y, z)$ from $W_2^1(\Omega_2)$ and $W_2^2(\Omega_2)$, respectively, that satisfy the boundary conditions (3.118). Let $D(L) = \overset{\circ}{W}{}_2^4(\Omega_1)$ and $D(K) = \overset{\circ}{W}{}_2^2(\Omega_2)$; we then obtain $D(L^{1/2}) = \overset{\circ}{W}{}_2^2(\Omega_1)$, $D(K^{1/2}) = \overset{\circ}{W}{}_2^1(\Omega_2)$. Using the notations described here, the problem (3.116)–(3.119) can be represented in the form (3.42), where, instead of w_0, w_1, θ_0, the initial data (3.119) appear.

Let n_1, n_2 be natural numbers $n = \{n_1, n_2\}$. Let us denote by H_1^n the linear neighbourhood of the function $\{w_{lkl}^{(1,3)} | 1 \le k,\ l \le n_1\}$, where $w_{kl}^{(1,3)}(x, y)$ are the basis functions defined by (3.68). Similarly, let us denote by H_2^n the linear neighbourhood of the function $\{w_{klm}^{(2)} | 1 \le k,\ l \le n_1,\ 1 \le m \le n_2\}$, where $w_{klm}^2(x, y, z)$ are the basis functions defined by (3.99).

The following conditions are applied to the given functions g_1, g_2, w_0, w_1, θ_0:

$$\begin{aligned} g_1 \in L_2(\Omega_1), \quad g_2 \in L_2(\Omega_2), \quad w_0 \in \overset{\circ}{W}{}_2^4(\Omega_1), \\ w_1 \in \overset{\circ}{W}{}_2^2(\Omega_1), \quad \theta_0 \in \overset{\circ}{W}{}_2^2(\Omega_2). \end{aligned} \tag{3.125}$$

Therefore, for the problem (3.116)–(3.119), all conditions required by Theorem 3.3 are satisfied and the convergence velocity of the Bubnov–Galerkin method is characterized by the inequalities (3.47), (3.48). We now show explicitly how the series s_n appears in these inequalities.

For arbitrary functions $w \in H_1^n$, $\theta \in \overset{\circ}{W}{}_2^2(\Omega_2)$, we have

$$\begin{aligned} &\left|(R_{H_2^n} N w, \theta)_{L_2(\Omega_2)}\right| \\ &= \frac{1}{\pi(1-\nu)} \left| \int_{\Omega_2} \sum_{k=[\frac{n_2}{2}]+1}^{\infty} \frac{(-1)^{k+1}}{k} \times \sin 2\pi k z \, \nabla^2 w \theta \, d\Omega_2 \right| \\ &= \frac{1}{2\pi^2(1-\nu)} \left| \int_{\Omega_2} \sum_{k=[\frac{n_2}{2}]+1}^{\infty} \frac{(-1)^{k+1}}{k^2} \times \cos 2\pi k z \, \nabla^2 w \frac{\partial \theta}{\partial z} \, d\Omega_2 \right| \\ &\le \frac{1}{2\pi^2(1-\nu)} \left[\int_{-\frac{1}{2}}^{\frac{1}{2}} \left(\sum_{k=[\frac{n_2}{2}]+1}^{\infty} \frac{(-1)^{k+1}}{k^2} \times \cos 2\pi k z \right)^2 dz \right]^{1/2} \\ &\quad \left[\int_{\Omega_1} (\nabla^2 w)^2 \, d\Omega_1 \right]^{1/2} \times \left[\int_{\Omega_2} \left(\frac{\partial \theta}{\partial z} \right)^2 d\Omega_2 \right]^{1/2} \\ &= \frac{1}{\pi^2} \left[\frac{3\beta(1+\nu)}{2} \sigma_{[\frac{n_2}{2}]+1} \right]^{1/2} \left\| L^{1/2} w \right\|_{L_2(\Omega_1)} \left\| K^{1/2} \theta \right\|_{L_2(\Omega_2)}, \end{aligned} \tag{3.126}$$

where

$$\sigma_m^{(4)} = \sum_{k=m}^{\infty} \frac{1}{k^4} \equiv \frac{\pi^4}{90} - \sum_{k=1}^{m-1} \frac{1}{k^4}, \tag{3.127}$$

and $\left[\frac{n^2}{2}\right]$ denotes the largest number not larger than $\frac{n^2}{2}$. In (3.47), (3.48), s_n is an arbitrary series of which the elements approach zero and satisfy the inequality (3.45). From (3.126), we conclude that we can take

$$s_n = \frac{1}{\pi^2}\left[\frac{3\beta(1+\nu)}{2}\sigma^{(4)}_{[\frac{n^2}{2}]+1}\right]^{1/2}. \tag{3.128}$$

Let us formulate an algorithm for calculating

$$\left\|L^{-1/2}R_{H_1^n}g_1\right\|^2_{H_1}, \quad \left\|K^{-1/2}R_{H_2^n}g_2\right\|^2_{H_2},$$

which occur in the righ-hand sides of the inequalities (3.47), (3.48). Taking into account (3.120), (3.122), (3.124), we obtain:

$$\left\|L^{-1/2}R_{H_1^n}g_1\right\|^2_{H_1} = 12(1-\nu^2)\left\|(\nabla^2)^{-1}R_{H_1^n}g_1\right\|^2_{L_2(\Omega_1)}, \tag{3.129}$$

$$\left\|K^{-1/2}R_{H_2^n}g_2\right\|_{H_2} = \frac{1}{\beta^{1/2}(1-\nu)^{1/2}}\left\|\left(-\lambda^2\frac{\partial^2}{\partial x^2} - \lambda^2\frac{\partial^2}{\partial y^2} - \frac{\partial^2}{\partial z^2}\right)^{-1/2}R_{H_2^n}q_2\right\|_{L_2(\Omega_2)}. \tag{3.130}$$

By expanding the function $(\nabla^2)^{-1}R_{H_1^n}q_1$ into a Fourier series with respect to the basis $\{w_{kl}^{(1,3)}\}$ and expanding the functions $\left(-\lambda^2\frac{\partial^2}{\partial x^2} - \lambda^2\frac{\partial^2}{\partial y^2} - \frac{\partial^2}{\partial z^2}\right)^{-1}R_{H_2^n}q_2$, $R_{H_2^n}q_2$ into Fourier series with respect to the basis $\{w_{klm}^{(2)}\}$, we obtain

$$\left\|(\nabla^2)^{-1}R_{H_1^n}q_1\right\|^2_{L_2(\Omega_1)} = \frac{4}{\pi^4}\left(\sum_{k,l=1}^{\infty} - \sum_{k,l=1}^{n_1}\right)(k^2+l^2)^{-2}\left(\int_{\Omega_1} q_1 w_{kl}^{(1,3)}\, d\Omega_1\right)^2, \tag{3.131}$$

$$\left\|(-\lambda^2\frac{\partial^2}{\partial x^2} - \lambda^2\frac{\partial^2}{\partial y^2} - \frac{\partial^2}{\partial z^2})^{-1/2}R_{H_2^n}q_2\right\|_{L_2(\Omega_2)}$$

$$= \left(\left(-\lambda^2 \frac{\partial^2}{\partial x^2} - \lambda^2 \frac{\partial^2}{\partial y^2} - \frac{\partial^2}{\partial z^2} \right)^{-1} R_{H_2^n} q_2, R_{H_2^n} q_2 \right)^{1/2}_{L_2(\Omega_2)}$$

$$= \frac{2\sqrt{2}}{\pi} \left[\left(\sum_{k,l=1}^{\infty} \sum_{m=1}^{\infty} - \sum_{k,l=1}^{n_1} \sum_{m=1}^{n_2} \right) (\lambda^2 k^2 + \lambda^2 l^2 + m^2)^{-1} \right.$$

$$\left. \times \left(\int_{\Omega_2} q_2 w_{klm}^{(2)} \, d\Omega_2 \right)^2 \right]^{1/2} . \qquad (3.132)$$

Taking into account (3.120)–(3.124), (3.128)–(3.130), we can obtain a more detailed estimate of the error (3.47), (3.48) introduced by the Bubnov–Galerkin method for the problem (3.116)–(3.119). The following is one of the estimates obtained from (3.48):

$$\max_{0 \le t \le T} \left\{ \int_{\Omega_1} \left[\frac{c}{\lambda^4} \left(\frac{\partial}{\partial t} \Delta w_n \right)^2 + \frac{1}{24(1-\nu^2)} \left(\nabla^2 \Delta w_n \right)^2 \right] d\Omega_1 \right.$$

$$\left. + \frac{1}{\beta(1-\nu)} \int_{\Omega_2} (\Delta \theta_n)^2 \, d\Omega_2 \right\}$$

$$\le \int_{\Omega_1} \left[\frac{c}{\lambda^4} (R_{H_1^n} w_1)^2 + \frac{1}{12(1-\nu^2)} (R_{H_1^n} \nabla^2 w_0)^2 - 2 R_{H_1^n} q_1 R_{H_1^n} w_0 \right] d\Omega_1$$

$$+ \frac{1}{\beta(1-\nu)} \int_{\Omega_2} (R_{H_1^n} \theta_0)^2 \, d\Omega_2 + 24(1-\nu^2) \left\| (\nabla^2)^{-1} R_{H_1^n} q_1 \right\|^2_{L_2(\Omega_1)}$$

$$+ \frac{\beta}{2(1-\nu)} \left\{ \frac{1}{\beta} \left\| (-\lambda^2 \frac{\partial^2}{\partial x^2} - \lambda^2 \frac{\partial^2}{\partial y^2} - \frac{\partial^2}{\partial z^2})^{-1/2} R_{H_2^n} q_2 \right\|_{L_2(\Omega_2)} \right. \qquad (3.133)$$

$$\left. + \frac{c_n}{\pi^2} \left[\frac{3(1-\nu)}{2} \sigma^{(4)}_{[\frac{n_2}{2}]+1} \right]^{1/2} \right\}^2 T,$$

where $\Delta w_n = w_n - w$, $\Delta \theta_n = \theta_n - \theta$, $\{w, \theta\}$ is the exact solution to the problem, $\{w_n, \theta_n\}$ is the approximate solution obtained using the Bubnov–Galerkin method, and

$$c_n = \left[\frac{\lambda^4}{c} \int_{\Omega_1} \left(q_1 - \frac{1}{12(1-\nu^2)} \nabla^2 \nabla^2 w_0 - \frac{1}{1-\nu} \nabla^2 \int_{-\frac{1}{2}}^{\frac{1}{2}} P_{H_2^n} \theta_0 \right. \right.$$

$$\left. \times z \, dz \right)^2 d\Omega_1 + \frac{1}{12(1-\nu)} \int_{\Omega_1} (\nabla^2 w_1)^2 \, d\Omega_1 + \frac{1}{\beta(1-\nu)}$$

$$\left. \times \int_{\Omega_2} \left(q_2 + \lambda^2 \frac{\partial^2 \theta_0}{\partial x^2} + \lambda^2 \frac{\partial^2 \theta_0}{\partial y^2} + \frac{\partial^2 \theta_0}{\partial z^2} + \beta z P_{H_1^n} \nabla^2 w_1 \right)^2 d\Omega_2 \right]^{1/2} . \qquad (3.134)$$

Furthermore,

$$\left\|(\nabla^2)^{-1} R_{H_1^n} q_1\right\|^2_{L_2(\Omega_1)}, \quad \left\|\left(-\lambda^2\frac{\partial^2}{\partial x^2} - \lambda^2\frac{\partial^2}{\partial y^2} - \frac{\partial^2}{\partial z^2}\right)^{-1/2} R_{H_1^n} q_2\right\|_{L_2(\Omega_2)}$$

are the quantities defined by (3.131), (3.132); $\sigma_m^{(4)}$ is the series defined by the relation (2.127); and $\left[\frac{n_2}{2}\right]$ denotes the largest number not larger than $\frac{n_2}{2}$.

Let us consider now a problem of the influence of symmetry of the data g_1, g_2, w_0, w_1, θ_0 on the theoretically guaranteed velocity of convergence of the Bubnov–Galerkin method applied to the problem (3.116)–(3.119). Let, for instance, the data of the problem exhibit the following symmetry:

$$g_1(x,y) = g_1(y,x) = g_1(-x,y) = g_1(x,-y), \tag{3.135}$$

$$\begin{aligned} g_2(x,y,z) &= g_2(y,x,z) = g_2(-x,y,z) \\ &= g_2(x,-y,z) = -g_2(x,y,-z), \end{aligned} \tag{3.136}$$

$$w_0(x,y) = w_0(y,x) = w_0(-x,y) = w_0(x,-y), \tag{3.137}$$

$$w_1(x,y) = w_1(y,x) = w_1(-x,y) = w_1(x,-y), \tag{3.138}$$

$$\begin{aligned} \theta_0(x,y,z) &= \theta_0(y,x,z) = \theta_0(-x,y,z) \\ &= \theta_0(x,-y,z) = -\theta_0(x,y,-z), \end{aligned} \tag{3.139}$$

where x, y, z are arbitrary numbers from $\left(-\frac{1}{2}, \frac{1}{2}\right)$. Let us denote by S_1 a set of functions $g_1(x,y)$ defined on Ω_1, satisfying the symmetry conditions (3.135), and let us denote by S_2 a set of functions $g_2(x,y,z)$ defined on Ω_2, satisfying the symmetry conditions (3.136). Let

$$H_1 = L_2(\Omega_1) \cap S_1, \quad H_2 = L_2(\Omega_2) \cap S_2. \tag{3.140}$$

The operators I_1, I_2, L, K, M, N for (3.116)–(3.119) with the symmetric data (3.135)–(3.139) are defined via the relations (3.121)–(3.123). However, we take

$$D(I_1) = L_2(\Omega_1) \cap S_1, \quad D(I_2) = L_2(\Omega_2) \cap S_2, \quad D(L) = \overset{\circ}{W}{}_2^4(\Omega_2) \cap S_2,$$
$$D(M) = \overset{\circ}{W}{}_2^2(\Omega_2) \cap S_2, \quad D(N) = \overset{\circ}{W}{}_2^2(\Omega_1) \cap S_1,$$

and use the notation (3.124). Therefore, the initial–boundary problem (3.116)–(3.119) with the symmetric data (3.135)–(3.139) can be reduced to the form (3.42), where the initial data (3.119) play the role of the elements w_0, w_1, θ_0.

The basis functions to be used in the Bubnov–Galerkin method for the problem (3.116)–(3.119) with the symmetric data (3.135)–(3.139) can be obtained in the following way:

$$\tilde{w}_{kl}^{(1)}(x,y)$$
$$= \cos(2k-1)\pi x \cos(2l-1)\pi y + \cos(2l-1)\pi x \times \cos(2k-1)\pi y, \quad (3.141)$$
$$k,l = 1,2,\dots, \ k \le l,$$
$$\tilde{w}_{klm}^{(2)}(x,y,z) = \tilde{w}_{kl}^{(1)}(x,y)\sin 2\pi m z, \quad k,l,m = 1,2,\dots, \ k \le l. \quad (3.142)$$

It is easy to check that the functions $\tilde{w}_{kl}^{(1)}(x,y)$ defined by the relations (3.141) create a full system in the space $L_2(\Omega_1) \cap S_1$, and that the functions $\tilde{w}_{klm}^{(2)}(x,y,z)$ defined by the relations (3.141) create a full system in the space $L_2(\Omega_2) \cap S_2$. Let n_1, n_2 be natural numbers $n = \{n_1, n_2\}$. Let us denote by H_1^n the linear neighbourhood of the function system $\{\tilde{w}_{kl}^{(1)} | 1 \le k \le l \le n_1\}$, and by H_2^n the linear neighbourhood of the function system $\{\tilde{w}_{klm}^{(2)} | 1 \le k \le l \le n_1, \ 1 \le m \le n_2\}$. We assume that the data g_1, g_2, w_0, w_1, θ_0 satisfy the symmetry conditions (3.135). All conditions of Theorem 3.3 are then satisfied, and the velocity of convergence of the Bubnov–Galerkin method for the present problem is characterized by the inequalities (3.47), (3.48). In these inequalities, the following series occurs:

$$s_n = \frac{1}{\pi^2}\left[\frac{3\beta(1+\nu)}{2}\sigma_{n_2+1}^{(4)}\right]^{1/2}, \quad (3.143)$$

where $\sigma_m^{(4)}$ is the series defined by the relation (3.127) (the relation (3.143) can be obtained in a manner analogous to that used for (3.128)). In order to calculate the quantities $\left\|L^{-1/2}R_{H_1^n}q_1\right\|_{H_1}^2$, $\left\|K^{-1/2}R_{H_2^n}q_2\right\|_{H_2}$, we again use the relations (3.129), (3.130), in which the following relations appear:

$$\left\|(\nabla^2)^{-1}R_{H_1^n}q_1\right\|_{L_2(\Omega_1)}^2 = \frac{4}{\pi^4}\left[\sum_{k,l=1}^{\infty} - \sum_{k,l=1}^{n_1}\left[(2k-1)^2+(2l-1)^2\right]^{-2}\right]$$
$$\times\left(\int_{\Omega_1} q_1(x,y)\cos(2k-1)\pi x\cos(2l-1)\pi y\,dx\,dy\right)^2,$$

$$\left\|\left(-\lambda^2\frac{\partial^2}{\partial x^2} - \lambda^2\frac{\partial^2}{\partial y^2} - \frac{\partial^2}{\partial z^2}\right)^{-1/2} R_{H_2^n}q_2\right\|_{L_2(\Omega_2)} \quad (3.144)$$

$$= \frac{2\sqrt{2}}{\pi}\left[\left(\sum_{k,l=1}^{\infty}\sum_{m=1}^{\infty} - \sum_{k,l=1}^{n_1}\sum_{m=1}^{n_2}\right)\left(\lambda^2(2k-1)^2+\lambda^2(2l-1)^2+4m^2\right)^{-1}\right.$$
$$\left.\times\left(\int_{\Omega_2} q_2(x,y,z)\cos(2k-1)\pi x\cos(2l-1)\pi y\sin 2\pi m z\,dx\,dy\,dz\right)^2\right]^{1/2}.$$
$$(3.145)$$

The relations (3.3), (3.145) can be obtained using an expression of the function $(\nabla^2)^{-1}R_{H_1^n}q_1$ into a Fourier series in the basis (3.141) and of the functions $\left(-\lambda^2\frac{\partial^2}{\partial x^2}-\lambda^2\frac{\partial^2}{\partial y^2}-\frac{\partial^2}{\partial z^2}\right)^{-1/2}R_{H_2^n}q_2$, $R_{H_2^n}q_2$ into Fourier series in the basis (3.142).

Using (3.121)–(3.124), (3.129), (3.130), (3.140), (3.143), we can obtain estimates of the errors (3.47), (3.48) introduced by the Bubnov–Galerkin method for the problem (3.135)–(3.139). The following is one of the estimates obtained from (3.48):

$$\max_{0\le t\le T}\left\{\int\limits_{\Omega_1}\left[\frac{c}{\lambda^4}\left(\frac{\partial}{\partial t}\Delta w_n\right)^2+\frac{1}{24(1-\nu^2)}\left(\nabla^2\Delta w_n\right)^2\right]d\Omega_1\right.$$
$$\left.+\frac{1}{\beta(1-\nu)}\int\limits_{\Omega_2}\left(\Delta\theta_n\right)^2 d\Omega_2\right\}$$
$$\le\int\limits_{\Omega_1}\left[\frac{c}{\lambda^4}\left(R_{H_1^n}w_1\right)^2+\frac{1}{12(1-\nu^2)}\left(R_{H_1^n}\nabla^2 w_0\right)^2-2R_{H_1^n}q_1R_{H_1^n}w_0\right]d\Omega_1$$
$$+\frac{1}{\beta(1-\nu)}\int\limits_{\Omega_2}\left(R_{H_1^n}\theta_0\right)^2 d\Omega_2+24(1-\nu^2)\left\|(\nabla^2)^{-1}R_{H_1^n}q_1\right\|^2_{L_2(\Omega_1)}$$
$$+\frac{\beta}{2(1-\nu)}\left\{\frac{1}{\beta}\left\|\left(-\lambda^2\frac{\partial^2}{\partial x^2}-\lambda^2\frac{\partial^2}{\partial y^2}-\frac{\partial^2}{\partial z^2}\right)^{-1/2}R_{H_2^n}q_2\right\|_{L_2(\Omega_2)}\right.$$
$$\left.+\frac{c_n}{\pi^2}\left[\frac{3(1-\nu)}{2}\sigma^{(4)}_{[\frac{n_2}{2}]+1}\right]^{1/2}\right\}^2 T, \tag{3.146}$$

where $\Delta w_n=w_n-w$; $\Delta\theta_n=\theta_n-\theta$; $\{w,\theta\}$ is the exact solution of the problem; $\{w_n,\theta_n\}$ is the approximate solution obtained using the Bubnov–Galerkin method; c_n is the series defined by the relation (3.134), where by H_1^n, H_2^n we mean the linear neighbourhoods of the systems of basis functions $\{\tilde{w}^{(1)}_{kl}|1\le k\le l\le n_1\}$, $\{\tilde{w}^{(2)}_{klm}|1\le k\le l\le n_1,\quad 1\le m\le n_2\}$;

$$\left\|(\nabla^2)^{-1}R_{H_1^n}q_1\right\|^2_{L_2(\Omega_1)},\quad \left\|\left(-\lambda^2\frac{\partial^2}{\partial x^2}-\lambda^2\frac{\partial^2}{\partial y^2}-\frac{\partial^2}{\partial z^2}\right)^{-1/2}R_{H_2^n}q_2\right\|_{L_2(\Omega_2)},$$

are the quantities defined by (3.3), (3.145), respectively; and $\sigma^{(4)}_m$ is the series defined by the relation (3.127).

Assuming that the symmetry conditions (3.135)–(3.139) are satisfied, we compare the estimate (3.146) with the estimate (3.133). For the same values of n_1 and n_2, the right-hand side of the inequality (3.146) is smaller than the right-hand side of the inequality (3.133) (the operators $R_{H_1^n}$, $R_{H_2^n}$ in the right-hand side of (3.146) correspond to the functions $L_2(\Omega_1)\cap S_1$,

$L_2(\Omega_2) \cap S_2$, with shorter remainders of the Fourier series than for the operators in the right hand-side of (3.133)). The number of basis functions used for the symmetric data, with the estimate (3.146) $\left(\frac{1}{2}(n_1+1\right)n_1$ basis functions in the series for the first component of the approximate solution $w_n(x,y,t)$ and $\frac{1}{2}(n_1+1)n_1n_2$ basis functions in the series for the second component of the approximate solution $\theta_n(x,y,z,t)\big)$, is also smaller than in the case without symmetry, with the estimate (3.133) $\big(n_1^2$ basis functions in the series for the first component of the approximate solution $w_n(x,y,z)$ and $n_1^2n_2$ basis functions in the series for the second component of the approximate solution $\theta_n(x,y,z,t)\big)$. Therefore, the estimate (3.146) of the error introduced by the Bubnov–Galerkin method is better in the case of symmetry.

The occurrence in the left-hand side of each of the inequalities (3.133), (3.146) of a term containing a harmonic operator gives us a possibility to estimate the error of the deflection function $w(x,y,t)$ in a unique norm because of the variables x, y, t, i.e. in a norm of the space $C(\overline{\Omega}_1 \times [0,T])$. Let us establish some following estimates. We denote by $\{\tilde{w}_n, \tilde{\theta}_n\}$ the exact solution of the intermediate initial–boundary value problem of the form (3.116)–(3.119) with free terms $P_{H_1^n}q_1$, $P_{H_1^n}q_2$ and initial data $P_{H_1^n}w_0$, $P_{H_1^n}w_1$, $P_{H_1^n}\theta_0$, where H_1^n denotes either a linear neighbourhood of the system of functions $\{\tilde{w}_{kl}^{(1)} | 1 \le k \le l \le n_1\}$ (if the input data of the problem (3.116)–(3.119) satisfy the symmetry conditions (3.135)–(3.139)) or a linear neighbourhood of the system of functions $\{\tilde{w}_{kl}^{(1,3)} | 1 \le k,\ l \le n_1\}$ (if the symmetry conditions (3.135)–(3.139) are not satisfied). We start with the obvious inequality

$$\left\|\Delta w_n\right\|_{C(\overline{\Omega}_1\times[0,T])} \le \left\|\tilde{w}_n - w\right\|_{C(\overline{\Omega}_1\times[0,T])} + \left\|w_n - \tilde{w}_n\right\|_{C(\overline{\Omega}_1\times[0,T])}, \tag{3.147}$$

where w is the first component of the exact solution $\{w,\theta\}$ to the basic problem (3.116)–(3.119), w_n is the first component of the approximate solution to the initial problem constructed with the use of the Bubnov–Galerkin method, and $\Delta w_n = w_n - w$. Considering $\{\tilde{w}_n, \tilde{\theta}_n\}$ as the approximate solution of the problem (3.116)–(3.119) and $\{w_n, \theta_n\}$ as the approximate solution to the intermediate problem, we obtain the corresponding estimates of the error introduced by the Bubnov–Galerkin methodeither of type (3.133) (if the data of the initial problem (3.116)–(3.119) are nonsymmetric) or of type (3.146) (if the data of the initial problem satisfy the symmetry conditions (3.135)–(3.139)). In the left hand sides of the inequalities obtained we omit all terms except those containing a harmonic operator, and the following inequalities are obtained:

$$\max_{0\le t\le T}\left\| \nabla^2\left[\tilde{w}_n(t) - w(t)\right]\right\|_{L_2(\Omega_1)} \le \alpha_n, \tag{3.148}$$

$$\max_{0\le t\le T}\left\| \nabla^2\left[w_n(t) - \tilde{w}_n(t)\right]\right\|_{L_2(\Omega_1)} \le \beta_n. \tag{3.149}$$

Let us introduce the following notation:

$$\sigma_0^* = \sum_{k,l=1}^{\infty} (k^2 + l^2)^{-2}, \quad \sigma_0^{**} = \sum_{k,l=1}^{\infty} \left[(2k-1)^2 + (2l-1)^2\right]^{-2}, \tag{3.150}$$

$$\sigma_m^* = \left(\sum_{k,l=1}^{\infty} - \sum_{k,l=1}^{m} \right) (k^2 + l^2)^{-2}, \tag{3.151}$$

$$\sigma_m^{**} = \left(\sum_{k,l=1}^{\infty} - \sum_{k,l=1}^{m} \right) \left[(2k-1)^2 + (2l-1)^2\right]^{-2}. \tag{3.152}$$

Let $w(x,y)$ be an arbitrary function of the class $\overset{\circ}{W}{}_2^2(\Omega_1)$; let $w_n(x,y)$ be an arbitrary function of the class $\overset{\circ}{W}{}_2^2(\Omega_1)$ satisfying the condition

$$P_{H_1^n} w_n = 0, \tag{3.153}$$

where H_1^n is a linear neighbourhood of the system of functions $\{w_{kl}^{(1,3)} | 1 \le k,\ l \le n_1\}$; let $w^{(s)}(x,y)$ be an arbitrary function of $\overset{\circ}{W}{}_2^2(\Omega_1)$ satisfying the symmetry conditions (3.137); and let $w_{(n)}^{(s)}(x,y)$ be an arbitrary function of $\overset{\circ}{W}{}_2^2(\Omega_1)$ satisfying both the condition (3.153), where H_1^n is the linear neighbourhood of the system of functions $\{\tilde{w}_{kl}^{(1)} | 1 \le k \le l \le n_1\}$, and the symmetry conditions (3.137). The following inequalities hold (these inequalities are similar to those for $W_2^2(\Omega_1) \to C(\overline{\Omega_1})$):

$$\left\| w \right\|_{C(\overline{\Omega}_1)} \le \frac{2}{\pi^2} \sqrt{\sigma_0^*} \left\| \nabla^2 w \right\|_{L_2(\Omega_1)}, \tag{3.154}$$

$$\left\| w_{(n)} \right\|_{C(\overline{\Omega}_1)} \le \frac{2}{\pi^2} \sqrt{\sigma_{n_1}^*} \left\| \nabla^2 w_{(n)} \right\|_{L_2(\Omega_1)}, \tag{3.155}$$

$$\left\| w^{(s)} \right\|_{C(\overline{\Omega}_1)} \le \frac{2}{\pi^2} \sqrt{\sigma_0^{**}} \left\| \nabla^2 w^{(s)} \right\|_{L_2(\Omega_1)}, \tag{3.156}$$

$$\left\| w_{(n)}^{(s)} \right\|_{C(\overline{\Omega}_1)} \le \frac{2}{\pi^2} \sqrt{\sigma_{n_1}^{**}} \left\| \nabla^2 w_{(n)}^{(s)} \right\|_{L_2(\Omega_1)}. \tag{3.157}$$

The inequalities (3.154)–(3.157) are obtained by the series expansion of the functions w, $\nabla^2 w$, $w_{(n)}$, $\nabla^2 w_{(n)}$ into Fourier series in the basis $\{w_{kl}^{(1,3)}\}$ and by the series expansion of the functions $w^{(s)}$, $\nabla^2 w^{(s)}$, $w_{(n)}^{(s)}$, $\nabla^2 w_{(n)}^{(s)}$ into Fourier series in the basis $\{\tilde{w}_{kl}^{(1)}\}$, with successive applications of the Cauchy–Buniakowski inequality. From (3.154)–(3.157), we obtain

$$\left\|w\right\|_{C(\overline{\Omega}_1\times[0,T])} \leq \frac{2}{\pi^2}\sqrt{\sigma_0^*}\max_{0\leq t\leq T}\left\|\nabla^2 w(t)\right\|_{L_2(\Omega_1)}, \tag{3.158}$$

$$\left\|w_{[n]}\right\|_{C(\overline{\Omega}_1\times[0,T])} \leq \frac{2}{\pi^2}\sqrt{\sigma_{n1}^*}\max_{0\leq t\leq T}\left\|\nabla^2 w_{[n]}(t)\right\|_{L_2(\Omega_1)}, \tag{3.159}$$

$$\left\|w^{[s]}\right\|_{C(\overline{\Omega}_1\times[0,T])} \leq \frac{2}{\pi^2}\sqrt{\sigma_0^{**}}\max_{0\leq t\leq T}\left\|\nabla^2 w^{[s]}(t)\right\|_{L_2(\Omega_1)}, \tag{3.160}$$

$$\left\|w_{[n]}^{[s]}\right\|_{C(\overline{\Omega}_1\times[0,T])} \leq \frac{2}{\pi^2}\sqrt{\sigma_{n1}^{**}}\max_{0\leq t\leq T}\left\|\nabla^2 w_{[n]}^{[s]}(t)\right\|_{L_2(\Omega_1)}, \tag{3.161}$$

where $w(x,y,t)$ is an arbitrary function of the class $C([0,T];\overset{\circ}{W}{}_2^2(\Omega_1))$, $w_{[n]}(x,y,t)$ is an arbitrary function of the class $C([0,T];\overset{\circ}{W}{}_2^2(\Omega_1))$ satisfying the condition (3.153), in which H_1^n is the linear neighbourhood of the system of functions $\{w_{kl}^{(1,3)}|1\leq k,\ l\leq n_1\}$; $w^{[s]}(x,y)$ is an arbitrary function of the class $C([0,T];\overset{\circ}{W}{}_2^2(\Omega_1))$ satisfying symmetry conditions of the type (3.137); and $w_{[n]}^{[s]}(x,y)$ is an arbitrary function of the class $C([0,T];\overset{\circ}{W}{}_2^2(\Omega_1))$ satisfying simultaneously the condition (3.153), in which H_1^n is the linear neighbourhood of the system of functions $\{\tilde{w}_{kl}^{(1)}|\leq k\leq l\leq n_1\}$, and the symmetry condition (3.137).

It is easy to check that the components $\tilde{w}_n$, $\tilde{\theta}_n$ of the exact solution to the intermediate problem are projections of the corresponding components w, θ of the exact solution to the initial problem (3.116)–(3.119) into H_1^n, and, in particular, that $P_{H_1^n}(\tilde{w}_n - w) = 0$. We now use the inequalities (3.158), (3.159) with $w = w_n - \tilde{w}_n$, $w_{[n]} = \tilde{w}_n - w$ when the symmetry conditions (3.135), (3.139) are not satisfied, and the inequalities (3.160), (3.161) with $w^{[s]} = w_n - \tilde{w}_n$, $w_{[n]}^{[s]} = \tilde{w}_n - w$ when the symmetry conditions (3.135)–(3.139) are satisfied, and we estimate from above the right-hand side of the inequality (3.147). Combining the estimates of the norm $\|w\|_{C(\overline{\Omega}_1\times[0,T])}$ obtained and the inequalities (3.148), (3.149), we obtain

$$\left\|\Delta w_n\right\|_{C(\overline{\Omega}_1\times[0,T])} \leq \frac{2}{\pi^2}\left(\sqrt{\sigma_{n1}^*}\alpha_n + \sqrt{\sigma_0^*}\beta_n\right) \tag{3.162}$$

when the symmetry conditions (3.135)–(3.139) are not satisfied, and

$$\left\|\Delta w_n\right\|_{C(\overline{\Omega}_1\times[0,T])} \leq \frac{2}{\pi^2}\left(\sqrt{\sigma_{n1}^{**}}\alpha_n + \sqrt{\sigma_0^{**}}\beta_n\right) \tag{3.163}$$

when the symmetry conditions (3.135)–(3.139) are satisfied.

The inequalities (3.162), (3.163) are used to estimate the errors occurring during calculation of a deflection function in an equally measured norm. We now present the estimates in a fully worked-out form. When the symmetry conditions (3.135)–(3.139) are not satisfied, we have

$$\left\|\Delta w_n\right\|_{C(\overline{\Omega}_1\times[0,T])} \leq \frac{2}{\pi^2}\sqrt{\sigma^*_{n_1}}\left\{\int\limits_{\Omega_1}\left[\frac{24(1-\nu^2)c}{\lambda^4}(R_{H_1^n}w_1)^2\right.\right.$$

$$\left.+2(R_{H_1^n}\nabla^2 w_0)^2 - 48(1-\nu)R_{H_1^n}q_1 R_{H_1^n}w_0\right] d\Omega_1$$

$$+\frac{24(1+\nu)}{\beta}\int\limits_{\Omega_2}(R_{H_1^n}\theta_0)^2\, d\Omega_2 + 576(1-\nu^2)^2\left\|(\nabla^2)^{-1}R_{H_1^n}q_1\right\|^2_{L_2(\Omega_1)}$$

$$\left.+\frac{12(1+\nu)}{\beta}\left\|\left(-\lambda^2\frac{\partial^2}{\partial x^2}-\lambda^2\frac{\partial^2}{\partial y^2}-\frac{\partial^2}{\partial z^2}\right)^{-1/2}R_{H_1^n}q_2\right\|^2_{L_2(\Omega_1)}T\right\}^{1/2}$$

$$+\frac{2}{\pi^2}\sqrt{\sigma_0^*}\left\{\frac{24(1-\nu)}{\beta}\int\limits_{\Omega_1}(R_{H_1^n}P_{H_1^n}\theta_0)^2 + 12\beta(1+\nu)\right.$$

$$\times\left[\frac{1}{\beta}\left\|\left(-\lambda^2\frac{\partial^2}{\partial x^2}-\lambda^2\frac{\partial^2}{\partial y^2}-\frac{\partial^2}{\partial z^2}\right)^{-1/2}R_{H_1^n}q_2\right\|_{L_2(\Omega_1)}\right.$$

$$\left.\left.+\frac{c_n}{\pi^2}\left(\frac{3(1-\nu^2)}{2}\sigma^{(4)}_{[\frac{n_2}{2}]+1}\right)^{1/2}\right]^2 T\right\}^{1/2}, \tag{3.164}$$

where H_1^n is the linear neighbourhood of the system of functions $\{w_{kl}^{(1,3)}|1\leq k,\ l\leq n_1\}$; H_n^2 is the linear neighbourhood of the system of functions $\{w_{klm}^{(2)}|1\leq k,\ l\leq n_1,\ 1\leq m\leq n_2\}$; σ^4_m, c_n, σ^*_m are the series defined by (3.127), (3.134), (3.151); $\left\|(\nabla^2)^{-1}R_{H_1^n}q_1\right\|^2_{L_2(\Omega_1)}$ is the quantity defined by the relation (3.131); and σ_0^* is the constant defined by the first relation of (3.150). We then have

$$\left\|\left(-\lambda^2\frac{\partial^2}{\partial x^2}-\lambda^2\frac{\partial^2}{\partial y^2}-\frac{\partial^2}{\partial z^2}\right)^{-1/2}R_{H_1^n}q_2\right\|^2_{L_2(\Omega_1)}$$

$$=\frac{8}{\pi^2}\left(\sum_{k,l=1}^{\infty}-\sum_{k,l=1}^{n_1}\right)\sum_{m=1}^{\infty}\left(\lambda^2k^2+\lambda^2l^2+m^2\right)^{-1}\left(\int\limits_{\Omega_2}q_2w^{(2)}_{klm}\,d\Omega_2\right)^2,$$

$$\left\|\left(-\lambda^2\frac{\partial^2}{\partial x^2}-\lambda^2\frac{\partial^2}{\partial y^2}-\frac{\partial^2}{\partial z^2}\right)^{-1/2}R_{H_1^n}P_{H_1^n}q_2\right\|_{L_2(\Omega_1)}$$

$$=\frac{2\sqrt{2}}{\pi}\left[\sum_{k,l=1}^{n_1}\sum_{m=1}^{\infty}\left(\lambda^2k^2+\lambda^2l^2+m^2\right)^{-1}\left(\int\limits_{\Omega_2}q_2w^{(2)}_{klm}\,d\Omega_2\right)^2\right]^{1/2}.$$

When the symmetry conditions (3.135)–(3.139) are satisfied, we have

$$\left\|\Delta w_n\right\|_{C(\overline{\Omega}_1\times[0,T])} \leq \frac{2}{\pi^2}\sqrt{\sigma_{n_1}^{**}}\Bigg\{\int\limits_{\Omega_1}\Bigg[\frac{24(1-\nu^2)c}{\lambda^4}(R_{H_1^n}w_1)^2$$

$$+2(R_{H_1^n}\nabla^2 w_0)^2 - 48(1-\nu)R_{H_1^n}q_1 R_{H_1^n}w_0\Bigg]\,d\Omega_1$$

$$+\frac{24(1+\nu)}{\beta}\int\limits_{\Omega_2}(R_{H_1^n}\theta_0)^2\,d\Omega_2 + 576(1-\nu^2)^2\left\|(\nabla^2)^{-1}R_{H_1^n}q_1\right\|^2_{L_2(\Omega_1)}$$

$$+\frac{12(1+\nu)}{\beta}\left\|\left(-\lambda^2\frac{\partial^2}{\partial x^2}-\lambda^2\frac{\partial^2}{\partial y^2}-\frac{\partial^2}{\partial z^2}\right)^{-1/2}R_{H_1^n}q_2\right\|^2_{L_2(\Omega_1)}T\Bigg\}^{1/2}$$

$$+\frac{2}{\pi^2}\sqrt{\sigma_0^*}\Bigg\{\frac{24(1+\nu)}{\beta}\int\limits_{\Omega_1}(R_{H_1^n}P_{H_1^n}\theta_0)^2 + 12\beta(1+\nu)$$

$$\times\Bigg[\frac{1}{\beta}\left\|\left(-\lambda^2\frac{\partial^2}{\partial x^2}-\lambda^2\frac{\partial^2}{\partial y^2}-\frac{\partial^2}{\partial z^2}\right)^{-1/2}R_{H_1^n}P_{H_1^n}q_2\right\|_{L_2(\Omega_2)}$$

$$+\frac{c_n}{\pi^2}\left(\frac{3(1-\nu^2)}{2}\sigma^{(4)}_{n_2+1}\right)^{1/2}\Bigg]^2 T\Bigg\}^{1/2}, \tag{3.165}$$

where H_1^n is the linear neighbourhood of the system of functions $\{w_{kl}^{(1)}|1\leq k\leq l\leq n_1\}$; H_2^n is the linear neighbourhood of the system of functions $\{\tilde{w}_{klm}^{(2)}|1\leq k\leq l\leq n_1,\ 1\leq m\leq n_2$; $\sigma_m^{(4)}$, c_n, σ_m^{**} are the series defined by the relations (3.127), (3.134), (3.152); $\|(\nabla^2)^{-1}R_{H_1^n}q_1\|^2_{L_2(\Omega_1)}$ is the quantity defined by the relation (3.3); and σ_0^{**} is the constant defined by the second relation of (3.150). We then have

$$\left\|\left(-\lambda^2\frac{\partial^2}{\partial x^2}-\lambda^2\frac{\partial^2}{\partial y^2}-\frac{\partial^2}{\partial z^2}\right)^{-1/2}R_{H_1^n}q_2\right\|^2_{L_2(\Omega_2)}$$

$$=\frac{8}{\pi^2}\left(\sum_{k,l=1}^{\infty}-\sum_{k,l=1}^{n_1}\right)\sum_{m=1}^{\infty}\left[\lambda^2(2k-1)^2+\lambda^2(2l-1)^2+4m^2\right]^{-1}$$

$$\times\left(\int\limits_{\Omega_2}q_2(x,y,z)\cos(2k-1)\pi x\cos(2l-1)\pi y\sin 2\pi mz\,dx\,dy\,dz\right)^2,$$

$$\left\|\left(-\lambda^2\frac{\partial^2}{\partial x^2}-\lambda^2\frac{\partial^2}{\partial y^2}-\frac{\partial^2}{\partial z^2}\right)^{-1/2}R_{H_1^n}P_{H_1^n}q_2\right\|_{L_2(\Omega_2)}$$

$$=\frac{2\sqrt{2}}{\pi}\Bigg[\sum_{k,l=1}^{n_1}\sum_{m=1}^{\infty}\left(\lambda^2(2k-1)^2+\lambda^2(2l-1)^2+4m^2\right)^{-1}$$

$$\times\left(\int\limits_{\Omega_2}q_2(x,y,z)\cos(2k-1)\pi x\cos(2l-1)\pi y\sin 2\pi mz\,dx\,dy\,dz\right)^2\Bigg]^{1/2}.$$

Remark 3.7 *The method presented above for obtaining an estimate of the error of a calculation of the deflection function (with use of the intermediate function* $\tilde{w}_n$*) in a uniformly distributed norm leads to a better estimate than does a direct application of the inclusion theorem* $W_2^2(\Omega_1) \to C(\overline{\Omega}_1)$ *to the inequalities (3.133), (3.146).*

Remark 3.8 *The method of obtaining the estimates (3.47), (3.48) can be extended without major difficulties to cases of coupled thermoelastic problems for shallow shells with a constant thickness in the framework of the Kirchhoff–Love model with boundary conditions of free support on ribs that cannot be stretched or compressed in the tangential plane.*

The computational examples showing the efficiency of the estimates described in this section are given in Chap. 4.

3.4 Coupled Problem of Thermoelasticity Within a Timoshenko-Type Model

In this section, a scheme leading to prior estimates given by (3.3), (3.19), (3.23), (3.39) of the errors introduced by the Bubnov–Galerkin method for a coupled problem of thermoelasticity of shallow shells, rectangular in plan and with variable thickness, in the framework of a kinematic model of Timoshenko type, is described.

Assuming that the conditions (2.64)–(2.68), (2.83) are satisfied, we consider the system of differential equations (2.20)–(2.23) with the homogeneous initial conditions (2.33)–(2.37), where u, v, w, Ψ_x, Ψ_y, θ are the sought functions, defined for $(x,y) \in \Omega_1$, $|z| \le \frac{1}{2}h(x,y)$, $t \in [0,T]$. We assume also that $\Omega_1 = (-\frac{a}{2}, \frac{a}{2}) \times (-\frac{b}{2}, \frac{b}{2})$, where a, b are positive constants. Any one of the sets of boundary conditions (2.48), (2.50), (2.52) is applied to the shell $\partial\Omega_2$, and on the surface of $\partial\Omega_2$ we apply a homogeneous condition of type (2.38).

We use the notation $H_1 = (L_2(\Omega_1))^5$, $H_2 = L_2(\Omega_2)$ and assume that the scalar product (3.98) is determined for the space $L_2(\Omega_2)$. The differential operator L in this case is defined on $D(L) = (\overset{\circ}{W}{}_2^2(\Omega_1))^5$ by the relation (2.72). The auxiliary operator L_0 with a known spectrum is sought in the form

$$L_0\overline{\omega} = \tilde{L}\overline{\omega} + P_L\overline{\omega}, \tag{3.166}$$

where $\overline{\omega} = \{u, v, w, \Psi_x, \Psi_y, \theta\}$, $\tilde{L}$ is the self-adjoint extension of the differential operator initially defined on $(\overset{\circ}{W}{}_2^2(\Omega_1))^5 \cap (C^2(\overline{\Omega}_1))^5$ in the sense of Friedrichs by the relation

$$\tilde{L}\overline{\omega} = \left\{ -\frac{\partial^2 u}{\partial x^2} - \frac{1-\nu}{2}\frac{\partial^2 u}{\partial y^2}, -\frac{\partial^2 v}{\partial y^2} - \frac{1-\nu}{2}\frac{\partial^2 v}{\partial x^2}, \right.$$
$$\left. -\nabla^2 w, -\frac{\partial^2 \Psi_x}{\partial x^2} - \frac{1-\nu}{2}\frac{\partial^2 \Psi_x}{\partial y^2}, -\frac{\partial^2 \Psi_y}{\partial y^2} - \frac{1-\nu}{2}\frac{\partial^2 \Psi_y}{\partial x^2} \right\},$$

and P_L is a positive constant to be found. The eigenvector functions $\overline{v}_{kl}^{(j)}$ $(j = 1, \dots, 5)$ of the operator L_0 have the following form:

$$\overline{v}_{kl}^{(j)}(x,y) = \left\{\sigma_i^j \omega_{kl}^{(1,1)}\left(\frac{x}{a}, \frac{y}{b}\right)\right\}_{i=1}^5, \ j = 1,4, \ k = 0,1,\dots, \ l = 1,2,\dots, \tag{3.167}$$

$$\overline{v}_{kl}^{(j)}(x,y) = \left\{\sigma_i^j \omega_{kl}^{(1,2)}\left(\frac{x}{a}, \frac{y}{b}\right)\right\}_{i=1}^5, \ j = 2,5, \ k = 1,2,\dots, \ l = 0,1,\dots, \tag{3.168}$$

$$\overline{v}_{kl}^{(3)}(x,y) = \left\{\sigma_i^3 \omega_{kl}^{(1,3)}\left(\frac{x}{a}, \frac{y}{b}\right)\right\}_{i=1}^5, \ k,l = 1,2,\dots, \tag{3.169}$$

for the boundary conditions (2.48),

$$\overline{v}_{kl}^{(j)}(x,y) = \left\{\sigma_i^j w_{kl}^{(1,3)}\left(\frac{x}{a}, \frac{y}{b}\right)\right\}_{i=1}^5, \ j = 1,2,3, \ k,l = 1,2,\dots, \tag{3.170}$$

$$\overline{v}_{kl}^{(4)}(x,y) = \left\{\sigma_i^4 \omega_{kl}^{(1,1)}\left(\frac{x}{a}, \frac{y}{b}\right)\right\}_{i=1}^5, \ k = 0,1,\dots, \ l = 1,2,\dots, \tag{3.171}$$

$$\overline{v}_{kl}^{(5)}(x,y) = \left\{\sigma_i^5 \omega_{kl}^{(1,2)}\left(\frac{x}{a}, \frac{y}{b}\right)\right\}_{i=1}^5, \ k = 1,2,\dots, \ l = 1,2,\dots, \tag{3.172}$$

for the boundary conditions (2.50), and

$$\overline{v}_{kl}^{(j)}(x,y) = \left\{\sigma_i^j \omega_{kl}^{(1,3)}\left(\frac{x}{a}, \frac{y}{b}\right)\right\}_{i=1}^5, \ j = 1,\dots,5, \ k,l = 1,2,\dots, \tag{3.173}$$

for the boundary conditions (2.52).

In (3.167)–(3.173), $\omega_{kl}^{(1,j)}$ $(j = 1,2,3)$ are the functions defined by the relations (3.67), (3.68); σ_i^j is the Kronecker delta. The eigenvalues $\lambda_{kl}^{(j)}$ of the operator L_0 (for all three cases (2.48), (2.50), (2.52)) are defined by the relations

$$\lambda_{kl}^{(1)} = \lambda_{kl}^{(4)} = \pi^2\left[\frac{k^2}{a^2} + \frac{(1-\nu)l^2}{2b^2}\right] + P_L, \tag{3.174}$$

$$\lambda_{kl}^{(2)} = \lambda_{kl}^{(5)} = \pi^2\left[\frac{(1-\nu)k^2}{2a^2} + \frac{l^2}{b^2}\right] + P_L, \tag{3.175}$$

$$\lambda_{kl}^{(3)} = \pi^2\left[\frac{k^2}{a^2} + \frac{l^2}{b^2}\right] + P_L. \tag{3.176}$$

Using a method similar to that described in [246] and the technique described in Sect. 3.2, it can be seen that for an appropriate choice of the constant P_L the operator (3.166) is convergent and forms an acute angle to the operator (2.72), and that estimates of types (3.93), (3.95), (3.96) hold.

The operators K and K_0 are defined by the relations (3.103), (3.100), and with an appropriate choice of p_k both of those operators are convergent and form an acute angle, and the estimates (3.104), (3.105) hold. The eigenfunctions and eigenvalues of the operator K_0 are defined by the relations (3.101), (3.102).

The operators I_1, M for the problem under consideration are defined by (2.70), (2.76). In accordance with (3.103), the operators I_1, N are defined by a differential equation obtained by multiplying both sides of (2.23) by $T_0^{-1}h(x,y)$, i.e. we take

$$I_2\theta = \frac{c_\varepsilon(1+\varepsilon)h}{T_0}\theta,$$

$$N\varpi = \frac{E\alpha_T h}{1-\nu}\left[\frac{\partial u}{\partial x} + \frac{\partial v}{\partial y} - (k_x+k_y)w + z\left(\frac{\partial \Psi_x}{\partial x} + \frac{\partial \Psi_y}{\partial y}\right)\right].$$

The free terms of (3.1), (3.2) can be obtained by the relations $g_1 = \{p_1, p_2, q_1, 0, 0\}$, $g_2 = T_0^{-1}hq_2$.

In an analogous way to that presented in Sect. 3.2, we obtain

$$\left\|I_1^{1/2}\right\|_{H_1\to H_1} = \rho^{1/2}\tilde{H}_0^{1/2}, \tag{3.177}$$

where

$$\tilde{H}_0 = \max_{(x,y)\in\overline{\Omega}_1}\left\{h(x,y), \frac{1}{12}h^3(x,y)\right\}.$$

We also obtain the second relation of (3.108), the inequalities (3.110), (3.111) and the inequality

$$\left\|MK^{-1/2}\right\|_{H_2\to H_1} \le \tilde{B}_1, \tag{3.178}$$

where $\tilde{B}_1$ is a positive constant. In accordance with (3.178), for the problem under consideration (unlike the coupled thermoelastic problem of a shallow shell within the Kirchhoff–Love kinematic model), the estimate (3.19), (3.39) hold, together with the estimates of the error introduced by the Bubnov–Galerkin method (3.3), (3.23).

Using the relation (3.177), the second relation of (3.108), inequalities of the form (3.96), (3.110), (3.111), and the inequalities (3.105), (3.178), we estimate from above the constants c_i $(i = 1, \dots, 5)$, c_2^*, c_3^*, d_j $(j = 1, 2, 3)$ and d_2^* defined by (3.5) and (3.40), as well as the values of e_1, e_2, e_1^* defined by (3.37), (3.38), (3.41). Then, substituting into the right-hand sides of (3.3), (3.19), (3.23), (3.39) the estimates of these constants and variables, we obtain prior estimates of the errors of the Bubnov–Galerkin method for the differential equations (2.20)–(2.23) with the homogeneous initial conditions

(2.33)–(2.37) and homogeneous boundary conditions. The following is one of the estimates obtained, and is a specific case of (3.19):

$$\max_{0\le t\le T}\Bigg\{\int_{\Omega_1}\Bigg[\rho h\left(\left(\frac{\partial}{\partial t}\Delta u_n\right)^2+\left(\frac{\partial}{\partial t}\Delta v_n\right)^2+\left(\frac{\partial}{\partial t}\Delta w_n\right)^2\right)$$
$$+\frac{1}{12}\rho h^3\left(\left(\frac{\partial}{\partial t}\Delta\Psi_{xn}\right)^2+\left(\frac{\partial}{\partial t}\Delta\Psi_{yn}\right)^2\right)+\frac{Eh}{1+\nu}\Bigg(\left(\frac{\partial}{\partial x}\Delta u_n\right.$$
$$\left.-k_x\Delta w_n\right)^2+\left(\frac{\partial}{\partial y}\Delta v_n-k_y\Delta w_n\right)^2+\frac{1}{2}\left(\frac{\partial}{\partial y}\Delta u_n+\frac{\partial}{\partial x}\Delta v_n\right)^2\Bigg)$$
$$+\frac{Eh\nu}{1-\nu^2}\left(\frac{\partial}{\partial x}\Delta u_n+\frac{\partial}{\partial y}\Delta v_n-(k_x+k_y)\Delta w_n\right)^2$$
$$+\frac{k^2Eh}{2(1+\nu)}\left(\left(\frac{\partial}{\partial x}\Delta w_n+\Delta\Psi_{xn}\right)^2+\left(\frac{\partial}{\partial y}\Delta w_n+\Delta\Psi_{yn}\right)^2\right)$$
$$+D(1-\nu)\Bigg(\left(\frac{\partial}{\partial x}\Delta\Psi_{xn}\right)^2+\left(\frac{\partial}{\partial y}\Delta\Psi_{yn}\right)^2+\frac{1}{2}\left(\frac{\partial}{\partial y}\Delta\Psi_{xn}\right.$$
$$\left.+\frac{\partial}{\partial x}\Delta\Psi_{yn}\right)^2\Bigg)+D\nu\left(\frac{\partial}{\partial x}\Delta\Psi_{xn}+\frac{\partial}{\partial y}\Delta\Psi_{yn}\right)^2\Bigg]\,d\Omega_1$$
$$+\frac{c_\varepsilon(1+\varepsilon)}{T_0}\int_{\Omega_2}(\Delta\theta_n)^2\,d\Omega_2+\frac{\lambda_q}{T_0}\int_0^t\int_{\Omega_2}|\nabla\Delta\theta_n|^2\,d\Omega_2\,d\tau\Bigg\}$$
$$\le 4A_LT^{1/2}(1+\alpha_L\beta_L^{-1})\big(M_g(T)\big)^{1/2}\Bigg\{\Bigg[\int_0^T\int_{\Omega_1}\frac{1}{h}(p_1^2+p_2^2+q_1^2)$$
$$\times d\Omega_1\,d\tau\Bigg]^{1/2}+\frac{\sqrt{2}}{2}e^{1/2}T^{1/2}\tilde{B}_1\Bigg[\frac{1}{\rho}\int_0^T\int_{\Omega_1}\frac{1}{h}(p_1^2+p_2^2+q_1^2)$$
$$\times d\Omega_1\,d\tau+\frac{T_0}{c_\varepsilon(1+\varepsilon)}\int_0^T\int_{\Omega_2}q_2^2\,d\Omega_2\,d\tau\Bigg]^{1/2}$$
$$+\rho^{1/2}\tilde{H}_0^{1/2}T^{1/2}\big(M_g(T)\big)^{1/2}\Bigg\}\tilde{\lambda}_{i(n)+1}^{-1/2}+A_k^2(1+\alpha_k\beta_k^{-1})^2$$
$$\times\Bigg\{\left(\int_0^T\int_{\Omega_2}hq_2^2\,d\Omega_2\,d\tau\right)^{1/2}+\big[c_\varepsilon(1+\varepsilon)T_0^{-1}H_0+B_2^2\big]^{1/2}$$
$$\times T^{1/2}\big(M_g(T)\big)^{1/2}\Bigg\}^2\tilde{æ}_{j(n)+1}^{-1}\equiv 0\big(\max(\tilde{\lambda}_{i(n)+1}^{-1/2},\tilde{æ}_{j(n)+1}^{-1})\big),$$

where $\Delta u_n = u_n - u$, $\Delta v_n = v_n - v$, $\Delta w_n = w_n - w$, $\Delta\Psi_{xn} = \Psi_{xn} - \Psi_x$, $\Delta\Psi_{yn} = \Psi_{yn} - \Psi_y$, $\Delta\theta_n = \theta_n - \theta$, $\{u, v, w, \Psi_x, \Psi_y, \theta\}$ is the exact ("classical") solution to the problem, and $\{u_n, v_n, w_n, \Psi_{xn}, \Psi_{yn}, \theta_n\}$ is the approximate solution obtained using the Bubnov–Galerkin approach. This approximate solution is given by

$$\overline{\omega}_n(x,y,t) = \{u_n(x,y,t), v_n(x,y,t), w_n(x,y,t), \Psi_{xn}(x,y,t), \Psi_{yn}(x,y,t)\}$$
$$= \sum_{\overline{v}_{kl}^{(j)} \in H_1^n} \alpha_{kl}^{(j)}(t)\overline{v}_{kl}^{(j)}(x,y),$$

$$\theta_n(x,y,z,t) = \sum_{\eta_{klm} \in H_2^n} \beta_{klm}(t)\eta_{klm}(x,y,z),$$

$$\tilde{\lambda}_{i(n)+1} = \min\{\lambda_{kl}^{(j)} : \overline{v}_{kl}^{(j)} \overline{\in} H_1^n\},$$

$$\tilde{æ}_{i(n)+1} = \min\{æ_{kl}^{(j)} : \eta_{klm} \overline{\in} H_2^n\},$$

where H_1^n, H_2^n are the subspaces of the spaces $(L_2(\Omega_1))^5$ and $L_2(\Omega_2)$, spanned by $i(n)$ basis vector functions $\overline{v}_{kl}^{(j)}$ and defined in relation to the mechanical boundary conditions of the problem under cosideration, either by the relations (3.167)–(3.169), by the relations (3.170)–(3.172), by the relation (3.173), or on $j(n)$ basis functions η_{klm} defined by the relations (3.101). The $\lambda_{kl}^{(j)}$, $æ_{klm}$ are the eigenvalues of the auxiliary operators L_0, K_0 defined by the relations (3.174)–(3.176), (3.102). $M_g(T)$ is the constant defined by the relation (3.113), in which $M[\cdot,\cdot,\cdot]$ is the value defined by the relation (3.114).

Remark 3.9 *If the thickness of the shell is constant and the shell edges are supported freely on elastic ribs that cannot be stretched or compressed, and if both the mechanical and the thermal loads are constant in time, then for the problem considered here the specific scheme for estimation of the errors (3.47), (3.48) of the Bubnov–Galerkin method is similar to that presented in Sect. 3.3.*

4 Numerical Investigations of the Errors of the Bubnov–Galerkin Method

In Chap. 3, many prior error estimates for the Bubnov–Galerkin method applied to coupled thermoelastic problems of shallow shells and plates were derived. The estimates related to the general case of a shell with a variable thickness subjected to time-varying mechanical and thermal loads possess an important theoretical meaning (they guarantee, for a wide class of problems, strong convergence of successive approximations to the exact solution with a velocity larger than the estimated velocity). However, in applications, when specific real problems have to be considered the estimates possess a more generalized meaning. This question is addressed, for instance, in the case of the estimates obtained in Sect. 3.3 for vibrations of a simply supported plate with constant thickness, subjected to mechanical and thermal loads that are constant in time. In this chapter, numerical results are given and the efficiency of the estimate used is verified.

4.1 Vibration of a Transversely Loaded Plate

We consider the initial–boundary value problem (3.116)–(3.119) (the bars over nondimensional quantities are omitted). The behaviour of a plate without a thermal load will be considered, and the following three groups of data will be used:

$$\begin{aligned} &q_1(x,y) \not\equiv 0, \quad w_0(x,y) = w_1(x,y) \equiv 0, \\ &q_2(x,y,z) = \theta_0(x,y,z) \equiv 0, \end{aligned} \tag{4.1}$$

$$\begin{aligned} &q_1(x,y) = w_1(x,y) \equiv 0, \quad w_0(x,y) \not\equiv 0, \\ &q_2(x,y,z) = \theta_0(x,y,z) \equiv 0, \end{aligned} \tag{4.2}$$

$$\begin{aligned} &q_1(x,y) = w_0(x,y) \equiv 0, \quad w_1(x,y) \not\equiv 0, \\ &q_2(x,y,z) = \theta_0(x,y,z) \equiv 0. \end{aligned} \tag{4.3}$$

It should be noted that the solution of an arbitrary initial–boundary value problem of the form (3.116)–(3.119) with nonhomogeneous mechanical and homogeneous thermal data (no thermal load) can be represented in the form of a sum of solutions of the three problems related to the data (4.1), (4.2), (4.3).

In this chapter, the notation used Sect. 3.3 is used here also.

In this section, the problem (3.116)–(3.119) with the data (4.1) is considered (the other variants of the data, i.e. (4.2) and (4.3), will be considered later). The problem relates to the vibrations of a transversely loaded simply supported plate. We assume that $q_1 \in L_2(\Omega_1)$, and the case of a symmetric load (3.135) will be considered separately.

We are going to solve the problem stated above using the Bubnov–Galerkin method by taking the basis functions either in the form (3.68), (3.99) (when the symmetry conditions (3.135) are not satisfied) or in the form (3.141), (3.142) (when the symmetry conditions are satisfied). Thus, the estimates (3.133), (3.146), (3.164) and (3.165) hold. Taking into account (4.1), the estimates (3.164), (3.165) read

$$\|w_n - w\|_{C(\overline{\Omega}_1 \times [0,T])} \leq \frac{96(1-\nu^2)}{\pi^4} \sqrt{\sigma^*_{n_1}} \left[\left(\sum_{k,l=1}^{\infty} - \sum_{k,l=1}^{n_1} \right) (k^2+l^2)^{-2} \times \left(\int_{\Omega_1} q_1 w_{kl}^{(1,3)} \, d\Omega_1 \right)^2 \right]^{1/2} + \frac{6\lambda^2}{\pi^4} \left[\frac{2\beta(1+\nu)(1-\nu^2)}{c} \times \sigma_0^* \sigma^{(4)}_{[\frac{n_2}{2}]+1} T \right]^{1/2} \|q_1\|_{L_2(\Omega_1)}, \tag{4.4}$$

$$\|w_n - w\|_{C(\overline{\Omega}_1 \times [0,T])} \leq \frac{96(1-\nu^2)}{\pi^4} \sqrt{\sigma^{**}_{n_1}} \left[\left(\sum_{k,l=1}^{\infty} - \sum_{k,l=1}^{n_1} \right) ((2k-1)^2 + (2l-1)^2)^{-2} \left(\int_{\Omega_1} q_1(x,y) \cos(2k-1)\pi x \cos(2l-1)\pi y \, dx\, dy \right)^2 \right]^{1/2} + \frac{6\lambda^2}{\pi^4} \left[\frac{2\beta(1+\nu)(1-\nu^2)\sigma_0^{**}\sigma^{(4)}_{0n_2+1} T}{c} \right]^{1/2} \|q_1\|_{L_2(\Omega_1)}. \tag{4.5}$$

In the left-hand side of each of the inequalities (3.133), (3.146), we omit the terms containing Δw_n, and the newly obtained inequalities, taking account of (4.1), have the form

$$\max_{0 \leq t \leq T} \|\theta_n(t) - \theta(t)\|_{L_2(\Omega_2)} \leq \frac{96(1-\nu)(1-\nu^2)}{\pi^4} \times \left[\left(\sum_{k,l=1}^{\infty} - \sum_{k,l=1}^{n_1} \right) (k^2+l^2)^{-2} \left(\int_{\Omega_1} q_1 w_{kl}^{(1,3)} \, d\Omega_1 \right)^2 + \frac{3\beta^2(1-\nu^2)\lambda^4}{4\pi^4 c} T \sigma^{(4)}_{[\frac{n_2}{2}]+1} \|q_1\|^2_{L_2(\Omega_1)} \right]^{1/2}, \tag{4.6}$$

$$\max_{0\leq t\leq T}\left\|\theta_n(t)-\theta(t)\right\|_{L_2(\Omega_2)} \leq \frac{96(1-\nu)(1-\nu^2)}{\pi^4}\left[\left(\sum_{k,l=1}^{\infty}-\sum_{k,l=1}^{n_1}\right)\right.$$

$$\times\left((2k-1)^2(2l-1)^2\right)^{-2}\left(\int_{\Omega_1} q_1(x,y)\cos(2k-1)\pi x\right.$$

$$\left.\left.\times\cos(2l-1)\pi y\,dx\,dy\right)^2+\frac{3\beta^2(1-\nu^2)\lambda^4}{4\pi^4 c}T\sigma_{n_2+1}^{(4)}\left\|q_1\right\|^2_{L_2(\Omega_1)}\right]. \tag{4.7}$$

In (4.4)–(4.7), $\{w,\theta\}$ is the exact solution to the problem under cosideration, and $\{w_n,\theta_n\}$ is its approximate solution constructed using the Bubnov–Galerkin method. The estimates (4.4), (4.6) correspond to a case of a nonsymmetric distribution of the transverse load, and the estimates (4.5), (4.7) correspond to a case of a symmetric distribution of the transverse load (3.135).

Let us remember that the plate under consideration is a square with side a and thickness h. For the purpose of computation we take $a = 7\times10^{-2}$ m and $h = 10^{-3}$ m, and therefore $\lambda = \dfrac{h}{a} = 1/70$. Two materials for a plate will be considered: aluminium (Al) and copper (Cu). The material properties, taken from [127], are

$$E=\begin{cases}6.85\times10^{10}\ \text{Pa} & \text{(Al)},\\ 11.2\times10^{10}\ \text{Pa} & \text{(Cu)},\end{cases}$$

$$\nu=\begin{cases}0.36 & \text{(Al)},\\ 0.37 & \text{(Cu)},\end{cases} \tag{4.8}$$

$$\rho=\begin{cases}2.70\times10^{3}\ \text{kg/m}^3 & \text{(Al)},\\ 8.93\times10^{3}\ \text{kg/m}^3 & \text{(Cu)},\end{cases}\quad \alpha_t=\begin{cases}2.28\times10^{-5}\ \text{K}^{-1} & \text{(Al)},\\ 1.66\times10^{-5}\ \text{K}^{-1} & \text{(Cu)},\end{cases}$$

$$\lambda_q=\begin{cases}207\ \text{W/(m K)} & \text{(Al)},\\ 398\ \text{W/(m K)} & \text{(Cu)},\end{cases}\quad c_p=\begin{cases}902\ \text{J/(kg K)} & \text{(Al)},\\ 386\ \text{J/(kg K)} & \text{(Cu)}.\end{cases}$$

The initial temperature of the plate is $T_0 = 293$ K. Using the known formula [90]

$$c_\varepsilon = c_\sigma - 3\alpha_T^2 E T_0/(1-2\nu),$$

where $c_\sigma = \rho c_p$ is the specific heat capacity for a constant stress tensor, we can calculate the specific heat for the two materials considered, for a constant strain tensor, as follows:

$$c_\varepsilon=\begin{cases}2.324\times10^{6}\ \text{J/(m}^3\ \text{K)} & \text{(Al)},\\ 3.343\times10^{6}\ \text{J/(m}^3\ \text{K)} & \text{(Cu)}.\end{cases}$$

The coupling coefficient is calculated using the second expression of (2.24):

$$\varepsilon=\begin{cases}3.407\times10^{-2} & \text{(Al)},\\ 2.262\times10^{-2} & \text{(Cu)}.\end{cases}$$

The values of the nondimensional parameters c, β for the materials are calculated using (3.115):

$$c = \begin{cases} 2.92 \times 10^{-10} & \text{(Al)}, \\ 1.08 \times 10^{-9} & \text{(Cu)}, \end{cases} \quad \beta = \begin{cases} 6.78 \times 10^{-3} & \text{(Al)}, \\ 4.20 \times 10^{-3} & \text{(Cu)}. \end{cases} \tag{4.9}$$

We consider time intervals related to the maximum period T^* of free shell vibrations $T^* = \dfrac{2\pi}{\omega^*}$, where ω^* is the minimum shell eigenfrequency obtained for uncoupled deformation and temperature fields. Finding ω^* from the equation governing elastic vibrations of a plate (without a thermal term), we obtain $T^* = \dfrac{2\sqrt{3c(1-\nu)^2}}{\pi\lambda^2}$; we obtain from this

$$T^* = \begin{cases} 0.08614 & \text{(Al)}, \\ 0.16496 & \text{(Cu)}. \end{cases} \tag{4.10}$$

As the transverse load we take $g_1(x, y) \equiv 1$, which is uniformly distributed over the plate surface. Therefore the symmetry conditions (3.135) are satisfied and the estimates for the Bubnov–Galerkin method (4.5), (4.7) hold. Substituting $\lambda = 1/70$, the values of ν, c, β given by (4.8), (4.9), and $g_1(x, y) \equiv 1$, we obtain

$$\left\| w_n - w \right\|_{C(\overline{\Omega_{11}} \times [0,T])} \le \Phi_1(n_1, n_2, T), \tag{4.11}$$

$$\max_{0 \le t \le T} \left\| \theta_n(t) - \theta(t) \right\|_{L_2(\Omega_2)} \le \Psi_1(n_1, n_2, T), \tag{4.12}$$

where

$$\Phi_1(n_1, n_2, T) = \begin{cases} 0.34766\sqrt{\sigma_{n_1}^{**}\sigma_{n_1}} + 4.9537 \times 10^{-2}\sqrt{\sigma_{n_2+1}^{(4)}T} & \text{(Al)}, \\ 0.34475\sqrt{\sigma_{n_1}^{**}\sigma_{n_1}} + 2.0262 \times 10^{-2}\sqrt{\sigma_{n_2+1}^{(4)}T} & \text{(Cu)}, \end{cases}$$

$$\Psi_1(n_1, n_2, T) = \begin{cases} \left(6.1140 \times 10^{-4}\sigma_{n_1} + 4.3939 \times 10^{-5}\sigma_{n_2+1}^{(4)}T\right)^{1/2} & \text{(Al)}, \\ \left(3.6970 \times 10^{-4}\sigma_{n_1} + 4.5207 \times 10^{-6}\sigma_{n_2+1}^{(4)}T\right)^{1/2} & \text{(Cu)}, \end{cases}$$

$$\sigma_m = \left(\sum_{k,l=1}^{\infty} - \sum_{k,l=1}^{m} \right) (2k-1)^2 (2l-1)^{-2} \left[(2k-1)^2 + (2l-1)^2 \right]^{-2}.$$

Here $\sigma_m^{(4)}$, σ_m^{**} are numerical series defined by the relations (3.172), (3.152); and $n = \{n_1, n_2\}$, where n_1, n_2 are natural numbers characterizing the number of terms taken in the components of the approximate solution w_n, θ_n (let us remember that w_n includes $\frac{1}{2}(n_1+1)n_1$ terms, whereas θ_n includes $\frac{1}{2}(n_1+1)n_1n_2$ terms).

Let us now introduce the relative errors of the Bubnov–Galerkin method $\delta_{n,T}w$, $\delta_{n,T}\theta$, which characterize the accuracy of computation of the deflection function in terms of a norm that is uniform with respect to the variables x, y, t, and the accuracy of computation of the thermal function in relation to the average squared norm with respect to the spatial coordinates:

$$\delta_{n,T}w = \frac{\|w_n - w\|_{C(\overline{\Omega}_1 \times [0,T])}}{\|w_n\|_{C(\overline{\Omega}_1 \times [0,T])}}, \tag{4.13}$$

$$\delta_{n,T}\theta = \frac{\|\theta_n(T) - \theta(T)\|_{L_2(\Omega_2)}}{\|\theta_n(T)\|_{L_2(\Omega_2)}}. \tag{4.14}$$

Let us introduce the following notation:

$$\Delta_{n,T}w = \frac{\Phi_1(n_1, n_2, T)}{\|w_n\|_{C(\Omega_1 \times [0,T])}}, \tag{4.15}$$

$$\Delta_{n,T}\theta = \frac{\Psi_1(n_1, n_2, T)}{\|\theta_n(T)\|_{L_2(\Omega_2)}}. \tag{4.16}$$

According to (4.11), (4.12) the inequalities $\delta_{n,T}w \leq \Delta_{n,T}w$, $\delta_{n,T}\theta \leq \Delta_{n,T}\theta$ hold, i.e. the quantities defined by the relations (4.15), (4.16) must be treated as theoretically guaranteed estimates from above of the relative errors $\delta_{n,T}w$, $\delta_{n,T}\theta$ of the Bubnov–Galerkin method, defined by the relations (4.13), (4.14).

The problem considered here has been solved numerically using the Bubnov–Galerkin method (the set of basis functions used changes in the intervals $1 \leq n_1$, $n_2 \leq 6$), with successive application of the Runge–Kutta method for solving the system of differential equations. The following computational scheme has been followed:

1. Using the maximal set of basis functions ($n_1 = n_2 = 6$), we calculate $w_n(0, 0, t)$ and $\theta_n(0, 0, 0.25, t)$ for $t = (i/160)T^*$, $i = 1, 2, \dots, 160$, where T^* is the maximum period of the eigenvibrations of the plate (the value of T^* was calculated from (4.10)).
2. We find the time value $T_1 = (i_1/160)T^*$ for which the maximum of the function $i \to |w_n(0, 0, (i/160)T^*)|$ is achieved, and then we calculate the values $w_n(x, y, T_1)$ for $x = j/40$, $y = k/40$, $j, k = 0, 1, \dots, 19$ (owing to the eveness of the function w_n in relation to x and y, we calculate only its values for $x, y \geq 0$, i.e. on a quadrant of the plate).
3. We find the value $T_2 = (i_2/160)T^*$ for which the first explicitly expressed extremum of the function $i \to (0, 0, 0.25, (i/160)T^*)$ appears, and calculate the values $\theta_n(x, y, z, T_2)$ for $x = j/40$, $y = k/40$, $z = l/40$, $j, k, l = 0, 1, \dots, 19$ (because the function θ is even with respect to x and y and odd with respect to z, only the calculations for $x, y, z \geq 0$ are

carried out, i.e. 1/8 of the plate is treated as a three-dimensional body). By "the first explicitly expressed extremum" of the function $i \to \theta(i)$ calculated on a set I of integers i, we mean that one of the two extrema of the functions $\max_{i \in I} \theta(i)$, $\min_{i \in I} \theta(i)$ which is achieved for the smaller value of the argument i.

4. Now we vary a number of basis functions in the intervals corresponding to $1 \le n_1,\ n_2 \le 6$ for each pair of n_1, n_2 and calculate the norm $\|w_n\|_{C(\overline{\Omega}_1[0,T_1])}$ according to the approximate formula

$$\|w_n\|_{C(\overline{\Omega}_1[0,T_1])} \approx \max_{\substack{1 \le i \le i_1 \\ 0 \le j,\ k \le 19}} \left| w_n \left(\frac{j}{40}, \frac{k}{40}, \frac{i}{160} T^* \right) \right| .$$

   We then calculate the guaranteed estimate $\Delta_{n,T_1} w$ using the formula (4.15).

5. The number of basis functions is varied in the intervals corresponding to $1 \le n_1,\ n_2 \le 6$ for each pair n_1, n_2, and for each pair of values n_1, n_2 we calculate the norm $\|\theta_n\|_{L_2(\Omega_2)}$ from the approximate formula

$$\|\theta\|_{L_2(\Omega_2)} \approx \left[\frac{1}{8000} \sum_{j,k,l=0}^{19} \left| \theta_n \left(\frac{j}{40}, \frac{k}{40}, \frac{i}{160} T_2 \right) \right|^2 \right]^{1/2} .$$

   We then calculate the guaranteed estimate $\Delta_{n,T_2} \theta$ of the relative error of the calculation of the thermal function $\delta_{n,T_2} \theta$, using the formula (4.16).

The following values of T_1, T_2 have been obtained:

$$T_1 = 0.5T^* = \begin{cases} 4.307 \times 10^{-2} & \text{(Al)}, \\ 8.248 \times 10^{-2} & \text{(Cu)}, \end{cases} \tag{4.17}$$

$$T_2 = 0.4T^* = \begin{cases} 3.446 \times 10^{-2} & \text{(Al)}, \\ 6.598 \times 10^{-2} & \text{(Cu)}. \end{cases} \tag{4.18}$$

In Tables 4.1 and 4.2, the theoretically guaranteed estimates $\Delta_{n,T_1} w$ of the relative error of a computation of the deflection function, in terms of a uniformly distributed norm, for all pairs n_1, n_2 $(1 \le n_1,\ n_2 \le 6)$ for aluminium and copper plates are given as percentages. It can be seen that the estimates of the errors introduced by the Bubnov–Galerkin method guarantee values of the relative error of a computation of the deflection of 1–2% (or even smaller). For instance, in the case of an aluminium plate (see Table 4.1), the relative error of a computation of the deflection function is less than 1.838% for $n_1 = 3$, $n_2 = 2$, less than 1.538% for $n_1 = 2$, $n_2 = 3$, and less than 0.856% for $n_1 = 3$, $n_2 = 4$. In the case of a copper plate (see Table 4.2), the relative error of a computation of the deflection function is less than 1.464% for $n_1 = n_2 = 2$, 0.73% for $n_1 = n_2 = 3$. When the maximum number of basis functions is used $(n_1 = n_2 = 6)$, the guaranteed estimate of the relative

Table 4.1. Estimates of relative error (as a percentage) of a computation of the deflection function in terms of a uniformly distributed norm, $\Delta_{n,T_1}w$; problem (3.116)–(3.119); data set (4.1); $\lambda = 1/70$; $q_1(x, y) \equiv 1$; Al ($\nu = 0.36$, $c = 2.92 \times 10^{-10}$, $\beta = 6.78 \times 10^{-3}$)

$n_2 \backslash n_1$	1	2	3	4	5	6
1	6.935	3.971	3.607	3.528	3.500	3.489
2	5.206	2.197	1.838	1.757	1.730	1.719
3	4.564	1.538	1.180	1.099	1.072	1.061
4	4.248	1.214	0.856	0.775	0.748	0.737
5	4.066	1.028	0.670	0.589	0.562	0.551
6	3.951	0.909	0.522	0.471	0.444	0.433

Table 4.2. Estimates of relative error (as a percentage) of a computation of the deflection function in terms of a uniformly distributed norm, $\Delta_{n,T_1}w$; problem (3.116)–(3.119); data set (4.1); $\lambda = 1/70$; $q_1(x, y) \equiv 1$; Cu ($\nu = 0.37$, $c = 1.08 \times 10^{-9}$, $\beta = 4.20 \times 10^{-3}$)

$n_2 \backslash n_1$	1	2	3	4	5	6
1	5.478	2.476	2.116	2.036	2.008	1.997
2	4.491	1.464	1.106	1.025	0.998	0.986
3	4.125	1.088	0.730	0.649	0.622	0.611
4	3.994	0.903	0.546	0.465	0.438	0.426
5	3.841	0.796	0.439	0.358	0.331	0.320
6	3.775	0.729	0.372	0.291	0.264	0.252

Table 4.3. Estimates of relative error (as a percentage) of a computation of the temperature function in terms of a mean square norm, $\Delta_{n,T_2}\theta$; problem (3.116)–(3.119); data set (4.1); $\lambda = 1/70$; $q_1(x, y) \equiv 1$; Al ($\nu = 0.36$, $c = 2.92 \times 10^{-10}$, $\beta = 6.78 \times 10^{-3}$)

$n_2 \backslash n_1$	1	2	3	4	5	6
1	180.50	66.35	53.20	51.06	50.56	50.40
2	172.56	49.08	29.72	25.82	24.84	24.53
3	171.31	45.12	22.68	17.26	15.77	15.27
4	170.94	43.81	19.95	13.48	11.51	10.82
5	170.79	43.26	18.72	11.59	9.22	8.33
6	170.72	42.99	18.09	10.55	7.88	6.82

error of a computation of the deflection function does not exceed 0.433% for a plate made from aluminium and does not exceed 0.242% for a plate made from copper.

In Tables 4.3 and 4.4, the theoretically guaranteed estimates $\Delta_{n,T_2}\theta$ of the relative error of a computation of the temperature function, in terms of a mean square norm with respect to the spatial variables, calculated for all pairs n_1, n_2 ($1 \le n_1,\ n_2 \le 6$) are given as percentages.

Table 4.4. Estimates of relative error (as a percentage) of a computation of the temperature function in terms of a mean square norm, $\Delta_{n,T_2}\theta$; problem (3.116)–(3.119); data set (4.1); $\lambda = 1/70$; $q_1(x, y) \equiv 1$; Cu ($\nu = 0.37$, $c = 1.08 \times 10^{-9}$, $\beta = 4.20 \times 10^{-3}$)

$n_2 \backslash n_1$	1	2	3	4	5	6
1	330.95	97.54	63.17	56.61	55.01	54.49
2	325.03	84.67	41.67	31.05	28.08	27.07
3	324.18	82.03	36.09	23.05	18.86	17.32
4	323.94	81.19	34.14	19.86	14.80	12.78
5	323.84	80.84	33.31	18.40	12.77	10.37
6	323.80	80.67	32.90	17.65	11.67	8.97

It can be seen that the estimates of the error of the Bubnov–Galerkin method guarantee that the relative error of a computation of the temperature function is of the order of 10–20% or less. For example, in the case of an aluminium plate (see Table 4.3), the relative error of a computation of the temperature function is less than 19.95% for $n_1 = 3$, $n_2 = 4$, less than 17.26% for $n_1 = 4$, $n_2 = 3$, and less than 9.22% for $n_1 = n_2 = 5$. According to Table 4.4, for the copper plate, the relative error of a calculation of the temperature function is less than 19.86% for $n_1 = n_2 = 4$, and less than 8.97% for $n_1 = n_2 = 6$.

To conclude, the estimates of the accuracy of the Bubnov–Galerkin method allow numerical study (in the framework of the model used here) of one form of interesting behaviour caused by coupling of the temperature and deformation fields. In spite of the lack of a thermal load, a temperature field has been created inside a vibrating plate, as a result of only the deformation process.

4.2 Vibration of a Plate with an Imperfection in the Form of a Deflection

Let us consider the problem (3.116)–(3.119) with the data set (4.2). This problem concerns the vibration of a simply supported plate with a given initial deflection distribution (imperfection), with neither a mechanical nor a thermal load. The coupling of the thermal and deformation fields, as well as the nonlinearity of the temperature field distribution through the plate thickness, are taken into account.

Let us remember that the plate considered here is square, with side a and thickness h. For the computations, we take $a = 7 \times 10^{-2}$ m, $h = 10^{-3}$ m and $\lambda = \dfrac{h}{a} = 1/70$. We consider two materials from which the plate may be made: aluminium (Al) and copper (Cu). The values of the parameters ν, c,

β for these materials have been taken from the relations (4.8), (4.9). Let us assume an initial deflection of the plate of the following form:

$$w_0(x,y) = \left(x^2 - \frac{1}{4}\right)\left(x^2 - \frac{5}{4}\right)\left(y^2 - \frac{1}{4}\right)\left(y^2 - \frac{5}{4}\right).$$

That is, we take the function w_0 to be the minimum-order polynomial which satisfies the boundary conditions of simple support (thus the problem is included in the class $\overset{\circ}{W}{}_2^4(\Omega_1)$).

The problem has been solved using the Bubnov–Galerkin method, with the basis functions $\tilde{w}_{kl}^{(1)}(x,y)$ $(1 \le k \le l \le n_1)$, $\tilde{w}_{klm}^{(2)}(x,y,z)$ $(1 \le k \le l \le n_1, 1 \le m \le n_2)$ in the form (3.141), (3.142). Because the symmetry conditions (3.137) are satisfied in the present case, the estimates (3.146), (3.165) hold. Taking into account the actual values of the parameters λ, ν, c, β and of the function $w_0(x,y)$, the estimates (3.146), (3.165) have the following form (similar to problem presented in Sect. 4.1 with a data set of type (4.1)):

$$\left\|w_n\right\|_{C(\overline{\Omega}_1\times[0,T_1])} < \Phi_2(n_1,n_2,T), \tag{4.19}$$

$$\max_{0\le t\le T}\left\|\theta_n(t)-\theta(t)\right\|_{L_2(\Omega_2)} < \Psi_2(n_1,n_2,T), \tag{4.20}$$

where $\{w,\theta\}$ is the exact solution of the problem, $\{w_n,\theta_n\}$ is the approximate solution obtained by the Bubnov–Galerkin method, and

$$\Phi_2(n_1,n_2,T) = 0.28658\Big[\sigma_{n_1}^{**}\big(0.47170\tilde{\sigma}_{n_1+1}^{(6)} + 0.94353\tilde{\sigma}_{n_1+1}^{(8)}$$
$$+0.47238\tilde{\sigma}_{n_1+1}^{(10)}\big)\Big]^{1/2} + \begin{cases} 9.2628\times 10^{-2}\sqrt{\sigma_{n_2+1}^{(4)}T} & \text{(Al)},\\ 3.8210\times 10^{-2}\sqrt{\sigma_{n_2+1}^{(4)}T} & \text{(Cu)},\end{cases}$$

$$\Psi_2(n_1,n_2,T) = \begin{cases} \big[1.9597\times 10^{-4}\tilde{\sigma}_{n_1+1}^{(6)} + 3.9198\times 10^{-4}\tilde{\sigma}_{n_1+1}^{(8)} & \\ \quad + 1.9624\times 10^{-4}\tilde{\sigma}_{n_1+1}^{(10)} + 1.5363\times 10^{-4}\tilde{\sigma}_{n_2+1}^{(4)}T\big]^{1/2} & \text{(Al)},\\ \big[1.2051\times 10^{-4}\tilde{\sigma}_{n_1+1}^{(6)} + 2.4105\times 10^{-4}\tilde{\sigma}_{n_1+1}^{(8)} & \\ \quad + 1.2068\times 10^{-4}\tilde{\sigma}_{n_1+1}^{(10)} + 1.5241\times 10^{-5}\tilde{\sigma}_{n_2+1}^{(4)}T\big]^{1/2} & \text{(Cu)},\end{cases}$$

$$\tilde{\sigma}_m^{(\alpha)} = \sum_{k=m}^{\infty}(2k-1)^{-\alpha}, \quad \alpha = 6,8,10. \tag{4.21}$$

Here $\sigma_m^{(4)}$ and σ_m^{**} are the numerical series defined by the relations (3.127) and (3.152), and $n = \{n_1, n_2\}$.

Let us consider the relative errors $\delta_{n,T}w$, $\delta_{n,T}\theta$ of the Bubnov–Galerkin method defined by the relations (4.13), (4.14), which characterize both the accuracy of the deflection function calculation, in terms of a norm uniformly distributed with respect to x, y, t, and the accuracy of the temperature function calculation, in terms of a mean square norm with respect to the spatial coordinates. Let us denote

$$\Delta_{n,T}w = \frac{\Phi_2(n_1, n_2, T)}{\|w_n\|_{C(\overline{\Omega}_1 \times [0,T])}}, \tag{4.22}$$

$$\Delta_{n,T}\theta = \frac{\Psi_2(n_1, n_2, T)}{\|\theta_n(T)\|_{L_2(\Omega_2)}}. \tag{4.23}$$

In accordance with (4.19), (4.20), the following inequalities hold: $\delta_{n,T}w < \Delta_{n,T}w$, $\delta_{n,T}\theta < \Delta_{n,T}\theta$, i.e. the quantities $\Delta_{n,T}w$, $\Delta_{n,T}\theta$ defined by (4.22), (4.23) can be considered as theoretically guaranteed estimates from above of the relative errors $\delta_{n,T}w$, $\delta_{n,T}\theta$ of the Bubnov–Galerkin method, defined by (4.13), (4.14).

The problem has been solved using the Bubnov–Galerkin method (the number of basis functions used was taken from the intervals $1 \le n_1$, $n_2 \le 6$), with successive application of the Runge–Kutta method for solving the ordinary differential equations. The numerical scheme presented in Sect. 4.1 was applied. Here we shall mention only that during performance of steps 1–3 of the numerical scheme, the following values of T_1, T_2 were obtained:

$$T_1 = 0.5T^* = \begin{cases} 4.307 \times 10^{-2} & \text{(Al)}, \\ 8.248 \times 10^{-2} & \text{(Cu)}, \end{cases} \tag{4.24}$$

$$T_2 = \begin{cases} 0.4T^* = 3.446 \times 10^{-2} & \text{(Al)}, \\ 0.375T^* = 6.186 \times 10^{-2} & \text{(Cu)}, \end{cases} \tag{4.25}$$

where T^* is the maximum period of free oscillations of the plate defined by (4.10).

In Tables 4.5 and 4.6, the theoretically guaranteed estimates $\Delta_{n,T_1}w$ from above of the relative error of a deflection function calculation using the uniform norm, for all pairs n_1, n_2 ($1 \le n_1$, $n_2 \le 6$) are given in percentages

Table 4.5. Estimates of relative error (as percentages) of deflection function calculations using the uniform norm $\Delta_{n,T_1}w$; problem (3.116)–(3.119); data set (4.2); $\lambda = 1/70$; $w_0(x,y) = \left(x^2 - \frac{1}{4}\right)\left(x^2 - \frac{5}{4}\right)\left(y^2 - \frac{1}{4}\right)\left(y^2 - \frac{5}{4}\right)$; Al ($\nu = 0.36$, $c = 2.92 \times 10^{-10}$, $\beta = 6.78 \times 10^{-3}$)

$n_2 \backslash n_1$	1	2	3	4	5	6
1	7.123	5.827	5.693	5.665	5.656	5.652
2	4.269	2.949	2.816	2.789	2.779	2.775
3	3.208	1.879	1.747	1.720	1.710	1.707
4	2.686	1.353	1.222	1.194	1.184	1.181
5	2.386	1.050	0.919	0.891	0.882	0.878
6	2.195	0.858	0.727	0.699	0.690	0.686

Table 4.6. Estimates of relative error (as percentages) of deflection function calculations using the uniform norm $\Delta_{n,T_1}w$; problem (3.116)–(3.119); data set (4.2); $\lambda = 1/70$; $w_0(x,y) = \left(x^2 - \frac{1}{4}\right)\left(x^2 - \frac{5}{4}\right)\left(y^2 - \frac{1}{4}\right)\left(y^2 - \frac{5}{4}\right)$; Cu ($\nu = 0.37$, $c = 1.08 \times 10^{-9}$, $\beta = 4.20 \times 10^{-3}$)

$n_2 \backslash n_1$	1	2	3	4	5	6
1	4.718	3.402	3.269	3.241	3.232	3.228
2	3.088	1.759	1.627	1.599	1.590	1.586
3	2.483	1.148	1.017	0.989	0.980	0.976
4	2.185	0.848	0.717	0.689	0.679	0.676
5	2.014	0.675	0.544	0.516	0.507	0.503
6	1.905	0.565	0.435	0.406	0.397	0.393

for aluminium and copper plates. Analysing the tables, one can conclude that the estimates given here guarantee a relative error of a computation of the deflection function of 1–2% or less. For instance, according to Table 4.5 (aluminium), the relative error has the following values: less than 1.879% for $n_1 = 2$, $n_2 = 3$, and less than 0.919% for $n_1 = 3$, $n_2 = 5$. Also, according to the results shown in Table 4.6 (copper), the relative error of a computation of the deflection function is less than 1.759% for $n_1 = n_2 = 4$, and less than 0.717% for $n_1 = 3$, $n_2 = 4$.

In Tables 4.7 and 4.8 (for aluminium and copper, respectively), the theoretically guaranteed estimates from above, $\Delta_{n,T_2}\theta$, of the relative error of a computation of the temperature function in terms of the mean square norm with respect to the spatial coordinates are given as percentages for all pairs n_1, n_2 in the intervals $1 \le n_1, n_2 \le 6$. An analysis of Tables 4.7 and 4.8 leads to the conclusion that the estimates given here of the error introduced by the Bubnov–Galerkin method guarantee a relative error of a computation of the temperature function of less than 10–15%. For instance, according to Table 4.7 (aluminium) the relative error of a temperature function computation is less than 14.61% for $n_1 = 2$, $n_2 = 3$ and less than 9.03% for

Table 4.7. Estimates of relative error (as percentages) of temperature function calculations using the mean square norm $\Delta_{n,T_2}\theta$; problem (3.116)–(3.119); data set (4.2); $\lambda = 1/70$; $w_0(x,y) = \left(x^2 - \frac{1}{4}\right)\left(x^2 - \frac{5}{4}\right)\left(y^2 - \frac{1}{4}\right)\left(y^2 - \frac{5}{4}\right)$; Al ($\nu = 0.36$, $c = 2.92 \times 10^{-10}$, $\beta = 6.78 \times 10^{-3}$)

$n_2 \backslash n_1$	1	2	3	4	5	6
1	55.80	42.36	41.72	41.63	41.62	41.61
2	41.80	21.59	20.34	20.17	20.14	20.13
3	38.63	14.61	12.70	12.43	12.38	12.36
4	37.58	11.57	9.03	8.66	8.57	8.55
5	37.14	10.06	7.00	6.51	6.39	6.36
6	36.93	9.25	5.77	5.16	5.02	4.97

Table 4.8. Estimates of relative error (as percentages) of temperature function calculations using the mean square norm $\Delta_{n,T_2}\theta$; problem (3.116)–(3.119); data set (4.2); $\lambda = 1/70$; $w_0(x,y) = \left(x^2 - \frac{1}{4}\right)\left(x^2 - \frac{5}{4}\right)\left(y^2 - \frac{1}{4}\right)\left(y^2 - \frac{5}{4}\right)$; Cu ($\nu = 0.37$, $c = 1.08 \times 10^{-9}$, $\beta = 4.20 \times 10^{-3}$)

$n_2 \backslash n_1$	1	2	3	4	5	6
1	79.87	43.71	41.57	41.29	41.23	41.21
2	70.69	24.71	20.77	20.22	20.09	20.05
3	68.85	18.98	13.47	12.59	12.40	12.34
4	68.27	16.77	10.12	8.92	8.64	8.55
5	68.03	15.78	8.38	6.88	6.51	6.39
6	67.91	15.28	7.39	5.63	5.17	5.03

$n_1 = 3$, $n_2 = 4$. Also according to Table 4.8 (copper), the relative error of a temperature function computation is less than 13.47% for $n_1 = n_2 = 3$ and less than 8.92% for $n_1 = n_2 = 4$.

To conclude, the estimates of the errors introduced by the Bubnov–Galerkin method given here confirm numerically that there is coupling between the temperature and deformation fields when, in the absence of any thermal interaction, a temperature field occurs owing to a deformation process.

4.3 Vibration of a Plate with a Given Variable Deflection Change

Let us consider the problem (3.116)–(3.119) with the data set (4.3). This problem concerns the vibration of a simply supported plate with an initial distribution of the velocity of change of the deflection, with neither a thermal nor a mechanical load, but with inclusion of coupling between the temperature and deformation fields and with a nonlinear distribution of the temperature field through the plate thickness.

Following Sects. 4.1, 4.2, we assume that the relative plate thickness is $\lambda = 1/70$ and the values of the parameters ν, c, β are defined by (4.8), (4.9). We assume also that the initial distribution of the velocity of deflection change over the plate surface has the form

$$w_1(x,y) = \left(x^2 - \frac{1}{4}\right)\left(y^2 - \frac{1}{4}\right),$$

i.e. we take the function w_1 to be equal to the polynomial of the lowest power which takes zero values on $\partial\Omega_1$ (it belongs to the class $\overset{\circ}{W}{}_2^2(\Omega_1)$).

We have solved the problem using the Bubnov—Galerkin method, with the basis functions $\tilde{w}_{kl}^{(1)}(x,y)$ $(1 \le k \le l \le n_1)$, $\tilde{w}_{klm}^{(2)}(x,y,z)$ $(1 \le k \le l \le$

n_1, $1 \le m \le n_2$) in the form (3.141), (3.142). The symmetry conditions (3.137) are satisfied, and therefore the estimates (3.146), (3.165) can be applied. Taking into account the actual values of the parameters λ, ν, c, β and of the function $w_1(x, y)$, the estimates (3.146), (3.165), obtained analogously to the method used in Sect. 4.1 for the data set (4.1), read as follows:

$$\left\| w_n - w \right\|_{C(\overline{\Omega}_1 \times [0,T_1])} < \Phi_3(n_1, n_2, T), \tag{4.26}$$

$$\max_{0 \le t \le T} \left\| \theta_n(t) - \theta(t) \right\|_{L_2(\Omega_2)} < \Psi_3(n_1, n_2, T), \tag{4.27}$$

where $\{w, \theta\}$ is the exact solution of the problem, $\{w_n, \theta_n\}$ is its approximate solution obtained by the Bubnov—Galerkin method, and

$$\Phi_3(n_1, n_2, T) = \begin{cases} 3.6532 \times 10^{-3} \sqrt{\sigma_{n_1}^{**} \tilde{\sigma}_{n_1+1}^{(6)}} + 9.0151 \times 10^{-4} \sqrt{\sigma_{n_2+1}^{(4)} T} & \text{(Al)}, \\ 6.9961 \times 10^{-3} \sqrt{\sigma_{n_1}^{**} \tilde{\sigma}_{n_1+1}^{(6)}} + 7.1094 \times 10^{-4} \sqrt{\sigma_{n_2+1}^{(4)} T} & \text{(Cu)}, \end{cases}$$

$$\Psi_3(n_1, n_2, T) = \begin{cases} \left[6.7507 \times 10^{-8} \tilde{\sigma}_{n_1+1}^{(6)} + 1.4552 \times 10^{-8} \sigma_{n_2+1}^{(4)} T \right]^{1/2} & \text{(Al)}, \\ \left[1.5226 \times 10^{-7} \tilde{\sigma}_{n_1+1}^{(6)} + 5.5652 \times 10^{-9} \sigma_{n_2+1}^{(4)} T \right]^{1/2} & \text{(Cu)}. \end{cases}$$

Here $\sigma_m^{(4)}$, σ_m^{**}, $\tilde{\sigma}_m^{(6)}$ are the series of numbers defined by the relations (3.127), (3.152), (4.21), and $n = \{n_1, n_2\}$.

Let us consider the relative errors of the Bubnov—Galerkin method $\delta_{n,T} w$, $\delta_{n,T} \theta$, defined by the relations (4.13), (4.14), characterizing the accuracy of computation of the deflection function in terms of a uniformly distributed norm with respect to x, y, t, and of the temperature function in terms of a mean square norm with respect to the space variables. We use the notations

$$\Delta_{n,T} w = \frac{\Phi_3(n_1, n_2, T)}{\left\| w_n \right\|_{C(\overline{\Omega}_1) \times [0,T])}}, \tag{4.28}$$

$$\Delta_{n,T} \theta = \frac{\Psi_3(n_1, n_2, T)}{\left\| \theta_n(T) \right\|_{L_2(\Omega_2)}}. \tag{4.29}$$

In accordance with (4.26), (4.27), the inequalities $\delta_{n,T} w < \Delta_{n,T} w$, $\delta_{n,T} \theta < \Delta_{n,T} \theta$ hold, i.e. the quantities $\Delta_{n,T} w$, $\Delta_{n,T} \theta$ defined by (4.28), (4.29) can be considered as theoretically guaranteed estimates from above of the relative errors of the Bubnov—Galerkin method $\delta_{n,T} w$, $\delta_{n,T} \theta$, defined by (4.13), (4.14).

The problem has been solved numerically using the Bubnov—Galerkin method (the number of basis functions used was taken from intervals $1 \le n_1$, $n_2 \le 6$), and the Runge–Kutta method was applied to solve the ordinary differential equations. A computational experiment, similar to that presented in Sect. 4.1, was performed. During the computations, we found

$$T_1 = 0.25T^* = \begin{cases} 2.153 \times 10^{-2} & \text{(Al)}, \\ 4.124 \times 10^{-2} & \text{(Cu)}, \end{cases} \tag{4.30}$$

$$T_2 = \begin{cases} 0.175T^* = 1.507 \times 10^{-2} & \text{(Al)}, \\ 0.15T^* = 2.474 \times 10^{-2} & \text{(Cu)}, \end{cases} \tag{4.31}$$

where T^* is the maximum period of free vibration of the plate, given by (4.10).

In Tables 4.9 and 4.10, the theoretically guaranteed estimates from above of the relative error of a computation of the deflection function, for all pairs n_1, n_2 in the intervals $1 \le n_1,\ n_2 \le 6$, are presented. It can be seen from the tables that these estimates guarantee a relative error of the deflection function values of the order of 1–2% or less. Conclusions similar to those presented in the previous sections can be easily derived.

To conclude, in this case also the estimates of the error introduced by the Bubnov—Galerkin method given here allow us to detect and follow the distribution of a temperature field inside a vibrating plate that is due only to a deformation process, without heat sources.

Table 4.9. Estimates of relative error (as percentages) of deflection function calculations using the uniform norm $\Delta_{n,T_1} w$; problem (3.116)–(3.119); data set (4.3); $\lambda = 1/70$; $w_1(x,y) = \left(x^2 - \frac{1}{4}\right)\left(y^2 - \frac{1}{4}\right)$; Al ($\nu = 0.36$, $c = 2.92 \times 10^{-10}$, $\beta = 6.78 \times 10^{-3}$)

$n_2 \backslash n_1$	1	2	3	4	5	6
1	6.904	4.563	4.305	4.251	4.232	4.225
2	4.786	2.413	2.158	2.103	2.085	2.077
3	3.999	1.614	1.360	1.305	1.287	1.279
4	3.611	1.221	0.968	0.912	0.894	0.887
5	3.389	0.995	0.742	0.687	0.688	0.661
6	3.247	0.851	0.599	0.543	0.525	0.517

Table 4.10. Estimates of relative error (as percentages) of deflection function calculations using the uniform norm $\Delta_{n,T_1} w$; problem (3.116)–(3.119); data set (4.3); $\lambda = 1/70$; $w_1(x,y) = \left(x^2 - \frac{1}{4}\right)\left(y^2 - \frac{1}{4}\right)$; Cu ($\nu = 0.37$, $c = 1.08 \times 10^{-9}$, $\beta = 4.20 \times 10^{-3}$)

$n_2 \backslash n_1$	1	2	3	4	5	6
1	5.115	2.747	2.492	2.437	2.418	2.411
2	3.908	1.522	1.268	1.213	1.195	1.187
3	3.459	1.067	0.814	0.758	0.740	0.732
4	3.239	0.843	0.590	0.535	0.516	0.509
5	3.112	0.714	0.461	0.406	0.388	0.380
6	3.031	0.632	0.380	0.324	0.306	0.298

5 Coupled Nonlinear Thermoelastic Problems

In this chapter, we formulate fundamental assumptions and relations similar to those presented in Chap. 2 for coupled linear thermoelasticity problems of shallow shells. A Timoshenko-type model including the inertial effect of rotation of shell elements is used. Both the generalized heat transfer equation and the equations governing vibration of a shell are formulated in Sect. 5.2, and then some special cases of these equations are analysed. In the next section, boundary and initial conditions are attached to the differential equations. In Sect. 5.4, the existence and uniqueness of a solution as well as the convergence of the Bubnov–Galerkin method, are rigorously discussed.

The question of solvability of the initial–boundary value problems of coupled thermoelasticity defined in Sects. 5.2 and 5.3, and the possibility of the approximate solution of these problems using the Bubnov–Galerkin method are addressed. The existence of a solution to the problem governed by the differential equations and initial and boundary conditions formulated earlier is proved. A few additional theorems are formulated.

5.1 Fundamental Relations and Assumptions

We consider again, as in Chap. 2, a shallow shell with a constant thickness h occupying an area Ω in plan, with a boundary $\partial\Omega_1$. We assume the shell material to be isotropic, homogeneous and elastic. We introduce an orthogonal system of coordinates x, y, z and, in the usual way [232], the coordinates x, y are tangent to the middle surface, and the coordinate z is normal to the middle surface, with its positive direction oriented toward the centre of curvature.

Let us denote the displacements of the middle surface along the axes x, y, z by $u(x, y, z)$, $v(x, y, z)$, $w(x, y, z)$, respectively. We denote the initial shell curvatures related to the coordinate axes x, y by k_x, k_y.

As the fundamental model, a Timoshenko-type model including the inertial effect of rotation of shell elements will be used [232]. It is assumed that shell fibres normal to the middle surface do not curve during a deformation process, but may cease to be normal to the middle surface. The deformation of fibres normal to the middle surface is defined by the condition of a plane stress state ($\sigma_{zz} = 0$). Within this model the following relations between deformations and displacements [232] hold:

$$\varepsilon_{11}^z = \varepsilon_{11} + z æ_{11}, \tag{5.1}$$

$$\varepsilon_{22}^z = \varepsilon_{22} + z æ_{22}, \tag{5.2}$$

$$\varepsilon_{12}^z = \varepsilon_{12} + z æ_{12} \qquad \left(-\frac{h}{2} \le z \le \frac{h}{2}\right), \tag{5.3}$$

$$\varepsilon_{13}^z = \varepsilon_{13}, \tag{5.4}$$

$$\varepsilon_{23}^z = \varepsilon_{23}, \tag{5.5}$$

where ε_{ij} $(i, j = 1, 2)$ are the tangential deformations of the middle surface, ε_{13}, ε_{23} are the shear deformations, $æ_{ij}$ $(i, j = 1, 2)$ are the bending deformations, and Ψ_x, Ψ_y are the rotation angles in the xz and yz planes, respectively. We have

$$\varepsilon_{11} = \frac{\partial u}{\partial x} - k_x w + \frac{1}{2}\left(\frac{\partial w}{\partial x}\right)^2 \qquad (x \leftrightarrow y), \tag{5.6}$$

$$\varepsilon_{12} = \frac{\partial u}{\partial y} + \frac{\partial v}{\partial x} + \frac{\partial w}{\partial x}\frac{\partial w}{\partial y}, \tag{5.7}$$

$$æ_{11} = \frac{\partial \Psi_x}{\partial x}, \quad æ_{22} = \frac{\partial \Psi_y}{\partial y}, \quad æ_{12} = \frac{\partial \Psi_x}{\partial y} + \frac{\partial \Psi_y}{\partial x}, \tag{5.8}$$

$$\varepsilon_{13} = \Psi_x + \frac{\partial w}{\partial x}, \quad \varepsilon_{23} = \Psi_y + \frac{\partial w}{\partial y}. \tag{5.9}$$

Note that the deformations ε_{11}, ε_{22}, ε_{12}, ε_{23}, ε_{13} are assumed to be small in comparison with one, i.e. the derivatives of deflections are neglected.

Experiment shows that the deformation of a body is related to its internal heat capacity and vice versa. Therefore, one can believe that an appropriate modelling of both the deformation and temperature field will result in a better understanding of real processes inside bodies. This effect is particularly important in polymers [110], where interesting elastic waves occur.

Let us assume that the shell has a temperature T_0 in its unstrained, undeformed state. Its temperature then starts to change because of the action of surface and mass forces, internal heat sources and heat exchange with the surrounding medium. We denote by $\theta(x, y, z, t) = T_1(x, y, z, t) - T_0$ the temperature increase at the point of the shell point x, y, z at the time moment t; $T_1(x, y, z)$ is the absolute temperature at this point. We assume $\left|\frac{\theta}{T_0}\right| \ll 1$, i.e. that the change of temperature θ is so small that it does not affect the elastic and thermal constants of the shell material. The following notation will be used in what follows: E, Young modulus; ν, Poisson's ratio; α_T, coefficient of thermal expansion; ρ, density; λ_q, heat transfer coefficient; c_ε, specific heat transfer coefficient for a constant deformation. Assuming that the shell is in a local quasi-equilibrium condition ([96]) and taking into account the relations of irreversible thermodynamic processes for small values of deformation and of temperature increase (see [179]), the following Duhamel–Neumann relations can be obtained:

$$\varepsilon_{ij}^{z} = \frac{1+\nu}{E}\sigma_{ij} - \nu\frac{(\sigma_{11}+\sigma_{22})}{E}\delta_{ij} + \alpha_T Q\delta_{ij}, \quad i,j = 1,2,3,$$
$$\delta_{ij} = \begin{cases} 0, & i \neq j \\ 1, & i = j \end{cases}. \tag{5.10}$$

These relations establish the links between the stresses σ_{ij} and deformations ε_{ij}^{z} in the condition of a plane stress state ($\sigma_{33} = 0$). We take

$$\sigma_{xx} = \sigma_{11},\ \sigma_{yy} = \sigma_{22},\ \sigma_{zz} = \sigma_{33},\ \sigma_{xy} = \sigma_{12},\ \sigma_{xz} = \sigma_{13},\ \sigma_{yz} = \sigma_{23}.$$

Solving (5.1) for σ_{ij}, we obtain

$$\sigma_{11} = \frac{E}{1-\nu^2}(\varepsilon_{11}^{z} + \nu\varepsilon_{22}^{z}) - \frac{E}{1-\nu}\alpha_T Q, \tag{5.11}$$

$$\sigma_{22} = \frac{E}{1-\nu^2}(\varepsilon_{22}^{z} + \nu\varepsilon_{11}^{z}) - \frac{E}{1-\nu}\alpha_T Q, \tag{5.12}$$

$$\sigma_{12} = \frac{E}{2(1+\nu)}\varepsilon_{12}^{z}, \tag{5.13}$$

$$\sigma_{13} = \frac{E}{2(1+\nu)}\varepsilon_{13}^{z}, \tag{5.14}$$

$$\sigma_{23} = \frac{E}{2(1+\nu)}\varepsilon_{23}^{z}. \tag{5.15}$$

Integrating the stresses (5.11)–(5.15) with respect to z and taking into account (5.1), the following longitudinal and transverse forces in the middle surface are found:

$$N_x = \int_{-\frac{h}{2}}^{\frac{h}{2}} \sigma_{11}\,dz = \frac{Eh}{1-\nu^2}\left\{\frac{\partial u}{\partial x} - k_x w + \frac{1}{2}\left(\frac{\partial w}{\partial x}\right)^2 + \nu\left[\frac{\partial v}{\partial y} - k_y w + \frac{1}{2}\left(\frac{\partial w}{\partial y}\right)^2\right]\right\} - \frac{E\alpha_T}{1-\nu}\int_{-\frac{h}{2}}^{\frac{h}{2}} Q\,dz, \tag{5.16}$$

$$N_y = \int_{-\frac{h}{2}}^{\frac{h}{2}} \sigma_{22}\,dz = \frac{Eh}{1-\nu^2}\left\{\frac{\partial v}{\partial y} - k_y w + \frac{1}{2}\left(\frac{\partial w}{\partial y}\right)^2 + \nu\left[\frac{\partial u}{\partial x} - k_x w + \frac{1}{2}\left(\frac{\partial w}{\partial x}\right)^2\right]\right\} - \frac{E\alpha_T}{1-\nu}\int_{-\frac{h}{2}}^{\frac{h}{2}} Q\,dz, \tag{5.17}$$

$$S = \int_{-\frac{h}{2}}^{\frac{h}{2}} \sigma_{12}\, dz = \frac{Eh}{2(1+\nu)} \left(\frac{\partial u}{\partial y} + \frac{\partial v}{\partial x} + \frac{\partial w}{\partial x} \frac{\partial w}{\partial y} \right), \tag{5.18}$$

$$Q_x = \int_{-\frac{h}{2}}^{\frac{h}{2}} \sigma_{12}\, dz = k^2 \frac{Eh}{2(1+\nu)} \left(\Psi_x + \frac{\partial w}{\partial x} \right), \tag{5.19}$$

$$Q_y = \int_{-\frac{h}{2}}^{\frac{h}{2}} \sigma_{23}\, dz = k^2 \frac{Eh}{2(1+\nu)} \left(\Psi_y + \frac{\partial w}{\partial y} \right), \tag{5.20}$$

where

$$\frac{1}{k^2} = \frac{1}{h} \int_{-\frac{h}{2}}^{\frac{h}{2}} f^2(z)\, dz.$$

The function $f(z)$ characterizes the distribution of tangential stresses through the shell thickness. In order to obtain the moments, the stresses are integrated with weight z over the thickness:

$$M_x = \int_{-\frac{h}{2}}^{\frac{h}{2}} \sigma_{11} z\, dz = \frac{Eh^3}{12(1-\nu^2)} \left(\frac{\partial \Psi_x}{\partial x} + \nu \frac{\partial \Psi_y}{\partial y} \right) - \frac{E\alpha_T}{1-\nu} \int_{-\frac{h}{2}}^{\frac{h}{2}} zQ\, dz, \tag{5.21}$$

$$M_y = \int_{-\frac{h}{2}}^{\frac{h}{2}} \sigma_{22} z\, dz = \frac{Eh^3}{12(1-\nu^2)} \left(\frac{\partial \Psi_y}{\partial y} + \nu \frac{\partial \Psi_x}{\partial x} \right) - \frac{E\alpha_T}{1-\nu} \int_{-\frac{h}{2}}^{\frac{h}{2}} zQ\, dz, \tag{5.22}$$

$$H = \int_{-\frac{h}{2}}^{\frac{h}{2}} \sigma_{12} z\, dz = \frac{Eh^3}{12(1-\nu^2)} \frac{(1-\nu)}{2} \left(\frac{\partial \Psi_x}{\partial y} + \frac{\partial \Psi_y}{\partial x} \right). \tag{5.23}$$

5.2 Differential Equations

From the equation governing the entropy balance and taking into account the coupling between the components of heat flow and the thermodynamic forces associated with the Fourier law, we obtain the generalized heat transfer equation (see [179]):

$$c_\varepsilon \frac{\partial \theta}{\partial t} - \lambda_q \Delta \theta = -\frac{E\alpha_T T_0}{(1-2\nu)} \frac{\partial}{\partial t} (\varepsilon_{11}^z + \varepsilon_{22}^z + \varepsilon_{33}^z) + g_2. \tag{5.24}$$

The underlined term links the temperature increase with the velocity of the volume change, and $g_2(x, y, z, t)$ is the quantity of heat transferred in a unit volume within a unit time interval.

In order to obtain the equations governing the vibration of a shell with respect to the displacements, we use the following differential equation, which governs the motion (including inertial forces) of the shell element $h\,dx\,dy$ in the x, y, z directions and the rotational inertia of the element with respect to x, y ([232]):

$$\frac{\partial N_x}{\partial x} + \frac{\partial S}{\partial y} + p_1 - \rho h \frac{\partial^2 u}{\partial t^2} = 0, \tag{5.25}$$

$$\frac{\partial S}{\partial x} + \frac{\partial N_y}{\partial y} + p_2 - \rho h \frac{\partial^2 v}{\partial t^2} = 0, \tag{5.26}$$

$$\begin{aligned} &\frac{\partial Q_x}{\partial x} + \frac{\partial Q_y}{\partial y} + k_x N_x + k_y N_y + \frac{\partial}{\partial x}\left(N_x \frac{\partial w}{\partial x} + S \frac{\partial w}{\partial y}\right) \\ &+ \frac{\partial}{\partial y}\left(S \frac{\partial w}{\partial x} + N_y \frac{\partial w}{\partial x}\right) + g_1 - \rho h \frac{\partial^2 w}{\partial t^2} = 0, \end{aligned} \tag{5.27}$$

$$\frac{\partial M_x}{\partial x} + \frac{\partial H}{\partial y} + Q_x - \rho \frac{h^3}{12} \frac{\partial^2 \Psi_x}{\partial t^2} = 0, \tag{5.28}$$

$$\frac{\partial H}{\partial x} + \frac{\partial M_y}{\partial y} + Q_y - \rho \frac{h^3}{12} \frac{\partial^2 \Psi_y}{\partial t^2} = 0, \tag{5.29}$$

where p_1, p_2, g_1 are the intensities of the external loads along the x, y, z axes; $\rho h \frac{\partial^2 u}{\partial t^2}\, dx\, dy$, $\rho h \frac{\partial^2 v}{\partial t^2}\, dx\, dy$, $\rho h \frac{\partial^2 w}{\partial t^2}\, dx\, dy$ are the components of the inertial force along the x, y, z axes; and $\rho \frac{h^3}{12} \frac{\partial^2 \Psi_x}{\partial t^2}$, $\rho \frac{h^3}{12} \frac{\partial^2 \Psi_y}{\partial t^2}$ are the moments of inertia around the x and y axes.

Substituting (5.16)–(5.23) into (5.25)–(5.29) and attaching the heat transfer equation (5.24), we obtain the full system of differential equations of thermoelasticity with respect to displacements, in the following form:

$$\begin{aligned} &\frac{Eh}{1-\nu^2} \frac{\partial}{\partial x}\left[\left(\frac{\partial u}{\partial x} - k_x w + \frac{1}{2}\left(\frac{\partial w}{\partial x}\right)^2\right) + \nu\left(\frac{\partial v}{\partial y} - k_y w + \frac{1}{2}\left(\frac{\partial w}{\partial y}\right)^2\right)\right] \\ &-\frac{E\alpha_T}{1-\nu} \frac{\partial}{\partial x}\left\{\int_{-\frac{h}{2}}^{\frac{h}{2}} \theta\, dz\right\} + \frac{Eh}{2(1+\nu)} \frac{\partial}{\partial y}\left[\frac{\partial u}{\partial y} + \frac{\partial v}{\partial x} + \frac{\partial w}{\partial x}\frac{\partial w}{\partial y}\right] \\ &+p_1 - \rho h \frac{\partial^2 u}{\partial t^2} = 0, \end{aligned} \tag{5.30}$$

$$\frac{Eh}{1-\nu^2}\frac{\partial}{\partial y}\left[\left(\frac{\partial v}{\partial y}-k_y w+\frac{1}{2}\left(\frac{\partial w}{\partial y}\right)^2\right)+\nu\left(\frac{\partial u}{\partial x}-k_x w+\frac{1}{2}\left(\frac{\partial w}{\partial x}\right)^2\right)\right]$$
$$-\frac{E\alpha_T}{1-\nu}\frac{\partial}{\partial y}\left\{\int_{-\frac{h}{2}}^{\frac{h}{2}}\theta\,dz\right\}+\frac{Eh}{2(1+\nu)}\frac{\partial}{\partial x}\left[\frac{\partial u}{\partial y}+\frac{\partial v}{\partial x}+\frac{\partial w}{\partial x}\frac{\partial w}{\partial y}\right]$$
$$+p_2-\rho h\frac{\partial^2 v}{\partial t^2}=0, \tag{5.31}$$

$$k^2\frac{Eh}{2(1+\nu)}\frac{\partial}{\partial x}\left[\Psi_x+\frac{\partial w}{\partial x}\right]+k^2\frac{Eh}{2(1+\nu)}\frac{\partial}{\partial y}\left[\Psi_y+\frac{\partial w}{\partial y}\right]+k_x\left[\frac{Eh}{1-\nu}\left\{\left(\frac{\partial u}{\partial x}\right.\right.\right.$$
$$\left.\left.\left.-k_x w+\frac{1}{2}\left(\frac{\partial w}{\partial x}\right)^2\right)+\nu\left(\frac{\partial v}{\partial y}-k_y w+\frac{1}{2}\left(\frac{\partial w}{\partial y}\right)^2\right)\right\}-\frac{E\alpha_t}{1-\nu}\int_{-\frac{h}{2}}^{\frac{h}{2}}\theta\,dz\right]$$
$$+k_y\left[\frac{Eh}{1-\nu^2}\left\{\left(\frac{\partial v}{\partial y}-k_y w+\frac{1}{2}\left(\frac{\partial w}{\partial y}\right)^2\right)+\nu\left(\frac{\partial u}{\partial x}-k_x w+\frac{1}{2}\left(\frac{\partial w}{\partial x}\right)^2\right)\right\}\right.$$
$$\left.-\frac{E\alpha_T}{1-\nu}\int_{-\frac{h}{2}}^{\frac{h}{2}}\theta\,dz\right]+\frac{\partial}{\partial x}\left[\frac{\partial w}{\partial x}\left\{\frac{Eh}{1-\nu^2}\left(\left[\frac{\partial u}{\partial x}-k_x w+\frac{1}{2}\left(\frac{\partial w}{\partial x}\right)^2\right]\right.\right.\right.$$
$$\left.\left.+\nu\left[\frac{\partial v}{\partial y}-k_y w+\frac{1}{2}\left(\frac{\partial w}{\partial y}\right)^2\right]\right)-\frac{E\alpha_T}{1-\nu}\int_{-\frac{h}{2}}^{\frac{h}{2}}\theta\,dz\right\}$$
$$\left.+\frac{\partial w}{\partial y}\left\{\frac{Eh}{2(1+\nu)}\left[\frac{\partial u}{\partial y}+\frac{\partial v}{\partial x}+\frac{\partial w}{\partial x}\frac{\partial w}{\partial y}\right]\right\}\right]$$
$$+\frac{\partial}{\partial y}\left[\frac{\partial w}{\partial y}\left\{\frac{Eh}{1-\nu^2}\left(\left[\frac{\partial v}{\partial y}-k_y w+\frac{1}{2}\left(\frac{\partial w}{\partial y}\right)^2\right]+\nu\left[\frac{\partial u}{\partial x}-k_x w+\frac{1}{2}\left(\frac{\partial w}{\partial x}\right)^2\right]\right)\right.\right.$$
$$\left.\left.-\frac{E\alpha_T}{1-\nu}\int_{-\frac{h}{2}}^{\frac{h}{2}}\theta\,dz\right\}+\frac{\partial w}{\partial x}\left\{\frac{Eh}{2(1+\nu)}\left[\frac{\partial u}{\partial y}+\frac{\partial v}{\partial x}+\frac{\partial w}{\partial x}\frac{\partial w}{\partial y}\right]\right\}\right]$$
$$+g_1-\rho h\frac{\partial^2 w}{\partial t^2}=0, \tag{5.32}$$

$$D\frac{\partial}{\partial x}\left[\frac{\partial \Psi_x}{\partial x}+\nu\frac{\partial \Psi_y}{\partial y}\right]-\frac{E\alpha_T}{1-\nu}\frac{\partial}{\partial x}\left\{\int\limits_{-\frac{h}{2}}^{\frac{h}{2}}\theta\,dz\right\}+D\frac{1-\nu}{2}\frac{\partial}{\partial y}\left[\frac{\partial \Psi_x}{\partial y}\right.$$

$$\left.+\frac{\partial \Psi_y}{\partial x}\right]-k^2\frac{Eh}{2(1+\nu)}\left(\Psi_x+\frac{\partial w}{\partial x}\right)-\rho\frac{h^3}{12}\frac{\partial^2 \Psi_x}{\partial t^2}=0, \tag{5.33}$$

$$D\frac{\partial}{\partial y}\left[\frac{\partial \Psi_y}{\partial y}+\nu\frac{\partial \Psi_x}{\partial x}\right]-\frac{E\alpha_T}{1-\nu}\frac{\partial}{\partial y}\left\{\int\limits_{-\frac{h}{2}}^{\frac{h}{2}}\theta\,dz\right\}+D\frac{1-\nu}{2}\frac{\partial}{\partial x}\left[\frac{\partial \Psi_x}{\partial y}\right.$$

$$\left.+\frac{\partial \Psi_y}{\partial x}\right]-k^2\frac{Eh}{2(1+\nu)}\left(\Psi_y+\frac{\partial w}{\partial y}\right)-\rho\frac{h^3}{12}\frac{\partial^2 \Psi_y}{\partial t^2}=0, \tag{5.34}$$

$$C_0\frac{\partial \theta}{\partial t}-\lambda_q\Delta\theta=-\frac{E\alpha_T T_0}{1-\nu}\frac{\partial}{\partial t}\left[\frac{\partial u}{\partial x}-k_x w+\frac{1}{2}\left(\frac{\partial w}{\partial x}\right)^2\right.$$

$$\left.+z\frac{\partial \Psi_x}{\partial x}+\frac{\partial v}{\partial y}-k_y w+\frac{1}{2}\left(\frac{\partial w}{\partial y}\right)^2+z\frac{\partial \Psi_y}{\partial y}\right]+g_2, \tag{5.35}$$

$$C_0=C_\varepsilon+\frac{E\alpha_T^2 T_0(1+\nu)}{(1-\nu)(1-2\nu)},\qquad D=\frac{Eh^3}{12(1-\nu^2)}.$$

The system of differential equations (5.30)–(5.35) governs the coupled dynamical problem of thermoelasticity taking into account the geometric nonlinearity. Let us emphasize that during the derivation of the equations, no limitations on the temperature distribution through the shell thickness have been assumed, and therefore a system of equations with different dimensions has been obtained. The variable θ occurring in the heat transfer equation (a parabolic-type equation) depends on the three spatial variables x, y, z and the time t. The variables Ψ_x, Ψ_y, w, u, v occurring in the system of equations governing the motion of a shell element (a hyperbolic-type equation) depend on the two spatial variables x, y and the time t.

Let us now consider some special cases of the equations above. First, we do not take transverse shear effects into account, i.e. we consider the coupled dynamical problem of thermoelasticity within the Kirchhoff–Love model. In this case the rotation angles Ψ_x, Ψ_y of the normal to the middle surface have the following form ([232]):

$$\Psi_x=-\frac{\partial w}{\partial x},\quad \Psi_y=-\frac{\partial w}{\partial y}. \tag{5.36}$$

The transverse forces and moments now have the form

$$Q_x=\frac{\partial M_x}{\partial x}+\frac{\partial H}{\partial y}-\rho\frac{h^3}{12}\frac{\partial^2}{\partial t^2}\left(\frac{\partial w}{\partial x}\right), \tag{5.37}$$

$$Q_y=\frac{\partial M_y}{\partial y}+\frac{\partial H}{\partial x}-\rho\frac{h^3}{12}\frac{\partial^2}{\partial t^2}\left(\frac{\partial w}{\partial y}\right), \tag{5.38}$$

$$M_x = -D\left[\frac{\partial^2 w}{\partial x^2} + \nu \frac{\partial^2 w}{\partial y^2}\right] - \frac{E\alpha_T}{1-\nu}\int\limits_{-\frac{h}{2}}^{\frac{h}{2}} \theta z \, dz, \tag{5.39}$$

$$M_y = -D\left[\frac{\partial^2 w}{\partial y^2} + \nu \frac{\partial^2 w}{\partial x^2}\right] - \frac{E\alpha_T}{1-\nu}\int\limits_{-\frac{h}{2}}^{\frac{h}{2}} \theta z \, dz, \tag{5.40}$$

$$H = -D(1-\nu)\frac{\partial^2 w}{\partial x \partial y}. \tag{5.41}$$

According to (5.36)–(5.41), the equations governing the motion of a shell element have now the form

$$\frac{\partial N_x}{\partial x} + \frac{\partial S}{\partial y} + p_1 - \rho\frac{h^3}{12}\frac{\partial^2 u}{\partial t^2} = 0,$$

$$\frac{\partial S}{\partial x} + \frac{\partial N_y}{\partial y} + p_2 - \rho\frac{h^3}{12}\frac{\partial^2 v}{\partial t^2} = 0,$$

$$\frac{\partial^2 M_x}{\partial x^2} + 2\frac{\partial^2 H}{\partial x\,\partial y} + \frac{\partial^2 M_y}{\partial y^2} + k_x N_x + k_y N_y$$

$$+\frac{\partial}{\partial x}\left(N_x\frac{\partial w}{\partial x} + S\frac{\partial w}{\partial y}\right) + \frac{\partial}{\partial y}\left(S\frac{\partial w}{\partial x} + N_y\frac{\partial w}{\partial y}\right)$$

$$+g_1 + \rho\frac{h^3}{12}\frac{\partial^2}{\partial t^2}\Big(\Delta w\Big) - \rho h\frac{\partial^2 w}{\partial t^2} = 0. \tag{5.42}$$

The volume deformation is governed by the following equation:

$$\varepsilon_{11}^z + \varepsilon_{22}^z + \varepsilon_{33}^z = \Bigg[\frac{\partial u}{\partial x} - k_x w + \frac{1}{2}\left(\frac{\partial w}{\partial x}\right)^2 + \frac{\partial v}{\partial y} - k_y w$$

$$+\frac{1}{2}\left(\frac{\partial w}{\partial y}\right)^2 - z\left(\frac{\partial^2 w}{\partial x^2} + \frac{\partial^2 w}{\partial y^2}\right)\Bigg]\frac{1-2\nu}{1-\nu} + \frac{1+\nu}{1-\nu}\alpha_T\theta. \tag{5.43}$$

Substituting (5.16)–(5.18) and (5.39)–(5.41) into (5.42), attaching the generalized equation of heat transfer (5.24) and taking into account (5.43), the following system of equations is obtained:

$$\frac{Eh}{1-\nu^2}\frac{\partial}{\partial x}\left[\left(\frac{\partial u}{\partial x} - k_x w + \frac{1}{2}\left(\frac{\partial w}{\partial x}\right)^2\right) + \nu\left(\frac{\partial v}{\partial y} - k_y w + \frac{1}{2}\left(\frac{\partial w}{\partial y}\right)^2\right)\right]$$

$$-\frac{E\alpha_T}{1-\nu}\frac{\partial}{\partial x}\left\{\int\limits_{-\frac{h}{2}}^{\frac{h}{2}} \theta \, dz\right\} + \frac{Eh}{2(1+\nu)}\frac{\partial}{\partial y}\left[\frac{\partial u}{\partial y} + \frac{\partial v}{\partial x} + \frac{\partial w}{\partial x}\frac{\partial w}{\partial y}\right]$$

$$+p_1 - \rho h\frac{\partial^2 u}{\partial t^2} = 0, \tag{5.44}$$

$$\frac{Eh}{1-\nu^2}\frac{\partial}{\partial y}\left[\left(\frac{\partial v}{\partial y}-k_y w+\frac{1}{2}\left(\frac{\partial w}{\partial y}\right)^2\right)+\nu\left(\frac{\partial u}{\partial x}-k_x w+\frac{1}{2}\left(\frac{\partial w}{\partial x}\right)^2\right)\right]$$
$$-\frac{E\alpha_T}{1-\nu}\frac{\partial}{\partial y}\left\{\int_{-\frac{h}{2}}^{\frac{h}{2}}\theta\,dz\right\}+\frac{Eh}{2(1+\nu)}\frac{\partial}{\partial x}\left[\frac{\partial u}{\partial y}+\frac{\partial v}{\partial x}+\frac{\partial w}{\partial x}\frac{\partial w}{\partial y}\right]$$
$$+p_2-\rho h\frac{\partial^2 v}{\partial t^2}=0, \tag{5.45}$$

$$-D\frac{\partial^2}{\partial x^2}\left[\frac{\partial^2 w}{\partial x^2}+\nu\frac{\partial^2 w}{\partial y^2}\right]-2D(1-\nu)\frac{\partial^2}{\partial x\partial y}\left[\frac{\partial^2 w}{\partial x\partial y}\right]-D\frac{\partial^2}{\partial y^2}\left[\frac{\partial^2 w}{\partial y^2}+\nu\frac{\partial^2 w}{\partial x^2}\right]$$
$$-\frac{E\alpha_T}{1-\nu}\Delta\left\{\int_{-\frac{h}{2}}^{\frac{h}{2}}\theta z\,dz\right\}+\rho\frac{h^3}{12}\frac{\partial^2}{\partial t^2}(\Delta w)$$
$$+k_x\left[\frac{Eh}{1-\nu}\left\{\left(\frac{\partial u}{\partial x}-k_x w+\frac{1}{2}\left(\frac{\partial w}{\partial x}\right)^2\right)+\nu\left(\frac{\partial v}{\partial y}-k_y w+\frac{1}{2}\left(\frac{\partial w}{\partial y}\right)^2\right)\right\}\right.$$
$$\left.-\frac{E\alpha_t}{1-\nu}\int_{-\frac{h}{2}}^{\frac{h}{2}}\theta\,dz\right]+k_y\left[\frac{Eh}{1-\nu^2}\left\{\left(\frac{\partial v}{\partial y}-k_y w+\frac{1}{2}\left(\frac{\partial w}{\partial y}\right)^2\right)\right.\right.$$
$$\left.\left.+\nu\left(\frac{\partial u}{\partial x}-k_x w+\frac{1}{2}\left(\frac{\partial w}{\partial x}\right)^2\right)\right\}-\frac{E\alpha_T}{1-\nu}\int_{-\frac{h}{2}}^{\frac{h}{2}}\theta\,dz\right]$$
$$+\frac{\partial}{\partial x}\left[\frac{\partial w}{\partial x}\left\{\frac{Eh}{1-\nu^2}\left(\left[\frac{\partial u}{\partial x}-k_x w+\frac{1}{2}\left(\frac{\partial w}{\partial x}\right)^2\right]+\nu\left[\frac{\partial v}{\partial y}-k_y w+\frac{1}{2}\left(\frac{\partial w}{\partial y}\right)^2\right]\right)\right.\right.$$
$$\left.\left.-\frac{E\alpha_T}{1-\nu}\int_{-\frac{h}{2}}^{\frac{h}{2}}\theta\,dz\right\}+\frac{\partial w}{\partial y}\left\{\frac{Eh}{2(1+\nu)}\left[\frac{\partial u}{\partial y}+\frac{\partial v}{\partial x}+\frac{\partial w}{\partial x}\frac{\partial w}{\partial y}\right]\right\}\right]$$
$$+\frac{\partial}{\partial y}\left[\frac{\partial w}{\partial y}\left\{\frac{Eh}{1-\nu^2}\left(\left[\frac{\partial v}{\partial y}-k_y w+\frac{1}{2}\left(\frac{\partial w}{\partial y}\right)^2\right]+\nu\left[\frac{\partial u}{\partial x}-k_x w+\frac{1}{2}\left(\frac{\partial w}{\partial x}\right)^2\right]\right)\right.\right.$$
$$\left.\left.-\frac{E\alpha_T}{1-\nu}\int_{-\frac{h}{2}}^{\frac{h}{2}}\theta\,dz\right\}+\frac{\partial w}{\partial x}\left\{\frac{Eh}{2(1+\nu)}\left[\frac{\partial u}{\partial y}+\frac{\partial v}{\partial x}+\frac{\partial w}{\partial x}\frac{\partial w}{\partial y}\right]\right\}\right]$$
$$+g_1-\rho h\frac{\partial^2 w}{\partial t^2}=0, \tag{5.46}$$

$$C_0\frac{\partial\theta}{\partial t}-\lambda_q\Delta\theta=-\frac{E\alpha_T T_0}{1-\nu}\frac{\partial}{\partial t}\left[\frac{\partial u}{\partial x}-k_x w+\frac{1}{2}\left(\frac{\partial w}{\partial x}\right)^2\right.$$
$$\left.-z\left(\frac{\partial^2 w}{\partial x^2}+\frac{\partial^2 w}{\partial y^2}\right)+\frac{\partial v}{\partial y}-k_y w+\frac{1}{2}\left(\frac{\partial w}{\partial y}\right)^2\right]+g_2. \tag{5.47}$$

Therefore, the system of differential equations (5.44)–(5.47) governs the coupled problem of thermoelasticity in the framework of the Kirchhoff–Love model, taking into account the geometrical nonlinearity and rotational inertia of the shell elements. It is clear that in the system (5.44)–(5.47), different types of equations (parabolic and hyperbolic) with different dimensions appear again.

The equations for linear problems discussed in Chap. 2 can be derived from the above system as a paricular case.

5.3 Boundary and Initial Conditions

The differential equations given above must be completed by initial and boundary conditions. We need to include the initial state of the shell and to formulate the conditions describing its influence on the surrounding medium. For the mathematical model of thermoelastic processes in shells used here, we take into account only mechanical and thermal phenomena between the shell and the neighbouring medium.

As the initial conditions, we take the distribution (for $t = 0$) of the displacements w, u, v, the angles of rotation of the normal Ψ_x, Ψ_y, the velocities of those displacments and angles $\frac{\partial w}{\partial t}, \frac{\partial u}{\partial t}, \frac{\partial v}{\partial t}, \frac{\partial \Psi_x}{\partial t}, \frac{\partial \Psi_y}{\partial t}$ and the temperature increase θ (the last condition is equivalent to introducing the temperature T_1) inside the shell:

$$\begin{aligned} w|_{t=0} &= \varphi_1(x,y), \quad u|_{t=0} = \varphi_2(x,y), \\ v|_{t=0} &= \varphi_3(x,y), \quad \theta|_{t=0} = \varphi_4(x,y,z), \end{aligned} \tag{5.48}$$

$$\Psi_x|_{t=0} = \varphi_5(x,y), \quad \Psi_y|_{t=0} = \varphi_6(x,y), \tag{5.49}$$

$$\left.\frac{\partial w}{\partial t}\right|_{t=0} = \Psi_1(x,y), \quad \left.\frac{\partial u}{\partial t}\right|_{t=0} = \Psi_2(x,y), \quad \left.\frac{\partial v}{\partial t}\right|_{t=0} = \Psi_3(x,y), \tag{5.50}$$

$$\left.\frac{\partial \Psi_x}{\partial t}\right|_{t=0} = \Psi_4(x,y), \quad \left.\frac{\partial \Psi_y}{\partial t}\right|_{t=0} = \Psi_5(x,y). \tag{5.51}$$

As an example of the mechanical conditions (boundary conditions) on the shell boundary, we present the following set, which will be used later (the formulation for other boundary conditions is known and can be found in the series of monographs [133, 232, 234]):

$$w|_{\partial\Omega_1} = f_1(x,y,t), \quad \left.\frac{\partial w}{\partial n}\right|_{\partial\Omega_1} = f_2(x,y,t), \tag{5.52}$$

$$u|_{\partial\Omega_1} = f_3(x,y,t), \quad v|_{\partial\Omega_1} = f_4(x,y,t), \tag{5.53}$$

$$\Psi_x|_{\partial\Omega_1} = f_5(x,y,t), \quad \Psi_y|_{\partial\Omega_1} = f_6(x,y,t), \tag{5.54}$$

$$(x,y) \in \partial\Omega_1, \ t \in [0,T],$$

where $[0, T]$ denotes the time interval during which the shell vibrations are investigated; $f_i(x, y, t)$, $i = 1, \ldots, 6$, are known functions of the coordinates and time; and $\frac{\partial}{\partial n}$ is the differential operator acting along an external normal to $\partial\Omega_1$. For $f_i = 0$, $i = 1, \ldots, 6$, these conditions correspond to a shell clamped on the contour $\partial\Omega_1$.

As the heat transfer conditions, one of the following ([123]) can be used:

1. A temperature distribution is assumed on the shell surface $\partial\Omega_2$:
$$\theta(x, y, z, t) = f_7(x, y, z, t), \quad t \in [0, T], \quad (x, y, z) \in \partial\Omega_2. \tag{5.55}$$
2. The density of the heat flow through the surface $\partial\Omega_2$ is given:
$$\frac{\partial\theta(x, y, z, t)}{\partial n} = f_8(x, y, z, t), \quad t \in [0, T], \quad (x, y, z) \in \partial\Omega_2,$$
where $\frac{\partial}{\partial n}$ denotes the differential operator along an external normal to $\partial\Omega_2$.
3. A condition of free heat exchange with the surrounding medium on the surface $\partial\Omega_2$ is specyfied:
$$\left(\frac{\partial}{\partial n} + \alpha\right) \theta(x, y, z, t) = f_9(x, y, z, t), \quad t \in [0, T], \quad (x, y, z) \in \partial\Omega_2,$$
where $\alpha = const$, $f_i(x, y, z, t)$, $i = 7, 8, 9$, are known functions of the coordinates and time.

5.4 On the Existence and Uniqueness of a Solution and on the Convergence of the Bubnov–Galerkin Method

We introduce the following notations and definitions.

Ω_1 is a bounded volume of the Euclidean space E_2 ($\overline{x} = (x, y)$ denotes a point of E_2), i.e. Ω_1 is a coupled open set; $\overline{\Omega_1} = \Omega_1 \cup \partial\Omega_1$, and $\partial\Omega_1$ is the boundary of Ω_1. $Q_1 = \Omega_1 \times (0, T)$ is a cylinder of height $T > 0$ in the space $R_3 = E_2 \times (-\infty < t < +\infty)$. $G = \partial\Omega_1 \times [0, T]$ is the surface of the cylinder Q_1. $\Omega_2 = \Omega_1 \times \left(-\frac{h}{2}, \frac{h}{2}\right)$ is a cylinder of height h in the space E_3 ($\overline{x} = (x, y, z)$ is a point of E_3). $\overline{\Omega_2} = \Omega_2 \cup \partial\Omega_2$, where $\partial\Omega_2$ is the boundary surface of Ω_2. $Q_2 = \Omega_2 \times (0, T)$ is a cylinder of height $T > 0$ in the space $R_4 = E_3 \times \{-\infty < t < +\infty\}$. $S = \partial\Omega_2 \times [0, T]$ is the surface of the cylinder Q_2.

In our investigations of nonstationary problems, we use the following functional space: $L^p(0, T; X)$ is the space of functions $f(t)$ with values in the space X, measured with respect to the Lebesgue measure dt of the interval $[0, T]$ and possessing the finite norm

$$\|f\| = \left[\int_0^T \|f\|_x^p\right]^{1/p},$$

where $\|f\|_x$ is the norm of elements in X. In this section, we use as X the spaces $\mathring{W}_n^m(\Omega_2)$, $\mathring{W}_n^m(\Omega_1)$, $W_n^m(\Omega_1)$, $L_n(\Omega_1)$, $L_n(\Omega_2)$.

The aim of this section is to investigate the solvability of the initial-boundary value problems of coupled thermoelasticity which were defined in the Sects. 5.2 and 5.3 and the possibility of their approximate solution using the Bubnov–Galerkin method. Firstly, we prove the existence of a solution to the problem defined by the equations (5.44)–(5.47), the initial conditions (5.48)–(5.51) and the boundary conditions (5.52)–(5.54). Using the notation defined above, the initial and boundary conditions have the following form (here we consider only homogeneous boundary conditions, i.e. $f_i = 0$, $i = 1, 2, 3, 4, 7$):

$$\begin{gathered} w\big|_{t=0} = \varphi_1(x,y), \quad u\big|_{t=0} = \varphi_2(x,y), \quad v\big|_{t=0} = \varphi_3(x,y), \\ (x,y) \in \partial\Omega_1, \quad t \in [0,T], \\ \left.\frac{\partial w}{\partial t}\right|_{t=0} = \Psi_1(x,y), \quad \left.\frac{\partial u}{\partial t}\right|_{t=0} = \Psi_2(x,y), \quad \left.\frac{\partial v}{\partial t}\right|_{t=0} = \Psi_3(x,y), \end{gathered} \tag{5.56}$$

$$w\big|_G = \left.\frac{\partial w}{\partial n}\right|_G = u\big|_G = v\big|_G = 0, \tag{5.57}$$

$$\theta\big|_{t=0} = \varphi_4(x,y,z), \quad (x,y,z) \in \Omega_2, \tag{5.58}$$

$$\theta\big|_S = 0. \tag{5.59}$$

Theorem 5.1 *Let $\partial\Omega_1$ be a sufficiently smooth boundary and let $p_1, p_2, g_1 \in L^2(0,T;(L_2(\Omega_1))$, $g_2 \in L^2(0,T;L_2(\Omega_2))$, $\varphi_1 \in \mathring{W}_2^2(\Omega_1)$, $\varphi_2, \varphi_3 \in \mathring{W}_2^1(\Omega_1)$, $\varphi_4 \in \mathring{W}_2^1(\Omega_2)$, $\Psi_1 \in \mathring{W}_2^1(\Omega_1)$, $\Psi_2, \Psi_3 \in L_2(\Omega_1)$. Then:*

1. *There exists at least one set of four functions $\{w, u, v, \theta\}$ that are a solution to the problem (5.44)–(5.47), (5.56)–*(5.59)*, and the following relations hold:*

$$w \in L^2(0,T;\mathring{W}_2^2(\Omega_1)), \quad \frac{\partial w}{\partial t} \in L^2(0,T;\mathring{W}_2^1(\Omega_1)), \tag{5.60}$$

$$u, v \in L^2(0,T;\mathring{W}_2^1(\Omega_1)), \quad \frac{\partial u}{\partial t}, \frac{\partial v}{\partial t} \in L^2(0,T;L_2(\Omega_1)), \tag{5.61}$$

$$\theta \in L^2(0,T;\mathring{W}_2^1(\Omega_2)). \tag{5.62}$$

2. *An approximate solution to the problem stated above can be found with the help of the Bubnov–Galerkin method in the form*

$$\begin{gathered} w_n = \sum_{i=1}^{n} g_{1i}(t)\chi_{1i}(x,y), \quad u_n = \sum_{i=1}^{n} g_{2i}(t)\chi_{2i}(x,y), \\ v_n = \sum_{i=1}^{n} g_{3i}(t)\chi_{2i}(x,y), \quad \theta_n = \sum_{i=1}^{n} g_{4i}(t)\chi_{3i}(x,y). \end{gathered} \tag{5.63}$$

Here $\{\chi_{1i}\}$ is a basis system in $\overset{\circ}{W}{}_2^2(\Omega_1)$, orthonormalized with the norm $\overset{\circ}{W}{}_2^1(\Omega_1)$, where the norm (in this case) has the form

$$\left\|a\right\|_{\overset{\circ}{W}{}_2^1(\Omega_1)} = \left[\int\limits_{\Omega_1} \left\{\rho h a^2 + \rho\frac{h^3}{12}\left[(\frac{\partial a}{\partial x})^2 + (\frac{\partial a}{\partial y})^2\right]\right\} d\Omega_1\right]^{1/2},$$

for an arbitrary element $a \in \overset{\circ}{W}{}_2^1(\Omega_1)$; $\{\chi_{2i}\}$ is a basis system in $\overset{\circ}{W}{}_2^1(\Omega_1)$, orthonormalized in $L_2(\Omega_1)$; $\{\chi_{3i}\}$ is a basis system in $\overset{\circ}{W}{}_2^1(\Omega_2)$, orthonormalized in $L_2(\Omega_2)$; and for arbitrary i, $\chi_{3i} = a_i(x,y)b_i(z)$ holds. In order to define the unknown $g_{ij}(t)$, $j = 1, \ldots, 4$, we obtain a system of ODEs solvable for arbitrary n on the whole interval $[0,T]$.

3. *The whole set of approximate solutions $\{w_n, u_n, v_n, \theta_n\}$ is weakly compact in the spaces corresponding to (5.60)–(5.62).*

Note that sufficient smoothness of $\partial\Omega_1$ is required by the theorems on inclusion and on representation of the basis functions in $\overset{\circ}{W}{}_2^1(\Omega_2)$ which are used in the proof.

Proof. The proof of the above theorem will be constructed here using the method of compactness (in the sense given in [147]). Fundamental investigations of this method and its application to problems of shells were performed by Vorovich and Morozov [164, 165, 235].

Step 1. Construction of the approximate solution using the Bubnov–Galerkin method.

Substituting the functions (5.63) into the system (5.44)–(5.47), we define the unknown $g_{ij}(t)$, $1, \ldots, 4$, by the following ordinary differential equations:

$$-\frac{Eh}{1-\nu^2}\left(\left[\left(\frac{\partial u_n}{\partial x} - k_x w_n + \frac{1}{2}\left(\frac{\partial w_n}{\partial x}\right)^2\right) + \nu\left(\frac{\partial v_n}{\partial y} - k_y w_n + \frac{1}{2}\left(\frac{\partial w_n}{\partial y}\right)^2\right)\right],\right.$$

$$\left.\frac{\partial \chi_{2l}}{\partial x}\right)_{L_2(\Omega_1)} + \frac{E\alpha_T}{1-\nu}\left(\int\limits_{-\frac{h}{2}}^{\frac{h}{2}} \theta_n\, dz,\ \frac{\partial \chi_{2l}}{\partial x}\right)_{L_2(\Omega_1)}$$

$$-\frac{Eh}{2(1+\nu)}\left(\left[\frac{\partial u_n}{\partial y} + \frac{\partial v_n}{\partial x} + \frac{\partial w_n}{\partial x}\frac{\partial w_n}{\partial y}\right],\ \frac{\partial \chi_{2l}}{\partial x}\right)_{L_2(\Omega_1)}$$

$$+ (p_1,\ \chi_{2l})_{L_2(\Omega_1)} = \rho h\left(\frac{\partial^2 u_n}{\partial t^2},\ \chi_{2l}\right)_{L_2(\Omega_1)}, \tag{5.64}$$

$$-\frac{Eh}{1-\nu^2}\left(\left[\left(\frac{\partial v_n}{\partial y}-k_y w_n+\frac{1}{2}\left(\frac{\partial w_n}{\partial y}\right)^2\right)+\nu\left(\frac{\partial u_n}{\partial x}-k_x w_n+\frac{1}{2}\left(\frac{\partial w_n}{\partial x}\right)^2\right)\right],\right.$$

$$\left.\frac{\partial \chi_{2l}}{\partial y}\right)_{L_2(\Omega_1)}+\frac{E\alpha_T}{1-\nu}\left(\int_{-\frac{h}{2}}^{\frac{h}{2}}\theta_n\,dz,\ \frac{\partial \chi_{2l}}{\partial y}\right)_{L_2(\Omega_1)}$$

$$-\frac{Eh}{2(1+\nu)}\left(\left[\frac{\partial u_n}{\partial y}+\frac{\partial v_n}{\partial x}+\frac{\partial w_n}{\partial x}\frac{\partial w_n}{\partial y}\right],\ \frac{\partial \chi_{2l}}{\partial x}\right)_{L_2(\Omega_1)}$$

$$+\left(p_1,\ \chi_{2l}\right)_{L_2(\Omega_1)}=\rho h\left(\frac{\partial^2 v_n}{\partial t^2},\ \chi_{2l}\right)_{L_2(\Omega_1)}, \tag{5.65}$$

$$-D\left(\left[\frac{\partial^2 w_n}{\partial x^2}+\nu\frac{\partial^2 w_n}{\partial y^2}\right],\ \frac{\partial^2 \chi_{1l}}{\partial x^2}\right)_{L_2(\Omega_1)}-2D(1-\nu)\left(\frac{\partial^2 w_n}{\partial x\partial y},\ \frac{\partial^2 \chi_{1l}}{\partial x\partial y}\right)_{L_2(\Omega_1)}$$

$$-D\left(\left[\frac{\partial^2 w_n}{\partial y^2}+\nu\frac{\partial^2 w_n}{\partial x^2}\right],\ \frac{\partial^2 \chi_{1l}}{\partial x^2}\right)_{L_2(\Omega_1)}$$

$$-\frac{E\alpha_T}{1-\nu}\left(\int_{-\frac{h}{2}}^{\frac{h}{2}}\theta_n z\,dz,\ \left[\frac{\partial^2 \chi_{1l}}{\partial x^2}+\frac{\partial^2 \chi_{1l}}{\partial y^2}\right]\right)_{L_2(\Omega_1)}$$

$$+\frac{Eh}{1-\nu^2}\left(\left[\left(\frac{\partial u_n}{\partial x}-k_x w_n+\frac{1}{2}\left(\frac{\partial w_n}{\partial x}\right)^2\right)+\nu\left(\frac{\partial v_n}{\partial y}-k_y w_n\right.\right.\right.$$

$$\left.\left.\left.+\frac{1}{2}\left(\frac{\partial w_n}{\partial y}\right)^2\right)\right],k_x\chi_{1l}\right)_{L_2(\Omega_1)}-\frac{E\alpha_T}{1-\nu}\left(\int_{-\frac{h}{2}}^{\frac{h}{2}}\theta_n\,dz,\ [k_x+k_y]\chi_{1l}\right)_{L_2(\Omega_1)}$$

$$+\frac{Eh}{1-\nu^2}\left(\left[\left(\frac{\partial v_n}{\partial y}-k_y w_n+\frac{1}{2}\left(\frac{\partial w_n}{\partial y}\right)^2\right)+\nu\left(\frac{\partial u_n}{\partial x}-k_x w_n\right.\right.\right.$$

$$\left.\left.\left.+\frac{1}{2}\left(\frac{\partial w_n}{\partial x}\right)^2\right)\right],k_y\chi_{1l}\right)_{L_2(\Omega_1)}-\frac{Eh}{1-\nu^2}\left(\frac{\partial w_n}{\partial x}\left[\frac{\partial u_n}{\partial x}-k_x w_n+\frac{1}{2}\left(\frac{\partial w_n}{\partial x}\right)^2\right.\right.$$

$$\left.\left.+\nu\left(\frac{\partial v_n}{\partial y}-k_y w_n+\frac{1}{2}\left(\frac{\partial w_n}{\partial y}\right)^2\right)\right],\ \frac{\partial \chi_{1l}}{\partial x}\right)_{L_2(\Omega_1)}$$

$$-\frac{Eh}{2(1+\nu)}\left(\frac{\partial w_n}{\partial y}\left[\frac{\partial u_n}{\partial y}+\frac{\partial v_n}{\partial x}+\frac{\partial w_n}{\partial x}\frac{\partial w_n}{\partial y}\right],\ \frac{\partial \chi_{1l}}{\partial x}\right)_{L_2(\Omega_1)}$$

$$-\frac{Eh}{1-\nu^2}\left(\frac{\partial w_n}{\partial y}\left[\frac{\partial v_n}{\partial y}-k_y w_n+\frac{1}{2}\left(\frac{\partial w_n}{\partial y}\right)^2+\nu\left(\frac{\partial u_n}{\partial x}-k_x w_n\right.\right.\right.$$

$$\left.\left.\left.+\frac{1}{2}\left(\frac{\partial w_n}{\partial x}\right)^2\right)\right],\ \frac{\partial \chi_{1l}}{\partial y}\right)_{L_2(\Omega_1)}$$

$$-\frac{Eh}{2(1+\nu)}\left(\left[\frac{\partial u_n}{\partial y}+\frac{\partial v_n}{\partial x}+\frac{\partial w_n}{\partial x}\frac{\partial w_n}{\partial y}\right],\ \frac{\partial \chi_{1l}}{\partial y}\right)_{L_2(\Omega_1)}$$

$$+\frac{E\alpha_T}{1-\nu}\left(\frac{\partial w_n}{\partial x}\int\limits_{-\frac{h}{2}}^{\frac{h}{2}}\theta_n\,dz,\ \frac{\partial \chi_{1l}}{\partial x}\right)_{L_2(\Omega_1)}$$

$$+\frac{E\alpha_T}{1-\nu}\left(\frac{\partial w_n}{\partial y}\int\limits_{-\frac{h}{2}}^{\frac{h}{2}}\theta_n\,dz,\ \frac{\partial \chi_{1l}}{\partial y}\right)_{L_2(\Omega_1)}+(g_1,\ \chi_{1l})_{L_2(\Omega_1)}$$

$$=\rho h\left(\frac{\partial^2 w_n}{\partial t^2},\ \chi_{1l}\right)_{L_2(\Omega_1)}-\frac{\rho h^3}{12}\left(\frac{\partial^2}{\partial t^2}\left[\frac{\partial^2 w_n}{\partial x^2}+\frac{\partial^2 w_n}{\partial y^2}\right],\ \chi_{1l}\right)_{L_2(\Omega_1)}, \tag{5.66}$$

$$-\lambda_q\left[\left(\frac{\partial\theta_n}{\partial x},\ \frac{\partial\chi_{3l}}{\partial x}\right)_{L_2(\Omega_2)}+\left(\frac{\partial\theta_n}{\partial y},\ \frac{\partial\chi_{3l}}{\partial y}\right)_{L_2(\Omega_2)}+\left(\frac{\partial\theta_n}{\partial z},\ \frac{\partial\chi_{3l}}{\partial z}\right)_{L_2(\Omega_2)}\right]$$

$$-\frac{E\alpha_T T_0}{1-\nu}\left(\frac{\partial}{\partial t}\left[\frac{\partial u_n}{\partial x}-k_x w_n+\frac{1}{2}\left(\frac{\partial w_n}{\partial x}\right)^2+\nu\left(\frac{\partial v}{\partial y}-k_y w_n+\frac{1}{2}\left(\frac{\partial w_n}{\partial y}\right)^2\right)\right.\right.$$

$$\left.\left.-z\left\{\frac{\partial^2 w_n}{\partial x^2}+\frac{\partial^2 w_n}{\partial y^2}\right\}\right],\ \chi_{3l}\right)_{L_2(\Omega_2)}+(g_2,\ \chi_{3l})_{L_2(\Omega_2)}=c_0\left(\frac{\partial\theta_n}{\partial t},\ \chi_{3l}\right)_{L_2(\Omega_2)}, \tag{5.67}$$

$$l=1,\dots,n.$$

We complete the system of equations (5.64)–(5.67) in the usual way [147, 235] with the following initial conditions:

$$g_{1l}(0)=(\varphi_1,\ \chi_{1l})_{L_2(\Omega_1)},\quad g_{2l}(0)=(\varphi_2,\ \chi_{2l})_{L_2(\Omega_1)},$$
$$g_{3l}(0)=(\varphi_3,\ \chi_{2l})_{L_2(\Omega_1)},$$
$$\frac{dg_{1l}(0)}{dt}=(\Psi_1,\ \chi_{1l})_{L_2(\Omega_1)},\quad \frac{dg_{2l}(0)}{dt}=(\Psi_2,\ \chi_{2l})_{L_2(\Omega_1)},$$
$$\frac{dg_{3l}(0)}{dt}=(\Psi_3,\ \chi_{2l})_{L_2(\Omega_1)},\quad g_{4l}(0)=(\varphi_4,\ \chi_{3l})_{L_2(\Omega_2)},\quad l=1,\dots,n. \tag{5.68}$$

We now prove[1] the solvability of (5.64)–(5.68) using Schauder's principle of a fixed point, which reads:

> If a fully continuous operator A maps a bounded closed convex manifold S of the Banach space E onto a part of the space, then a fixed point of this mapping exists, i.e. $x \in S$ such that $Ax = x$.

[1] In general, the problem stated here is related to the proof of the Peano theorem about the existence of solutions to the Cauchy problem for a system of ordinary differential equations. However, for problems related to the theory of shells, a detailed proof of this theorem is not known to the authors.

First, taking into account the orthogonality property of the basis functions $\{\chi_{ji}\}$, $j = 1,2,3$, in the spaces $L_2(\Omega_1)$, $L_2(\Omega_2)$, $\overset{\circ}{W}{}_2^1(\Omega_1)$, we transform (5.64)–(5.68) to the form

$$\frac{d^2 g_{1l}(0)}{dt^2} = f_{1l}(\overline{g_1}, \overline{g_2}, \overline{g_3}, \overline{g_4}) + (g_1,\ \chi_{1l})_{L_2(\Omega_1)},$$
$$\frac{d^2 g_{2l}(0)}{dt^2} = \frac{1}{\rho h} f_{2l}(\overline{g_1}, \overline{g_2}, \overline{g_3}, \overline{g_4}) + \frac{1}{\rho h}(p_1,\ \chi_{2l})_{L_2(\Omega_1)},$$
$$\frac{d^2 g_{3l}(0)}{dt^2} = \frac{1}{\rho h} f_{3l}(\overline{g_1}, \overline{g_2}, \overline{g_3}, \overline{g_4}) + \frac{1}{\rho h}(g_1,\ \chi_{2l})_{L_2(\Omega_1)},$$
$$\frac{d^2 g_{4l}(0)}{dt^2} = \frac{1}{c_0} f_{4l}(\overline{g_1}, \overline{g_1}, \overline{g_2}, \overline{g_3}, \overline{g_4}) + \frac{1}{c_0}(g_2,\ \chi_{3l})_{L_2(\Omega_2)}, \tag{5.69}$$
$$l = 1,2,\dots,n,$$

where $\overline{g}_i = (g_{i1}, g_{i2}, \dots, g_{in})$, $\overline{g}_i = \dfrac{d\overline{g}_i}{dt}$, $i = 1,\dots,4$, and f_{il} denotes the left-hand sides of (5.69). It is not difficult to see that all f_{il} are continuous with respect to $g_{ij}(t)$, $\dot{g}_{ij}(t)$, $i = 1,\dots,4$, $j = 1,2,3,4,\dots,n$, for all $-\infty < g_{ij}(t) < +\infty$, $-\infty < \dot{g}_{ij}(t) < +\infty$ and for arbitrary $t \in [0,T]$, and therefore it is sufficient to consider only the terms in which g_{ij} and $\dot{g}_{ij}$ appear in a nonlinear way. Let us consider the following

$$R = \int_{\Omega_1} \frac{\partial w_n}{\partial y}\frac{\partial v_n}{\partial y}\frac{\partial \chi_{1l}}{\partial y}\, d\Omega_1 = \int_{\Omega_1} \left(\sum_{i=1}^{n} g_{1i}\frac{\partial \chi_{1i}}{\partial y} \sum_{j=1}^{n} g_{3j}\frac{\partial \chi_{2i}}{\partial y} \right) \frac{\partial \chi_{1l}}{\partial y}\, d\Omega_1$$
$$= \sum_{i,j=1}^{n} g_{1i}g_{3j} \int_{\Omega_1} \frac{\partial \chi_{1i}}{\partial y}\frac{\partial \chi_{1l}}{\partial y}\frac{\partial \chi_{2j}}{\partial y}\, d\,\Omega_1 = \sum_{i,j=1}^{n} g_{1i}g_{3j}a_{ilj},$$

where a_{ilj} are arbitrary constants. It is obvious that the function R is continuous on the set of variables $g_{1i}g_{3j}$. The rest of the nonlinear terms can be analysed similarly. Let us now move from the system (5.69) to an equivalent form using integral equations:

$$g_{1l}(t) = g_{1l}(0) + \int_0^t \dot{g}_{1l}(\tau)\, d\tau,$$

$$\dot{g}_{1l}(t) = \dot{g}_{1l}(0) + \int_0^t f_{1l}(\overline{g}_1, \overline{g}_2, \overline{g}_3, \overline{g}_4)\, d\tau + \int_0^t \int_{\Omega_1} g_1 \chi_{1l}\, d\Omega_1\, d\tau,$$

$$g_{jl}(t) = g_{jl}(0) + \int_0^t g_{jl}(\tau)\, d\tau, \quad j = 2,3,$$

$$\dot{g}_{jl}(t) = \dot{g}_{jl}(0) + \frac{1}{\rho h}\int_0^t f_{jl}(\overline{g}_1, \overline{g}_2, \overline{g}_3, \overline{g}_4)\, d\tau + \frac{1}{\rho h}\int_0^t \int_{\Omega_1} p_k \chi_{2l}\, d\Omega_1\, d\tau,$$

$$g_{4l}(t) = g_{4l}(0) + \frac{1}{c_0}\int_0^t f_{4l}(\overline{g}_1, \overline{g}_1, \overline{g}_2, \overline{g}_3, \overline{g}_4)\, d\tau + \frac{1}{c_0}\int_0^t \int_{\Omega_2} g_2 \chi_{3l}\, d\Omega_2\, d\tau,$$

$$l = 1, \dots, n, \quad k = 1, 2. \tag{5.70}$$

We take the right-hand sides of the system (5.70) as the operator A used in the Schauder principle. This operator is defined in the cube

$$D_0 = \Big\{ \overline{g}_i, \overline{g}_i, g_4 \big| \, |g_{ij}(t) - g_{ij}(0)| \le b, \quad |g_{4j}(t) - g_{4j}(0)| \le b,$$
$$|\dot{g}_{ij}(t) - \dot{g}_{ij}(0)| \le b, \quad i = 1, 2, 3, \quad j = 1, \dots, n \Big\}$$

of the space $C[0, T]$ of the continuous functions g_{ij}, $\dot{g}_{ij}$, g_{4j}, with the norm

$$\left\| g_{ij}, \dot{g}_{ij} \right\|_C = \max_{t \in [0,T]} \left\{ |g_{ij}|, |\dot{g}_{ij}| \right\},$$

$$\left\| g_{4j}, 0 \right\|_C = \max_{t \in [0,T]} \left\{ |g_{4j}|, 0 \right\}.$$

In order to show that the operator A defined in this way is fully continuous in D_0, we need to show that:

(a) A is continuous on D_0,
(b) A is compact on D_0, i.e. A maps an arbitrary bounded set from D_0 to a compact one.

It is clear that the operator A will be continuous if all its components are continuous. We consider only the component f_{1l}, because the proof for other components works similarly. Continuity of f_{1l} implies that $\forall \varepsilon, \exists \delta$, and if $|\overline{g}_i - \overline{g}'_i| < \delta$, then

$$|f_{1l}(\overline{g}_1, \overline{g}_2, \overline{g}_3, \overline{g}_4) - f_{1l}(\overline{g}'_1, \overline{g}'_2, \overline{g}'_3, \overline{g}'_4)| < \varepsilon. \tag{5.71}$$

Let us consider an arbitrary uniformly convergent series $\{\overline{g}^m\} \to \overline{g}^0$, $m \to \infty$, $\overline{g}^m = \{\overline{g}_1^m, \overline{g}_2^m, \overline{g}_3^m, \overline{g}_4^m\}$, i.e. $\forall \varepsilon_1, \exists N(\varepsilon_1)$ such that for $m \ge N(\varepsilon_1)$ the inequality

$$\left| \overline{g}^m(t) - \overline{g}^0(t) \right| < \varepsilon_1, \quad \forall t \in [0, T],$$

holds. Choosing N large enough for ε_1 to be smaller than δ, we obtain from (5.71):

$$\left| f_{1l}(\overline{g}^m(t)) - f_{1l}(\overline{g}^0(t)) \right| < \varepsilon, \ \forall t \in [0, T]. \tag{5.72}$$

Using (5.72), we then obtain

$$\left\| \dot{g}_{1l}^m(t) - \dot{g}_{1l}^0(t) \right\|_c = \max_{t \in [0,T]} \left| \int_0^t \left[f_{1l}(\overline{g}^m(t)) - f_{1l}(\overline{g}^0(t)) \right] d\tau \right|$$

$$\le \max_{t \in [0,T]} \int_0^t \left| f_{1l}(\overline{g}^m(t)) - f_{1l}(\overline{g}^0(t)) \right| d\tau \le \varepsilon T. \tag{5.73}$$

The inequality (5.73) proves the continuity of one of the components of the operator A. Thus, the condition (a) has been proved. We shall now give a proof of the condition (b) using this component. We use the Arcel theorem [238]. First, let us observe that because f_{il}, $i = 1, \dots, 4$, are continuous functions on D_0, they also achieve their maximal values there. Defining

$$M = \max_{l=1,\dots,n} \left[\max_{\overline{g},\dot{\overline{g}} \in D_0} \{|f_{1l}|, |f_{2l}|, |f_{3l}|, |f_{4l}|\} \right],$$

we obtain inequalities corresponding to the conditions of the Arcel theorem.

For an arbitrary $\overline{g}(t) \in D_0$, we have

$$\left| \dot{g}_{1l}(0) + \int_0^t f_{1l}(\overline{g})\, d\tau + \int_0^t \int_{\Omega_1} g_1 \chi_{1l}\, d\Omega_1\, d\tau \right| \leq |\dot{g}_{1l}(0)|$$

$$+ \int_0^t |f_{1l}(\overline{g})|\, d\tau + \frac{1}{2} \int_0^t \|g_1\|^2_{L_2(\Omega_1)}\, d\tau + \frac{1}{2} \int_0^t \|\chi_{1l}\|^2_{L_2(\Omega_1)}\, d\tau$$

$$\leq |\dot{g}_{1l}(0)| + MT + \frac{1}{2} \int_0^T \|g_1\|^2_{L_2(\Omega_1)}\, d\tau + \frac{T}{2} \|\chi_{1l}\|^2_{L_2(\Omega_1)} . \tag{5.74}$$

For arbitrary $t_1, t_2 \in [0, T]$ $(t_1 > t_2)$, we have

$$\Bigg| \dot{g}_{1l}(0) + \int_0^{t_2} f_{1l}(\overline{g})\, d\tau + \int_0^{t_2} \int_{\Omega_1} g_1 \chi_{1l}\, d\Omega_1\, d\tau - \dot{g}_{1l}(0) - \int_0^{t_1} f_{1l}(\overline{g})\, d\tau$$

$$- \int_0^{t_1} \int_{\Omega_1} g_1 \chi_{1l}\, d\Omega_1\, d\tau \Bigg| = \left| \int_{t_1}^{t_2} f_{1l}(\overline{g})\, d\tau + \int_{t_1}^{t_2} \int_{\Omega_1} g_1 \chi_{1l}\, d\Omega_1\, d\tau \right|$$

$$\leq M(t_2 - t_1) + \left[\int_{t_1}^{t_2} \|g_1\|^2_{L_2(\Omega_1)}\, d\tau \right]^{1/2} \left[\int_{t_1}^{t_2} \|\chi_{1l}\|^2_{L_2(\Omega_1)}\, d\tau \right]^{1/2}$$

$$\leq M(t_2 - t_1) + \left[\int_0^T \|g_1\|^2_{L_2(\Omega_1)}\, d\tau \right]^{1/2} \sqrt{t_2 - t_1}\, \|\chi_{1l}\|^2_{L_2(\Omega_1)} . \tag{5.75}$$

The inequalities (5.74), (5.75) show that the set of patterns of the component of the operator A under consideration is equally bounded and equally continuous (the proof is analogous for the other components). Taking the largest constant from those which equally bound each of the operator components (i.e. we use estimates in the sense of (5.74)), we obtain in accordance with the Arcel theorem, the result that the operator A is compact. Thus, the operator is fully continuous.

According to the definition given above, the set D_0 is convex and closed. In order to finish our treatment of the solvability of (5.64)–(5.68), we use the Schauder principle and show that the operator maps the set D_0 into itself. The last observation is true if each of the components of the operator A maps D_0 into itself. A proof of the latter requirement is given here using f_{1l} as an example:

$$|\dot{g}_{1l}(t) - \dot{g}_{1l}(0)| = \left| \int_0^t f_{1l}(\overline{g})\, d\tau + \int_0^t \int_{\Omega_1} g_1 \chi_{1l}\, d\Omega_1\, d\tau \right|$$

(we use the Cauchy inequality)

$$\leq Mt + \frac{\varepsilon}{2} \int_0^t \left\| g_1 \right\|^2_{L_2(\Omega_1)} d\tau + \frac{1}{2\varepsilon} \int_0^t \left\| \chi_{1l} \right\|^2_{L_2(\Omega_1)} d\tau$$

$$\leq \left(M + \frac{1}{2\varepsilon} \left\| \chi_{1l} \right\|^2_{L_2(\Omega_1)} \right) t + \frac{\varepsilon}{2} \int_0^T \left\| g_1 \right\|^2_{L_2(\Omega_1)} d\tau, \quad 0 \leq t \leq T, \tag{5.76}$$

where ε is an arbitrary positive number. From (5.76), we conclude that the component maps D_0 into itself when

$$\left(M + \frac{1}{2\varepsilon} \left\| \chi_{1l} \right\|^2_{L_2(\Omega_1)} \right) t + \frac{\varepsilon}{2} \int_0^T \left\| g_1 \right\|^2_{L_2(\Omega_1)} d\tau = b. \tag{5.77}$$

Now, letting ε be small enough that the inequality

$$b - \frac{\varepsilon}{2} \int_0^T \left\| g_1 \right\|^2_{L_2(\Omega_1)} d\tau > 0$$

is satisfied, we find the upper boundary of the time interval $[0, t_0]$ for which the condition of mapping D_0 into itself is satisfied:

$$t_0 = \frac{b - \frac{\varepsilon}{2} \int_0^T \left\| g_1 \right\|^2_{L_2(\Omega_1)} d\tau}{M + \frac{1}{2\varepsilon} \left\| \chi_{1l} \right\|^2_{L_2(\Omega_1)}}. \tag{5.78}$$

In particular, if we take

$$\varepsilon = \frac{b}{\int\limits_0^T \left\| g_1 \right\|^2_{L_2(\Omega_1)} d\tau}$$

in (5.78), then we obtain

$$t_0 = \frac{\frac{b}{2}}{M + \frac{1}{2b}\left(\int\limits_0^T \left\| g_1 \right\|^2_{L_2(\Omega_1)} d\tau\right) \left\| \chi_{1l} \right\|^2_{L_2(\Omega_1)}}$$
$$= \frac{b}{2M + \frac{1}{b}\left\| \chi_{1l} \right\|^2_{L_2(\Omega_1)} \int\limits_0^T \left\| g_1 \right\|^2_{L_2(\Omega_1)} d\tau}. \tag{5.79}$$

Analysis of (5.76) and (5.79) leads to the conclusion that during the investigation of the other components only the $\|\chi_{1l}\|^2_{L_2(\Omega_1)} \int\limits_0^T \|g_1\|^2_{L_2(\Omega_1)} d\tau$ term can be changed, and therefore we take

$$c_1 = \max_{\substack{l=1,\dots,n \\ k=1,2}} \left\{ \left\| \chi_{1l} \right\|^2_{L_2(\Omega_1)} \int\limits_0^T \left\| g_1 \right\|^2_{L_2(\Omega_1)} d\tau; \int\limits_0^T \left\| p_k \right\|^2_{L_2(\Omega_1)} d\tau; \right.$$
$$\left. \int\limits_0^T \left\| g_2 \right\|^2_{L_2(\Omega_2)} d\tau \right\},$$

while the systems of functions $\{\chi_{2l}\}$, $\{\chi_{3l}\}$ are orthonormalized in $L_2(\Omega_1)$, $L_2(\Omega_2)$. Therefore

$$t_0 = \frac{b^2}{2Mb + c_1}. \tag{5.80}$$

It should be emphasized that owing to its definition, the quantity M is a function of $\overline{g}(0)$, $\overline{g}(0)$, b. Therefore, the following lemma has been proved.

Lemma 5.1 *If $t \in [0, T_0]$, where $T_0 = \min\{T, t_0\}$, then there exists in a cube of the space $C[0,T]$ at least one solution of the problem (5.64)–(5.68).*

Later in a typical way [237], we shall extend the solution to the whole interval $[0,T]$ using prior estimates.

Step 2. Prior estimates. In order to obtain prior estimates we multiply (5.64) by $\dot{g}_{1l}(t)$, (5.65) by $\dot{g}_{2l}(t)$, (5.66) by $\dot{g}_{3l}(t)$ and (5.67) by $\dot{g}_{4l}(t)$. We then sum each of the resulting expressions with respect to l, and add the first three equations. We obtain

$$\frac{1}{2}\frac{d}{dt}\left\{\rho h\left\|\frac{\partial w_n(t)}{\partial t}\right\|^2_{L_2(\Omega_1)} + \frac{\rho h^3}{12}\left\|\frac{\partial w_n(t)}{\partial t}\right\|^2_{\overset{\circ}{W}{}^1_2(\Omega_1)} + \rho h\left\|\frac{\partial u_n(t)}{\partial t}\right\|^2_{L_2(\Omega_1)}\right.$$

$$+\left\|\frac{\partial v_n(t)}{\partial t}\right\|^2_{L_2(\Omega_1)} + D\left\|w_n\right\|^2_{\overset{\circ}{W}{}^2_2(\Omega_1)} + \int\limits_{\Omega_1}\left(\frac{Eh}{1-\nu^2}\left[\frac{\partial u_n}{\partial x} - k_x w_n\right.\right.$$

$$\left.+\frac{1}{2}\left(\frac{\partial w_n}{\partial x}\right)^2\right]^2 + \frac{Eh}{1-\nu^2}\left[\frac{\partial v_n}{\partial y} - k_y w_n + \frac{1}{2}\left(\frac{\partial w_n}{\partial y}\right)^2\right]^2$$

$$+\frac{2\nu Eh}{1-\nu^2}\left[\frac{\partial u_n}{\partial x} - k_x w_n + \frac{1}{2}\left(\frac{\partial w_n}{\partial x}\right)^2\right]\left[\frac{\partial v_n}{\partial y} - k_y w_n\right.$$

$$\left.\left.\left.+\frac{1}{2}\left(\frac{\partial w_n}{\partial y}\right)^2\right] + \frac{Eh}{2(1+\nu)}\left[\frac{\partial u_n}{\partial y} - \frac{\partial v_n}{\partial x} + \frac{\partial w_n}{\partial x}\frac{\partial w_n}{\partial y}\right]^2\right) d\Omega_1\right\}$$

$$-\frac{E\alpha_t}{1-\nu}\int\limits_{\Omega_1}\int\limits_{-\frac{h}{2}}^{\frac{h}{2}}\theta_n\frac{\partial}{\partial t}\left[\frac{\partial u_n}{\partial x} - k_x w_n + \frac{1}{2}\left(\frac{\partial \Omega_n}{\partial x}\right)^2 + \frac{\partial v_n}{\partial y} - k_y w_n\right.$$

$$\left.+\frac{1}{2}\left(\frac{\partial \Omega_n}{\partial y}\right)^2 - z\left(\frac{\partial^2 w_n}{\partial x^2} + \frac{\partial^2 w_n}{\partial y^2}\right)\right] dz\, d\Omega_1$$

$$= \int\limits_{\Omega_1}\left(g_1\frac{\partial w_n}{\partial t} + p_1\frac{\partial u_n}{\partial t} + p_2\frac{\partial v_n}{\partial t}\right) d\Omega_1, \tag{5.81}$$

$$\frac{1}{2}\frac{d}{dt}\left\{\frac{c_0}{T_0}\left\|\theta_n\right\|^2_{L_2(\Omega_2)}\right\} + \frac{\lambda_q}{T_0}\left\|\theta_n\right\|^2_{\overset{\circ}{W}{}^1_2(\Omega_2)}$$

$$+\frac{E\alpha_t}{1-\nu}\int\limits_{\Omega_2}\frac{\partial}{\partial t}\left[\frac{\partial u_n}{\partial x} - k_x w_n + \frac{1}{2}\left(\frac{\partial w_n}{\partial x}\right)^2 + \frac{\partial v_n}{\partial y}\right.$$

$$\left.-k_y w_n + \frac{1}{2}\left(\frac{\partial w_n}{\partial y}\right)^2 - z\left(\frac{\partial^2 w_n}{\partial x^2} + \frac{\partial^2 w_n}{\partial y^2}\right)\right] d\Omega_2$$

$$= \int\limits_{\Omega_2} g_2\theta_n\, d\Omega_2. \tag{5.82}$$

Summing (5.81) and (5.82), we obtain the following (note that in the left-hand sides, the terms containing temperature cancel):

$$\frac{1}{2}\frac{d}{dt}\left\{\rho h\left(\left\|\frac{\partial w_n(t)}{\partial t}\right\|^2_{L_2(\Omega_1)}+\left\|\frac{\partial u_n(t)}{\partial t}\right\|^2_{L_2(\Omega_1)}+\left\|\frac{\partial v_n(t)}{\partial t}\right\|^2_{L_2(\Omega_1)}\right)\right.$$
$$+\frac{\rho h^3}{12}\left\|\frac{\partial w_n(t)}{\partial t}\right\|^2_{\overset{\circ}{W}{}^1_2(\Omega_1)}+D\left\|w_n\right\|^2_{\overset{\circ}{W}{}^1_2(\Omega_1)}+\frac{c_0}{T_0}\left\|\theta_n\right\|^2_{L_2(\Omega_2)}$$
$$+\int\limits_{\Omega_1}\left(\frac{Eh}{1-\nu^2}\left[\frac{\partial u_n}{\partial x}-k_x w_n+\frac{1}{2}\left(\frac{\partial w_n}{\partial x}\right)^2\right]^2+\left[\frac{\partial v_n}{\partial y}-k_y w_n\right.\right.$$
$$\left.+\frac{1}{2}\left(\frac{\partial w_n}{\partial y}\right)^2\right]^2+2\nu\left[\frac{\partial u_n}{\partial x}-k_x w_n+\frac{1}{2}\left(\frac{\partial w_n}{\partial x}\right)^2\right]^2\left[\frac{\partial v_n}{\partial y}\right.$$
$$\left.-k_y w_n+\frac{1}{2}\left(\frac{\partial w_n}{\partial y}\right)^2\right]+\frac{Eh}{2(1+\nu)}\left[\frac{\partial u_n}{\partial x}+\frac{\partial v_n}{\partial x}+\frac{\partial w_n}{\partial x}\right.$$
$$\left.\left.\left.\times\frac{\partial w_n}{\partial y}\right]^2\right)d\Omega_1\right\}+\frac{\lambda_q}{T_0}\left\|\theta_n\right\|^2_{\overset{\circ}{W}{}^1_2(\Omega_2)}=\int\limits_{\Omega_1}\left(g_1\frac{\partial w_n}{\partial t}+p_1\frac{\partial u_n}{\partial t}\right.$$
$$\left.+p_2\frac{\partial v_n}{\partial t}\right)d\Omega_1+\int\limits_{\Omega_2}g_2\theta_n\,d\Omega_2. \tag{5.83}$$

Integrating (5.83) with respect to t in the interval from 0 to τ, $\tau\leq T$, and then transforming the right-hand side of (5.83) by use of the Cauchy inequality, we obtain

$$\frac{1}{2}\rho h\left(\left\|\frac{\partial u_n(\tau)}{\partial t}\right\|^2_{L_2(\Omega_1)}+\left\|\frac{\partial v_n(\tau)}{\partial t}\right\|^2_{L_2(\Omega_1)}+\left\|\frac{\partial w_n(\tau)}{\partial t}\right\|^2_{L_2(\Omega_1)}\right)$$
$$+\frac{1}{2}\frac{\rho h^3}{12}\left\|\frac{\partial w_n(\tau)}{\partial t}\right\|^2_{\overset{\circ}{W}{}^1_2(\Omega_1)}+\frac{1}{2}D\left\|w_n(\tau)\right\|^2_{\overset{\circ}{W}{}^2_2(\Omega_1)}+\frac{\lambda_q}{T_0}\int\limits_0^\tau\left\|\theta_n\right\|^2_{\overset{\circ}{W}{}^1_2(\Omega_2)}dt$$
$$+\frac{1}{2}\frac{c_0}{T_0}\left\|\theta_n\right\|^2_{L_2(\Omega_2)}$$
$$\leq\frac{1}{2}\rho h\left(\left\|\frac{\partial u_n(0)}{\partial t}\right\|^2_{L_2(\Omega_1)}+\left\|\frac{\partial v_n(0)}{\partial t}\right\|^2_{L_2(\Omega_1)}+\left\|\frac{\partial w_n(0)}{\partial t}\right\|^2_{L_2(\Omega_1)}\right)$$
$$+\frac{1}{2}\frac{\rho h^3}{12}\left\|\frac{\partial w_n(0)}{\partial t}\right\|^2_{\overset{\circ}{W}{}^1_2(\Omega_1)}+\frac{1}{2}D\left\|w_n(0)\right\|^2_{\overset{\circ}{W}{}^2_2(\Omega_1)}+\frac{1}{2}\frac{c_0}{T_0}\left\|\theta_n(0)\right\|^2_{L_2(\Omega_2)}$$
$$+\frac{1}{2}\frac{1}{\rho h}\int\limits_0^\tau\int\limits_{\Omega_1}\left[(g_1)^2+(p_1)^2+(p_2)^2\right]d\Omega_1\,dt+\frac{1}{2}\int\limits_0^\tau\left\{\rho h\left(\left\|\frac{\partial u_n}{\partial t}\right\|^2_{L_2(\Omega_1)}\right.\right.$$

$$+\left\|\frac{\partial v_n}{\partial t}\right\|^2_{L_2(\Omega_1)} + \left\|\frac{\partial w_n}{\partial t}\right\|^2_{L_2(\Omega_1)}\Bigg) + \frac{\rho h^3}{12}\left\|\frac{\partial w_n}{\partial t}\right\|^2_{\overset{\circ}{W}{}^1_2(\Omega_1)}$$

$$+D\left\|w_n\right\|^2_{\overset{\circ}{W}{}^2_2(\Omega_1)}\Bigg\} dt + \frac{1}{2}\frac{T_0}{c_0}\int_0^{\tau}\int_{\Omega_2}(g_2)^2\, d\Omega_2 + \frac{1}{2}\frac{c_0}{T_0}\int_0^{\tau}\left\|\theta_n\right\|^2_{L_2(\Omega_2)} dt$$

$$+\int_{\Omega_1}\Bigg(\frac{Eh}{1-\nu^2}\Bigg\{\left[\frac{\partial u_n(0)}{\partial x} - k_x w_n(0) + \frac{1}{2}\left(\frac{\partial w_n(0)}{\partial x}\right)^2\right]^2$$

$$+\left[\frac{\partial v_n(0)}{\partial y} - k_y w_n(0) + \frac{1}{2}\left(\frac{\partial w_n(0)}{\partial y}\right)^2\right]^2 + 2\nu\left[\frac{\partial u_n(0)}{\partial x}\right.$$

$$\left.-k_x w_n(0) + \frac{1}{2}\left(\frac{\partial w_n(0)}{\partial x}\right)^2\right]\left[\frac{\partial v_n(0)}{\partial y} - k_y w_n(0) + \frac{1}{2}\left(\frac{\partial w_n(0)}{\partial y}\right)^2\right]\Bigg\}$$

$$+\frac{Eh}{2(1+\nu)}\left[\frac{\partial u_n(0)}{\partial y} + \frac{\partial v_n(0)}{\partial x} + \frac{\partial w_n(0)}{\partial x}\frac{\partial w_n(0)}{\partial y}\right]^2\Bigg) d\Omega_1$$

$$\leq \frac{1}{2}\rho h\left(\left\|\frac{\partial u_n(0)}{\partial t}\right\|^2_{L_2(\Omega_1)} + \left\|\frac{\partial v_n(0)}{\partial t}\right\|^2_{L_2(\Omega_1)} + \left\|\frac{\partial w_n(0)}{\partial t}\right\|^2_{L_2(\Omega_1)}\right)$$

$$+\frac{1}{2}\frac{\rho h^3}{12}\left\|\frac{\partial w_n(0)}{\partial t}\right\|^2_{\overset{\circ}{W}{}^1_2(\Omega_1)} + \frac{1}{2}D\left\|w_n(0)\right\|^2_{\overset{\circ}{W}{}^2_2(\Omega_1)} + \frac{1}{2}\frac{c_0}{T_0}\left\|\theta_n(0)\right\|^2_{L_2(\Omega_2)}$$

$$+\frac{1}{2}\int_0^T\left[\left(\left\|g_1\right\|^2_{L_2(\Omega_1)} + \left\|p_1\right\|^2_{L_2(\Omega_1)} + \left\|p_2\right\|^2_{L_2(\Omega_1)}\right) + \frac{T_0}{c_0}\left\|g_2\right\|^2_{L_2(\Omega_2)}\right] dt$$

$$+\frac{Eh}{1-\nu^2}c_2\left(\left\|\frac{\partial u_n(0)}{\partial x}\right\|^2_{L_2(\Omega_1)} + \left\|\frac{\partial v_n(0)}{\partial y}\right\|^2_{L_2(\Omega_1)}\right.$$

$$\left.+\frac{1-\nu}{2}\left\|\frac{\partial u_n(0)}{\partial y}\right\|^2_{L_2(\Omega_1)} + \frac{1-\nu}{2}\left\|\frac{\partial v_n(0)}{\partial x}\right\|^2_{L_2(\Omega_1)} + 1 + \left\|w_n(0)\right\|^4_{\overset{\circ}{W}{}^2_2(\Omega_1)}\right)$$

$$+\frac{1}{2}\int_0^{\tau}\Bigg\{\frac{1}{2}\rho h\left(\left\|\frac{\partial u_n(t)}{\partial t}\right\|^2_{L_2(\Omega_1)} + \left\|\frac{\partial v_n(t)}{\partial t}\right\|^2_{L_2(\Omega_1)} + \left\|\frac{\partial w_n(t)}{\partial t}\right\|^2_{L_2(\Omega_1)}\right)$$

$$+\frac{\rho h^3}{12}\left\|\frac{\partial w_n(t)}{\partial t}\right\|^2_{\overset{\circ}{W}{}^1_2(\Omega_1)} + \frac{c_0}{T_0}\left\|\theta_n(t)\right\|^2_{L_2(\Omega_2)} + D\left\|w_n(t)\right\|^2_{\overset{\circ}{W}{}^2_2(\Omega_1)}\Bigg\} dt, \quad (5.84)$$

$$c_2 = const > 0.$$

In order to obtain the chain of inequalities (5.84), the following inequalities and inclusion theorems have been used (in addition to the Cauchy inequality):

(a)

$$\int\limits_{\Omega_1}\left(\left[\frac{\partial u}{\partial x}-k_x w+\frac{1}{2}\left(\frac{\partial w}{\partial x}\right)^2\right]^2+\left[\frac{\partial v}{\partial y}-k_y w+\frac{1}{2}\left(\frac{\partial w}{\partial y}\right)^2\right]^2\right.$$

$$\left.+2\nu\left[\frac{\partial u}{\partial x}-k_x w+\frac{1}{2}\left(\frac{\partial w}{\partial x}\right)^2\right]\left[\frac{\partial v}{\partial y}-k_y w+\frac{1}{2}\left(\frac{\partial w}{\partial y}\right)^2\right]^2\right)d\Omega_1$$

$$\geq\int\limits_{\Omega_1}(1-\nu)\left(\left[\frac{\partial u}{\partial x}-k_x w+\frac{1}{2}\left(\frac{\partial w}{\partial x}\right)^2\right]^2+\left[\frac{\partial v}{\partial y}-k_y w+\frac{1}{2}\left(\frac{\partial w}{\partial y}\right)^2\right]^2\right)d\Omega_1\geq 0,$$

$$0<\nu<\frac{1}{2};$$

(b) $\forall u\in \mathring{W}_2^1(\Omega_1)$, $\Omega_1\subset E_2$, the following applies

$$\int\limits_{\Omega_1}u^4\,d\Omega_1\leq c\left[\int\limits_{\Omega_1}\left\{\left(\frac{\partial u}{\partial x}\right)^2+\left(\frac{\partial u}{\partial y}\right)^2\right\}d\Omega_1\right]^2.$$

Using one of the integrals as an example, we illustrate below how the estimates in the chain (5.84) have been obtained:

$$\int\limits_{\Omega_1}\left[\frac{\partial v_n(0)}{\partial y}-k_y w_n(0)+\frac{1}{2}\left(\frac{\partial w_n(0)}{\partial y}\right)^2\right]^2 d\Omega_1$$

(using the Cauchy inequality)

$$\leq\int\limits_{\Omega_1}\left[\left(\frac{\partial v_n(0)}{\partial y}-k_y w_n(0)\right)^2+\left(\frac{\partial w_n(0)}{\partial y}\right)^4\right]d\Omega_1$$

$$\leq\int\limits_{\Omega_1}\left[4\left(\frac{\partial w_n(0)}{\partial y}\right)^2+4k_y^2 w_n^2(0)+\left(\frac{\partial w_n(0)}{\partial y}\right)^4\right]d\Omega_1$$

(completing the second term in the integral to the full norm in the space $\mathring{W}_2^2(\Omega_1)$, and using the inequality (b) for the third term)

$$\leq c\left(\left\|\frac{\partial v_n(0)}{\partial y}\right\|^2_{L_2(\Omega_1)}+\|w_n(0)\|^2_{\mathring{W}_2^2(\Omega_1)}+\|w_n(0)\|^4_{\mathring{W}_2^2(\Omega_1)}\right)$$

$$\leq c\left(\left\|\frac{\partial v_n(0)}{\partial y}\right\|^2_{L_2(\Omega_1)}+\|w_n(0)\|^4_{\mathring{W}_2^2(\Omega_1)}+\frac{1}{2}\right).$$

The remaining terms of (5.84) are estimated in an analogous way, and c_2 is expected to be the largest constant from all c.

Furthermore, let us observe that the arbitrary elements $u_n(0)$, $v_n(0)$, $w_n(0)$, $\theta_n(0)$ occurring in (5.84) can be considered as elements of convergent series with a corresponding norm, i.e. they approximate the initial conditions and, as the elements of convergent series, they are bounded by a constant independent of the number n. Finally, taking into account the conditions applied to g_1, g_2, p_1, p_2 in (5.84), we obtain:

$$\begin{aligned}
&\frac{1}{2}\rho h\left(\left\|\frac{\partial u_n(\tau)}{\partial t}\right\|^2_{L_2(\Omega_1)}+\left\|\frac{\partial v_n(\tau)}{\partial t}\right\|^2_{L_2(\Omega_1)}+\left\|\frac{\partial w_n(\tau)}{\partial t}\right\|^2_{L_2(\Omega_1)}\right)\\
&\quad+\frac{1}{2}\rho h^3\left\|\frac{\partial w_n(\tau)}{\partial t}\right\|^2_{\overset{\circ}{W}{}^1_2(\Omega_1)}+\frac{1}{2}D\left\|w_n(\tau)\right\|^2_{\overset{\circ}{W}{}^2_2(\Omega_1)}\\
&\quad+\frac{\lambda_q}{T_0}\int_0^T\left\|\theta_n\right\|^2_{\overset{\circ}{W}{}^1_2(\Omega_2)}dt+\frac{1}{2}\frac{c_0}{T_0}\left\|\theta_n(\tau)\right\|^2_{L_2(\Omega_2)}\\
&\le c+\frac{1}{2}\int_0^\tau\left\{\frac{1}{2}\rho h\left(\left\|\frac{\partial u_n(t)}{\partial t}\right\|^2_{L_2(\Omega_1)}+\left\|\frac{\partial v_n(t)}{\partial t}\right\|^2_{L_2(\Omega_1)}\right.\right.\\
&\quad\left.+\left\|\frac{\partial w_n(t)}{\partial t}\right\|^2_{L_2(\Omega_1)}\right)+\frac{\rho h^3}{12}\left\|\frac{\partial w_n(t)}{\partial t}\right\|^2_{\overset{\circ}{W}{}^1_2(\Omega_1)}\\
&\quad\left.+\frac{c_0}{T_0}\left\|\theta_n(t)\right\|^2_{L_2(\Omega_2)}+D\left\|w_n(t)\right\|^2_{\overset{\circ}{W}{}^2_2(\Omega_1)}\right\}dt. \qquad (5.85)
\end{aligned}$$

Rejecting the term $\frac{\lambda_q}{T_0}\int_0^T\left\|\theta_n\right\|^2_{\overset{\circ}{W}{}^1_2(\Omega_2)}dt$ and using Gronwill's lemma[2] [140], we obtain

$$\begin{aligned}
&\rho h\left(\left\|\frac{\partial u_n(t)}{\partial t}\right\|^2_{L_2(\Omega_1)}+\left\|\frac{\partial v_n(t)}{\partial t}\right\|^2_{L_2(\Omega_1)}+\left\|\frac{\partial w_n(t)}{\partial t}\right\|^2_{L_2(\Omega_1)}\right)\\
&+\frac{\rho h^3}{12}\left\|\frac{\partial w_n(t)}{\partial t}\right\|^2_{\overset{\circ}{W}{}^1_2(\Omega_1)}+D\left\|w_n(t)\right\|^2_{\overset{\circ}{W}{}^2_2(\Omega_1)}+\frac{c_0}{T_0}\left\|\theta_n(t)\right\|^2_{L_2(\Omega_2)}\le c, \qquad (5.86)\\
&\forall t\in[0,T].
\end{aligned}$$

Returning to (5.85) and (5.83), with the help of (5.86) we obtain

$$\frac{\lambda_q}{T_0}\int_0^T\left\|\theta_n\right\|^2_{\overset{\circ}{W}{}^1_2(\Omega_2)}dt\le c, \qquad c=const>0. \qquad (5.87)$$

[2] Gronwill's lemma may be stated as follows. Let $v(t)$ be a nonnegative function continuous on $[a,b]$, let $c\ge 0$ be a constant, and in addition let $v(t)\le c+\int_a^t v(\sigma)\,d\sigma$, $a\le t\le b$. Then $v(t)\le c_1$, where $c_1=\exp(\int_a^t d\sigma)$, and if $c=0$ then $v=0$.

(The symbol c is used here mean various constants). From (5.83), using (5.86), (5.87), we obtain

$$\int\limits_{\Omega_1}\Bigg[\left(\frac{\partial u_n(t)}{\partial x}-k_xw_n(t)+\frac{1}{2}\left(\frac{\partial w_n(t)}{\partial x}\right)^2\right)^2+\left(\frac{\partial v_n(t)}{\partial y}\right.$$

$$\left.-k_yw_n(t)+\frac{1}{2}\left(\frac{\partial w_n(t)}{\partial y}\right)^2\right)^2+2\nu\left(\frac{\partial u_n(t)}{\partial x}-k_xw_n(t)\right.$$

$$\left.+\frac{1}{2}\left(\frac{\partial w_n(t)}{\partial x}\right)^2\right)\left(\frac{\partial v_n(t)}{\partial y}-k_yw_n(t)+\frac{1}{2}\left(\frac{\partial w_n(t)}{\partial y}\right)^2\right)$$

$$+\frac{1-\nu}{2}\left(\frac{\partial u_n(t)}{\partial y}+\frac{\partial v_n(t)}{\partial x}+\frac{\partial w_n(t)}{\partial x}\frac{\partial w_n(t)}{\partial y}\right)^2\Bigg]d\Omega_1\le c. \qquad (5.88)$$

From (5.88), using the Cauchy inequality, we obtain

$$\int\limits_{\Omega_1}\left(\frac{\partial u_n(t)}{\partial x}\right)^2 d\Omega_1\le c,\quad \int\limits_{\Omega_1}\left(\frac{\partial v_n(t)}{\partial y}\right)^2 d\Omega_1\le c,$$

$$\int\limits_{\Omega_1}\left(\frac{\partial u_n(t)}{\partial y}+\frac{\partial v_n(t)}{\partial x}\right)^2 d\Omega_1\le c. \qquad (5.89)$$

Furthermore, using (5.89) and Korn's inequality [161, 162], we obtain

$$\int\limits_{\Omega_1}\left[\left(\frac{\partial u_n}{\partial x}\right)^2+\left(\frac{\partial u_n}{\partial y}\right)^2+\left(\frac{\partial v_n}{\partial x}\right)^2+\left(\frac{\partial v_n}{\partial y}\right)^2\right]d\Omega_1$$

$$\le c\int\limits_{\Omega_1}\left[\left(\frac{\partial u_n}{\partial x}\right)^2+\left(\frac{\partial v_n}{\partial y}\right)^2+\left(\frac{\partial u_n}{\partial y}+\frac{\partial v_n}{\partial x}\right)^2\right]d\Omega_1\le c, \qquad (5.90)$$

$$c=const>0.$$

Remark 5.1 *With the help of the prior estimates (5.86), (5.90) it is possible to extend the solution of the system (5.70) to the whole interval* $[0,T]$. *Owing to the orthonormalization of the basis functions in the spaces* $L_2(\Omega_1)$, $\overset{\circ}{W}{}_2^1(\Omega_1)$, $L_2(\Omega_2)$, *we obtain*

$$c\ge\left\|\frac{\partial u_n(t)}{\partial t}\right\|^2_{L_2(\Omega_1)}=\sum_{i=1}^n(\dot g_{2i}(t))^2,\quad \forall t\in[0,T]. \qquad (5.91)$$

Because $\|u_n\|^2_{L_2(\Omega_1)}\le c$, $\|u_n\|^2_{\overset{\circ}{W}{}_2^1(\Omega_1)}\le c$, *we have*

$$c\ge\|u_n(t)\|^2_{L_2(\Omega_1)}=\sum_{i=1}^n(g_{2i}(t))^2,\quad \forall t\in[0,T], \qquad (5.92)$$

and estimates for other functions can be obtained analogously. The estimates (5.91), (5.92) show that $g_{2i}(t)$ and $\dot{g}_{2i}(t)$ lie in a certain sphere R of the space $C[0,T]$ and, in what follows, a maximal value of M with respect to all $g_{2i}(t)$, $\dot{g}_{2i}(t) \in R$ can be obtained. Substituting the value of $\max\{M\}$ *into (5.80) we obtain the time interval $[0,t_0]$ on which the lemma on the solvability of (5.70) guarantees a solution for arbitrary initial conditions $\overline{g}(T_0)$, $\overline{g}(T_0) \in R$. Therefore, the whole interval $[0,T]$ is covered by a finite number of intervals with length t_0, which implies that the system (5.70) has a solution in the whole interval $[0,T]$.*

Step 3. Limiting transition. It is known that an arbitrary bounded set in a Hilbert space is weakly compact. Thus, using the prior estimates (5.86), (5.87), (5.90) obtained in the step 2, we can separate out the following series with the following properties:

$$\begin{aligned}
\frac{\partial u_{n_k}}{\partial t} &\to \frac{\partial u}{\partial t} \quad \text{weak in} \quad L_2(0,T;L_2(\Omega_1)),\\
\frac{\partial v_{n_k}}{\partial t} &\to \frac{\partial v}{\partial t} \quad \text{weak in} \quad L_2(0,T;L_2(\Omega_1)),\\
\frac{\partial w_{n_k}}{\partial t} &\to \frac{\partial w}{\partial t} \quad \text{weak in} \quad L_2(0,T;\overset{\circ}{W}{}_2^1(\Omega_1)),\\
u_{n_k} &\to u \quad \text{weak in} \quad L_2(0,T;\overset{\circ}{W}{}_2^1(\Omega_1)),\\
v_{n_k} &\to v \quad \text{weak in} \quad L_2(0,T;\overset{\circ}{W}{}_2^1(\Omega_1)),\\
w_{n_k} &\to w \quad \text{weak in} \quad L_2(0,T;\overset{\circ}{W}{}_2^2(\Omega_1)),\\
\theta_{n_k} &\to \theta \quad \text{weak in} \quad L_2(0,T;\overset{\circ}{W}{}_2^1(\Omega_1)).
\end{aligned}$$

Let us construct, for the arbitrary functions $\alpha_e(t) \in C^1([0,T])$, $\alpha_e(T) = 0$, the following operations:

$$\Phi_1 = \sum_{i=1}^{n} \alpha_i(t) \otimes \chi_{1i}(x,y), \;\; \Phi_2 = \sum_{i=1}^{n} \alpha_i(t) \otimes \chi_{2i}(x,y),$$

$$\Phi_3 = \sum_{i=1}^{n} \alpha_i(t) \otimes \chi_{3i}(x,y).$$

We multiply each of the equations (5.64)–(5.67) by $\alpha_l(t)$, sum them over l from 1 to n, and then integrate with respect to t. We obtain

$$-\rho h \int_0^T \left(\frac{\partial u_n}{\partial t}, \frac{\partial \Phi_2}{\partial t} \right)_{L_2(\Omega_1)} dt + \frac{Eh}{1-\nu^2} \int_0^T \left(\varepsilon_{11}^n + \nu\varepsilon_{22}^n, \frac{\partial \Phi_2}{\partial x} \right)_{L_2(\Omega_1)} dt$$

$$+ \frac{Eh}{2(1+\nu)} \int_0^T \left(\varepsilon_{12}^n, \frac{\partial \Phi_2}{\partial y} \right)_{L_2(\Omega_1)} dt - \frac{E\alpha_T}{1-\nu} \int_0^T \left(\int_{-\frac{h}{2}}^{\frac{h}{2}} \theta_n \, dz, \frac{\partial \Phi_2}{\partial x} \right)_{L_2(\Omega_1)} dt$$

$$= \int_0^T (p_1, \Phi_2)_{L_2(\Omega_1)}\, dt + \left(\frac{\partial u_n}{\partial t}, \Phi_2 \right)_{L_2(\Omega_1)} \bigg|_{t=0},$$

$$-\rho h \int_0^T \left(\frac{\partial v_n}{\partial t}, \frac{\partial \Phi_2}{\partial t} \right)_{L_2(\Omega_1)} dt + \frac{Eh}{1-\nu^2} \int_0^T \left(\varepsilon_{22}^n + \nu \varepsilon_{11}^n, \frac{\partial \Phi_2}{\partial y} \right)_{L_2(\Omega_1)} dt$$

$$+ \frac{Eh}{2(1+\nu)} \int_0^T \left(\varepsilon_{12}^n, \frac{\partial \Phi_2}{\partial x} \right)_{L_2(\Omega_1)} dt - \frac{E\alpha_T}{1-\nu} \int_0^T \left(\int_{-\frac{h}{2}}^{\frac{h}{2}} \theta_n \, dz, \frac{\partial \Phi_2}{\partial y} \right)_{L_2(\Omega_1)} dt$$

$$= \int_0^T (p_2, \Phi_2)_{L_2(\Omega_1)}\, dt + \left(\frac{\partial v_n}{\partial t}, \Phi_2 \right)_{L_2(\Omega_1)} \bigg|_{t=0},$$

$$-\rho h \int_0^T \left(\frac{\partial w_n}{\partial t}, \frac{\partial \Phi_1}{\partial t} \right)_{L_2(\Omega_1)} dt - \frac{\rho h^3}{12} \int_0^T \left(\frac{\partial^2 w_n}{\partial t\, \partial x}, \frac{\partial^2 \Phi_1}{\partial x\, \partial t} \right)_{L_2(\Omega_1)} dt$$

$$-\frac{\rho h^3}{12} \int_0^T \left(\frac{\partial^2 w_n}{\partial t\, \partial y}, \frac{\partial^2 \Phi_1}{\partial y\, \partial t} \right)_{L_2(\Omega_1)} dt + D \int_0^T (w_n, \Phi_1)_{\overset{\circ}{W}{}_2^2(\Omega_1)} dt$$

$$+ \frac{E\alpha_T}{1-\nu} \int_0^T \left(\int_{-\frac{h}{2}}^{\frac{h}{2}} \theta_n z \, dz, \frac{\partial^2 \Phi_1}{\partial x^2} + \frac{\partial^2 \Phi_1}{\partial y^2} \right)_{L_2(\Omega_1)} dt$$

$$-\frac{Eh}{1-\nu^2} \int_0^T (\varepsilon_{11}^n + \nu \varepsilon_{22}^n, k_x \Phi_1)_{L_2(\Omega_1)}\, dt - \frac{Eh}{1-\nu^2} \int_0^T (\varepsilon_{22}^n + \nu \varepsilon_{11}^n, k_y \Phi_1)_{L_2(\Omega_1)}\, dt$$

$$+ \frac{E\alpha_T}{1-\nu 2} \int_0^T \left(\int_{-\frac{h}{2}}^{\frac{h}{2}} \theta_n \, dz, [k_x + k_y]\Phi_1 \right)_{L_2(\Omega_1)} dt + \frac{Eh}{1-\nu^2} \int_0^T \left(\frac{\partial w_n}{\partial x} \left(\varepsilon_{11}^n \right.\right.$$

$$\left.\left. + \nu \varepsilon_{22}^n, \frac{\partial \Phi_1}{\partial x} \right) \right)_{L_2(\Omega_1)} dt + \frac{Eh}{1-\nu^2} \int_0^T \left(\frac{\partial w_n}{\partial y} \left(\varepsilon_{22}^n + \nu \varepsilon_{11}^n, \frac{\partial \Phi_1}{\partial y} \right) \right)_{L_2(\Omega_1)} dt$$

$$+ \frac{Eh}{2(1+\nu)} \int_0^T \left(\frac{\partial w_n}{\partial y} \varepsilon_{12}^n, \frac{\partial \Phi_1}{\partial x} \right)_{L_2(\Omega_1)} dt + \frac{Eh}{2(1+\nu)} \int_0^T \left(\frac{\partial w_n}{\partial x} \varepsilon_{12}^n, \frac{\partial \Phi_1}{\partial y} \right)_{L_2(\Omega_1)} dt$$

$$-\frac{E\alpha_T}{1-\nu} \int_0^T \left(\frac{\partial w_n}{\partial x} \theta_n, \frac{\partial \Phi_1}{\partial x} \right)_{L_2(\Omega_1)} dt - \frac{E\alpha_T}{1-\nu} \int_0^T \left(\frac{\partial w_n}{\partial y} \theta_n, \frac{\partial \Phi_1}{\partial y} \right)_{L_2(\Omega_1)} dt$$

$$= \int_0^T (g_1, \Phi_1)_{L_2(\Omega_1)}\, dt + \left(\frac{\partial w_n}{\partial x}, \Phi_1\right)_{L_2(\Omega_1)}\Bigg|_{t=0} + \left(\frac{\partial^2 w_n}{\partial t\, \partial x}, \frac{\partial \Phi_1}{\partial x}\right)_{L_2(\Omega_1)} dt\Bigg|_{t=0}$$

$$+ \left(\frac{\partial^2 w_n}{\partial t\, \partial y}, \frac{\partial \Phi_1}{\partial y}\right)_{L_2(\Omega_1)} dt\Bigg|_{t=0}, \tag{5.93}$$

$$-c_0 \int_0^T \left(\theta_n, \frac{\partial \Phi_3}{\partial t}\right)_{L_2(\Omega_1)} dt - \lambda_q \int_0^T (\theta_n, \Phi_3)_{\overset{\circ}{W}{}_2^1(\Omega_2)}\, dt$$

$$= -\frac{E\alpha_T T_0}{1-\nu} \int_0^T \Bigg\{ -\left(\frac{\partial u_n}{\partial t}, \frac{\partial \Phi_3}{\partial x}\right)_{L_2(\Omega_2)} - \left([k_x - k_y]\frac{\partial w_n}{\partial t}, \Phi_3\right)_{L_2(\Omega_2)}$$

$$+ \left(\frac{\partial^2 w_n}{\partial t\, \partial x}\frac{\partial w_n}{\partial x} + \frac{\partial^2 w_n}{\partial t\, \partial y}\frac{\partial w_n}{\partial y}, \Phi_3\right)_{L_2(\Omega_2)} - \left(\frac{\partial v_n}{\partial t}, \frac{\partial \Phi_3}{\partial y}\right)_{L_2(\Omega_2)}$$

$$+ \left(\frac{\partial^2 w_n}{\partial t\, \partial x}, z\frac{\partial \Phi_3}{\partial x}\right)_{L_2(\Omega_2)} + \left(\frac{\partial^2 w_n}{\partial t\, \partial y}, z\frac{\partial \Phi_3}{\partial y}\right)_{L_2(\Omega_2)} \Bigg\} dt$$

$$+ \int_0^T (g_2, \Phi_3)_{L_2(\Omega_2)}\, dt + (\theta_n, \Phi_3)_{L_2(\Omega_2)}\Bigg|_{t=0},$$

$$\varepsilon_{11}^n = \frac{\partial u_n}{\partial x} - k_x w_n + \frac{1}{2}\left(\frac{\partial w_n}{\partial x}\right)^2, \quad \varepsilon_{22}^n = \frac{\partial v_n}{\partial y} - k_y w_n + \frac{1}{2}\left(\frac{\partial w_n}{\partial y}\right)^2,$$

$$\varepsilon_{12}^n = \frac{\partial u_n}{\partial y} + \frac{\partial v_n}{\partial y} + \frac{\partial w_n}{\partial x}\frac{\partial w_n}{\partial y}.$$

We take the limit of (5.93) in accordance with (5.92). It is clear that attention needs to be focused only on the nonlinear terms, and therefore we consider some of the more typical nonlinear terms in detail. Before proceeding let us note that the transition to the limit for the nonlinear terms is carried out using the following lemmas:

Lemma 5.2 *An arbitrary series that is weakly convergent in* $\overset{\circ}{W}{}_2^{1,1}(Q_T)$*,* $Q_T = \Omega \times (0,T)$*, is strongly compact in the space* $L_4(Q_T)$ *when* Ω *is a bounded area of* E_2*.*

Lemma 5.3 *Let* $a(x,y,t) \in \overset{\circ}{W}{}_2^{2,1}(Q_T)$*,* $Q_T = \Omega \times (0,T)$*,* $\Omega \subset E_2$*. In this case* $\frac{\partial a}{\partial x}, \frac{\partial a}{\partial y} \in L_4(Q_T)$*, and* $\|a\|_{L_2(\Omega)}$ *is a continuous function of time.*

Lemma 5.4 *If the series* $\{a_n(x,y,t)\}$ *is weakly convergent in* $\overset{\circ}{W}{}_2^{2,1}(Q_T)$*,* $Q_T = \Omega \times (0,T)$*,* $\Omega \subset E_2$*, to a certain function* $a_0 \in \overset{\circ}{W}{}_2^{2,1}(Q_T)$*, then the* *series* $\left\{\dfrac{\partial a_n}{\partial x}\right\}$, $\left\{\dfrac{\partial a_n}{\partial y}\right\}$ *are strongly convergent in the space* $L_4(Q_T)$ *to the functions* $\dfrac{\partial a_0}{\partial x}$, $\dfrac{\partial a_n}{\partial y}$, *respectively.*

In these lemmas, the space $\overset{\circ}{W}{}_m^{l_1,l_2}(Q_T)$ has been used, which is the subspace of $W_m^{l_1,l_2}(Q_T)$ consisting of those elements which are equal to zero on $S_T = \partial\Omega \times (0,T)$, $\overline{\Omega} = \partial\Omega \cup \Omega$. $W_m^{l_1,l_2}(Q_T)$ is the Banach space of the functions $u(\overline{x},t)$, $\overline{x} = (x,y)$, belonging to $L_m(Q_T)$, together with their derivatives with respect to $\overline{x}$ of order l_1, and their derivatives with respect to t of order l_2. In this space, the norm can be defined in the following way:

$$\left\|u\right\|_{W_m^{l_1,l_2}} = \left\{\int\limits_{Q_T}\left(\sum_{k=0}^{l_1}\sum_{(k)}\left|D_{\overline{x}}^k u\right|^m + \sum_{k=1}^{l_2}\left|D_t^k u\right|^m\right) d\overline{x}\,dt\right\}^{1/m},$$

where $D_{\overline{x}}^k$ and D_t^k denote the derivatives with respect to the corresponding variables.

Let us consider the limiting transition in the following terms (by w_0 and θ_0 we mean the weak limits of the series $\{w_n\}$ and $\{\theta_n\}$ in the sense of (5.92)):

1.
$$\int\limits_0^T\int\limits_{\Omega_1}\left(\frac{\partial w_n}{\partial x}\frac{1}{2}\left(\frac{\partial w_n}{\partial x}\right)^2\frac{\partial \Phi_1}{\partial x} - \frac{\partial w_0}{\partial x}\frac{1}{2}\left(\frac{\partial w_0}{\partial x}\right)^2\frac{\partial \Phi_1}{\partial x}\right) d\Omega_1\,dt$$
$$= \frac{1}{2}\int\limits_0^T\int\limits_{\Omega_1}\left[\left(\frac{\partial w_n}{\partial x} - \frac{\partial w_0}{\partial x}\right)\left(\frac{\partial w_n}{\partial x}\right)^2\frac{\partial \Phi_1}{\partial x}\right.$$
$$\left. + \left(\frac{\partial w_n}{\partial x} - \frac{\partial w_0}{\partial x}\right)\left(\frac{\partial w_n}{\partial x} + \frac{\partial w_0}{\partial x}\right)\frac{\partial w_0}{\partial x}\frac{\partial \Phi_1}{\partial x}\right] d\Omega_1\,dt$$

(using the Holder inequality)

$$\leq \frac{1}{2}\int\limits_0^T\left\{\left[\int\limits_{\Omega_1}\left(\frac{\partial w_n}{\partial x} - \frac{\partial w_0}{\partial x}\right)^2\left(\frac{\partial \Phi_1}{\partial x}\right)^2 d\Omega_1\right]^{1/2}\left[\int\limits_{\Omega_1}\left(\frac{\partial w_n}{\partial x}\right)^4 d\Omega_1\right]^{1/2}\right.$$
$$+\left[\int\limits_{\Omega_1}\left(\frac{\partial w_n}{\partial x} - \frac{\partial w_0}{\partial x}\right)^2\left(\frac{\partial w_n}{\partial x} - \frac{\partial w_0}{\partial x}\right)^2 d\Omega_1\right]^{1/2}$$
$$\left.\times\left[\int\limits_{\Omega_1}\left(\frac{\partial w_{0n}}{\partial x}\right)^2\left(\frac{\partial \Phi_1}{\partial x}\right)^2 d\Omega_1\right]^{1/2}\right\} dt$$

$$\leq \frac{1}{2}\left\|\frac{\partial w_n}{\partial x}-\frac{\partial w_0}{\partial x}\right\|_{L_4(Q_1)}\left\|\frac{\partial w_n}{\partial x}\right\|_{L_4(Q_1)}\left\|\frac{\partial \Phi_1}{\partial x}\right\|_{L_4(Q_1)}$$

$$+\frac{1}{2}\left\|\frac{\partial w_n}{\partial x}-\frac{\partial w_0}{\partial x}\right\|_{L_4(Q_1)}\left[8\int_0^T\int_{\Omega_1}\left(\left(\frac{\partial w_n}{\partial x}\right)^4\right.\right.$$

$$\left.\left.+\left(\frac{\partial w_0}{\partial x}\right)^4\right)d\Omega_1\,dt\right]^{1/4}\left\|\frac{\partial w_0}{\partial x}\right\|_{L_4(Q_1)}\left\|\frac{\partial \Phi_1}{\partial x}\right\|_{L_4(Q_1)}.$$

It is clear that, owing to Lemma 5.3, all of the integrals in the expression above exist and are bounded, whereas, owing to Lemma 5.4, the whole expression approaches zero.

2.
$$\int_0^T\int_{\Omega_1}\left\{\frac{\partial w_n}{\partial x}\left[\int_{-\frac{h}{2}}^{\frac{h}{2}}\theta_n\,dz\right]\frac{\partial \Phi_1}{\partial x}-\frac{\partial w_0}{\partial x}\left[\int_{-\frac{h}{2}}^{\frac{h}{2}}\theta\,dz\right]\frac{\partial \Phi_1}{\partial x}\right\}d\Omega_1\,dt$$

$$=\int_0^T\int_{\Omega_1}\left(\frac{\partial w_n}{\partial x}-\frac{\partial w_0}{\partial x}\right)\frac{\partial \Phi_1}{\partial x}\left[\int_{-\frac{h}{2}}^{\frac{h}{2}}\theta_n\,dz\right]d\Omega_1\,dt$$

$$+\int_0^T\int_{\Omega_1}\frac{\partial w_0}{\partial x}\frac{\partial \Phi_1}{\partial x}\left[\int_{-\frac{h}{2}}^{\frac{h}{2}}(\theta_n-\theta_0)\,dz\right]d\Omega_1\,dt$$

(using again the Holder inequality)

$$\leq\left\{\int_0^T\int_{\Omega_1}\left(\frac{\partial w_n}{\partial x}-\frac{\partial w_0}{\partial x}\right)^2\left(\frac{\partial \Phi_1}{\partial x}\right)^2 d\Omega_1\,dt\right\}^{1/2}\left\{\int_0^T\int_{\Omega_1}\left[\int_{-\frac{h}{2}}^{\frac{h}{2}}\theta_n\,dz\right]^2 d\Omega_1\,dt\right\}^{1/2}$$

$$+\int_0^T\int_{\Omega_1}\int_{-\frac{h}{2}}^{\frac{h}{2}}\frac{\partial w_0}{\partial x}\frac{\partial \Phi_1}{\partial x}(\theta_n-\theta_0)\,dz\,d\Omega_1\,dt$$

$$\leq\left\|\frac{\partial w_n}{\partial x}-\frac{\partial w_0}{\partial x}\right\|_{L_4(Q_1)}\left\|\frac{\partial \Phi_1}{\partial x}\right\|_{L_4(Q_1)}(h)^{1/2}\left\|\theta_n\right\|_{L_2(Q_1)}$$

$$+\int_0^T\int_{\Omega_2}(\theta_n-\theta_0)\frac{\partial w_0}{\partial x}\frac{\partial \Phi_1}{\partial x}\,d\Omega_2\,dt.$$

The first term in the last expression approaches zero according to Lemmas 5.3, 5.4, whereas the second approaches zero owing to the convergence of the series $\{\theta_n\}$ to θ_0 in $L^2(0,T;L_2(\Omega_2))$, or $\dfrac{\partial w_1}{\partial x}\dfrac{\partial \Phi_1}{\partial x} \in L_2(\Omega_2)$.

3.

$$\int_0^T \int_{\Omega_1} \int_{-\frac{h}{2}}^{\frac{h}{2}} \left(\frac{\partial^2 w_n}{\partial t\,\partial x}\frac{\partial w_n}{\partial x}\Phi_3 - \frac{\partial^2 w_0}{\partial t\,\partial x}\frac{\partial w_0}{\partial x}\Phi_3 \right) dz\, d\Omega_1\, dt$$

$$= \int_0^T \int_{\Omega_1} \int_{-\frac{h}{2}}^{\frac{h}{2}} \left\{ \left(\frac{\partial w_n}{\partial x} - \frac{\partial w_{0n}}{\partial x} \right) \frac{\partial^2 w_n}{\partial t\,\partial x}\Phi_3 + \left(\frac{\partial^2 w_n}{\partial t\,\partial x} - \frac{\partial^2 w_0}{\partial t\,\partial x} \right) \frac{\partial w_0}{\partial x}\Phi_3 \right\} dz\, d\Omega_1\, dt$$

Note that owing to the special form of the function Φ_3 in the first and the second term, we initially integrate the components of Φ_3 along z. As a result, Φ_3 will contain only components depending on the two spatial variables x, y and on the time t. Therefore

$$\Phi_3^1 \in \overset{\circ}{W}{}_2^{1,1}(Q_1), \quad \Phi_3^1 = \sum_{i=1}^{n} \alpha_i(t) \otimes a_i(x,y) c_i, \quad c_i = \int_{-\frac{h}{2}}^{\frac{h}{2}} b_i(z)\, dz,$$

and Lemma 5.2 is applicable to Φ_3.

Taking into account the above considerations, we extend the inequality as follows:

$$\le \left\| \frac{\partial^2 w_n}{\partial t\,\partial x} \right\|_{L_2(Q_1)} \left\| \frac{\partial w_n}{\partial x} - \frac{\partial w_0}{\partial x} \right\|_{L_4(Q_1)} \|\Phi_3\|_{L_4(Q_1)}$$
$$+ \int_0^T \int_{\Omega_2} \left(\frac{\partial^2 w_n}{\partial t \partial x} - \frac{\partial^2 w_0}{\partial t \partial x} \right) \frac{\partial w_0}{\partial x} \Phi_3^1\, d\Omega_2\, dt.$$

The first term approaches zero because of Lemma 5.4, and the second term approaches zero either because of a weak convergence of $\dfrac{\partial^2 w_n}{\partial t\,\partial x}$ in $L_2(Q_1)$ or because $\dfrac{\partial w}{\partial x}\Phi_3^1 \in L_2(Q_1)$.

For the remaining nonlinear terms, the limiting transition is performed in a similar way.

Let us focus on the initial conditions. The expressions

$$\left(\frac{\partial w_n}{\partial t} - \Psi_1, \Phi_1 \right)_{L_2(Q_1)} \Bigg|_{t=0}, \qquad (\theta_n - \theta_0, \Phi_3)_{L_2(Q_2)} \Big|_{t=0},$$

and others approach zero when the approximate solution to the problem is constructed. The rest of the initial conditions must be understood in the usual way [140], i.e.

$$\lim_{t\to 0} \|w_0 - \varphi_1\|_{L_2(Q_1)} = 0,\ \lim_{t\to 0} \|u_0 - \varphi_2\|_{L_2(Q_1)} = 0,\ \lim_{t\to 0} \|v_0 - \varphi_3\|_{L_2(Q_1)} = 0,$$

where w_0, v_0, u_0 is the solution to the problem under consideration, and the existence of the limits results from a theorem about traces of the space $W_2^{1,1}(Q_1)$ [140] and a theorem on inclusion that states $W_2^{2,1} \subset W_2^{1,1}(Q_1)$.

This step ends the proof of the theorem about the existence of a generalized solution (in the sense of (5.60)–(5.62)) to the problem.

Let us now formulate a new theorem. For a class of sufficiently smooth initial data, the solution to the problem (5.42)–(5.47) is unique. This holds also for nonlinear problem if, in (5.42)–(5.47), the nonlinear terms (5.56)–(5.59) are neglected (and in addition, the term related to the rotational inertia of the shell element $\rho\frac{h^3}{12}\frac{\partial^2}{\partial t^2}(\Delta w)$ is ignored).

Theorem 5.2 *Let $\partial\Omega_1$ satisfy the conditions of Theorem 5.1 and, in addition*

$$p_1, p_2, g_1, \frac{\partial p_1}{\partial t}, \frac{\partial p_2}{\partial t}, \frac{\partial g_1}{\partial t} \in L^2(0,T;L_2(\Omega_1)),$$
$$g_2, \frac{\partial g_2}{\partial t} \in L^2(0,T;L_2(\Omega_2)),$$
$$\varphi_1 \in \overset{\circ}{W}{}_2^2(\Omega_1) \cap W_2^4(Q_1), \quad \Psi_1 \in \overset{\circ}{W}{}_2^2(\Omega_1),$$
$$\varphi_2, \varphi_3 \in \overset{\circ}{W}{}_2^1(\Omega_1) \cap W_2^2(Q_1), \quad \varphi_4 \in \overset{\circ}{W}{}_2^1(\Omega_2) \cap W_2^2(Q_2),$$
$$\Psi_2, \Psi_3 \in \overset{\circ}{W}{}_2^1(\Omega_1).$$

Then, a unique choice of the four functions $\{w, u, v, \theta\}$ that is the solution to the problem (5.56)–(5.59) exists, and these functions satisfy the following conditions:

$$w, \frac{\partial w}{\partial t} \in L^2(0,T;\overset{\circ}{W}{}_2^2(\Omega_1)), \quad \frac{\partial^2 w}{\partial t^2} \in L^2(0,T;L_2(\Omega_1)),$$
$$u, \frac{\partial u}{\partial t} \in L^2(0,T;\overset{\circ}{W}{}_2^1(\Omega_1)), \quad \frac{\partial^2 u}{\partial t^2} \in L^2(0,T;L_2(\Omega_1)),$$
$$v, \frac{\partial v}{\partial t} \in L^2(0,T;\overset{\circ}{W}{}_2^1(\Omega_1)), \quad \frac{\partial^2 v}{\partial t^2} \in L^2(0,T;L_2(\Omega_1)),$$
$$\theta \in L^2(0,T;\overset{\circ}{W}{}_2^1(\Omega_2)), \quad \frac{\partial \theta}{\partial t} \in L^2(0,T;\overset{\circ}{W}{}_2^1(\Omega_2)).$$

Proof. The proof of this theorem again relies on the Bubnov–Galerkin method, and therefore verifies its convergence. Because of the uniqueness of the solution, all series of approximate solutions (of the type given in Theorem

5.1) converge to the exact solution of the problem. Since a method for investigating the smoothness of the generalized solution is known [139, 147] and the principal elements of the proof have already been investigated carefully in the treatment of Theorem 5.1, only the fundamental steps of the proof will be outlined.

Step 1. We construct the system of ordinary differential equations related to the problem (5.56)–(5.59) according to the Faedo–Galerkin method. This system coincides with the system (5.64)–(5.67) if both the term responsible for rotational inertia and the nonlinear terms are ignored. We look for the following approximate solution:

$$w_n = \sum_{i=1}^{n} g_{5i}(t)\chi_{4i}(x,y), \quad u_n = \sum_{i=1}^{n} g_{6i}(t)\chi_{5i}(x,y),$$

$$v_n = \sum_{i=1}^{n} g_{7i}(t)\chi_{5i}(x,y), \quad \theta_n = \sum_{i=1}^{n} g_{8i}(t)\chi_{6i}(x,y,z),$$

where $\{\chi_{4i}\}$ is a full system in $\overset{\circ}{W}{}_2^2(\Omega_1) \cap W_2^4(Q_1)$, orthonormalized in $L_2(\Omega_1)$; $\{\chi_{5i}\}$ is a full system in $\overset{\circ}{W}{}_2^1(\Omega_1) \cap W_2^2(Q_1)$, orthonormalized in $L_2(\Omega_1)$; and $\{\chi_{6i}\}$ is a full system in $\overset{\circ}{W}{}_2^1(\Omega_2) \cap W_2^2(Q_2)$, orthonormalized in $L_2(\Omega_2)$ (it should be noted that a special representation for the function χ_{6i} is not needed now). Therefore, the solvability of the system is evident and, moreover, owing to the linear set of ordinary differential equations, this solution is unique for arbitrary n.

Step 2. First prior estimate. In the same way as described in the treatment of Theorem 5.1, we obtain the prior estimates (5.86), (5.87), (5.90) (of course, without the term related to the rotational inertia).

Step 3. Second prior estimate. We differentiate each of the equations of the linearized system (5.64)–(5.67) with respect to t (this is possible because $(\frac{\partial p_1}{\partial t}, \frac{\partial p_2}{\partial t}, \frac{\partial g_1}{\partial t}) \in L^2(Q_1)$, $\frac{\partial g_2}{\partial t} \in L^2(Q_2)$ owing to the conditions of the theorem). We multiply each of the resulting equations by $\frac{\partial^2 g_{il}}{\partial t^2}$, $j = 5, \dots, 8$, and sum them over l from 1 to n. Because the problem is a linear one, we can repeat the procedure outlined above for the first prior estimate to construct a second prior estimate of the form (5.86), (5.87), (5.90). The only difference is that now, in each expression within the norm symbols, we have time derivatives:

$$\rho h\left(\left\| \frac{\partial^2 u_n(t)}{\partial t^2} \right\|^2_{L_2(\Omega_1)} + \left\| \frac{\partial^2 v_n(t)}{\partial t^2} \right\|^2_{L_2(\Omega_1)} + \left\| \frac{\partial^2 w_n(t)}{\partial t^2} \right\|^2_{L_2(\Omega_1)} \right)$$

$$+D \left\| \frac{\partial^2 w_n(t)}{\partial t^2} \right\|^2_{\overset{\circ}{W}{}_2^2(\Omega_1)} + \frac{C_0}{T_0} \left\| \frac{\partial \theta_n(t)}{\partial t} \right\|^2_{L_2(\Omega_2)} \le C, \quad t \in [0,T],$$

$$\frac{\lambda_q}{T_0}\int_0^T \left\| \frac{\partial \theta_n(t)}{\partial t} \right\|^2_{\overset{\circ}{W}{}_2^1(\Omega_2)} dt \leq C,$$

$$\left\| \frac{\partial u_n(t)}{\partial t} \right\|^2_{\overset{\circ}{W}{}_2^1(\Omega_1)} + \left\| \frac{\partial v_n(t)}{\partial t} \right\|^2_{\overset{\circ}{W}{}_2^1(\Omega_1)} \leq C. \tag{5.94}$$

Step 4. Limiting transition. On the basis of the first and the second (5.94) prior estimates, we pass to the limit as in the treatment of Theorem 5.1. In this case, the initial conditions are completed by the following relations:

$$\|w(x,y,t) - \Psi_1\|_{L_2(\Omega_1)} \to 0, \quad t \to 0,$$
$$\|u(x,y,t) - \Psi_2\|_{L_2(\Omega_1)} \to 0, \quad t \to 0,$$
$$\|v(x,y,t) - \Psi_3\|_{L_2(\Omega_1)} \to 0, \quad t \to 0,$$

and the proof is analogous to that in the case of Theorem 5.1.

Step 5. Uniqueness. It is not difficult to check that in the framework of the theorem under consideration the following "energetical" equality holds:

$$\frac{1}{2}\left\{\rho h\left(\left\|\frac{\partial u}{\partial t}\right\|^2_{L_2(\Omega_1)} + \left\|\frac{\partial v}{\partial t}\right\|^2_{L_2(\Omega_1)} + \left\|\frac{\partial w}{\partial t}\right\|^2_{L_2(\Omega_1)}\right) + D\|w\|^2_{\overset{\circ}{W}{}_2^2(\Omega_1)}\right.$$
$$+\int_{\Omega_1}\frac{Eh}{1-\nu^2}\left(\left[\frac{\partial u}{\partial x} - k_x w\right]^2 + \left[\frac{\partial v}{\partial y} - k_y w\right]^2 + 2\nu\left[\frac{\partial u}{\partial x} - k_x w\right]\left[\frac{\partial v}{\partial y} - k_y w\right]\right.$$
$$\left.\left. + \frac{1-\nu}{2}\left[\frac{\partial u}{\partial y} + \frac{\partial v}{\partial x}\right]^2\right) d\Omega_1 + \frac{C_0}{T_0}\|\theta\|^2_{L_2(\Omega_2)}\right\}\Bigg|_{t=0}^{t=t_1} + \frac{\lambda_q}{T_0}\int_0^{t_1}\|\theta\|^2_{\overset{\circ}{W}{}_2^1(\Omega_2)}\,dt$$
$$= \int_0^{t_1}\left[\int_{\Omega_1}\left(p_1\frac{\partial u}{\partial t} + p_2\frac{\partial v}{\partial t} + g_1\frac{\partial w}{\partial t}\right) d\Omega_1 + \int_{\Omega_1} p_2\theta\, d\Omega_2\right] dt.$$

In order to obtain this equality, we proceed in the following stages:

1. The nonlinear terms in the system (5.44)–(5.47) are neglected.
2. Equation (5.44) is multiplied by a function equal to zero for $t \in [t_1, T]$ and to $\dfrac{\partial u}{\partial t}$ for $t \in [0, t_1]$.
3. Equation (5.45) is multiplied by a function equal to zero for $t \in [t_1, T]$ and to $\dfrac{\partial v}{\partial t}$ for $t \in [0, t_1]$.
4. Equation (5.46) (without the rotational-inertia term) is multiplied by a function equal to zero for $t \in [t_1, T]$ and to $\dfrac{\partial w}{\partial t}$ for $t \in [0, t_1]$.
5. Equation (5.47) is multiplied by a function equal to zero for $t \in [t, T]$ and to θ for $t \in [0, t_1]$;

6. Each of the equations obtained is integrated with respect to time and transformed by integration by parts.
7. The resulting equations are summed.

As a result, we obtain the "energetical" equation. Applying the Cauchy inequality and the Gronwill lemma to the "energetical" equation, we obtain

$$\left\{\rho h\left(\left\|\frac{\partial u}{\partial t}\right\|^2_{L_2(\Omega_1)}+\left\|\frac{\partial v}{\partial t}\right\|^2_{L_2(\Omega_1)}+\left\|\frac{\partial w}{\partial t}\right\|^2_{L_2(\Omega_1)}\right)+D\left\|w\right\|^2_{\overset{\circ}{W}{}^2_2(\Omega_1)}\right.$$
$$\left.+\frac{C_0}{T_0}\left\|\theta\right\|^2_{L_2(\Omega_2)}\right\}\Bigg|_{t=t_1}\le e^{t_1}c_3,$$

$$\frac{Eh}{1-\nu^2}\int\limits_{\Omega_1}\left[\left(\frac{\partial u}{\partial x}-k_x w\right)^2+\left(\frac{\partial v}{\partial y}-k_y w\right)^2+2\nu\left(\frac{\partial u}{\partial x}-k_x w\right)\left(\frac{\partial v}{\partial y}-k_y w\right)\right.$$
$$\left.+\frac{1-\nu}{2}\left(\frac{\partial u}{\partial y}+\frac{\partial v}{\partial x}\right)^2\right]d\Omega_1\Bigg|_{t=t_1}\le C_3+\frac{1}{2}\int\limits_0^{t_1}\left\{\rho h\left(\left\|\frac{\partial u}{\partial t}\right\|^2_{L_2(\Omega_1)}+\left\|\frac{\partial v}{\partial t}\right\|^2_{L_2(\Omega_1)}\right.\right.$$
$$\left.\left.+\left\|\frac{\partial w}{\partial t}\right\|^2_{L_2(\Omega_1)}\right)+\frac{C_0}{T_0}\left\|\theta\right\|^2_{L_2(\Omega_2)}\right\}dt,$$

$$\frac{\lambda_q}{T_0}\int\limits_0^{t_1}\left\|\theta\right\|^2_{\overset{\circ}{W}{}^1_2(\Omega_2)}dt\le C_3+\frac{1}{2}\int\limits_0^{t_1}\left\{\rho h\left(\left\|\frac{\partial u}{\partial t}\right\|^2_{L_2(\Omega_1)}+\left\|\frac{\partial v}{\partial t}\right\|^2_{L_2(\Omega_1)}\right.\right.$$
$$\left.\left.+\left\|\frac{\partial w}{\partial t}\right\|^2_{L_2(\Omega_1)}\right)+\frac{C_0}{T_0}\left\|\theta\right\|^2_{L_2(\Omega_2)}\right\}dt,\tag{5.95}$$

$$C_3=\rho h\left(\left\|\frac{\partial u}{\partial t}\right\|^2_{L_2(\Omega_1)}+\left\|\frac{\partial v}{\partial t}\right\|^2_{L_2(\Omega_1)}+\left\|\frac{\partial w}{\partial t}\right\|^2_{L_2(\Omega_1)}\right)+\frac{C_0}{T_0}\left\|\theta\right\|^2_{L_2(\Omega_2)}$$
$$+\frac{Eh}{1-\nu^2}\int\limits_{\Omega_1}\left[\left(\frac{\partial u}{\partial x}-k_x w\right)^2+\left(\frac{\partial v}{\partial y}-k_y w\right)^2+2\nu\left(\frac{\partial u}{\partial x}-k_x w\right)\left(\frac{\partial v}{\partial y}-k_y w\right)\right.$$
$$\left.+\frac{1-\nu}{2}\left(\frac{\partial u}{\partial y}+\frac{\partial v}{\partial x}\right)^2\right]d\Omega_1\Bigg|_{t=0}+\int\limits_0^{t_1}\left\{\frac{1}{\rho h}\left(\left\|p_1\right\|^2_{L_2(\Omega_1)}+\left\|p_2\right\|^2_{L_2(\Omega_1)}\right.\right.$$
$$\left.\left.+\left\|g_1\right\|^2_{L_2(\Omega_1)}\right)+\frac{T_0}{C_0}\left\|g_2\right\|^2_{L_2(\Omega_2)}\right\}dt.\tag{5.96}$$

Let us consider now two solutions $\{w_1, u_1, v_1, \theta_1\}$ and $\{w_2, u_2, v_2, \theta_2\}$. It is clear that $\{w_1-w_2, u_1-u_2, v_1-v_2, \theta_1-\theta_2\}$ is also a solution to the problem, where we must take

$$p_1=p_2=g_1=g_2=\varphi_1=\varphi_2=\varphi_3=\varphi_4=\Psi_1=\Psi_2=\Psi_3=0,$$

which means that $c_3 = 0$. Substituting these values into (5.95), we obtain

$$w_1 = w_2, \quad u_1 = u_2, \quad v_1 = v_2, \quad \theta_1 = \theta_2,$$

which proves the uniqueness of the solution. This ends the proof.

Two new theorems related to the solvability of the problem (5.48)–(5.54) (resulting from linearization of (5.44)–(5.47)) will now be formulated for homogeneous boundary conditions. The proofs are omitted as they repeat fully the proofs of Theorems 5.1 and 5.2.

Theorem 5.3 *Let $\partial\Omega_1$ satisfy the conditions of Theorem 5.1, and let*

$$p_1, p_2, g_1 \in L^2(0,T; L_2(\Omega_1)), \quad g_2 \in L^2(0,T; L_2(\Omega_2)), \quad \varphi_i \in \overset{\circ}{W}{}_2^1(\Omega_1),$$
$$i = 1,2,3,5,6, \quad \varphi_4 \in \overset{\circ}{W}{}_2^1(\Omega_2), \quad \Psi_j \in L_2(\Omega_1), \quad j = 1, \dots, 5.$$

Then:

1. *There exists at least one set of six functions $\{w, u, v, \Psi_x, \Psi_y, \theta\}$ that is the solution to (5.48)–(5.54) (the linearized form of (5.44)–(5.47)), and the following conditions are satisfied:*

$$w, u, v, \Psi_x, \Psi_y \in L^2(0,T; \overset{\circ}{W}{}_2^1(\Omega_1)), \quad \theta \in L^2(0,T; \overset{\circ}{W}{}_2^1(\Omega_2)),$$
$$\frac{\partial w}{\partial t}, \frac{\partial u}{\partial t}, \frac{\partial v}{\partial t}, \frac{\partial \Psi_x}{\partial t}, \frac{\partial \Psi_y}{\partial t} \in L^2(0,T; L_2(\Omega_1)). \tag{5.97}$$

2. *An approximate solution to the specified problem can be found with the help of the Bubnov–Galerkin method. A set of approximate solutions is weakly compact in the spaces corresponding to (5.97), and each of the weak limits is a solution to the problem.*

Theorem 5.4 *Let the conditions of Theorem 5.3 be satisfied, and let*

$$\frac{\partial p_1}{\partial t}, \frac{\partial p_2}{\partial t}, \frac{\partial g_1}{\partial t} \in L^2(0,T; L_2(\Omega_1)), \quad \frac{\partial g_2}{\partial t} \in L^2(0,T; L_2(\Omega_2)),$$
$$\varphi_1, \varphi_2, \varphi_3, \varphi_5, \varphi_6 \in \overset{\circ}{W}{}_2^1(\Omega_1) \cap W_2^2(Q_1), \quad \varphi_4 \in \overset{\circ}{W}{}_2^1(\Omega_2) \cap W_2^2(Q_2),$$
$$\Psi_1, \Psi_2, \Psi_3, \Psi_4, \Psi_5 \in \overset{\circ}{W}{}_2^1(\Omega_1).$$

Then there exists a unique set of six functions $\{w, u, v, \Psi_x, \Psi_y, \theta\}$ that is the solution to the problem (5.48)–(5.54). *These functions satisfy the following conditions:*

$$w, u, v, \Psi_x, \Psi_y, \frac{\partial w}{\partial t}, \frac{\partial u}{\partial t}, \frac{\partial v}{\partial t}, \frac{\partial \Psi_x}{\partial t}, \frac{\partial \Psi_y}{\partial t} \in L^2(0,T; \overset{\circ}{W}{}_2^1(\Omega_1)),$$
$$\frac{\partial^2 w}{\partial t^2}, \frac{\partial^2 u}{\partial t^2}, \frac{\partial^2 v}{\partial t^2}, \frac{\partial^2 \Psi_x}{\partial t^2}, \frac{\partial^2 \Psi_y}{\partial t^2} \in L^2(0,T; L_2(\Omega_1)),$$
$$\theta \in L^2(0,T; \overset{\circ}{W}{}_2^1(\Omega_2)), \quad \frac{\partial \theta}{\partial t} \in L^2(0,T; \overset{\circ}{W}{}_2^1(\Omega_2)).$$

Remark 5.2 *The contour bounding a shell that is rectangular or elliptic in plan can serve as a boundary contour satisfying the conditions of the above theorems.*

Note that the method presented above for proving Theorems 5.1, 5.2, 5.3 and 5.4 can be applied also in other cases of a simply supported shell with displacements equal to zero on its contour.

To conclude, in this chapter we have stated and proved some mathematical theorems governing the process of thermoelastic deformation of a shallow shell with an arbitrary distribution of the temperature field through the shell thickness. This means that the initial–boundary value problems corresponding to the models used here are solvable. However, in order to judge how well those models correspond to the real processes occurring in shells, one needs to compare the numerical results with real experimental data.

Finally, it should be emphasized that the method of construction of an approximate solution used in Theorem 5.1 generalizes a known (in the theory of thermoelasticity) rule for approximation of the temperature function by an arbitrary polynomial with respect to the variable z [122].

6 Theory with Physical Nonlinearities and Coupling

In this chapter, a theory of shells with physical nonlinearities and coupling is outlined. In Sect. 6.1, the fundamental assumptions and relations are introduced.

In Sect. 6.2, the variational equations of physically nonlinear coupled problems are derived. The equations are then reformulated into equations in terms of displacements (Sect. 6.3), which are suitable for a series of other thermoelastic problems.

6.1 Fundamental Assumptions and Relations

Similarly to previous chapters, a shallow shell with a constant thickness h having a shape Ω in plane and a boundary $d\Omega_1$ is considered. The shell material is assumed to be isotropic and physically nonlinear. An orthogonal system of coordinates x_1, x_2, x_3 is used here instead of the orthogonal system of coordinates x, y, z used in Chap. 5. As earlier, the x_1, x_2 coordinate axes coincide with the directions of the principal curvatures of the middle surface of the shell, and the x_3 coordinate is oriented towards the centre of curvature of the middle surface and is normal to the middle surface. The displacements of the points of the middle surface along x_1, x_2 and x_3 are denoted by $u_1(x_1, x_2, t)$, $u_2(x_1, x_2, t)$ and $w(x_1, x_2, t)$, respectively. The initial curvatures with respect to x_1, x_2 are denoted by k_1, k_2, respectively.

We shall now to derive the governing equations on the basis of the kinematic Kirchhoff–Love model, for which the following relations between the deformation of the middle surface and an arbitrary surface hold [231]:

$$\varepsilon_{ij}^{z} = \varepsilon_{ij} + x_3 æ_{ii} \quad (i, j = 1, 2) \tag{6.1}$$

$$\left(-\frac{h}{2} \le x_3 \le \frac{h}{2}\right).$$

Here ε_{ij} $(i, j = 1, 2)$ are the tangential deformations of the middle surface, and $æ_{ij}$ are its bending deformations, and the following relations hold:

$$\varepsilon_{ii} = \frac{\partial u_i}{\partial x_i} - k_1 w + \frac{1}{2}\left(\frac{\partial w}{\partial x_i}\right)^2 \quad (i = 1, 2), \tag{6.2}$$

$$\varepsilon_{12} = \frac{\partial u_1}{\partial x_2} + \frac{\partial u_2}{\partial x_1} + \frac{\partial w}{\partial x_1}\frac{\partial w}{\partial x_2}, \tag{6.3}$$

$$æ_{ii} = -\frac{\partial^2 w}{\partial {x_i}^2} \quad (i = 1, 2), \quad æ_{12} = -2\frac{\partial^2 w}{\partial x_1 \partial x_2}. \tag{6.4}$$

It should be noted that the deformations ε_{ij} are considered to be small in comparison with one. Thus, the derivatives of deformations can be neglected in comparison with the deformations.

The physical relations are assumed to be of the following form [112]:

$$\sigma_{ii} = 3Kæ(e_0)e_0 + 2G\gamma(e_i)(e_{ii} - e_0) \quad (i = 1, 2, 3), \tag{6.5}$$

$$\sigma_{ii} = G\gamma(e_1)e_{12}, \tag{6.6}$$

where K, G are the moduli of volume compression and shear, $e_0 = \frac{1}{3}(e_{11}^z + e_{22}^z + e_{33}^z)$ is the average deformation, $\gamma(e_1)$ is the shear function, $æ(e_0)$ is the elongation function, and

$$e_i = \frac{\sqrt{2}}{3}\sqrt{(e_{11}^z - e_{22}^z)^2 + (e_{22}^z - e_{33}^z)^2 + (e_{33}^z - e_{11}^z)^2 + \frac{3}{2}(e_{12}^z)^2}$$

is the deformation intensity (in accordance with [112], we take $æ(e_0) \equiv 1$).

Using the hypothesis of a plane stress–strain state ($\sigma_{33} = 0$) and taking into account the influence of temperature via the Duhamel–Neumann hypothesis (similarly to previous chapters), we obtain

$$\sigma_{11} = \frac{E\gamma(e_i)}{(1+\nu)(1-\tilde{\mu})}\left[\varepsilon_{11}^z + \tilde{\mu}\varepsilon_{22}^z - (1+\tilde{\mu})\alpha_t\theta\right], \tag{6.7}$$

$$\sigma_{22} = \frac{E\gamma(e_i)}{(1+\nu)(1-\tilde{\mu})}\left[\varepsilon_{22}^z + \tilde{\mu}\varepsilon_{11}^z - (1+\tilde{\mu})\alpha_t\theta\right], \tag{6.8}$$

$$\sigma_{12} = \frac{E\gamma(e_i)}{2(1+\nu)}\varepsilon_{12}^z, \tag{6.9}$$

where E is Young's modulus, ν is Poisson's ratio, $\theta(x_1, x_2, x_3, t) = T_1(x_1, x_2, x_3, t) - T_0$ is the temperature increase at the point x_1, x_2, x_3 at the time moment t, and $T_1(x_1, x_2, x_3, t)$ is the absolute temperature at this point at time t. We assume that $\left|\frac{\theta}{T_0}\right| \ll 1$, i.e. the temperature increase is so small that all of the thermoelastic parameters can be treated as constant (independent of temperature). The quantity α_t is the coefficient of linear thermal expansion, and

$$\tilde{\mu} = \frac{3 - 2g\gamma(e_i)}{2(3 + g\gamma(e_i))}, \tag{6.10}$$

where $g = \frac{G}{K}$.

We introduce (similarly to previous chapters) the following integral stress characteristics:

$$T_{ij} = \int\limits_{-\frac{h}{2}}^{\frac{h}{2}} \sigma_{ij}\, dx_3, \quad M_{ij} = \int\limits_{-\frac{h}{2}}^{\frac{h}{2}} \sigma_{ij}\, x_3 dx_3. \tag{6.11}$$

Substituting (6.1) into (6.7)–(6.9) and then into (6.11), we obtain after a series of transformations

$$T_{ii} = \frac{Eh}{1-\nu^2}(\varepsilon_{ii} + \nu\varepsilon_{jj}) + \Delta T_{ii} - N_T - \Delta_T T, \tag{6.12}$$

$$M_{ii} = \frac{Eh^3}{12(1-\nu^2)}(æ_{ii} + \nu æ_{jj}) + \Delta M_{ii} - M_T - \Delta_T M \tag{6.13}$$

$$(i, j = 1, 2, \quad i \neq j),$$

$$T_{12} = \frac{Eh}{2(1+\nu)}\varepsilon_{12} + \Delta T_{12}, \tag{6.14}$$

$$M_{12} = \frac{Eh^3}{24(1+\nu)}æ_{12} + \Delta M_{12}, \tag{6.15}$$

$$N_T = \frac{E\alpha_t}{1-\nu}\int\limits_{-\frac{h}{2}}^{\frac{h}{2}} \theta\, dx_3, \quad M_T = \frac{E\alpha_t}{1-\nu}\int\limits_{-\frac{h}{2}}^{\frac{h}{2}} \theta x_3\, dx_3, \tag{6.16}$$

where

$$\Delta T_{ii} = \frac{Eh}{1+\nu}\left[\varepsilon_{ii}p_1^1 + \varepsilon_{jj}p_1^2 + h(æ_{ii}p_2^1 + æ_{jj}p_2^2)\right] \tag{6.17}$$

$$\Delta M_{ii} = \frac{Eh^2}{1+\nu}\left[\varepsilon_{ii}p_2^1 + \varepsilon_{jj}p_2^2 + h(æ_{ii}p_3^1 + æ_{jj}p_3^2)\right] \tag{6.18}$$

$$(i, j = 1, 2, \quad i \neq j),$$

$$\Delta T_{12} = \frac{Eh}{2(1+\nu)}(\varepsilon_{12}p_1^3 + hæ_{12}p_2^3), \tag{6.19}$$

$$\Delta M_{12} = \frac{Eh^2}{2(1+\nu)}(\varepsilon_{12}p_2^3 + hæ_{12}p_2^3), \tag{6.20}$$

$$\Delta_T T = E\alpha_T W_1, \quad \Delta_T M = E\alpha_T W_2, \tag{6.21}$$

$$P_k^n = \frac{1}{h^k}\int\limits_{-\frac{h}{2}}^{\frac{h}{2}} F_n x_3^{k-1}\, dx_3 \tag{6.22}$$

$$(n, k = 1, 2, 3),$$

$$W_i = \int\limits_{-\frac{h}{2}}^{\frac{h}{2}} F_i \theta x_3^{i-1}\, dx_3 \quad (i = 1, 2), \tag{6.23}$$

$$F_1 = \frac{(1-\nu)\gamma(e_i) - (1-\tilde{\mu})}{(1-\nu)(1-\tilde{\mu})}, \tag{6.24}$$

$$F_2 = \frac{\tilde{\mu}(1-\nu)\gamma(e_i) - \nu(1-\tilde{\mu})}{(1-\nu)(1-\tilde{\mu})}, \tag{6.25}$$

$$F_3 = \gamma(e_i) - 1, \tag{6.26}$$

$$F_4 = \frac{(1+\tilde{\mu})(1-\nu)\gamma(e_i) - (1+\nu)(1-\tilde{\mu})}{(1-\nu^2)(1-\tilde{\mu})}. \tag{6.27}$$

Solving (6.12) and (6.14) for the deformations of the middle surface $\varepsilon_{11}, \varepsilon_{22}, \varepsilon_{12}$, and (6.13), (6.15) for æ$_{ii}$, æ$_{12}$, we obtain

$$\varepsilon_{ii} = \frac{1}{Eh}\Big[(T_{ii} - \nu T_{jj}) - (\Delta T_{ii} - \nu \Delta T_{jj}) + (1-\nu)(N_T + \Delta_T T)\Big], \tag{6.28}$$

$$\varepsilon_{12} = \frac{2(1+\nu)}{Eh}(T_{12} - \Delta T_{12}), \tag{6.29}$$

$$\text{æ}_{ii} = \frac{12}{Eh^3}\Big[(M_{ii} - \nu M_{jj}) - (\Delta M_{ii} - \nu \Delta M_{jj}) + (1-\nu)(M_T + \Delta_T M)\Big] \tag{6.30}$$

$$\text{æ}_{12} = \frac{24(1+\nu)}{Eh^3}(M_{12} - \Delta M_{12}) \tag{6.31}$$

$$(i, j = 1, 2, \quad i \neq j).$$

The theory of small elasto-plastic deformations yields

$$G\gamma(e_i) = \frac{1}{3}\frac{\sigma_i(e_i)}{e_i}, \tag{6.32}$$

$$\gamma(e_i) = \frac{\sigma_i(e_i)}{3Ge_i}, \tag{6.33}$$

where $\sigma_i(e_i)$ characterizes the relation between the stress intensity and the deformation intensity.

6.2 Variational Equations of Physically Nonlinear Coupled Problems

We apply the Biot variational principle [47]

$$\delta V_1 + \delta D + \delta K = \iint\limits_{\Omega_2} (\mathbf{F}_0\, \delta \mathbf{u}_0 - \theta \mathbf{n}\, \delta \mathbf{S})\, d\Omega_2, \tag{6.34}$$

where $\delta V_1, \delta D, \delta K$ are the variations of the generalized free vibrations of the dissipative function and of the kinetic energy, respectively; $\mathbf{F}_0, \mathbf{u}_0$ are the surface force and displacement; $\mathbf{n}$ is the external normal; and Ω_2 is the surface of a bounded body. The change of the entropy vector $\mathbf{S}$ is related to the increase of the entropy φ in a volume unit [47] by

$$\varphi = -div\,\mathbf{S} = -\sum_i \frac{\partial S_i}{\partial x_i}. \tag{6.35}$$

Using the results presented in [47], we obtain

$$\delta V_1 = \iiint_V \left(\sum_{ij} \delta_{ij} \delta e_{ij} + \theta\, \delta\varphi \right) dV, \tag{6.36}$$

$$\delta D = \iiint_V \frac{T_0}{R_T} \mathbf{S}\, \delta\mathbf{S}\, dV, \tag{6.37}$$

where R_T is the thermal conductivity, and V denotes the volume occupied by the shell.

Substituting (6.35) into (6.36) and integrating by parts, we obtain

$$\iiint_V \theta\, \delta\varphi\, dV \equiv \iiint_V grad\,\theta\, \delta\mathbf{S}\, dV - \iint_{\Omega_2} \theta \mathbf{n}\, \delta\mathbf{S}\, d\Omega_2. \tag{6.38}$$

The surface integrals in (6.38) and (6.34) cancel each other, and from (6.38) we obtain

$$\iiint_V \theta\, \delta\varphi\, dV + \delta D \equiv \iiint_V \left(grad\,\theta + \frac{T_0}{R_T}\mathbf{S} \right) \delta\mathbf{S}\, dV. \tag{6.39}$$

From (6.38), we can deduce the following vector equation:

$$grad\,\theta + \frac{T_0}{R_T}\mathbf{S} = 0. \tag{6.40}$$

Taking the divergence of (6.40) and taking into account a result given in [47]

$$\varphi = \frac{c\theta}{T_0} + \beta e, \tag{6.41}$$

we obtain

$$\nabla^2 \theta = \frac{T_0}{R_T} \frac{\partial}{\partial t} \left(\frac{c\theta}{T_0} + \beta e \right), \tag{6.42}$$

where c is the specific heat capacity, $e = \varepsilon_{11}^z + \varepsilon_{22}^z + \varepsilon_{33}^z$, ∇^2 is the three-dimensional Laplace operator, $\beta = k\alpha_t$, and t denotes time.

Therefore, a heat transfer equation which includes coupling between the temperature and the deformation has been obtained.

In order to reduce the order of the integrals occurring in (6.34), we integrate the terms that are not used with respect to $-\frac{h}{2} \le x_3 \le \frac{h}{2}$. Neglecting the terms $\ddot{u}_1, \ddot{u}_2$, and taking into account (6.1), (6.11) and the condition $\sigma_{33} = 0$, we obtain

$$\delta K = -\rho h \iint_{\Omega} \ddot{w}\, \delta w \, dx_1 dx_2, \tag{6.43}$$

$$\iiint_V \sum_{i,j} \sigma_{ij}\, \delta e_{ij}\, dV = \iint_{\Omega_2} (T_{11}\, \delta\varepsilon_{11} + T_{22}\, \delta\varepsilon_{22} + T_{12}\, \delta\varepsilon_{12})\, dx_1 dx_2 + \iint_{\Omega_2} (M_{11}\, \delta æ_{11} + M_{22}\, \delta æ_{22} + M_{12}\, \delta æ_{12})\, dx_1 dx_2, \tag{6.44}$$

$$\iint_{\Omega} (\mathbf{F}_0\, \delta \mathbf{u}_0)\, d\Omega = \iint_{\Omega} (q - \rho h \varepsilon \dot{w})\, \delta w\, dx_1 dx_2 + \int T_{11}^0\, \delta u_1\, dx_2 + \int T_{22}^0\, \delta u_2\, dx_1 + \int T_{12}^0\, \delta u_1\, dx_1 + \int T_{12}^0\, \delta u_2\, dx_2, \tag{6.45}$$

where ε is a damping coefficient; q denotes the intensity of the transverse load; T_{11}^0, T_{22}^0, are given longitudinal loads along x_1, x_2, respectively; and T_{12}^0 is a given shear load.

In order to obtain the initial differential equation in a hybrid form, we transform the first term in the right-hand side of the relations (6.44) to the form

$$\begin{aligned} &\iint_{\Omega_2} (T_{11}\, \delta\varepsilon_{11} + T_{22}\, \delta\varepsilon_{22} + T_{12}\, \delta\varepsilon_{12})\, dx_1 dx_2 \\ &\quad = \delta \iint_{\Omega_2} (T_{11}\varepsilon_{11} + T_{22}\varepsilon_{22} + T_{12}\varepsilon_{12})\, dx_1 dx_2 \\ &\quad - \iint_{\Omega_2} (\varepsilon_{11}\, \delta T_{11} + \varepsilon_{22}\, \delta T_{22} + \varepsilon_{12}\, \delta T_{12})\, dx_1 dx_2. \end{aligned} \tag{6.46}$$

We introduce also the stress function F (Airy's stress function) in the form

$$T_{ii} = \frac{\partial F}{\partial x_{jj}^2}, \quad T_{12} = \frac{\partial^2 F}{\partial x_1 \partial x_2} \quad (i, j = 1, 2, \quad i \ne j), \tag{6.47}$$

as well as the known differential operators

$$\nabla_k^2(\cdot) = K_2 \frac{\partial^2(\cdot)}{\partial {x_1}^2} + K_1 \frac{\partial^2(\cdot)}{\partial {x_2}^2}, \tag{6.48}$$

$$L((\cdot),(\cdot)) = \frac{\partial^2(\cdot)}{\partial {x_1}^2} \frac{\partial^2(\cdot)}{\partial {x_2}^2} - 2 \frac{\partial^2(\cdot)}{\partial x_1 \partial x_2} \frac{\partial^2(\cdot)}{\partial x_1 \partial x_2} + \frac{\partial^2(\cdot)}{\partial {x_2}^2} \frac{\partial^2(\cdot)}{\partial {x_1}^2}. \tag{6.49}$$

We substitute the values of the deformations (6.2), (6.3) into the first integral on the right-hand side of (6.46), and substitute the values of $æ_{ij}$ from (6.4) and ε_{ij} from (6.28), (6.29) into the second integral on the right-hand side of each of (6.44), (6.46). In the transformed expression (6.44), we integrate by parts and take the variations, taking into account both (6.47) and the rule that the contour integrals in (6.45) cancel with the contour integrals obtained via transformation of (6.46). We move from the variational principle (6.34) to the following variational equation:

$$
\begin{aligned}
&\iiint_V (grad\,\theta + \frac{T_0}{R_T}\mathbf{S})\,\delta\mathbf{S}\,dV - \iint_\Omega \Big\{\Big\langle \nabla_k^2\, w + \frac{1}{2}L(w,w)\\
&+\frac{1}{Eh}\Big[\nabla^2\nabla^2 F - (\Delta T_{11} - \nu\Delta T_{22})_{x_2x_2} - (\Delta T_{22} - \nu\Delta T_{11})_{x_1x_1}\\
&+2(1+\nu)(\Delta T_{12})_{x_1x_2} + (1-\nu)\nabla^2(N_T + \Delta_T T)\Big]\Big\rangle\delta F + \Big[\nabla_k^2 F\\
&+\frac{1}{2}L(w,F) - D\nabla^2\nabla^2 w - \nabla^2(M_T + \Delta_T M) + (\Delta M_{11})_{x_1x_1}\\
&+(\Delta M_{22})_{x_2x_2} + 2(\Delta M_{12})_{x_1x_2} - \rho h(\ddot{w} + \varepsilon\dot{\omega}) + q\Big]\delta w\Big\}\,dx_1dx_2\\
&-\int\Big[k_1 w - \frac{1}{2}\Big(\frac{\partial w}{\partial x_1}\Big)^2 + \varepsilon_{11}\Big](\delta F)_{x_2}\Big|_{x_2=0}^{x_2=b}dx_1\\
&-\int\Big[k_2 w - \frac{1}{2}\Big(\frac{\partial w}{\partial x_2}\Big)^2 + \varepsilon_{22}\Big](\delta F)_{x_1}\Big|_{x_1=0}^{x_1=a}dx_2 + \int\Big[k_2 w\frac{\partial w}{\partial x_2}\\
&+\frac{\partial^2 w}{\partial x_1{}^2}\frac{\partial w}{\partial x_2} - \frac{\partial\varepsilon_{12}}{\partial x_1}\frac{\partial\varepsilon_{11}}{\partial x_2}\Big]\delta F\Big|_{x_2=0}^{x_2=b}dx_1 + \int\Big[k_2 w\frac{\partial w}{\partial x_1}\\
&+\frac{\partial^2 w}{\partial x_2{}^2}\frac{\partial w}{\partial x_1} - \frac{\partial\varepsilon_{12}}{\partial x_2}\frac{\partial\varepsilon_{22}}{\partial x_1}\Big]\delta F\Big|_{x_1=0}^{x_1=a}dx_2 + \Big(\varepsilon_{12} - \frac{\partial w}{\partial x_1}\frac{\partial w}{\partial x_2}\Big)\\
&\times\delta F\Big|_{x_1=0}^{x_1=a}\Big|_{x_2=0}^{x_2=b} - 2M_{12}\,\delta w\Big|_{\substack{x_1=0\\x_2=0}}^{\substack{x_1=a\\x_2=b}} - \int M_{11}(\delta w)_{x_1}\Big|_{x_1=0}^{x_1=a}dx_2\\
&-\int M_{22}(\delta w)_{x_2}\Big|_{x_2=0}^{x_2=b}dx_1 + \int\Big(\frac{\partial M_{11}}{\partial x_1} + 2\frac{\partial M_{12}}{\partial x_2} + T_{11}\frac{\partial w}{\partial x_1}\\
&+T_{12}\frac{\partial w}{\partial x_2}\Big)\delta w\Big|_{x_1=0}^{x_1=a}dx_2 + \int\Big(\frac{\partial M_{22}}{\partial x_2} + 2\frac{\partial M_{12}}{\partial x_1} + T_{22}\frac{\partial w}{\partial x_2}\\
&+T_{12}\frac{\partial w}{\partial x_2}\Big)\delta w\Big|_{x_2=0}^{x_2=b}dx_1 = 0.
\end{aligned}
\tag{6.50}
$$

The subscripts x_i in (6.50) denote differentiation with respect to the coordinate x_i. The functional (6.50) has been written for a rectangular space $\Omega \in \{x_1x_2|0 \le x_1 \le a,\ 0 \le x_2 \le b\}$.

Equating to zero the expressions accompanying $\delta\mathbf{S}$, δw and δF, we obtain from the variational equation (6.50) a system of three differential equations with different dimensions: a three-dimensional heat transfer equation with coupling of deformation and temperature, and the two-dimensional equations for the shell motion.

The system of equations obtained reads

$$\begin{aligned}
&\frac{\partial^2\theta}{\partial {x_1}^2} + \frac{\partial^2\theta}{\partial {x_2}^2} + \frac{\partial^2\theta}{\partial {x_3}^2} = \frac{1}{2}\dot{\theta} + \frac{T_0 E\alpha_T}{3(1-2\nu)R_T}\dot{e},\\
&D\nabla^2\nabla^2 w - L(w,F) - \nabla_k^2 F + \nabla^2(M_T + \Delta_T M) - (\Delta M_{11})_{x_1x_1}\\
&-(\Delta M_{22})_{x_2x_2} - 2(\Delta M_{12})_{x_1x_2} - q + \rho h(\ddot{w} + \varepsilon\dot{w}) = 0,\\
&\frac{1}{Eh}\nabla^2\nabla^2 F = -\nabla_k^2 w - \frac{1}{2}L(w,w) + \frac{1}{Eh}\big[(\Delta T_{11} - \nu\Delta T_{22})_{x_2x_2}\\
&+(\Delta T_{22} - \nu\Delta T_{11})_{x_1x_1} - 2(1+\nu)(\Delta T_{12})_{x_1x_2}\\
&-(1-\nu)\nabla^2(N_T + \Delta_T T)\big],
\end{aligned} \tag{6.51}$$

where $\alpha = \dfrac{R_T}{\rho}$ is the thermal conductivity coefficient.

6.3 Equations in Terms of Displacements

For some series of problems (for example, in order to investigate thermoelastic stress wave transitions), it is much more convenient to use the governing equations expressed in terms of the displacements u_1, u_2 and w. However, omission of $\ddot{u}_1, \ddot{u}_2$ is not permitted in this case.

The general form of the heat transfer equation is not changed, but we need to substitute (6.1), (6.2) and (6.4) in the expression $e = \varepsilon_{11}^z + \varepsilon_{22}^z$.

In the variational principle (6.34), some terms are transformed to another form; in particular,

$$\delta K = -\rho h \iint\limits_{\Omega_2} \left(\sum_{i=1}^{2} \ddot{u}_i\,\delta u_i + \ddot{w}\,\delta w \right) dx_1 dx_2. \tag{6.52}$$

The expression (6.43) does not change, but one needs to substitute ε_{ij} as represented by (6.2) and (6.3) into the transformed equation (6.46). Then, constructing the standard variational equation including the influence of damping, we obtain

$$\iiint\limits_V (grad\,\theta + \frac{T_0}{R_T}\mathbf{S})\,\delta\mathbf{S}\,dV - \iint\limits_{\Omega} \Bigg\{ \sum_{\substack{i,j=1\\ i\neq j}}^{2} \left(\frac{\partial T_{ii}}{\partial x_i} + \frac{\partial T_{ij}}{\partial x_j} \right.$$

$$\left. - \rho h\ddot{u}_i \right)\delta u_i + \Bigg\langle \sum_{\substack{i,j=1\\ i\neq j}}^{2} \left[\frac{\partial}{\partial x_i}\left(T_{ii}\frac{\partial w}{\partial x_i} + T_{ij}\frac{\partial w}{\partial x_j} \right) k_i T_{ii} \right.$$

$$+\frac{\partial^2 M_{ii}}{\partial {x_i}^2}+\frac{\partial^2 M_{ii}}{\partial x_i \partial x_j}\Bigg]+q-\rho h\ddot{w}\Big\rangle\delta w\Bigg\}\,dx_1dx_2$$

$$+\int\sum_{\substack{i,j=1\\ i\neq j}}^{2}\left\{\left[T_{ii}\frac{\partial w}{\partial x_i}+T_{ij}\left(\frac{\partial w}{\partial x_i}+\frac{\partial w}{\partial x_j}\right)+T_{jj}\frac{\partial w}{\partial x_j}\right]\delta w\right. \tag{6.53}$$

$$\left.+M_{ii}(\delta w)_{x_i}+M_{ij}(\delta w)_{x_j}\right\}dx_j=0.$$

From the variational equation (6.53), we obtain the system of differential equations governing the shell dynamics,

$$\begin{aligned}&\frac{\partial T_{ii}}{\partial x_i}+\frac{\partial T_{ij}}{\partial x_j}=\rho h\ddot{u}_i \quad (1\leftrightarrow 2)\\&\frac{\partial^2 M_{11}}{\partial {x_1}^2}+2\frac{\partial^2 M_{12}}{\partial x_i\partial x_j}+\frac{\partial^2 M_{22}}{\partial {x_2}^2}+K_1T_{11}\\&+K_2T_{22}+\frac{\partial}{\partial x_1}\left(T_{11}\frac{\partial w}{\partial x_1}+T_{12}\frac{\partial w}{\partial x_2}\right)\\&+\frac{\partial}{\partial x_2}\left(T_{12}\frac{\partial w}{\partial x_1}+T_{22}\frac{\partial w}{\partial x_2}\right)+q=\rho h\ddot{w},\end{aligned} \tag{6.54}$$

and the heat transfer equation,

$$\begin{aligned}&\frac{\partial^2\theta}{\partial {x_1}^2}+2\frac{\partial^2\theta}{\partial {x_2}^2}+\frac{\partial^2\theta}{\partial {x_3}^2}+\frac{1}{2}\dot{\theta}+\frac{T_0E\alpha_T}{3(1-2\nu)R_T}\left[\frac{\partial\dot{u}_1}{\partial x_1}+\frac{\partial\dot{u}_2}{\partial x_2}\right.\\&\left.-(K_1+K_2)\dot{w}-x_3\nabla^2\dot{w}+\frac{1}{2}\left(\left(\frac{\partial\dot{w}}{\partial x_1}\right)^2+\left(\frac{\partial\dot{w}}{\partial x_2}\right)^2\right)\right].\end{aligned} \tag{6.55}$$

Using (6.12)–(6.17), (6.2) and (6.13), the following equations in terms of displacements are obtained:

$$\frac{\partial}{\partial x_1}\left[\left(\frac{1}{1-\nu}+p_1^1\right)\left(\frac{\partial u_1}{\partial x_1}-k_1w+\frac{1}{2}(\frac{\partial w}{\partial x_1})^2\right)\right.$$

$$+\left(\frac{\nu}{1-\nu}+p_1^2\right)\left(\frac{\partial u_2}{\partial x_2}-k_2w+\frac{1}{2}\left(\frac{\partial w}{\partial x_2}\right)^2\right)$$

$$\left.-h\left(\frac{\partial^2 w}{\partial {x_1}^2}p_2^1+\frac{\partial^2 w}{\partial {x_2}^2}p_2^2\right)\right]+\frac{\partial}{\partial x_2}\left[\frac{p_1^3+1}{2}\right.$$

$$\left.\times\left(\frac{\partial u_1}{\partial x_1}+\frac{\partial u_2}{\partial x_1}+\frac{\partial w}{\partial x_1}\frac{\partial w}{\partial x_1}\right)-h\frac{\partial^2 w}{\partial x_1\partial x_2}p_2^3\right]$$

$$= \frac{1+\nu}{E}\left[\rho \ddot{u}_i + \frac{1}{h}\frac{\partial}{\partial x_1}(N_T + \Delta_T T)\right] \quad (1 \leftrightarrow 2),$$

$$\sum_{\substack{i,j=1 \\ i \neq j}}^{2} \left\langle \left(\frac{\partial u_i}{\partial x_i} - k_i w + \frac{1}{2}\left(\frac{\partial w}{\partial x_i}\right)^2\right) k_i \left(\frac{1}{1-\nu} + p_1^1\right)\right.$$

$$+\left(\frac{\partial u_j}{\partial x_j} - k_j w + \frac{1}{2}\left(\frac{\partial w}{\partial x_j}\right)^2\right) k_i \left(\frac{\nu}{1-\nu} + p_1^2\right)$$

$$-h\left(\frac{\partial^2 w}{\partial {x_i}^2} p_2^1 + \frac{\partial^2 w}{\partial {x_i}^2} p_2^2\right) + \frac{\partial^2}{\partial {x_i}^2}\left[h\left(\frac{\partial u_i}{\partial x_i} - k_i w\right.\right.$$

$$\left. + \frac{1}{2}\left(\frac{\partial w}{\partial x_i}\right)^2\right) p_2^1 + \left(\frac{\partial u_j}{\partial x_j} - k_j w + \frac{1}{2}\left(\frac{\partial w}{\partial x_j}\right)^2\right) p_2^2$$

$$\left. -h^2 \frac{\partial^2 w}{\partial {x_i}^2}\left(\frac{1}{12(1-\nu)} + p_3^1\right) - h^2 \frac{\partial^2 w}{\partial {x_j}^2}\left(\frac{1}{12(1-\nu)} + p_3^2\right)\right]$$

$$+\frac{\partial^2}{\partial x_i \, \partial x_j}\left[\frac{h}{2}\left(\frac{\partial u_i}{\partial x_j} + \frac{\partial u_j}{\partial x_i} + \frac{\partial w}{\partial x_i}\frac{\partial w}{\partial x_j}\right) p_2^3\right.$$

$$\left. -h^2 \frac{\partial^2 w}{\partial x_i \partial x_j}\left(\frac{1}{12(1-\nu)} + p_3^3\right)\right] + \frac{\partial}{\partial x_i}\left\{\frac{\partial w}{\partial x_i}\right.$$

$$\times\left[\left(\frac{\partial u_i}{\partial x_i} - k_i w + \frac{1}{2}\left(\frac{\partial w}{\partial x_i}\right)^2\right)\left(p_1^1 + \frac{1}{1-\nu}\right) + \left(\frac{\partial u_j}{\partial x_j}\right.\right.$$

$$\left. -k_j w + \frac{1}{2}\left(\frac{\partial w}{\partial x_j}\right)^2\right)\left(p_1^2 + \frac{\nu}{1-\nu}\right) - h\left(\frac{\partial^2 w}{\partial {x_i}^2} p_2^1\right.$$

$$\left.\left. + \frac{\partial^2 w}{\partial {x_j}^2} p_2^2\right)\right] + \frac{\partial w}{\partial x_j}\left[\left(\frac{\partial u_i}{\partial x_j} + \frac{\partial u_j}{\partial x_i} + \frac{\partial w}{\partial x_i}\frac{\partial w}{\partial x_j}\right)\right.$$

$$\left.\left.\left. \times \frac{1+p_1^3}{2} - \frac{h}{2}\frac{\partial^2 w}{\partial x_i \, \partial x_j} p_2^3\right]\right\}\right\rangle = \frac{1+\nu}{Eh}\left\{\rho h \ddot{w} + q\right.$$

$$+\sum_{i=1}^{2}\left[k_i (N_T + \Delta_T T) + \frac{\partial}{\partial x_i}\left(\frac{\partial w}{\partial x_i}(N_T + \Delta_T T)\right)\right.$$

$$\left.\left. + \frac{\partial^2}{\partial {x_i}^2}(M_T + \Delta_T M)\right]\right\}, \tag{6.56}$$

where $\Delta_T N, \Delta_T M, p_n^k$ $(n = 1, 2, 3,\ k = 1, 2, 3)$ are nonlinear terms defined by the relations (6.16)–(6.27).

The boundary and initial conditions have been formulated in Sects. 2.3 and 5.3. The mechanical boundary conditions corresponding to certain types of support of the shell edge are cited below (because they will be used later):

$$w = \frac{\partial w}{\partial x_1} = \varepsilon_{22} = \frac{\partial \varepsilon_{22}}{\partial x_1} - \frac{\partial \varepsilon_{12}}{\partial x_2} = 0, \quad x_1 = const, \tag{6.57}$$

$$\begin{aligned} M_{11} = F = \frac{\partial F}{\partial x_1} = 0, \quad x_1 = const, \\ \frac{\partial M_{11}}{\partial x_1} + 2\frac{\partial M_{12}}{\partial x_1} = 0, \quad x_1 = const. \end{aligned} \tag{6.58}$$

The last condition is equivalent to the conditions

$$T_{11}\frac{\partial w}{\partial x_1} + T_{12}\frac{\partial w}{\partial x_2} = 0, \quad x_1 = const, \tag{6.59}$$

$$w = M_{11} = \varepsilon_{22} = \frac{\partial \varepsilon_{22}}{\partial x_1} - \frac{\partial \varepsilon_{12}}{\partial x_2} = 0, \quad x_1 = const, \tag{6.60}$$

$$w = \frac{\partial w}{\partial x_1} = F = \frac{\partial F}{\partial x_1} = 0, \quad x_1 = const, \tag{6.61}$$

$$w = M_{11} = F = \frac{\partial F}{\partial x_1} = 0, \quad x_1 = const. \tag{6.62}$$

In addition to the boundary conditions given above and those given in Sect. 2.3, there exist many more possibilities.

7 Nonlinear Problems of Hybrid-Form Equations

This chapter is devoted to the analysis of some nonlinear problems governed by the hybrid form of the differential equations obtained earlier.

In Sect. 7.1, a method of solution for nonlinear coupled problems is addressed. The finite-difference method is used to discretize the derivatives in order to obtain a system of ordinary differential and algebraic equations. A procedure for solution of the initial–boundary value problem is described. In Sect. 7.2, a relaxation method which reduces the problem to a Cauchy problem of ordinary differential equations is presented. Some advantages and disadvantages of the method are discussed. In addition, the high efficiency of the method is illustrated using examples of problems of plates and shells with physical and physical–geometrical nonlinearities. In Sect. 7.3, numerical investigations are described, and the reliability of the solution of the three-dimensional nonstationary heat transfer equation obtained using the relaxation method (later refered to as the "set-up" method) is shown. In addition, a solution to a coupled thermoelastic problem of a flexible shallow shell without any prior assumptions about the temperature distribution through the thickness is given.

The use of the three-dimensional heat transfer equation is addressed. The efficiency of the method used here when applied to the solution of integral–differential equations with different dimensions (three-dimensional equations related to the Kirchhoff–Love model) and of different type (heat transfer equations and the hyperbolic equations of shell theory) is demonstrated.

In Sect. 7.4, the vibration of an isolated shell subjected to an impulse load that is constant in time and uniformly distributed is addressed.

Many computational results are reported and discussed. In Sect. 7.5, the dynamic stability of shells under thermal shock is analysed. Many computational examples are included and many conclusions are derived. In Section 7.6, the influence of coupling and rotational inertia on stability is analysed. Theorems related to the existence of a solution of a nonlinear coupled thermoelastic problem within the Kirchhoff–Love model, including the rotational inertia of shell elements, are formulated and proved. Also, a numerical algorithm is proposed for solving the problem, and many numerical examples are given. In the next section, results of numerical tests are provided. In Sect. 7.8, the influence of two control parameters, i.e. damping and excitation amplitude, is investigated. Also, some tools to analyse the regular and chaotic behaviour of a plate are introduced. Section 7.9 is devoted to the analysis of spatial–temporal symmetric chaos. Nonsymmetric dissipative oscillations of a square isotropic plate subjected to a longitudinal, one-sided periodic load are reported in Sect. 7.10. The scenarios leading to chaos are illustrated. The last section, Sect. 7.11, deals with solitary waves in thin flexible plates within the kinematic Kirchhoff–Love model.

7.1 Method of Solution for Nonlinear Coupled Problems

We transform the system of differential equations (6.51) to the following non-dimensional form:

$$\begin{aligned}
&\lambda_1^2\frac{\partial^2\theta}{\partial {x_1}^2}+\lambda_2^2\frac{\partial^2\theta}{\partial {x_2}^2}+\frac{\partial^2\theta}{\partial {x_3}^2}=\dot{\theta}+\beta\dot{e},\\
&\frac{1}{12(1-\nu^2)}\left(\lambda^{-2}\frac{\partial^4 w}{\partial {x_1}^4}+2\frac{\partial^4 w}{\partial x_1^2\,\partial x_2^2}+\lambda^2\frac{\partial^4 w}{\partial {x_2}^4}\right)-L(w,F)\\
&-\nabla_k^2 F+\lambda^{-1}(M_T+\Delta_T M)_{x_1x_1}+\lambda(M_T+\Delta_T M)_{x_2x_2}\\
&-\lambda^{-1}(\Delta M_{11})_{x_1x_1}-\lambda(\Delta M_{22})_{x_2x_2}-2(\Delta M_{12})_{x_1x_2}\\
&-q+æ(\ddot{w}+\varepsilon\dot{w})=0,\\
&\lambda^{-2}\frac{\partial^4 F}{\partial {x_1}^4}+2\frac{\partial^4 F}{\partial x_1^2\,\partial x_2^2}+\lambda^2\frac{\partial^4 F}{\partial {x_2}^4}+(1-\nu)\big[\lambda^{-1}(N_T\\
&+\Delta_T T)_{x_1x_1}+\lambda(N_T+\Delta_T T)_{x_2x_2}\big]+\nabla_k^2 w+\frac{1}{2}L(w,w)\\
&-\lambda^{-1}(\Delta T_{22}-\nu\Delta T_{11})_{x_1x_1}-\lambda(\Delta T_{11}-\nu\Delta T_{22})_{x_2x_2}\\
&+2(1+\nu)(\Delta T_{12})_{x_1x_2}=0.
\end{aligned}\tag{7.1}$$

The notation used in (7.1) is described in Chap. 6.

As the initial conditions we take

$$\begin{aligned}
&w\big|_{t=0}=\varphi_1,\quad \theta\big|_{t=0}=\varphi_2,\\
&w\big|_{t=0}=\Psi_1,
\end{aligned}\tag{7.2}$$

and as the boundary conditions we take[1]

$$\begin{aligned}
&w=M_{11}=\varepsilon_{22}=\frac{\partial\varepsilon_{22}}{\partial x_1}-\frac{\partial\varepsilon_{12}}{\partial x_2}=0,\quad x_1=const,\\
&w=M_{22}=\varepsilon_{11}=\frac{\partial\varepsilon_{11}}{\partial x_2}-\frac{\partial\varepsilon_{12}}{\partial x_1}=0,\quad x_2=const,
\end{aligned}\tag{7.3}$$

[1] It is possible to apply any boundary conditions with the algorithm described here.

$$\frac{\partial \theta}{\partial x_3}\Big|_{x_3=0.5} = q_T, \quad \frac{\partial \theta}{\partial x_3}\Big|_{x_3=-0.5} = 0,$$
$$\frac{\partial \dot{\theta}}{\partial h}\Big|_{\Gamma} = 0. \tag{7.4}$$

The boundary conditions (7.4) characterize a thermal shock with an intensity q_T on the surface $x_3 = 0.5$ (the other surfaces of the shell are thermally isolated). We take $t \in [0, t_1]$, where t_1 is the observation time of the behaviour of the shell; $\Omega_1 = (0,1) \times (0,1)$ is the space in which the independent variables x_1, x_2 vary, which is referred to as the middle surface; $\overline{\Omega}_1 = \Omega_1 \cup \partial\Omega_1$; $\partial\Omega_1$ is the edge of the middle surface; $Q_1 = \Omega_1 \times (0, t_1)$; $\Gamma = \partial\Omega_1 \times [0, t_1]$; $\Omega_2 = \Omega_1 \times (-\frac{h}{2}, \frac{h}{2})$; h is the constant shell thickness; $\overline{\Omega}_2 = \Omega_2 \cup \partial\Omega_2$; $\partial\Omega_2$ is the surface surrounding the shell volume in three-dimensional space; $Q_2 = \Omega_2 \times (0, t_1)$; and $S = \partial\Omega_2 \times (0, t_1)$. The following nondimensional parameters have been used:

$$\overline{x}_1 = \frac{x_1}{a},\ \overline{x}_2 = \frac{x_2}{b},\ \overline{x}_3 = \frac{x_3}{h},\ \overline{w} = \frac{w}{h},\ \lambda_1 = \frac{h}{a},\ \lambda_2 = \frac{h}{b},\ \lambda = \frac{a}{b},$$
$$\overline{k}_1 = k_1\frac{a^2}{h},\ \overline{k}_2 = k_2\frac{b^2}{h},\ \overline{F} = \frac{F}{Eh^3},\ \overline{\varepsilon} = \varepsilon\frac{h^2}{\alpha},\ \overline{t} = t\frac{\alpha}{h^2},$$
$$æ = \frac{a^2 b^2 \rho \alpha^2}{Eh^6},\quad \overline{q}_T = q_T\frac{ab\alpha_T}{hk_t},\quad \overline{\theta} = \theta\frac{\alpha_T ab}{h^2}, \tag{7.5}$$
$$\Delta\overline{T}_{ij} = \Delta T_{ij}\frac{ab}{Eh^3},\quad \Delta\overline{M}_{ij} = \Delta M_{ij}\frac{ab}{Eh^4},$$
$$\beta = \frac{T_0 E\alpha_T}{3(1-2\nu)\rho h^2}.$$

In (7.1)–(7.4), the bars over nondimensional expressions have been omitted.

The first equation of (7.1) is three-dimensional and of parabolic type, whereas the second and third ones are two-dimensional. In addition, they are also integral-type equations. In order to reduce the partial differential equations to ordinary differential equations with respect to time, the finite-difference method is used to discretize the derivatives along the spatial coordinates x_1, x_2, x_3 with an $O(h^2)$ approximation. This method results in ODEs for w and θ, and a system of algebraic equations for the Airy function F.

The following difference operators have been used in relation to the spatial coordinates x_1, x_2, x_3, with a mesh of uniform spacings h_1, h_2, h_3, respectively:

$$\Lambda_i y = \frac{1}{h_i^2}\left[y(x_i - h_i) - 2y(x_i) + y(x_i + h)\right]_{i=1,2,3},$$
$$\Lambda_{ij} y = \frac{1}{4h_i h_j}\big[y(x_i + h_i, x_j + h_j) + y(x_i - h_i, x_j - h_j)$$
$$- y(x_i + h_i, x_j - h_j) - y(x_i - h_i, x_j + h_j)\big],$$

$$
\begin{aligned}
\Lambda_i^2 y &= \frac{1}{h_i^4}\big[y(x_i-2h_i)-4y(x_i-h_i)+6y_i(x_i) \\
&\quad -4y(x_i+h_i)+y(x_i+2h_i)\big], \\
\Lambda_{ij}^2 y &= \frac{1}{h_i^2 h_j^2}\big[y(x_i-h_i,x_j-h_j)-2y(x_i-h_i,x_j) \\
&\quad +y(x_i-h_i,x_j+h_j)-2y(x_i,x_j-h_j)+4y(x_i,x_j) \\
&\quad -2y(x_i,x_j+h_j)+y(x_i+h_i,x_j-h_j)-2y(x_i+h_i,x_j) \\
&\quad +y(x_i+h_i,x_j+h_j)\big].
\end{aligned} \tag{7.6}
$$

Using the difference operators (7.6) and introducing a new variable $\dfrac{dw}{dt} = \dot{w}$, the system of ODEs is reduced to first-order ODEs with respect to time and to algebraic equations:

$$
\begin{aligned}
\left[1+\frac{1+\nu}{1-\nu}\beta\right]\frac{d\theta_{ijk}}{dt} &= (1-2\nu)\beta\frac{d}{dt}\Big[\lambda^{-1}\Lambda_1 F_{ij} \\
&\quad +\lambda\Lambda_2 F_{ij}-(\Delta T_{11}+\Delta T_{22}-2N_T-2\Delta_T T)_{ij} \\
&\quad -\frac{x_{3k}}{1-\nu}(\lambda^{-1}\Lambda_1 w_{ij}+\lambda\Lambda_2 w_{ij})\Big] \\
&(i=0,\dots,n;\quad j=0,\dots,m;\quad k=0,\dots,p,)
\end{aligned} \tag{7.7}
$$

$$
\frac{dw_{ij}}{dt} = \dot{w}_{ij},
$$

$$
\begin{aligned}
\frac{d\dot{w}_{ij}}{dt}+\varepsilon\dot{w}_{ij} &= \big(-1/æ\big)\big[A(w)+B(w,F) \\
&\quad +C(M_T,\Delta_T M,\Delta M_{11},\Delta M_{22},\Delta M_{12})\big] \\
&\quad +\lambda_1^2\Lambda_1\theta_{ijk}+\lambda_2^2\Lambda_2\theta_{ijk}+\Lambda_3\theta_{ijk},
\end{aligned} \tag{7.8}
$$

$$
D(F) = E(w)+G(N_T,\Delta_T T,\Delta T_{11},\Delta T_{22},\Delta T_{12}). \tag{7.9}
$$

In (7.8) and (7.9), the following notation has been used:

$$
\begin{aligned}
A(w) &= \frac{1}{12(1-\nu^2)}(\lambda^{-2}\Lambda_1^2 w_{ij}+2\Lambda_{12}^2 w_{ij}+\lambda^2\Lambda_2^2 w_{ij}), \\
B(w,F) &= K_1\Lambda_2 F_{ij}+K_2\Lambda_1 F_{ij}+\Lambda_1 w_{ij}\Lambda_2 F_{ij} \\
&\quad +\Lambda_2 w_{ij}\Lambda_1 F_{ij}-\Lambda_{12} w_{ij}\Lambda_{12} F_{ij}, \\
&C(M_T,\Delta_T M,\Delta M_{11},\Delta M_{22},\Delta M_{12}) = \lambda^{-1}\Lambda_1(M_T \\
&+\Delta_T M-\Delta M_{11})_{ij}+\lambda\Lambda_2(M_T+\Delta_T M-\Delta M_{22})_{ij}-\Lambda_{12}(\Delta M_{12})_{ij}, \\
D(F) &= 12(1-\nu^2)A(F), \\
E(w) &= -K_1\Lambda_2 w_{ij}-K_2\Lambda_1 w_{ij}-\Lambda_1 w_{ij}\Lambda_2 w_{ij}+(\Lambda_{12}w_{ij})^2, \\
&G(N_T,\Delta_T T,\Delta T_{11},\Delta T_{22},\Delta T_{12}) = -(1-\nu)\lambda^{-1}\Lambda_1(N_T \\
&\quad +\Delta_T T)_{ij}+\lambda(1-\nu)\Lambda_2(N_T+\Delta_T T)_{ij}+\lambda^{-1}\Lambda_1(\Delta T_{22} \\
&\quad -\nu\Delta T_{11})_{ij}+\lambda\Lambda_2(\Delta T_{11}-\nu\Delta T_{22})_{ij}-\frac{1+\nu}{2}\Lambda_{12}(\Delta T_{12})_{ij}.
\end{aligned} \tag{7.10}
$$

The following relation has been used in the first equation of (7.1) in order to describe the volume extension e:

$$e = (1-2\nu)\big[\lambda^{-2}F_{x_1x_1} + \lambda F_{x_2x_2} - \Delta T_{11} - \Delta T_{22} + 2(N_T + \Delta_T T)\big] - \frac{1-2\nu}{1-\nu}x_3(\lambda^{-1}w_{x_1x_1} + \lambda w_{x_2x_2}) + \frac{1+\nu}{1-\nu}\theta. \tag{7.11}$$

This leads to (7.7).

In the procedure for solution of the coupled problem the time derivative in the right-hand side of (7.7) is approximated by a one-sided finite-difference relation:

$$\frac{dy_{ijk}}{dt} = \frac{1}{12h_t}(3y_{ijk}^{e-4} - 16y_{ijk}^{e-3} + 36y_{ijk}^{e-2} + 25y_{ijk}^{e}) + O(h_t^4), \tag{7.12}$$

where h_t is a time interval.

We need to attach to the difference equations (7.7)–(7.10) a group of boundary conditions, which must be formulated using a central finite-difference relation (it is necessary to find the values of the fundamental functions on nodes at the edge of the shell).

A procedure for solution of the initial–boundary value problem has been constructed in the following way.

Using the results found at the previous time step, w_{ij}, N_{Tij}, ΔT_{11ij}, $\Delta_T T_{ij}$, ΔT_{22ij}, ΔT_{12ij}, the system of linear algebraic equations (7.9) is solved for Airy's function F_{ij}. The values of F_{ij} obtained and values of N_{Tij}, $\Delta_T T_{ij}$, M_{Tij}, $\Delta_T M_{ij}$, ΔM_{11ij}, ΔM_{22ij}, ΔM_{12ij} found at the previous time step are substituted into the right-hand sides of (7.7)–(7.8). The ODEs are then integrated with respect to time using a method which will be explained later.

Using the temperature distribution in the shell volume found above, and integrating over the thickness with the help of Simpson's rule,

$$\int_{-\frac{1}{2}}^{\frac{1}{2}} \theta(x_1,x_2,x_3)\,dx_3 = \frac{1}{6\Delta h}\Big[(\theta_0 + \theta_{2n} + 4(\theta_1 + \cdots + \theta_{2n-1}) + 2(\theta_2 + \cdots + \theta_{2n-2})\Big], \tag{7.13}$$

the temperature terms N_T and M_T defined by (6.16) are found. They are used in later calculations.

Then, for each point of the shell volume, the deformation intensity is calculated:

$$e_i = \frac{\sqrt{2}}{3}\left[(e_{11}-e_{22})^2 + (e_{22}-e_{33})^2 + (e_{22}-e_{11})^2 + \frac{3}{2}e_{12}^2\right]^{1/2}, \tag{7.14}$$

where the e_{ij} are calculated using the formulas given in the previous chapter. A dependence $\sigma_i(e_i)$ is assumed (this depends on the shell material used), and σ_1 is defined at each point of the shell volume. The shear function $\gamma(e_i)$ calculated from (6.32) is used for calculation of the nonlinear terms (6.16)–(6.26), which are used during the second step of the algorithm.

In the first time step, owing to the initial conditions (7.2) $N_T = M_T = 0$, $\gamma \equiv 1$, all nonlinear terms are equal to zero. Use of N_T and M_T obtained in the previous step allows us to solve the integral–differential equations (7.1) as a system of differential equations.

We emphasize that we need to solve the system of algebraic equations (7.9) for each time step, but the matrix of this system remains constant. The Gauss method can be used to solve (7.9), and a transformation to a triangular form of the matrix is performed only once during the computations, in the first step of the calculation. The choice of the Gauss method possesses many advantages with respect to both high accuracy and a relatively short computational time. For instance, the use of a relaxation method requires computations five time longers to solve the problem under discussion.

The use of the three-dimensional heat transfer equation (7.7) is essential, because it makes any additional hypothesis about the temperature distribution through the thickness unnecessary.

During the integration of the ODEs (7.7), (7.8), a very careful verification of each computational step is needed. In addition, an economical choice of the integration method used for the ODEs (7.7), (7.8) is important. Consideration of the approaches proposed in this book has led the authors to apply a combination of the explicit and implicit Adams methods, known in the literature as the Adams–Bashford or prediction–correction method.

The methods used can be represented using the following formula:

$$P_k(EC_{k+1})^n E, \tag{7.15}$$

where P_k means prediction, i.e. finding a solution using the explicit Adams method; k denotes the order of accuracy; E means to computation of the right-hand sides of the ODEs; C_{k+1} means correction, i.e. improvement of the accuracy using the implicit Adams method, to an accuracy of order $(k+1)$ (here the solution obtained after the operation P_k is used as an initial approximation); and n denotes the required number of iterations.

When the operations $P_k(EC_{k+1})$ have been carried out, it is recommended to compute once more the right-hand sides of the ODEs in order to use them during the next steps of computation.

The right-hand sides of the ODEs are computed $(n+1)$ times, where n is independent of k; the Runge–Kutta method of the kth order needs computation of the right-hand sides to be performed k times.

In order to compare the accuracy and computation times of the methods, a few first-order ODEs have been solved using the Runge–Kutta method of the fourth order, and the Adams method in accordance with the schemes P_3EC_4E and P_5EC_6E. Results have been obtained for some equations with known solution, including $y = e^x \sin 10x$, and $y = \exp\left(2\frac{x^{11}}{11} - \frac{x^{12}}{12}\right)$. For

$x \in [0;3]$, the following results were obtained: for $h_t = 0.1$, the Runge–Kutta method gives more accurate results; for $h_t = 0.01$, the Runge–Kutta method gives a solution similar to that obtained by the scheme P_5EC_6E; for $h_t = 0.002$, the Adams methods give solutions with better accuracy in comparison with the Runge–Kutta method. Note that a solution obtained by the scheme $P_kEC_{k+1}E$ with $k = 3$ does not differ (practically) from that obtained with $k = 5$, but a test of the scheme with $k = 3$ in comparison with the fourth order Runge–Kutta method under conditions of a shortage of computational resources achieved about 45% better accuracy. This was observed for all model solutions used.

On the basis of these obtained results, the scheme (7.15) with $k = 3$ was used to integrate the system (7.7)–(7.9). This scheme, with one iteration, achieved 40% of the computational time of the fourth-order Runge–Kutta method, when applied to the nonlinear coupled problem.

7.2 Relaxation Method

Current methods applied to the solution of nonlinear problems of shells (with geometrical and physical nonlinearities) generally use a projection onto a system of nonlinear differential or nonlinear algebraic equations, with successive linearization using either Newton–Raphson or other iterative methods (for instance, the relaxation method). Nowadays there exists a wide literature devoted to such methods and their algorithms.

In 1963 Fedosev [76] proposed a dynamical approach to solve a problem related to the stability of shells. From a mathematical point of view, this method is called the "set-up" method [86]. The main idea of this method is that the solution of the nonlinear partial differential equations is reduced to a Cauchy problem of ODEs which is linear in time. This means that this method linearizes the nonlinear equations and decreases their dimension.

We discuss briefly an advantage of this method here. From a mathematical point of view, the "set-up" method can be treated as an iterative method to solve nonlinear algebraic equations, where each time step provides a new approximation to the exact solution. Like all iterative methods, this one is characterized by a high accuracy of computation. In addition, it does not have the common disadvantage of iterative methods of a high sensivity to the choice of the initial approximation. Additionally, the "set-up" method not only gives a very simple rule for obtaining nonunique solutions of static problems, but also allows one to find the stable and unstable branches of the equilibrium position of the system under cosideration and to capture all process of the jumping behaviour of a shell.

In the process of solution of homogeneous equations via traditional methods, in order to obtain a nontrivial solution one needs to introduce an artificial excitation (in the theory of shells this corresponds, for instance, to a small transverse load, a small curvature or some other initial imperfection).

However, this influences (sometimes significantly) the results obtained. In the case of the "set-up" method, the initial conditions play the role of the initial excitations, and small changes of these conditions do not influence the static solution obtained. Another advantage of the method is related to its simple realization, because nowadays there are many effective algorithms and programs devoted to solution of the Cauchy problem.

Let us clarify the method for obtaining unstable solutions using an example of an arbitrary nonlinear algebraic equation,

$$f(x) = 0. \tag{7.16}$$

We construct two differential equations from for (7.16),

$$c\left(\frac{d^2x}{dt^2} + \varepsilon\frac{d\dot{x}}{dt}\right) = \pm f(x). \tag{7.17}$$

In order to obtain the complete set of roots of (7.16), we need to solve the two differential equations (7.17). However, in practice we can do this in the following way.

We take an arbitrary initial condition (the initial approximation) $x_{in} = x_0$ for (7.17) with the positive right-hand side, and solving the equation we obtain x_1. Then we take the initial condition $x_0 = x_1 + \delta$ and solve (7.17) with the negative value of the right-hand side, and so on. As an example, we consider the solution of the equation

$$f(x) = x^4 - 12x^3 + 47x^2 - 60x = 0. \tag{7.18}$$

This equation has four real roots $x_1 = 0$, $x_2 = 3$, $x_3 = 4$, $x_4 = 5$. We have solved this equation by the "set-up" (SU) and Newton (N) methods. In Table 7.1 the roots found by both methods, the number of iterations (I) needed to obtain the solution and the initial approximations are reported.

Table 7.1. Comparison of SU and Newton methods

Exact values	$x_1 = 0$	$x_2 = 3$	$x_3 = 4$	$x_4 = 5$
Initial approximation (x_{in})	$x_1 = 2.5$	$x_2 = 3.5$	$x_3 = 4.5$	$x_4 = 10$
SU	$x_1 = 0$	$x_2 = 3$	$x_3 = 4$	$x_4 = 5$
N	$x_1 = 5$	$x_2 = 5$	$x_3 = 3$	$x_4 = 5$
Iteration number I(SU)	68	54	87	60
Iteration number I(N)	7	12	9	10

It can be seen that the SU method, in contrast to the Newton method, finds more exactly a stable root of the equation (7.18) close to the initial approximation.

Let us consider one more (transcendental) equation:

$$f(x) = \arctan x = 0. \tag{7.19}$$

This equation has a root $x = 0$. It is known that for (7.19) the Newton method approaches this root only for initial values such that $|x| < 1.39$, but the "set-up" method approaches this root for $\forall x_{in}$. For instance, taking $x_{in} = 2$, the root $x = 0$ is found after 33 iterations.

We now illustrate the high efficiency of the "set-up" method using a series of nonlinear problems of the theory of plates and shells, which are characterized by a wide range of properties with respect to different nonlinearity types (geometrical, physical and geometrical–physical) as well as different models(Kirchhoff–Love or Timoshenko).

We formulate now a nonlinear equation governing the dynamics of a shallow shell including transverse shear effects, in hybrid form:

$$
\begin{aligned}
&\nabla_k^2 F + L(w,F) + \frac{2}{3}\left[\lambda_1\left(\frac{\partial \gamma_x}{\partial x_1} + \frac{\partial^2 w}{\partial {x_1}^2}\right) + \lambda_2\left(\frac{\partial \gamma_y}{\partial x_2} + \frac{\partial^2 w}{\partial {x_2}^2}\right)\right] \\
&\qquad\qquad + q - \frac{\partial^2 w}{\partial t^2} - \varepsilon\frac{\partial w}{\partial t} = 0, \\
&\frac{1}{5}\left[\lambda^{-2}\frac{\partial^2 \gamma_x}{\partial {x_1}^2} + A_{1212}\frac{\partial^2 \gamma_y}{\partial {y_2}^2} + (A_{1122} + A_{1212})\frac{\partial^2 \gamma_y}{\partial x_1 \partial x_2}\right] \\
&\qquad - \frac{1}{20}\left[\lambda^{-1}\frac{\partial^2 w}{\partial {x_1}^2} + (A_{1122} + A_{1212})\frac{\partial^2 w}{\partial x_1 \partial x_2}\right] \\
&\qquad\qquad - 2\lambda_1\left(\gamma_x + \frac{\partial w}{\partial x_1}\right) = 0 \quad (x_1 \leftrightarrow x_2), \\
&\qquad\qquad \nabla^4 F + \nabla_x^2 w + \frac{1}{2}L(w,w) = 0.
\end{aligned}
\tag{7.20}
$$

The symbol $(x_1 \leftrightarrow x_2)$ denotes that a third equilibrium equation can be obtained from the second equation by a cyclic change of indices. The system (7.20) has been transformed to a nondimensional form as earlier. The equations (7.20) include several geometrical and physical–geometrical quantities; in particular, $k_{x_1} = \dfrac{a^2}{R_{\gamma_1}(2h)}$ and $k_{x_2} = \dfrac{a^2}{R_{\gamma_2}(2h)}$ (geometrical parameters characterizing the curvatures $\left(\dfrac{1}{R_{x_1}} \text{ and } \dfrac{1}{R_{x_2}}\right)$ of the shell and its size $(a, b, 2h)$), and $\lambda_1 = \dfrac{G_{13}}{A_{1111}}\left(\dfrac{b}{2h}\right)^2$ and $\lambda_2 = \dfrac{G_{23}}{A_{1111}}\left(\dfrac{a}{2h}\right)^2$ are the physical–geometrical parameters characterizing the effect of transverse shear on the solution.

The following boundary conditions are attached to (7.20):

$$
w = M_n = \gamma_{\overline{n}} = T_n = T_{\overline{n}}\big|_{\Gamma} = 0,
$$

where $n = x$, $\overline{n} = y$ define the shell boundary; $t \in [0, T]$, where T is the observation time; $\Omega_1 = (0;1) \times (0;1)$ is the space in which the independent parameters x_1 and x_2 vary, defining the middle surface; $\overline{\Omega}_1 = \Omega_1 \cup \partial\Omega_1$, where $\partial\Omega_1$ is the contour (edge) of the middle surface; $Q_1 = \Omega_1 \times (0;T)$; and $\Gamma = \partial\Omega_1 \times [0;1]$.

The following initial conditions are applied:

$$w\big|_{t=0} = \frac{\partial w}{\partial t}\big|_{t=0} = 0. \tag{7.21}$$

The problem defined by (7.20)–(7.21) is reduced to a Cauchy problem using higher approximations of the Bubnov-Galerkin method:

$$\begin{aligned} w &= \sum_{i,j=1}^{N} A_{ij}(t) \sin i\pi x_1 \sin j\pi x_2, \\ F &= \sum_{i,j=1}^{N} B_{ij}(t) \sin i\pi x_1 \sin j\pi x_2, \\ \gamma_{x_1} &= \sum_{i,j=1}^{N} C_{ij}(t) \cos i\pi x_1 \sin j\pi x_2, \\ \gamma_{x_2} &= \sum_{i,j=1}^{N} D_{ij}(t) \sin i\pi x_1 \cos j\pi x_2. \end{aligned} \tag{7.22}$$

As a result we obtain a system of N ODEs for $A_{ij}(t)$ and a system of algebraic equations for $B_{ij}(t)$, $C_{ij}(t)$ and $D_{ij}(t)$, which fortunately can be solved in closed form. The integrals over the middle surface $\Omega_1 = (0;1) \times (0;1)$ were computed analytically for each of the problems considered. The system of ODEs for $A_{ij}(t)$ with the initial conditions (7.21) was solved using the Runge–Kutta method with an automatic choice of the integration step.

Let us analyse the influence of the damping coefficient ε appearing in the first equation of the system (7.20) on the value of the dynamic critical load, when an impulse load of infinitesimal duration is applied (Fig. 7.1). We definite $\varepsilon = \varepsilon_{cr}$ as $\min \varepsilon$, when the dynamic critical load $q = q^d_{cr}$ is equal to the static critical load $q = q^s_{cr}$ (i.e. $q^s_{cr} = q^d_{cr}$). The criterion for a dynamic load given in [7] is used here. Plots of $q(w)$ and $t(w)$ are shown in the Fig. 7.2 for a cylindrical shell ($k_x = 0$, $k_{x_2} = 48$, $\lambda_1 = \lambda_2 = 1000$). Curve 2 corresponds to $\varepsilon = 3$, whereas curve 1 corresponds to a static solution for $q^d_{cr} < q^s_{cr}$. Both

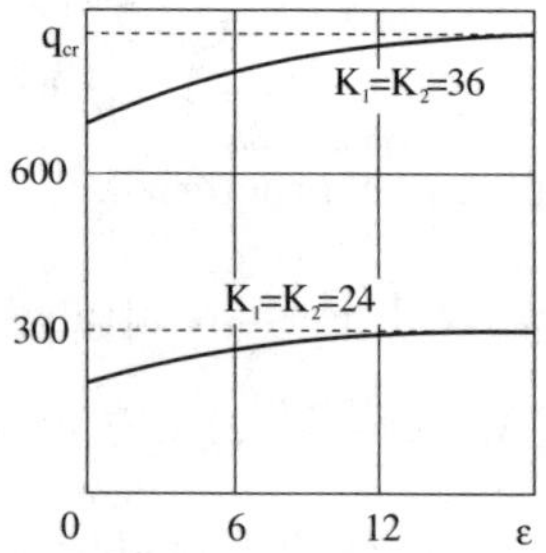

Fig. 7.1. Influence of damping coefficient ε on the dynamic critical load

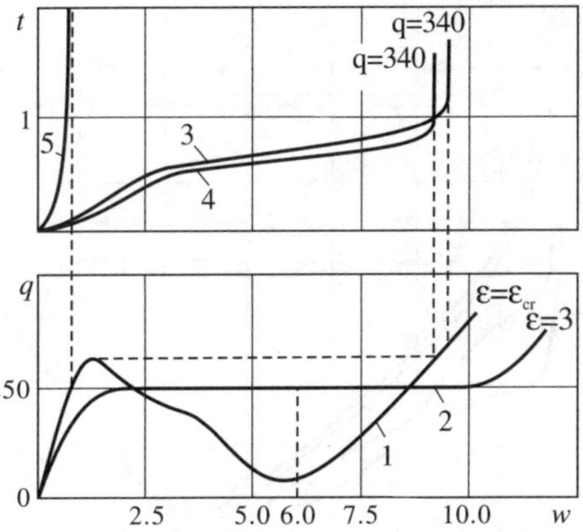

Fig. 7.2. Dependences $q(w)$, $t(w)$ for a cylindrical shell ($N = 5$)

parts of the figure need to be analysed simultaneously. For $\varepsilon = \varepsilon_n$ and for a given load q, the deflection approaches a constant value with increase of time. Having a set of values $\{t(w_i), q_i)\}$ one can construct a dependence $q(w)$. Curves 3, 4, 5 for $t(w)$, were obtained for $q = 255, 320$ and 350, with $\varepsilon = \varepsilon_{cr}$. From Fig. 7.1 it can be seen that for $\varepsilon \to \varepsilon_{cr}$, $q^d_{cr} \to q^s_{cr}$, for an impulse load that is uniformly distributed over the shell surface and constant in time (the results were obtained from (7.22) with $N = 5$). A further increase of the number of terms in (7.22) does not change the fundamental functions or their derivatives up to second order. The efficiency and high accuracy of the Bubnov–Galerkin method when applied to both static and dynamic problems were demonstrated long ago (see, for instance [133]).

In solving nonlinear problems of composite shells exhibiting weak shear stiffness within the Timoshenko-type theory, an increase in the number of modes of equilibrium for a given load occurs, and local stability loss is observable. The stable equilibrium configuration was chosen here from the set of various possible configurations using the "set-up" method. Plots of $q(w)$ for $k_{x_1} = k_{x_2} = 24$, $\lambda = 1$, $\lambda_1 = \lambda_2 = 40, 20, 10, 5$ (corresponding to curves 1, 2, 3, 4, respectively) are presented in Fig. 7.3 (obtained from (7.22) with $N = 7$). The $q(w)$ curves were obtained using a static method, i.e. a large set of nonlinear algebraic equations for A_{ij}, B_{ij}, C_{ij} and D_{ij} was solved using the Newton–Raphson method for a fixed value of the deflection in the centre of the shell. The arrows indicate external static loads which have been detected through the "set-up" method (the other equilibrium configurations are not physically realized).

In practice, one can find shallow shell structures with values of k_{x_1} and k_{x_2} even greater than 500. However, in the literature, only problems related to $k_{x_1}, k_{x_2} < 50$ have been addressed. The reason is clear. When the geometric parameters are increased, the number of possible equilibrium configurations increases greatly. This requires one to increase the number of terms in the approximating functions (and nodes) used when either variational or finite-

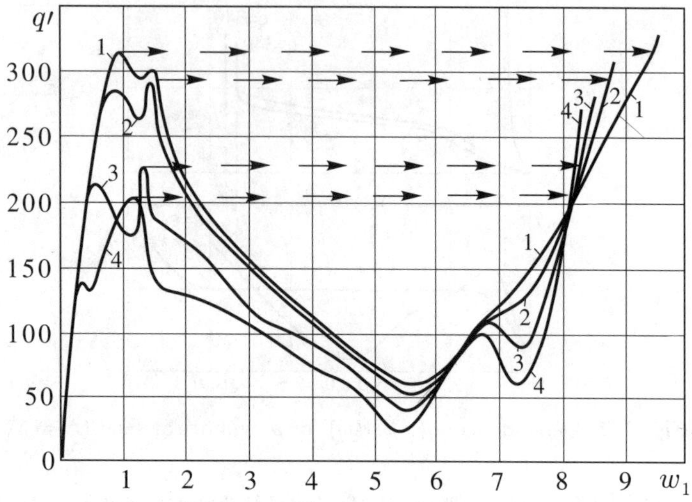

Fig. 7.3. Dependences $q(w)$, $t(w)$ for a cylindrical shell ($N = 7$)

difference methods are applied to static problems, in order to properly describe the complex surface shape. The convergence of the solution obtained for the nonlinear algebraic equations dramatically decreases, because their order increases.

The "set-up" method avoids the above disadvantages. We do not need to solve any system of equations at all, and in particular, not a nonlinear one. There is no problem in analysing shells with $24 \leq k_{x_1} = k_{x_2} \leq 500$ (we have taken $N = 15$ in (7.21)). The numerical results agree fully with experiment for $k_{x_1} = k_{x_2} = 409.1$. The difference in q^s_{cr} is 15%, and qualitatively the picture of stability loss is the same [218].

In Fig. 7.4, contours of equal relative deflection $w(x_1, x_2)$, $0 \leq x_1,\ x_2 \leq 0.5$, for $k_{x_1} = k_{x_2} = 409.1$ are drawn. Cases a–e are related to the pre-critical state, whereas f corresponds to a post-critical stability loss. With an increase of k_{x_1} and k_{x_2}, the shell loses its stability first in the corner zones, and during "jump" behaviour, holes appear on the axial curves, which is also indicated by experimental data.

We now illustrate the high efficiency of the "set-up" method using an example of a physically and geometrycally nonlinear problem of a plate. We recall (7.1) without the temperature terms:

$$\begin{aligned}
&\frac{1}{12(1-\nu^2)}\nabla^4 w - L(w,F) - \nabla_k^2 F - \lambda^{-1}(\Delta M_{11})_{x_1x_1} \\
&- 2(\Delta M_{12})_{x_1x_2} - \lambda(\Delta M_{22})_{x_2x_2} = q - æ(\ddot{w} + \varepsilon\dot{w}), \\
&\nabla^4 F + \nabla_k^2 w + \frac{1}{2}L(w,w) - \lambda^{-1}(\Delta T_{22} - \nu\Delta T_{11})_{x_1x_1} \\
&+ 2(1+\nu)(\Delta T_{12})_{x_1x_2} - \lambda(\Delta T_{11} - \nu\Delta T_{22})_{x_2x_2} = 0.
\end{aligned} \tag{7.23}$$

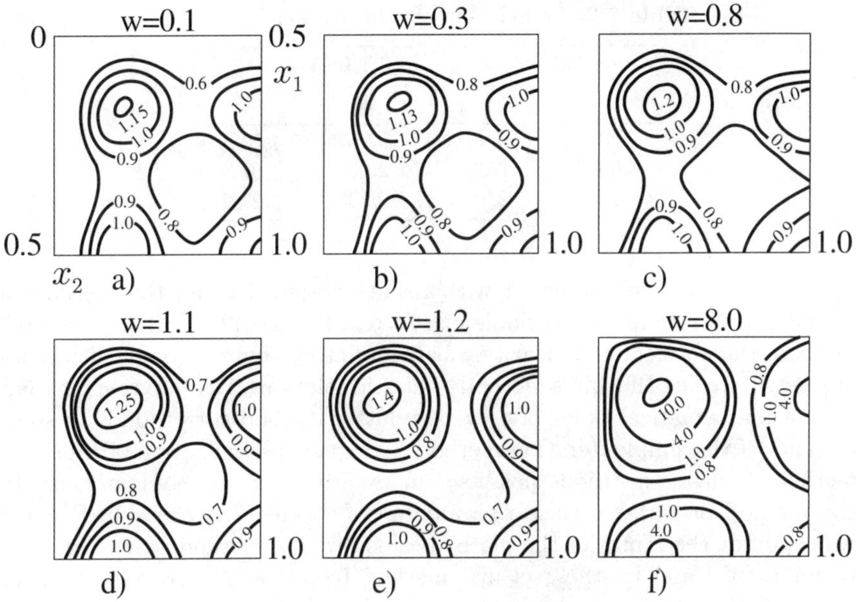

Fig. 7.4a–f. Diagrams for equal relative deflections $w(x_1, x_2)$

The construction of a solution to (7.23) using the "set-up" method has been described in Sect. 7.1. We consider the geometrically and physically nonlinear problem of bending of a square plate ($\lambda = 1$). We assume that a uniformly distributed load with intensity q on the surface $x_3 = -0.5$. The physical parameters of the plate material (AMC alloy) have been taken to be as follows: $E = 69$ GPa, $\mu = 0.3$, $\rho = 2800$ kg/m^3, $\sigma_i(e_i)$, and

$$\begin{aligned} &\sigma_i = E_1 e_i, \quad e_i \leq e_s, \\ &\sigma_i = E_1 e_s + E_2(e_i - e_s), \quad e_i > e_s. \end{aligned} \tag{7.24}$$

For the shear function, we obtain

$$\begin{aligned} &\gamma(e_i) = \frac{\sigma_i(e_i)}{3Ge_i}, \\ &\gamma \equiv 1, \quad e_i \leq e_s, \\ &\gamma = \frac{E_1}{E} + \left(1 - \frac{E_1}{E}\right)\left(\frac{e_s}{e_i}\right), \quad e_i > e_s, \\ &\frac{E_1}{E} = 0.57735, \quad e_s = 0.98 \cdot 10^{-3}. \end{aligned} \tag{7.25}$$

The geometrical parameters of the plate were taken to be as follows $a = b = 0.1$ m, $\dfrac{a}{h} = 50$. The boundary conditions were defined by (6.58), and the initial conditions were $w\big|_{t=0} = 0$, $\dot{w}\big|_{t=0} = 0$.

Table 7.2. Comparison of computational methods

q	$w(0.5; 0.5)$		
	1	2	3
23.4	0.96	0.95	1.04%
60.0	2.03	2.00	1.96%
70.3	2.16	2.23	3.24%

The results were compared with results obtained using the method of variational iterations and variable elastic parameters [242]. In the "set-up" method, the physical nonlinearity was taken into account through the theory of plasticity using Iliushin's method of elastic solutions. Here we remark that, from a mathematical point of view, the latter method corresponds to some variants of the simple iteration method. A numerical realization of Iliushin's method of elastic solutions possesses many advantages in comparison with the method of variable elastic parameters. In Table 7.2, the values of the deflection at the centre of the plate obtained by the method described in [7] (column "1") and by the "set-up" method (column "2") are reported, for three values of the transverse-load parameter. Column 3 reports the difference as a percentage.

First of all, it can be concluded from Table 7.2 that the results obtained by these two methods are similar. Some of the difference is caused by the specific structures of two algorithms. The static approach relies on a reduction from partial differential equations (PDEs) to ODEs as a first approximation and

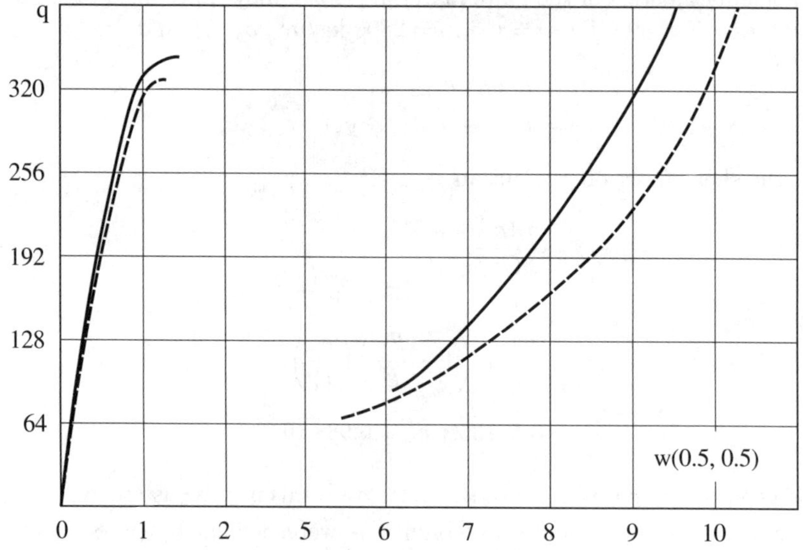

Fig. 7.5. Dependence $q(w(0.5; 0.5))$

on application of the variable-elasticity method. The dynamical approach is based on the "set-up" method and the method of elastic solutions.

We consider now the stability of a geometrically nonlinear elastic shell. The following parameters were chosen $k_{x_1} = k_{x_2} = 24$, $E = 69$ GPa, $\mu = 0.3$, $a = b = 0.1$ m, $h = 0.83 \times 10^{-3}$ m, $\sigma_i(e_i)$ a defined by (7.24), $E_1/E = 0.4478$, $e_s = 1.35 \times 10^{-3}$. The boundary conditions were governed by (6.58) and zero-value initial conditions were applied.

The dependence $q(w(0.5; 0.5))$ obtained is shown in Fig. 7.5.

7.3 Numerical Investigations and Reliability of the Results Obtained

In the previous section we have shown that reliable solutions of problems of shallow-shell theory with both geometrical and geometrical–physical nonlinearities can be obtained when the method described in Sect. 7.1 is applied. In this section, we show the reliability of the solution to the three-dimensional nonstationary heat transfer equation obtained when the "set-up" method is used. In addition, we also discuss and illustrate (using the technique described here) a solution to a coupled thermoelastic problem of a flexible shallow shell without any prior assumptions about the temperature distribution through the thickness, i.e. using the three-dimensional heat transfer equation. As a matter of fact, we need to demonstrate the efficiency of the method when it is applied to the solution of integral–differential equations with different dimensions (three-dimensional heat transfer equations and two-dimensional equations related to the Kirchhoff–Love model) and of different types (heat transfer equations and hyperbolic equations of shell the theory).

In the equations governing the shell motion, a temperature does not appear directly, but its integral characteristics (thermal forces and moments) are used; these appear because of the reduction of the three-dimensional heat transfer equation to a two-dimensional one.

The solution to the three-dimensional heat transfer equation has been sought using the finite-difference method. A number of researchers have emphasized that neglecting the nonlinear form of the temperature variation through the shell thickness greatly influences the solution to a nonstationary thermoelastic problem of a shell or a plate. Therefore, a solution to the three-dimensional nonstationary heat transfer equation without any introduction of hypotheses about the temperature distribution through the thickness possesses important practical meaning.

Let us recall the first equation of the system (7.1):

$$\frac{\partial \theta}{\partial t} = \lambda_1^2 \frac{\partial^2 \theta}{\partial {x_1}^2} + \lambda_2^2 \frac{\partial^2 \theta}{\partial {x_2}^2} + \frac{\partial^2 \theta}{\partial {x_3}^2} + w_0. \tag{7.26}$$

In this given equation, there is no term representing the shell geometry; however, the validity of this approach has been proved [121–123]. Using the finite-difference method along the spatial coordinates with an $O(h^2)$ approximation, we obtain the following ODEs with respect to time:

$$\frac{d}{dt}\frac{\partial\theta_{ijk}}{\partial t} = \lambda_1 \Lambda_1 \theta_{ijk} + \lambda_2 \Lambda_2 \theta_{ijk} + \Lambda_3 \theta_{ijk} + w_{0ijk}. \tag{7.27}$$

We need to attach to the system (7.27) the initial and boundary conditions formulated earlier, represented in a suitable (finite-difference) form.

We consider a numerical example (second type of boundary conditions) and convective heat transfer (third type of boundary conditions). The space was divided into $(6 \times 6 \times 6)$, $(8 \times 8 \times 8)$ and $(12 \times 12 \times 12)$ parts. The results obtained were compared with the analytical solution proposed by Kovalenko [121], showing very good agreement.

The results of a problem related to a thermal shock on a shell surface which is thermally isolated on its other sides are given in Table 7.3. The results where the surface was partitioned into $(12 \times 12 \times 12)$ and $(8 \times 8 \times 8)$ parts differ from the analytical results by no more than 0.2% and 0.5%, respectively.

We consider one of the methods of reduction of the three-dimensional problem to a two-dimensional one. In the case of linear problems of thermoelasticity, application of the operator method for thin-walled structures has been proposed [188], where thin-walled conditions from the point of view of heat transfer theory have also been proposed. Let us begin with the first invariant of the deformation tensor:

$$e = e_{11} + e_{22} + e_{33}. \tag{7.28}$$

Table 7.3. Comparison of results obtained for different mesh sizes

Solution method	Coordinates	Time $t = 0.2$	$t = 0.4$	$t = 0.6$
Exact	$z = 0.5$	67.19	97.01	124.06
	$z = 0$	21.06	47.66	74.26
	$z = -0.5$	8.18	31.55	57.71
$(12 \times 12 \times 12)$	$z = 0.5$	67.04	96.92	123.97
	$z = 0$	20.98	47.58	74.17
	$z = -0.5$	8.16	31.49	57.62
$(8 \times 8 \times 8)$	$z = 0.5$	66.86	96.80	123.87
	$z = 0$	20.89	47.48	74.07
	$z = -0.5$	8.14	31.41	57.53
$(4 \times 4 \times 4)$	$z = 0.5$	65.87	96.17	123.33
	$z = 0$	20.38	46.96	73.56
	$z = -0.5$	8.08	32.00	57.03

The quantity e_{33} can be found from the condition of a plane strain state ($\sigma_{33} = 0$), and it reads

$$e_{33} = -\frac{\nu}{E}(\sigma_{11} + \sigma_{22}) + \alpha_T \theta. \tag{7.29}$$

We shall now formulate the heat transfer equations in the absence of physical nonlinearities. Substituting σ_{11} and σ_{22} into (7.29), for a physically linear body, we obtain

$$e_{33} = -\frac{\nu}{1-\nu}(e_{11} + e_{22}) + \frac{1+\nu}{1-\nu}\alpha_T \theta. \tag{7.30}$$

Substituting (7.29) into (7.28), we obtain the first invariant of the deformation tensor:

$$\begin{aligned} e &= \frac{1-2\nu}{1-\nu}(e_{11} + e_{22}) + \frac{1+\nu}{1-\nu}\alpha_T \theta \\ &= \frac{1-2\nu}{1-\nu}\left[\frac{1-\nu}{Eh}(T_{11} + T_{22}) + 2\alpha_T N_T \right. \\ &\left. + z(æ_{11} + æ_{22})\right] + \frac{1+\nu}{1-\nu}\alpha_T \theta. \end{aligned} \tag{7.31}$$

Substituting (7.31) into the heat transfer equation, we obtain

$$\begin{aligned} &K_T \frac{\partial \theta}{\partial t} - K_T \left(\frac{\partial^2 \theta}{\partial {x_1}^2} + \frac{\partial^2 \theta}{\partial {x_2}^2} + \frac{\partial^2 \theta}{\partial {x_3}^2} \right) \\ &= \frac{E \alpha_T T_0}{1-2\nu} \frac{\partial}{\partial t}(e_{11} + e_{22} + e_{33}) + W_0, \end{aligned} \tag{7.32}$$

and we introduce the following notation:

$$\alpha_1 = \frac{K_T}{C}, \; \gamma_0 = \frac{\alpha_T E T_0}{C(1-2\nu)}, \; \beta = \frac{\alpha_1}{1 + \alpha_T \gamma_0 \frac{1+\nu}{1-\nu}},$$

$$\gamma = \frac{\gamma_0(1-2\nu)}{(1-\nu)\left(1 + \gamma_0 \alpha_T \frac{1+\nu}{1-\nu}\right)}, \; D^2 = \nabla^2 - \frac{1}{\beta}\frac{\partial}{\partial t},$$

$$D_1^2 = \frac{1}{\beta}\frac{\partial}{\partial t}, \; M = æ_{11} + æ_{22}, \; N = \frac{1-\nu}{Eh}(T_{11} + T_{22}) + 2\alpha_T N_T,$$

$$N_T = \frac{1}{h}\int_{-\frac{h}{2}}^{\frac{h}{2}} \theta \, dx_3. \tag{7.33}$$

In operator form, the heat transfer equation reads

$$\frac{\partial^2 \theta}{\partial {x_3}^2} + D^2 \theta - \gamma D_1^2 (N + x_3 M) = 0. \tag{7.34}$$

The solution to (7.34) obtained using the symbolic method reads

$$\theta = \frac{\sin x_3 D}{D}\left\{\left.\frac{\partial\theta}{\partial x_3}\right|_{x_3=0} - \gamma\frac{D_1^2}{D^2}M\right\} + \cos x^3 D\left\{\left.\theta\right|_{x_3=0} - \gamma\frac{D_1^2}{D^2}N\right\} + \gamma\frac{D_1^2}{D^2}(N+zM). \tag{7.35}$$

Integrating with respect to x_3, we obtain

$$N_T = \frac{1}{h}\int_{-\frac{h}{2}}^{\frac{h}{2}} \theta\, dx_3 = \frac{\sin\frac{h}{2}D}{\frac{h}{2}D}\left\{\left.\theta\right|_{x_3=0} - \gamma\frac{D_1^2}{D^2}N\right\} + \frac{1}{\gamma}\frac{D_1^2}{D^2}N, \tag{7.36}$$

$$M_T = \frac{12}{h^3}\int_{-\frac{h}{2}}^{\frac{h}{2}} \theta x_3\, dx_3 = \frac{12}{h^2D^2}\frac{\sin\frac{h}{2}D}{\frac{h}{2}D}\left\{\left.\frac{\partial\theta}{\partial x_3}\right|_{x_3=0} - \gamma\frac{D_1^2}{D^2}M\right\} - \frac{12}{h^2D^2}\cos\frac{h}{2}D\left\{\left.\frac{\partial\theta}{\partial x_3}\right|_{x_3=0} - \gamma\frac{D_1^2}{D^2}M\right\} - \gamma\frac{D_1^2}{D^2}M. \tag{7.37}$$

Removing $\left.\theta\right|_{x_3=0}$ and $\left.\frac{\partial\theta}{\partial x_3}\right|_{x_3=0}$ from (7.37) and (7.36) and substituting into (7.35), we obtain the temperature θ expressed in terms of the integral characteristics:

$$\theta(x_1,x_2,x_3,t) = \frac{\frac{h}{2}D}{\sin\frac{h}{2}D}\cos x_3 D N_T + \frac{h^3D^2}{24}\frac{\sin x_3 D}{\sin\frac{h}{2}D - \frac{h}{2}D\cot\frac{h}{2}D}M_T + \gamma\frac{D_1^2}{D^2}\left\{\frac{\frac{h}{2}D\cos zD}{\sin\frac{h}{2}D} - 1\right\}N - \frac{\gamma}{3}\frac{D_1^2}{D^2}\left\{\frac{\frac{h^3}{8}D^3\cot\frac{h}{2}D}{1-\frac{h}{2}\cot\frac{h}{2}D} - 3\right\}M. \tag{7.38}$$

After the boundary conditions on the surfaces

$$x_3 = \pm\frac{h}{2}: \left.\frac{\partial\theta}{\partial x_3}\right|_{x_3=-0.5} = 0 \quad \text{and} \quad \left.\frac{\partial\theta}{\partial x_3}\right|_{x_3=+0.5} = 0$$

are satisfied, the following equations are obtained:

$$
\begin{aligned}
&-\frac{h}{2}D^2N_T+\frac{h^3D^3}{24}\frac{\cot\frac{h}{2}D}{1-\frac{h}{2}D\cot\frac{h}{2}D}M_T+\gamma\frac{h}{2}D_1^2N \\
&-\frac{\gamma}{3}\frac{D_1^2}{D^2}\left\{\frac{\frac{h^3}{8}D^3\cot\frac{h}{2}D}{1-\frac{h}{2}\cot\frac{h}{2}D}-3\right\}M=0,
\end{aligned}
\tag{7.39}
$$

$$
\begin{aligned}
&\frac{h}{2}D^2N_T+\frac{h^3D^3}{24}\frac{\cot\frac{h}{2}D}{1-\frac{h}{2}D\cot\frac{h}{2}D}M_T+\gamma\frac{h}{2}D_1^2N \\
&-\frac{\gamma}{3}\frac{D_1^2}{D^2}\left\{\frac{\frac{h^3}{8}D^3\cot\frac{h}{2}D}{1-\frac{h}{2}\cot\frac{h}{2}D}-3\right\}M=0.
\end{aligned}
\tag{7.40}
$$

These equations yield an equation for the thermal stresses N_T,

$$
D^2N_T-\gamma D_1^2N=0,
\tag{7.41}
$$

and an equation for the thermal moments M_T,

$$
h^2D^2M_T-12M_T-\gamma h^2D_1^2N=0.
\tag{7.42}
$$

The equation (7.42) has been obtained by keeping only two terms of the series of $\cot\frac{h}{2}D$, which corresponds to a cubic form of the temperature distribution (7.38) through the thickness.

In terms of the nondimensional parameters introduced earlier, (7.41) and (7.42) read

$$
\begin{aligned}
&\lambda_1\lambda_2\left(\lambda^{-1}\frac{\partial^2N_T}{\partial {x_1}^2}+\lambda\frac{\partial^2N_T}{\partial {x_2}^2}\right)-\frac{\alpha_T^2ET_0}{C}\left(\frac{\partial}{\partial t}\left(\lambda^{-1}\frac{\partial^2F}{\partial {x_1}^2}\right.\right. \\
&\left.\left.+\lambda\frac{\partial^2F}{\partial {x_2}^2}\right)\right)=\left(1+\gamma_0\alpha_T\frac{1+\nu}{1-\nu}\right)\left(1+\frac{2\gamma_0\alpha_T}{1+\gamma_0\alpha_t\frac{1+\nu}{1-\nu}}\right)\frac{\partial N_T}{\partial t},
\end{aligned}
\tag{7.43}
$$

$$
\begin{aligned}
&\lambda_1\lambda_2\left(\lambda^{-1}\frac{\partial^2M_T}{\partial {x_1}^2}+\lambda\frac{\partial^2M_T}{\partial {x_2}^2}\right)-12M_T \\
&+\frac{\alpha_T^2ET_0}{C(1-\nu)}\frac{\partial}{\partial t}\left(\lambda^{-1}\frac{\partial^2w}{\partial {x_1}^2}+\lambda\frac{\partial^2w}{\partial {x_2}^2}\right) \\
&=\left(1+\gamma_0\alpha_T\frac{1+\nu}{1-\nu}\right)\frac{\partial M_T}{\partial t}.
\end{aligned}
\tag{7.44}
$$

We need to attach to these equations the equations governing the motion of the shell and the continuity equation:

$$\begin{aligned}
&\frac{1}{12(1-\nu^2)}\nabla^2\nabla^2 w - \nabla_k^2 F - L(w,F)\\
&+\frac{a^2b^2\rho\alpha_1^2}{Eh^3}\frac{\partial^2 w}{\partial t^2} + \frac{1}{12(1-\nu^2)}\left(\lambda^{-1}\frac{\partial^2 M_T}{\partial {x_1}^2} + \lambda\frac{\partial^2 M_T}{\partial {x_2}^2}\right) = q, \qquad (7.45)\\
&\nabla^2\nabla^2 F = -\lambda^{-1}\frac{\partial^2 N_T}{\partial {x_1}^2} - \lambda\frac{\partial^2 N_T}{\partial {x_2}^2} - \nabla_k^2 w - \frac{1}{2}L(w,w).
\end{aligned}$$

As a result, we have obtained a system of four PDEs (7.43)–(7.45) governing the shell motion with coupling of both the deformation and the temperature fields. Unlike the integral–differential set of equations obtained in the previous chapter, the system above is only differential and has one dimension. It s approximate, because in the series of $\cot\frac{h}{2}D$ only the first two terms are taken, which in practice leads to a cubic distribution of temperature through the shell thickness. However, the system remains a hybrid one, i.e. parabolic and hyperbolic.

We need to attach to the system (7.43)–(7.45) the boundary and initial conditions. In order to solve the system (similarly to the integral–differential system (7.1)), we have applied a finite-difference method with respect to the spatial coordinates with an $O(h^2)$ approximation. This finally results in ODEs for w_{ij}, N_{Tij}, M_{Tij} and a system of algebraic equations for F_{ij}. The system of algebraic equations was (again) solved using the Gauss method for every time step, and for the system of differential equations, the Runge–Kutta method with an automatically chosen step of integration was chosen. Unlike the situation for the "set-up" method, where the values of the thermal loads occurring in the three-dimensional heat transfer equation must be taken from the previous step of the integration in time, we do not need to take these values from the previous step now.

In order to estimate the influence of the assumption that we have made, the stress–strain state of a shell made from AMC alloy with the boundary conditions (6.59) has been analysed.

It has been assumed also that on the surfaces $x_1 = 0$, $x_2 = 0$, $x_1 = 1$, $x_2 = 1$, the following boundary condition of the first type for temperature has been used: $\theta = 0$, i.e. $N_T = M_T = 0$. At the initial time instant $t = 0$ a heat load, extending over an infinite time, uniformly distributed over the shell surface, was applied, with $q = 81$ and $k_{x_1} = k_{x_2} = 24$. The results obtained are presented in Table 7.4.

Method 1 in Table 7.4 corresponds to the solution of the system (7.43)–(7.45). Method 2 corresponds to the solution of the integral–differential equations. The results obtained indicate the very high efficiency of the approaches presented here for the solution of coupled thermoelastic problems of flexible shells subjected to transverse impulse-type loads.

Table 7.4. Comparison of results obtained for different methods

	Time							
	$t = 0.1$		$t = 0.2$		$t = 0.3$		$t = 0.4$	
Method / Function	1	2	1	2	1	2	1	2
$w(0.5; 0.5)$	0.99	0.99	1.78	1.78	0.85	0.85	0.02	0.02
$w(0.25; 0.25)$	0.68	0.68	1.07	1.07	0.68	0.68	0.03	0.03
$F(0.5; 0.5)$	0.38	0.38	0.63	0.63	0.33	0.33	0.00	0.00
$F(0.25; 0.25)$	0.15	0.15	0.23	0.23	0.14	0.14	0.004	0.004
$w_{x_1 x_2}(0.5; 0.5)$	–2.34	–2.34	–12.86	–12.86	0.90	0.94	–0.75	–0.74
$w_{x_1 x_2}(0.25; 0.25)$	–8.94	–8.94	–10.52	–10.52	–9.78	–9.77	–0.36	–0.37
$F_{x_1 x_2}(0.5; 0.5)$	–4.63	–4.63	–8.46	–8.46	–3.44	–3.44	–0.03	–0.02
$F_{x_1 x_2}(0.25; 0.25)$	–1.83	–1.83	–2.38	–2.38	–1.93	–1.93	–0.12	–0.12
$N_T(0.5; 0.5)$	0.124	0.126	0.228	0.330	0.092	0.093	0.001	0.001
$M_T(0.5; 0.5)$	–0.101	–0.101	0.298	0.310	0.353	0.344	0.080	0.084

7.4 Vibration of Isolated Shell Subjected to Impulse

We consider now a problem related to the dynamical response to an impulse load with intensivity q, constant in time and uniformly distributed, suddenly applied to the surface $x_3 = -0.5$. The problem is formulated in Sect. 7.1. The boundary conditions of the heat transfer equation (7.1) now correspond to thermal isolation of all shell surfaces (7.4). The boundary conditions associated with (7.3) correspond to free support on flexible ribs that are nonstretchable in their tangential plane. The initial conditions are $w\big|_{t=0} = \dot{w}\big|_{t=0} = \theta\big|_{t=0} = 0$.

As in earlier cases, all results presented from now on correspond to nondimensional quantities (except when it is specifically mentioned otherwise). We use the following notation: coupled geometrically nonlinear problem, GN(c); uncoupled geometrically nonlinear problem, GN; coupled geometrically and physically nonlinear problem, GPN(c); ucoupled geometrically and physically nonlinear problem, GPN. These problems are also referred to as problems 1, 2, 3 and 4, respectively; this latter notation is used in the figures.

In Fig. 7.6, the deflection in the shell centre versus time under the precritical load $q = 220$ is shown for these four problems. For both physically linear and nonlinear problems, the inclusion of coupling results in a decrease of the oscillation amplitude, which agrees with results obtained a long time ago (see, for instance, [139,176]). In the first period of oscillation, the decrease of the amplitude for problems 1 and 2 (GN(c) and GN) is 3.8%, whereas this decrease if 6% in the case of problems 3 and 4 (GPN(c) and GPN). In the second period of oscillation, the damping (caused by energy dissipation) is more pronounced and is 13% for problems 1 and 2, and 21% for problems 3 and 4.

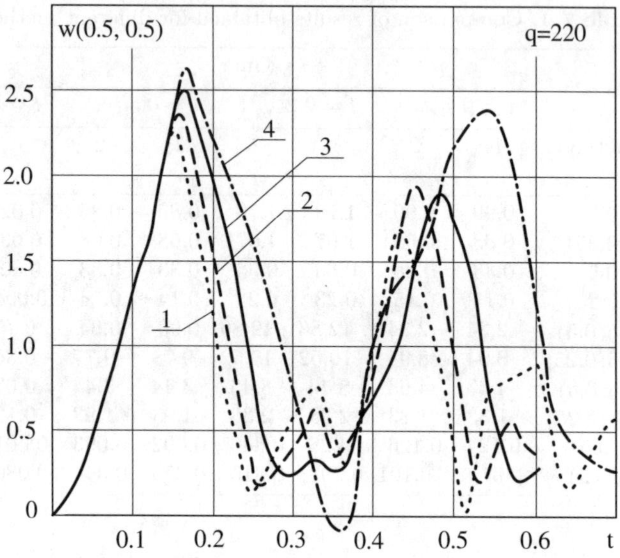

Fig. 7.6. Deflection in the shell centre for $q = 220$

In Figs. 7.7 and 7.8, the dependences of $F''_{x_1x_1}$ and $w''_{x_1x_1}$ on time in the shell centre are reported. These functions vary with time in a way that depends on the mathematical model used. In spite of the fact that the shell is very thin, the stress–strain state of the shell has an important influence on the effects of physical nonlinearity. For $t = 0.19 - 0.24$, the extremal stresses increase by 9.7% (for the uncoupled problems 2 and 4) and by 13%

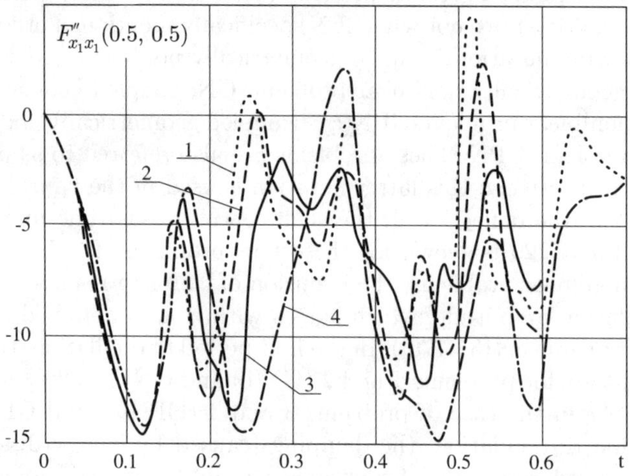

Fig. 7.7. Dependence of $F''_{x_1x_1}$ in the shell centre for $q = 220$

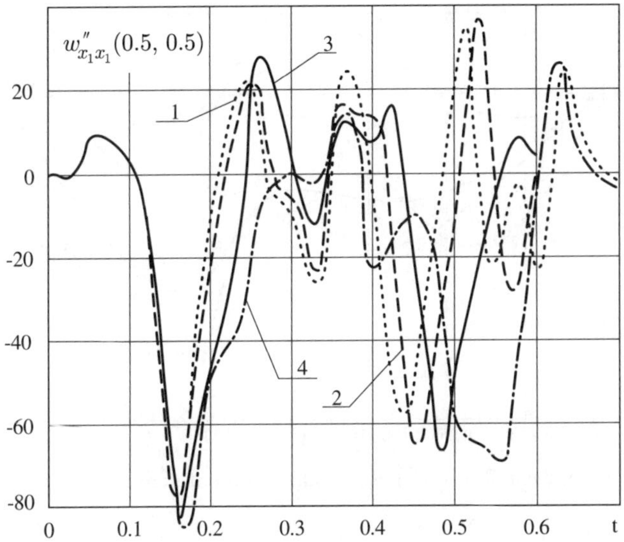

Fig. 7.8. Dependence of $W''_{x_1x_1}$ in the shell centre for $q = 220$

(for coupled problems). The inclusion of coupling leads to a decrease (in the time interval considered) in the maximum load (in terms of modulus) $F''_{x_1x_2}$ of 9.2% (for physically linear problems) and of 6.9% for the physically nonlinear problems. At the time moments $t = 0.25, 0.36, 0.48$ the $F''_{x_1x_1}$ values corresponding to the extreme deflection values differ by 2–2.5 times. Therefore, when the physical nonlinearity and the coupling are taken into account, the stress distribution changes in time significantly, and both their maximal and minimal values of the stresses are changed.

The distribution of $F''_{x_1x_1}$ along the axis of symmetry for the time moments $t = 0.16$ and $t = 0.36$ is shown in Figs. 7.9 and 7.10. It is apparent

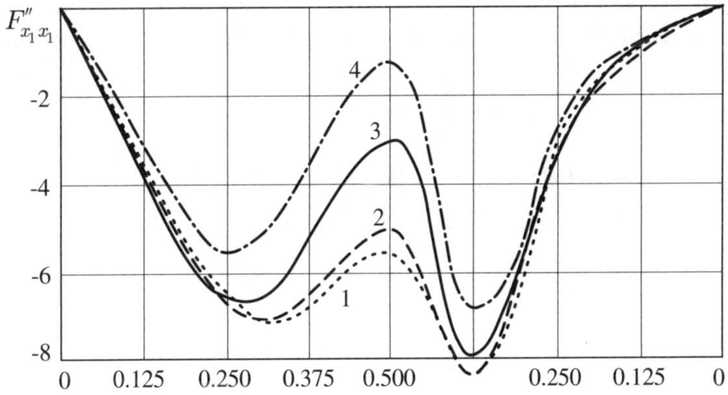

Fig. 7.9. Distribution of $F''_{x_1x_1}$ along axis of symmetry for $t = 0.16$

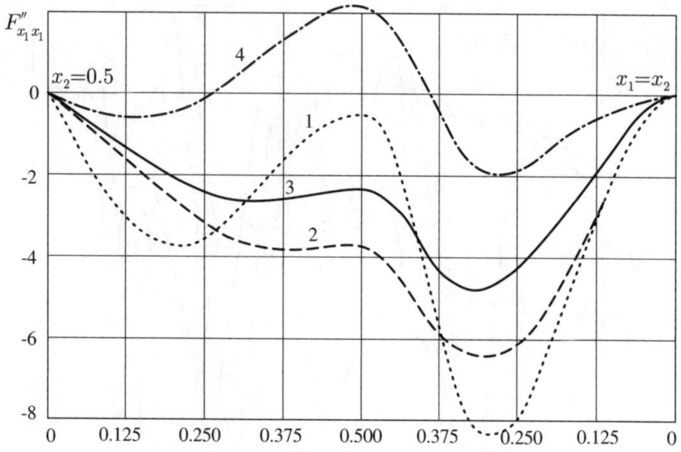

Fig. 7.10. Distribution of $F''_{x_1x_1}$ along axis of symmetry for $t = 0.36$

that the results obtained with different mathematical models have significant differences for $t = 0.16$, and these differences are more evident for $t = 0.36$. In the latter case, the solution corresponding to problem 4 differs from those for problems 1–3 not only quantitatively but also in the sign of the stresses.

In spite of the thinness of the plate, the influence of physical nonlinearity is also important. The maximal deflection in the shell centre obtained taking account of physical nonlinearity increases during the first period of oscillation by 9.9% (for the coupled problem) and 11% (for the uncoupled problems).

Consequently, in order to obtain a reliable estimate of the stress–strain state of a thin shallow shell with the geometrical and physical parameters considered above during oscillations caused by a load close to the critical value, one needs to consider the coupling of deformation and temperature. Although the deflections change only slightly, the stresses occurring in the shell can be arbitrary large owing to coupling effects.

In Fig. 7.11, the temperature distribution at the shell centre through the thickness of the shell is presented for problems 1 and 3. The maximum

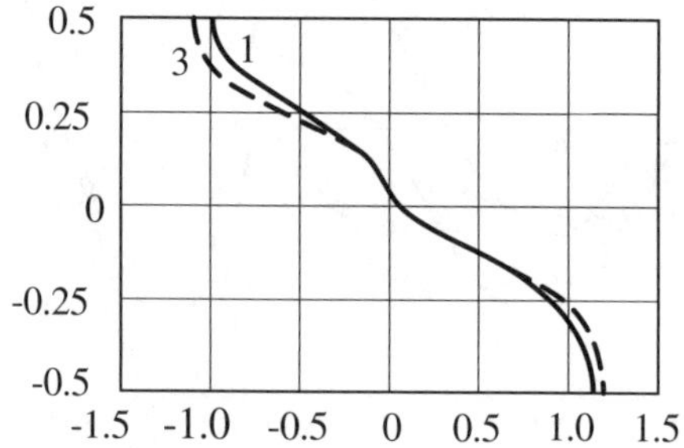

Fig. 7.11. Distribution of temperature in a shell centre along thickness

temperature rise of the shell due to the mechanical work is 4.6° in dimensional units for the physically nonlinear problem, and of 4.3° for the physically linear one. This heating is observed when the shell reaches its maximum deflection at $t = 0.16$ on the surface $x_3 = -0.5$ (Fig. 7.9 corresponds to this case). The coupling between deformation and temperature leads to a change of the temperature distribution in the shell volume. In Fig. 7.11, a negative increase of temperature can also be seen in certain parts of the shell, which is in agreement with the theoretical [188] and experimental [44] data obtained earlier.

The temperature distribution parallel to the middle surface of the shell at the same time moment $t = 0.16$ is shown in Figs. 7.12 and 7.13 for problems

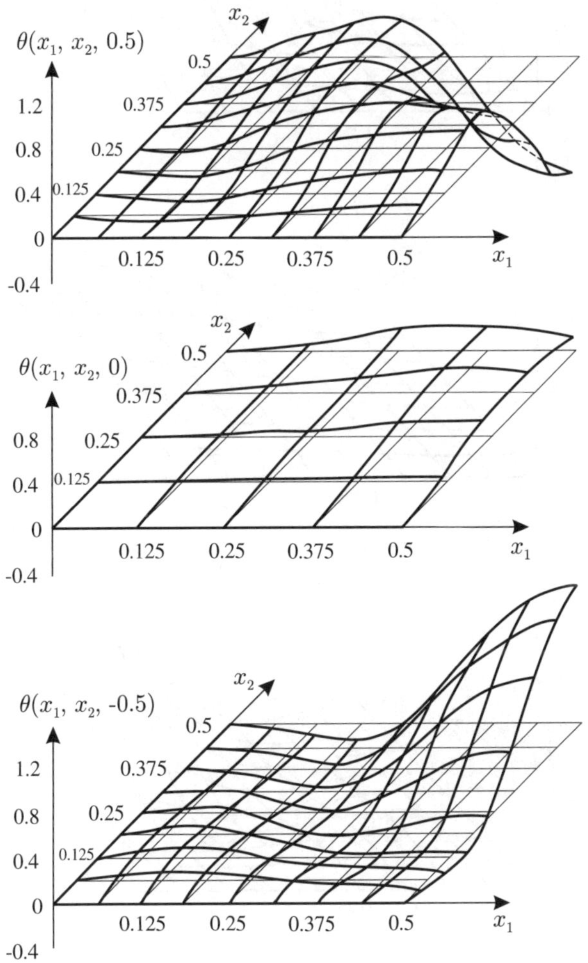

Fig. 7.12. Distribution of temperature for coupled geometrically nonlinear problem ($t = 0.16$)

1 and 3, respectively, for three different surfaces. Qualitatively, figures are similar. Inclusion of physical nonlinearity increases the self-heating of the shell only slihghty. Here also, a negative temperature increase, as well as a sudden temperature decrease in passing through the shell thickness, is observed. The maximum decrease is observed at $t = 0.16$ in the shell centre, as can be seen from the Figs. 7.12 and 7.13. For problem 1 this decrease is equal to

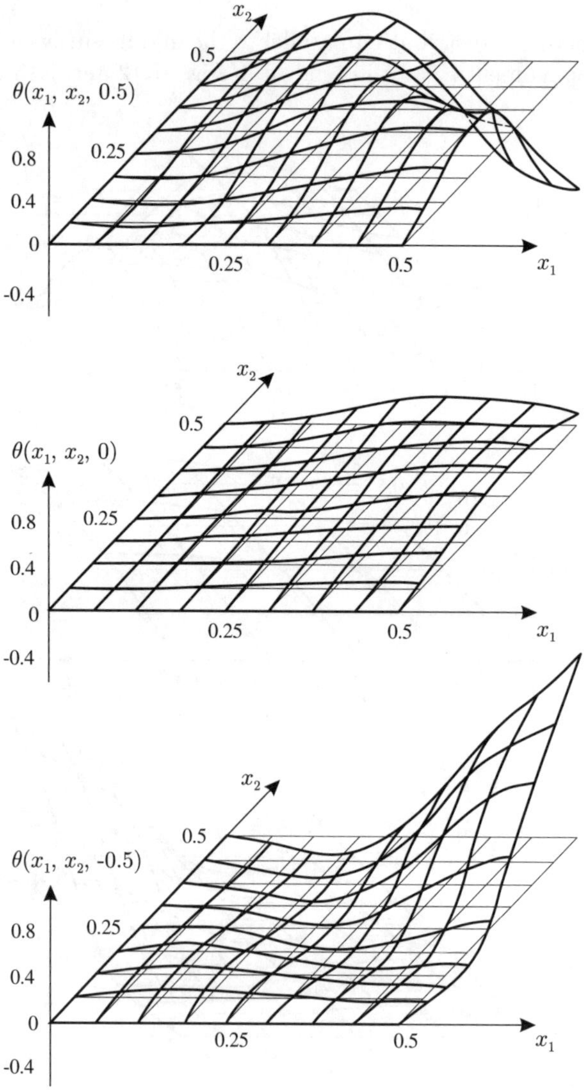

Fig. 7.13. Distribution of temperature for coupled geometrically and physically nonlinear problem ($t = 0.16$)

7.4°, whereas for problem 3 it is equal to 8.3°. The time moment $t = 0.16$ corresponds to the maximum shell deflections.

In Fig. 7.14, the temperature distribution parallel to the middle surface for problem 3 at $t = 0.26$ is reported (at this time $w(0.5, 0.5) = 0.54$). Figures 7.14 are again qualitatively similar to each other (the temperature distribution in the surfaces $x_3 = \pm 0.5$ at $t = 0.26$ is shown). The sudden temperature decrease at the shell centre reaches 4.9° at $t = 0.26$. Note that

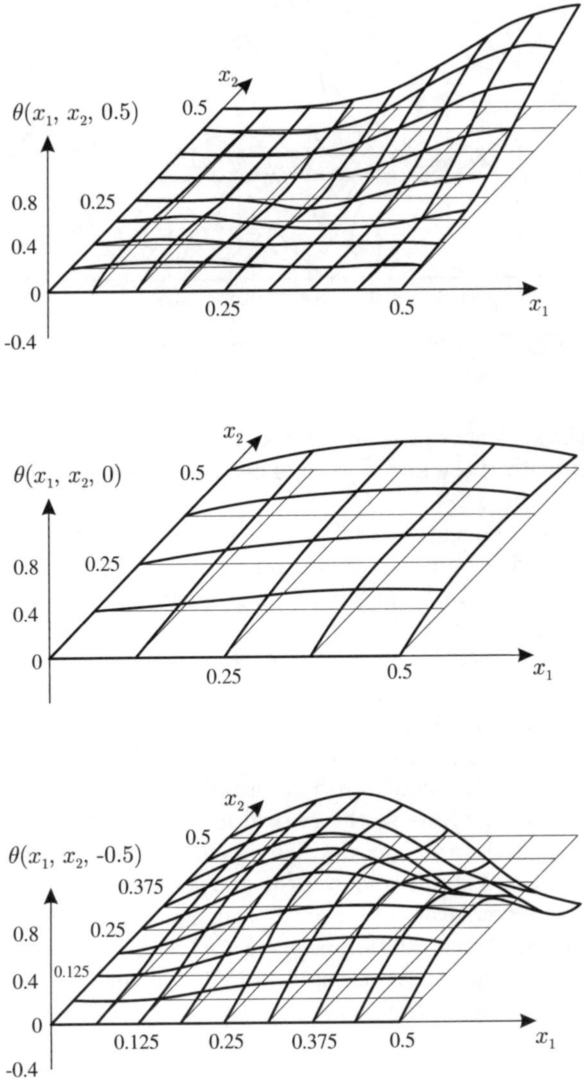

Fig. 7.14. Distribution of temperature for coupled geometrically and physically nonlinear problem ($t = 0.26$)

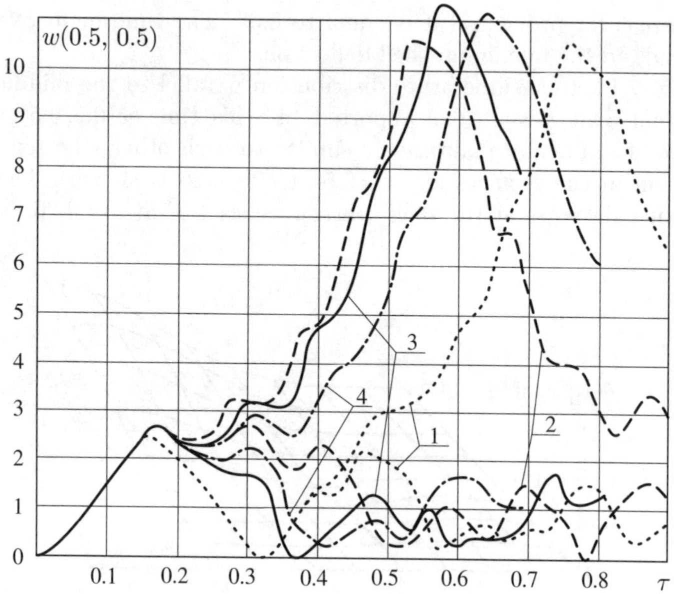

Fig. 7.15. Shell centre deflection for q_{cr} and $q_{cr} - 1$

the parts of the shell that heat up and cool down here changed their positions in comparison with those for $t = 0.16$. This observation is in agreement with the results given in [44].

We shall now consider the effect of coupling between the deformation and temperature fields and physical nonlinearity on the critical values of the stress–strain state for a flexible shell with geometric parameters $k_{x_1} = k_{x_2} = 24$.

In Fig. 7.15, the shell centre deflection versus time is reported for problems 1–4, for the critical value of the load q_{cr} and for pre-critical load that is one unit less. Therefore, an increase of q by 1, which corresponds to less then 0.5% of q_{cr}, corresponds to a deflection increase of more than three times. The maximum deflection values are greater than twice the camber of the shell. As the load is increased above the critical load, the time needed to achieve the maximum deflection decreases. The q values obtained are in agreement with the criteria of Vol'mir and Kantor and are referred to as the critical ones. The following critical values have been found: 230 for problems 1 and 2, 225 for problem 3 and 222 for problem 4. The Sho–Sung–Roth criterion has been applied in the case presented in Fig. 7.16. In this figure, the curve labelled 1 shows the deflection versus time at the shell centre for problem 4, with $q = 222$. Curve 2 shows the corresponding result for $q = 233$, curve 3 for $q = 224$.

The influence of coupling on the value of the critical load is not of significance. For the physically nonlinear, problems with the given geometrical

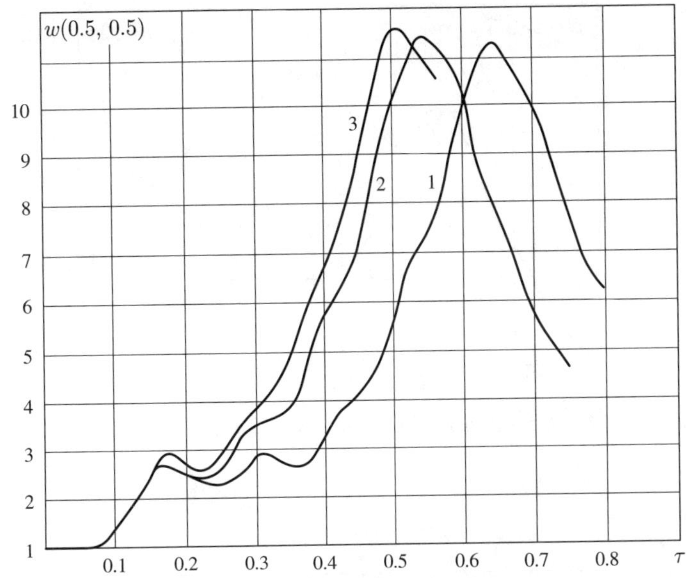

Fig. 7.16. Deflection at the shell centre for the uncoupled geometrically and physically nonlinear problem (curve 1 for $q = 222$, 2 for $q = 223$, 3 for $q = 224$)

and physical shell parameters, coupling increases q_{cr} by only 1.4%. The effect cannot be detected at all for problems 1 and 2. The increase of q_{cr} in the coupled problems can be explained by the part played by energy dissipation in the shell.

When we take into account the physical nonlinearity of the material, we find that q_{cr} for problem 4 is 3% smaller than for problem 2, and q_{cr} for problem 2 is 2.2% smaller than for problem 1. To conclude, the influence of physical nonlinearities on q_{cr} for a shell with the given parameters is also of small significance.

On the other hand, the choice of mathematical model can lead to qualitatively different results for a load close to the critical value. Let us consider, for instance, the stress–strain state obtained for problems 3 and 4 with $q = 223$ at the time moments $t = 0.16$ and $t = 0.36$. Although at $t = 0.16$ the shell deflection (Fig. 7.17) and the stress state differ only slightly, at $t = 0.36$ (Figs. 7.18–7.20) they differ qualitatively.

The shell centre deflection obtained from the solution to problem 4 (GPN) is 7% less than in the coupled doubly nonlinear problem at $t = 0.16$. The difference in $w''_{x_1x_1}$ at this chosen time moment for problem 3 and 4 (GN(c) and GPN) at the shell centre is almost negligible. However, a large difference has been observed in $F''_{x_1x_1}$. For problem 4, $F''_{x_1x_1}(0,0)$ is close to zero, whereas for problem 3 $F''_{x_1x_1}(0,0) = -2.5$. It should be noted that in this case also, the coupling between deformation and temperature is mainly exhibited in the stresses.

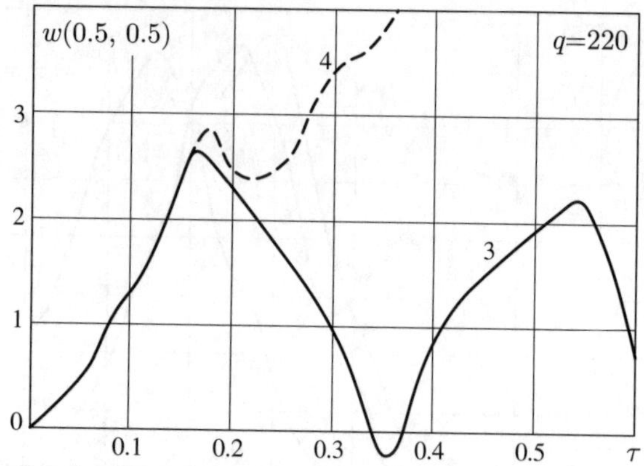

Fig. 7.17. Deflection at the shell centre for $q = 223$ and $t = 0.16$

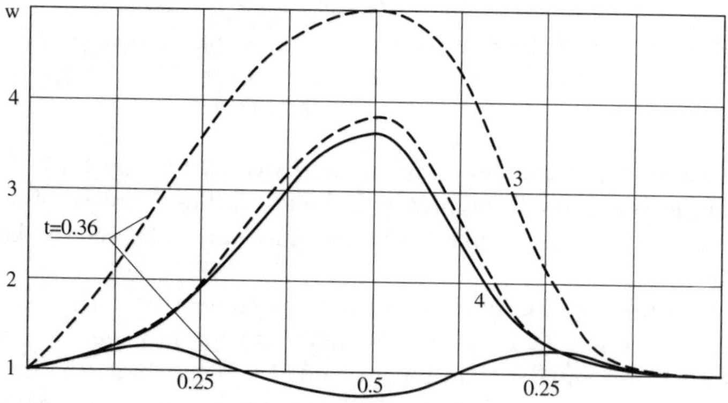

Fig. 7.18. Distribution of w for $t = 0.36$

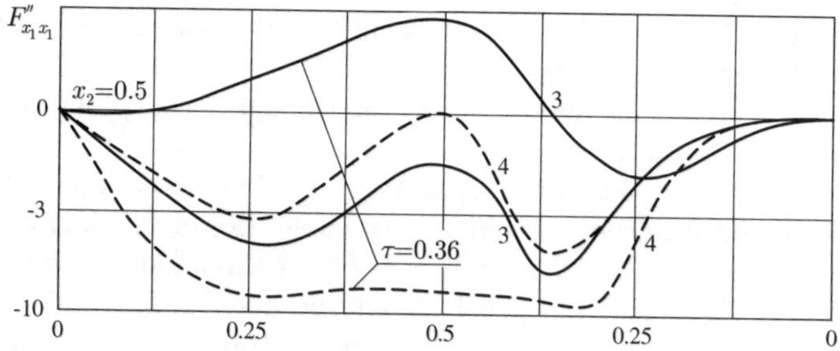

Fig. 7.19. Distribution of $F''_{x_1x_1}$ for $t = 0.36$

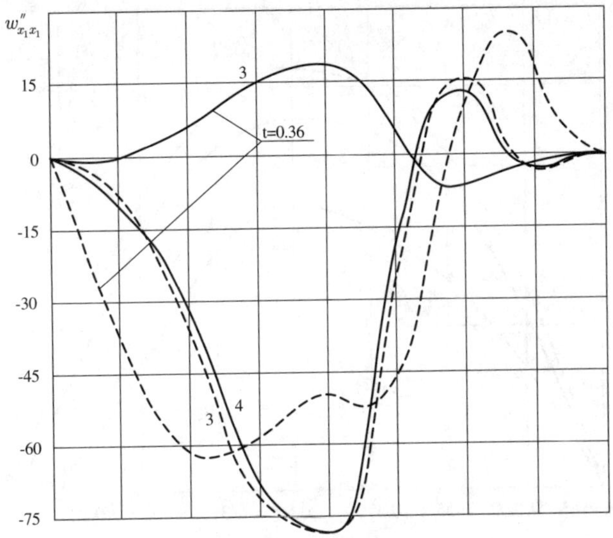

Fig. 7.20. Distribution of $W''_{x_1x_1}$

The time instant $t = 0.36$, for problem 3 with $q = 223$, corresponds to a case where the shell achieves negative deflection values during pre-critical vibrations. Note that at the same time moment, jumping behaviour occurs in the case of problem 4, and the maximum deflection here is of the order of four times the thickness. In what follows, the values of $F''_{x_1x_1}$ and $w''_{x_1x_1}$ not only have quantitative differences but also have different signs in a large part of the shell.

The jumping behaviour of the shell at the critical load has a time variation that depends on the model used. In Fig. 7.15, the maximum deflections are achieved at the critical load at $t = 0.76$ (problem 1), $t = 0.53$ (problem 2), $t = 0.56$ (problem 3) and $t = 0.64$ (problem 4). Therefore, if we compare the stress–strain states of the shell at the critical load at the same time moment, we can again obtain qualitatively different results. This statement is proved by the graphs shown in Figs. 7.21–7.24, where the stress–strain states corresponding to problems 1–4 at $t = 0.53$ are reported. The maximum deflections of the point on the middle surface (0.5,0.5) are 10.5 (problem 2), 9.2 (problem 3), 6.9 (problem 4) and 3.2 (problem 1). The different values of deflections correspond to different stress–strain states (the functions shown in Figs. 7.21–7.24 often differ from each other by a factor of several times at the same characteristic points).

The values of $w''_{x_1x_1}$, $F''_{x_1x_1}$ and σ_{11} for problems 1 and 2 differ not only in magnitude but also in sign over a large part of the shell. The values of $w''_{x_1x_1}$ for problems 3 and 4 differ at the shell centre by a factor of almost two, and at the point $(0.125, 0.125)$ they differ by almost of three in terms of absolute values; and in addition, they have different signs. Although the values of

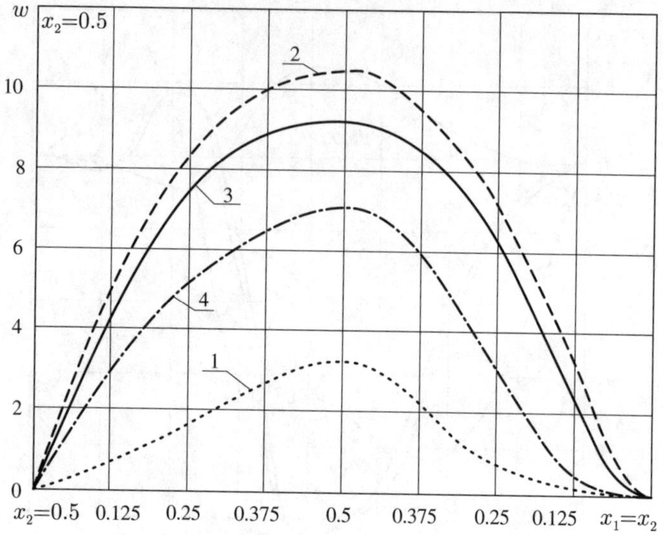

Fig. 7.21. Distribution of w for $t = 0.53$

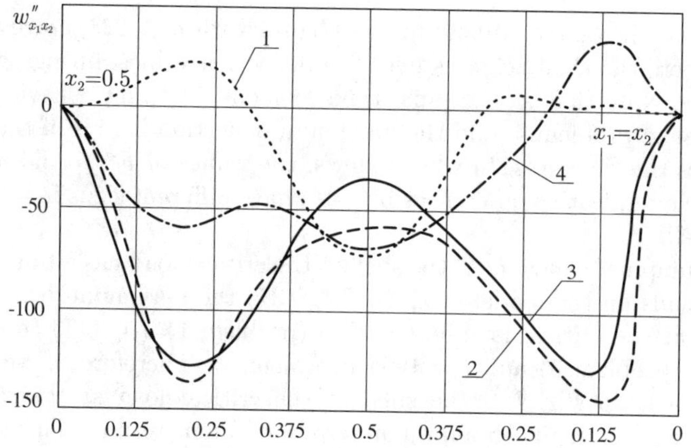

Fig. 7.22. Distribution of $w''_{x_1x_2}$ for $t = 0.53$

$F''_{x_1x_1}$ for those problems coincide at the point $(0.5, 0.5)$, at other points the stresses very often have different signs and their values differ significantly. The stresses σ_{11} obtained at q_{cr} are very close to each other along the $x_1 = 0.5$ axis for problems 3 and 4, but they often differ in sign.

To conclude, it has been shown that the stress–strain states obtained with different models can differ significantly, either quantitatively or qualitatively.

The discussion up to this point has illustrated the fact that taking account of coupling and of physical nonlinearities has a significant influence

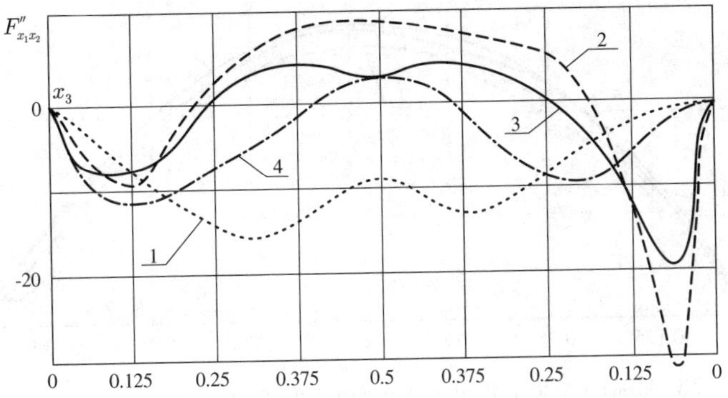

Fig. 7.23. Distribution of $F''_{x_1x_2}$ for $t = 0.53$

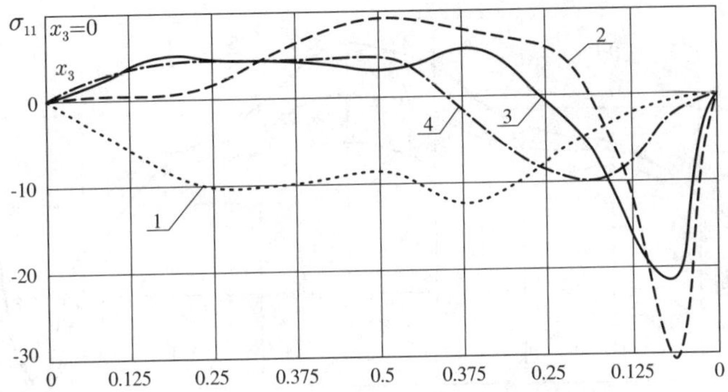

Fig. 7.24. Distribution of σ_{11} for $t = 0.53$

on the stress–strain state of a shell. The comparison has been carried out for problems 1–4 under the same pre-critical load. Now we are going to show how coupling and physical nonlinearities affect the stress–strain state of a shell during dynamic stability loss. The stress–strain state will be analysed at the time moment characteristic of the four types of problems discussed, i.e. when the maximum deflections are achieved. The various functions corresponding to this time value are reported in Figs. 7.25–7.31. As earlier, on the left in the figures the functions are plotted along one of the symetry axes of the shell crossing its centre ($x_2 = 0.5$, for instance). On the right, the function is plotted along a diagonal through the origin at $(0, 0)$. The data are reported for only one-quarter of the middle surface of the shell (because of symmetry).

The shell deflections shown in Fig. 7.25 at the time of maximum deflection differ only slightly (the values of $w(0.5, 0.5)$ for problems 1 and 2 are 6–7% less than for the physically nonlinear problems). However, the deformations of the middle surface differ significantly: $\varepsilon_{11}(0.5, 0.5)$ increases by 77% for the physically linear problem and by 57% for the physically nonlinear problem when coupling is assumed. In the neighbourhood of the corner, for given

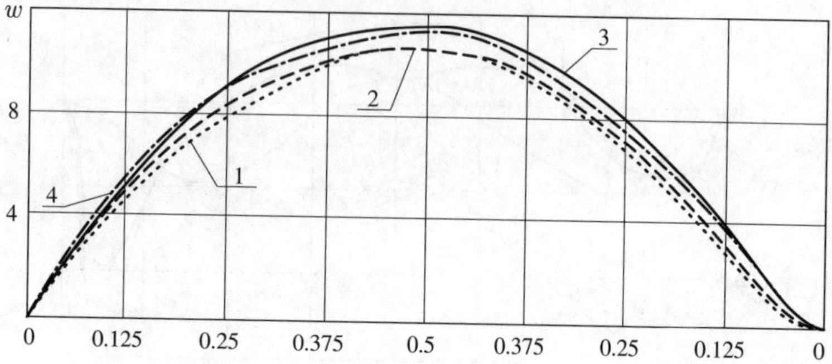

Fig. 7.25. Distribution of w at the time moment of maximum deflection

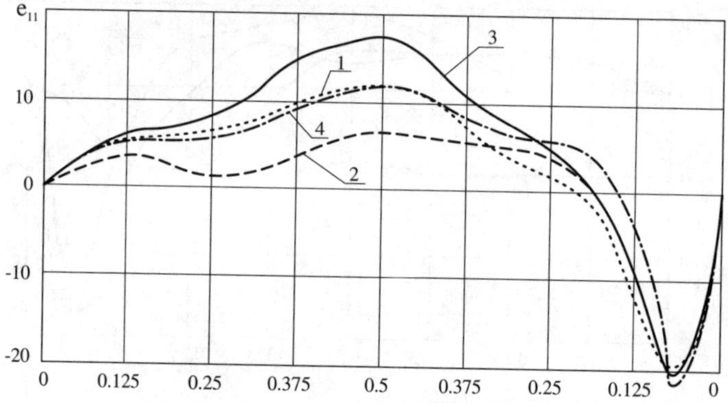

Fig. 7.26. Distribution of ε_{11}

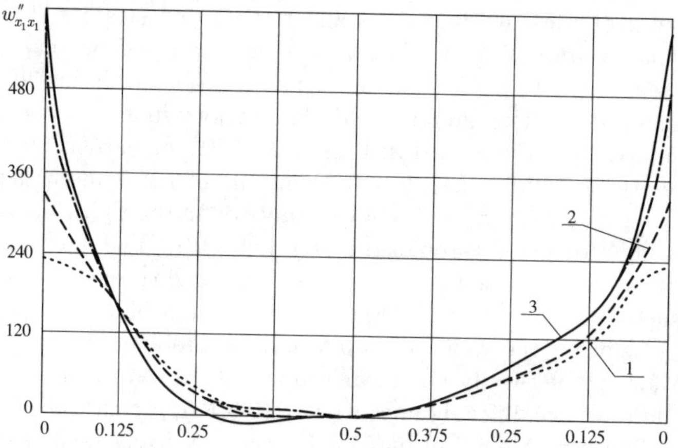

Fig. 7.27. Distribution of $w''_{x_1 x_1}$

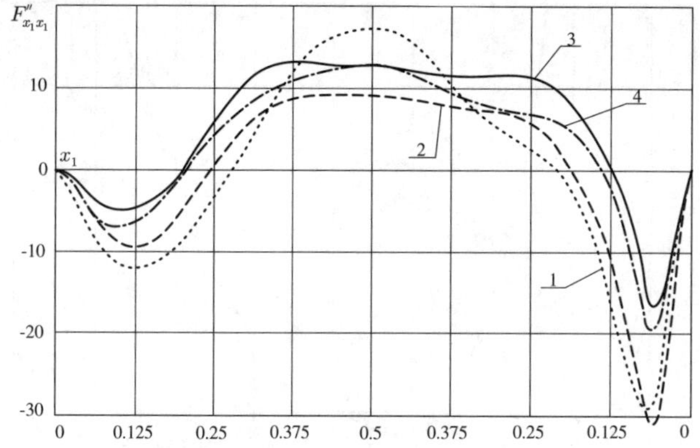

Fig. 7.28. Distribution of $F''_{x_1x_1}$

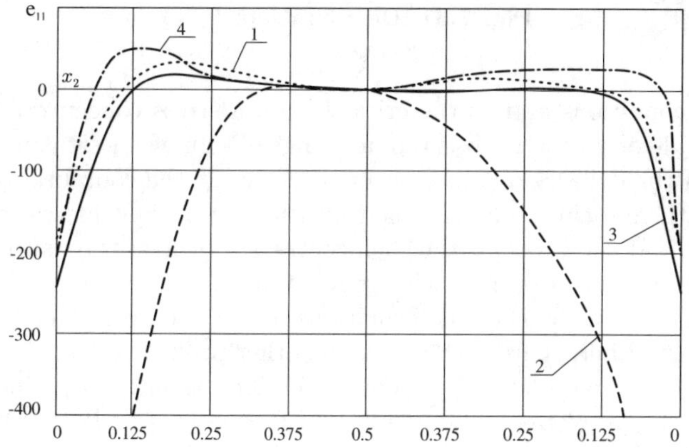

Fig. 7.29. Distribution of ε_{12}

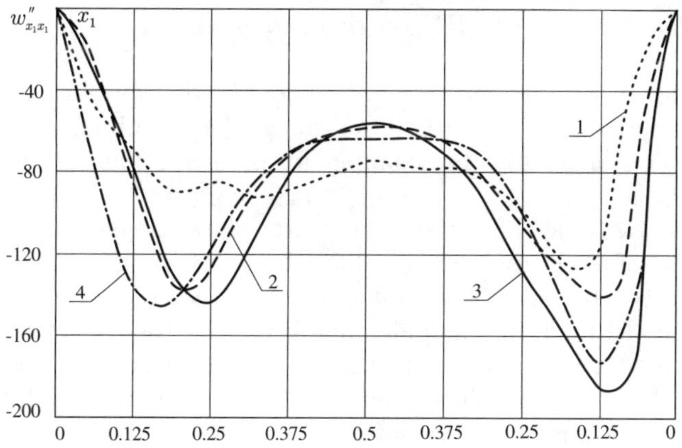

Fig. 7.30. Distribution of $w''_{x_1x_2}$

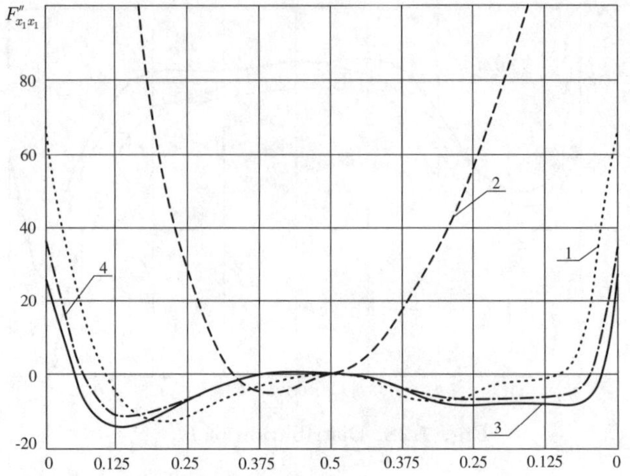

Fig. 7.31. Distribution of $F''_{x_1x_2}$

boundary conditions and at the critical load, a stress concentration and an increase of deformation is observed: a change of sign of ε_{11} sign and a notable increase of the deflection values occur. Performing the computations for the simple mathematical model (problem 2) leads to a slight increase of the results (by 4.8%). However, at the shell centre, ε_{11} decreases by approximately a factor of two in comparison with problem 3.

Calculations of the shear deformations of the middle surface for problem 2 give results different from those for the other problems (Fig. 7.29). A similar behaviour occurs for $F''_{x_1x_2}$ (Fig. 7.31). The values of ε_{12} and $F''_{x_1x_2}$ in the corner for problems 2 and 1 differ by factors of 4 and 4.7, respectively. Coupling has an important influence on $F''_{x_1x_2}$ and ε_{12}. For instance, at the point $(0.125, 0.125)$ $F''_{x_1x_2}$ values differ by more than 30% for problems 3 and 4, and the difference in the ε_{12} values is even larger close to a corner. The calculations of ε_{12} and $F''_{x_1x_2}$ without coupling give lower values of those functions.

A similar picture is seen in Fig. 7.27 for the distribution of $w''_{x_1x_2}$ over the middle surface. Close to $(0, 0)$ the values of $w''_{x_1x_2}$ obtained for different models differ significantly (for example, for problems 3 and 4 they differ by 14.4%).

In Figs. 7.28 and 7.30, the distribution of the second derivative with respect to x_1 of the load function and the deflection along the symmetry axes of the shell is reported. The calculations show that the use of a physically nonlinear model leads to a notable improvement of the results obtained: for $F''_{x_1x_1}$ at the centre of the shell, problems 3 and 4 give the same results, but for problem 2 we obtain a value of $F''_{x_1x_1}(0, 0)$ that is less by 26%, and for problem 1, a value that is greater by 40%. At the point $(0.0625, 0.0625)$, where stress concentration has been observed, the results for problem 2 are larger by 1.6 times than those for problem 4 (taking coupling into account decreases the stress by 15% at this point).

It turns out that the maximum values of $w''_{x_1 x_1}$ are located close to the corner, for all problems. The value of $w''_{x_1 x_1}(0.125, 0.125)$ is 7% larger for problem 3 than for problem 4. In contrast, in the physically linear problems, the coupling decreases the stresses concentration at this point. The difference between the results for problems 1 and 3 is 37%, whereas for problems 2 and 4 it is 18%.

Now we are able to formulate the following conclusions. Calculations for a load in the neighbourhood of the critical load can lead to qualitatively different results for different models. When coupling is taken into account, the value of the critical load increases owing to energy dissipation, which leads to heating of the shell. Inclusion of both coupling and physical nonlinearity of the material leads to changes of the times of jumping processes of the shell, and a comparison of the stress–strain states obtained with different mathematical models for the same time can also exhibit qualitative differences. The coupling and physical nonlinearity greatly influence the stress–strain state of the shell. When a simple mathematical model is used (without coupling or physical nonlinearity), it can lead to results improved theories (for instance, in the case of shear deformations).

The coupling between deformation and temperature helps us to analyse the influence of mechanical loads on the self-heating of the shell (remember that all shell surfaces are assumed to be thermally isolated).

In Figs. 7.32–7.34, the temperature distribution through the shell thickness at the centre for problem 3 at $t = 0.3, 0.5, 0.56$ and for problem 1 at $t = 0.54, 0.72, 0.76$ is shown. The time moments for the two problems have been chosen in order to obtain similar shell deflections in problems 3 and 1. The physical nonlinearity changes the temperature distribution θ obtained earlier. The maximum shell heating is observed at the time of maximum deflection close to the corner, i.e. in the place of maximum stress (deformation)

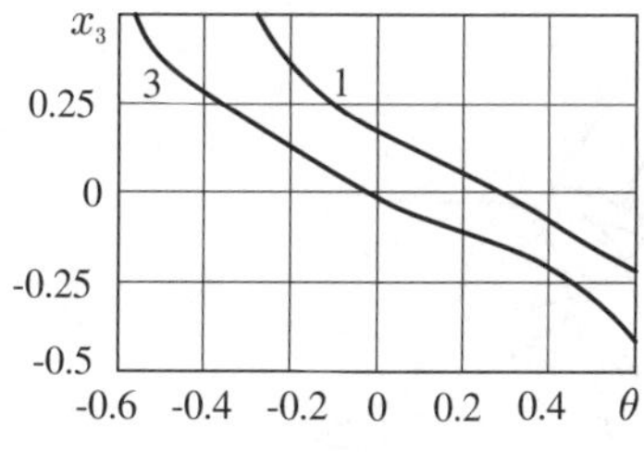

Fig. 7.32. Temperature distribution through the shell thickness at the centre of the shell: problem 1, $t = 0.54$; problem 3, $t = 0.3$

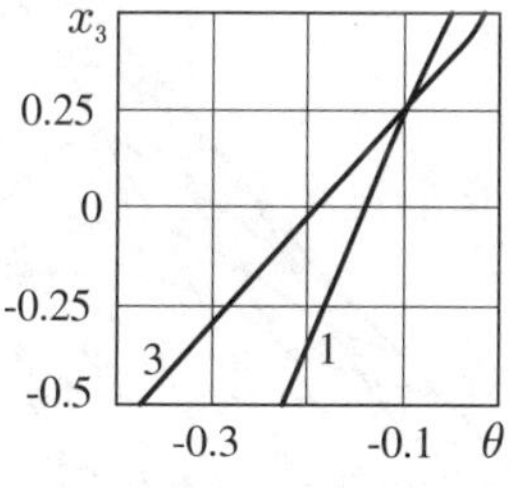

Fig. 7.33. Temperature distribution through the shell thickness at the centre of the shell: problem 1, $t = 0.72$; problem 3, $t = 0.5$

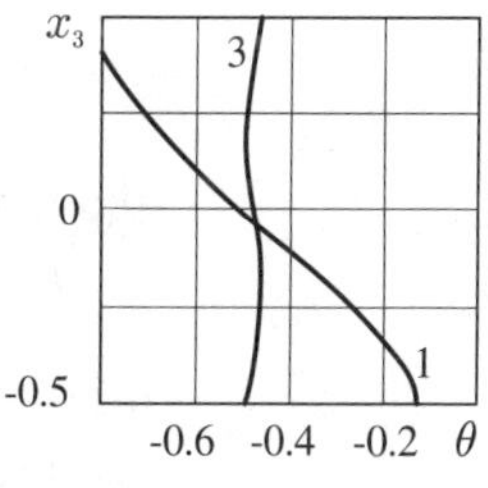

Fig. 7.34. Temperature distribution through the shell thickness at the centre of the shell problem 1, $t = 0.76$; problem 3, $t = 0.56$

concentration. The distribution of θ over the middle surface of the shell at times $t = 0.3, 0.5, 0.56$ for problem 3 and at $t = 0.76$ for problem 1 are shown in Figs. 7.35–7.38. The maximum self-heating is 12.1° for problem 3 and 8.3° for problem 1. Qualitatively, the temperature distributions over the middle surface for problem 3 (Fig. 7.37) and for the problem 1 (Figure 7.38) are similar (the maximal self-heating is observed in the surface $x_3 = 0.5$ both cases).

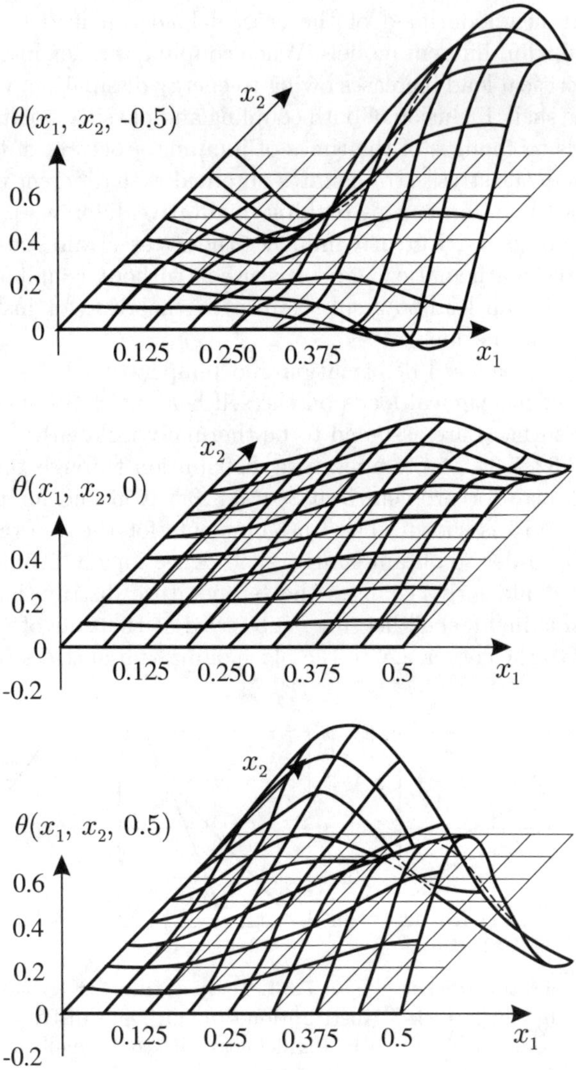

Fig. 7.35. Distribution of temperature over the middle surface of the shell, problem 3, $t = 0.3$

It is easy to check that the areas where θ increases on one of the shell surfaces correspond to areas where θ decreases on the opposite surface. Therefore, the temperature change across the thickness at the point $(0.0625, 0.0625)$ reaches 19.5° in problem 3, whereas at the point $(0.125, 0.125)$ it reaches 13.2° in problem 1.

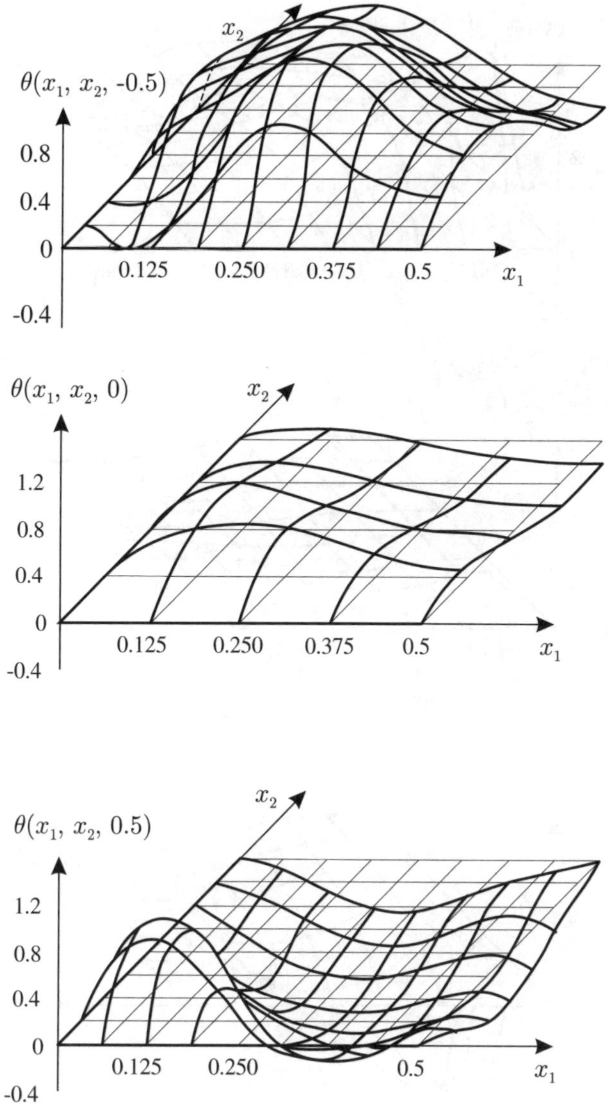

Fig. 7.36. Distribution of temperature over the middle surface of the shell, problem 3, $t = 0.5$

The shift of the area of maximum shell heating from the centre of the shell to the corner is caused by the increase in the load and deflection and by the type of jumping behaviour during stability loss. Incidentally, for a post-critical load ($q = 230$, problem 3), a value of $\theta = -0.78$ was reached at the shell centre on the surface $x_3 = 0.5$ (and $\theta = -0.46$ for $q = 225$), and the change of the solution for θ at the point $(0.0625, 0.0625)$ reached 19.9°.

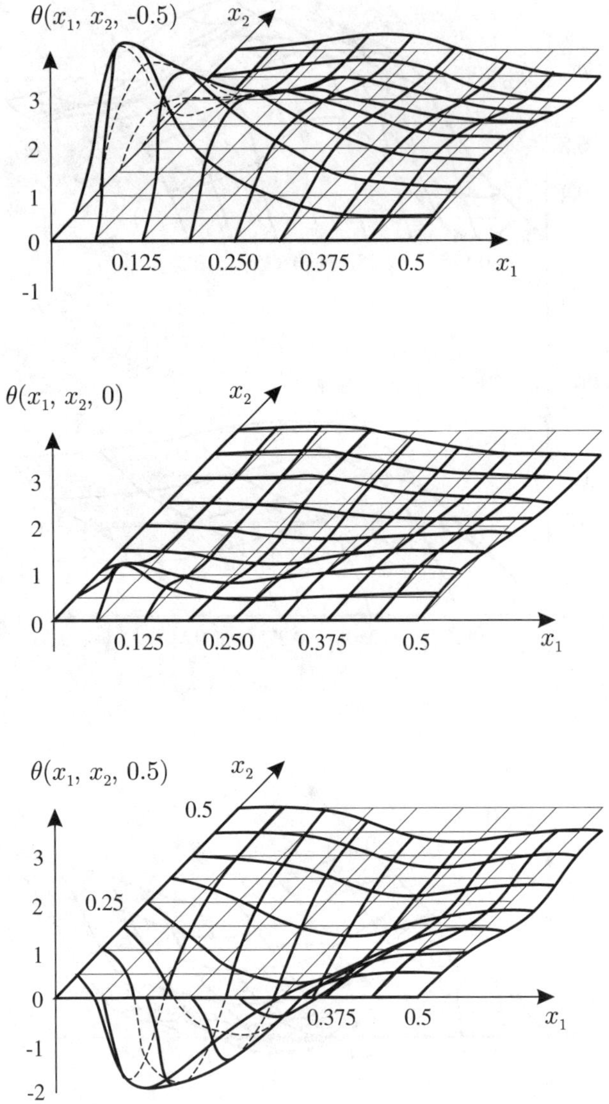

Fig. 7.37. Distribution of temperature over the middle of the shell, problem 3, $t = 0.56$

In Fig. 7.39, the temperature distribution θ through the shell thickness at the point $(0.0625, 0.0625)$, at the time moments $t = 0.1, 0.3, 0.5, 0.56, 0.6, 0.625$ is shown for problem 3. Curves 1–6 correspond to these times in increasing order, and curve 2 (not shown) almost coincides with the axis $\theta = 0$. The distributions of θ presented in Fig. 7.39 are sometimes (curves 1–3) close to linear, and sometimes (curves 4–6) very different from linear. Therefore,

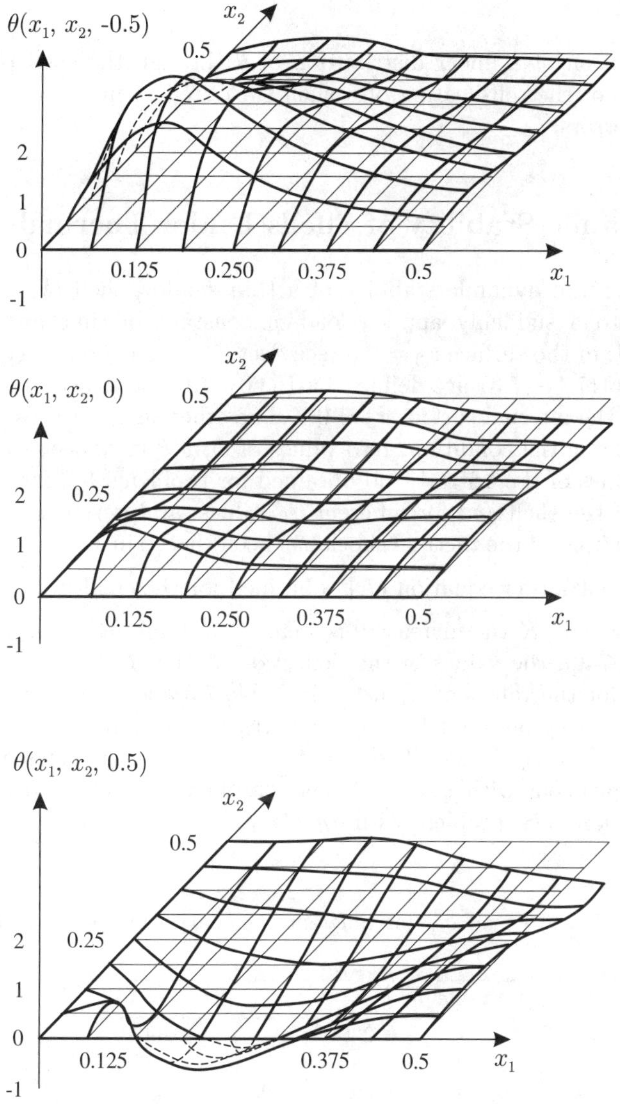

Fig. 7.38. Distribution of temperature over the middle shell, problem 1, $t = 0.76$

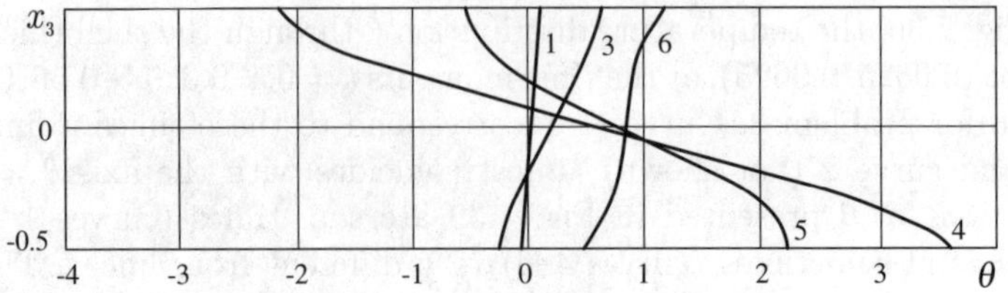

Fig. 7.39. Distribution of temperature θ through shell thickness at the point $(0.0625, 0.0625)$, at the time moments $t = 0.1, 0.3, 0.5, 0.56, 0.6, 0.625$

the assumption of a linear distribution of θ through the shell thickness in an analysis of the self-heating of a shell subjected to mechanical loads will introduce errors.

7.5 Dynamic Stability of Shells Under Thermal Shock

We consider the dynamic stability of a thin shallow shell ($k_1 = k_2 = 24$) subjected to a suddenly applied load q, constant in time and uniformly distributed, in the surface $x_3 = 0.5$ (see Sect. 7.1). The boundary conditions attached to (7.1)–(7.3) are defined by (6.60). At $t = 0$, a thermal shock is applied on the internal surface $x_3 = 0.5$. The other shell surfaces are isolated (7.4). As the initial condition, zero values of $w, \dot{w}, \theta$ have been taken.

The values of the critical load obtained for problems 1–4 for full thermal isolation of the shell and for different q_T values, are presented in Table 7.5. The intensity q_T of the thermal shock has been chosen in such a way as to satisfy the heat transfer equation (7.1) obtained for the condition $\dfrac{T - T_0}{T_0} \ll 1$, i.e. for $T_0 = 320° \, K$ the increase θ is treated as small up to 100°.

In Fig. 7.40, the values of the deflection at the shell centre versus time are shown for the values of q_{cr} listed in Table 7.5 and for certain pre-critical loads differing by one unit from q_{cr}. Curves 1–4 refer to problems 1–4 with a thermal shock $q_T = 25$. "Problem 5" refers to the uncoupled physically nonlinear problem with $q_T = 12$, and "problem 6" refers to the coupled physically nonlinear problem with $q_T = 12$.

Table 7.5. Critical load q_{cr} for different boundary conditions of heat transfer equation

Problem	$q_T = 0$	$q_T = 12$	$q_T = 25$
1	230	–	260
2	230	–	257
3	225	236	245
4	222	232	244

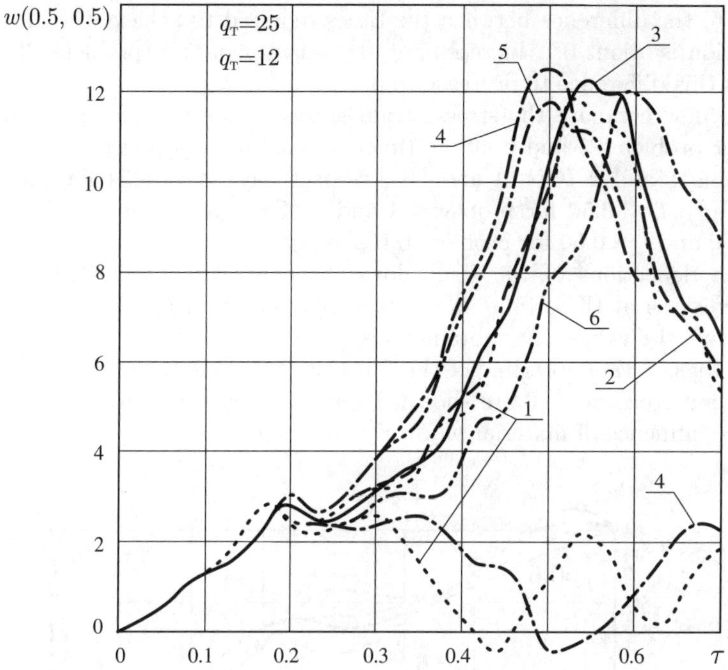

Fig. 7.40. Deflection of the shell centre for problems 1–4 with a thermal shock $q_T =$ 25, and for uncoupled and coupled physically nonlinear problems with a thermal shock $q_T = 12$

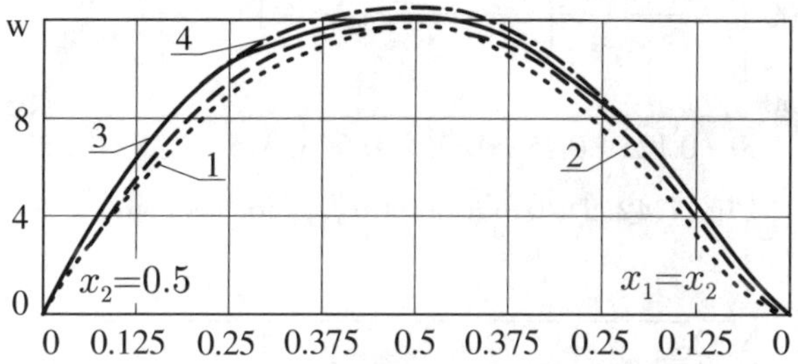

Fig. 7.41. Shell deflections for problem 1, $t = 0.53$; problem 2, $t = 0.53$; problem 3, $t = 0.54$; problem 4, $t = 0.50$

It is observed that when q_T is increased, the influence of physical nonlinearities on the value of the critical loads increases. For $q_T = 0$, the difference between the values of q_T for problems 1 and 3 is 2.2% and the difference for problems 2 and 4 is 3.5%. For $q_T = 25$, the corresponding differences are 6.1% and 5.3%, respectively.

The influence of coupling onto "jumping" phenomenon greatly decreases with increase of q_T. Curves 1 and 2 in Fig. 7.40 almost overlap. For problems

5 and 6, the difference between the times required to achieve the maximum deflection is about 0.1. Increasing q_T by a factor of two (problems 3 and 4) causes this difference to decrease to 0.04.

We now compare the stress–strain states of the shell for problems 1–4 and for problems 5 and 6 at the times of maximum deflection: $t = 0.53$ for problems 1 and 2 (GN(c) and GN, respectively); $t = 0.54$ for problem 3 (GPN(c)); $t = 0.50$ for problems 4 and 5 (GPN problems for different q_T values); and $t = 0.60$ for problem 6 (GPN(c)).

The deflections of the shell along its symmetry axes obtained in for problems 1–4 at these time moments are presented in Fig. 7.41. For all four problems, the values of w obtained are close to each other.

In Figs. 7.42–7.46, some of the functions characterizing the stress-strain states for problems 1–6 are shown. It can be seen that with an increase of q_T, the influence of material physical nonlinearity increases and the effect of

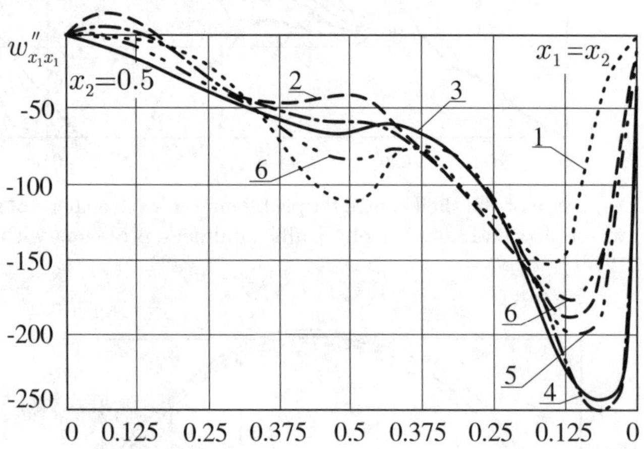

Fig. 7.42. Distribution of $w''_{x_1x_1}$ for problems 1–6

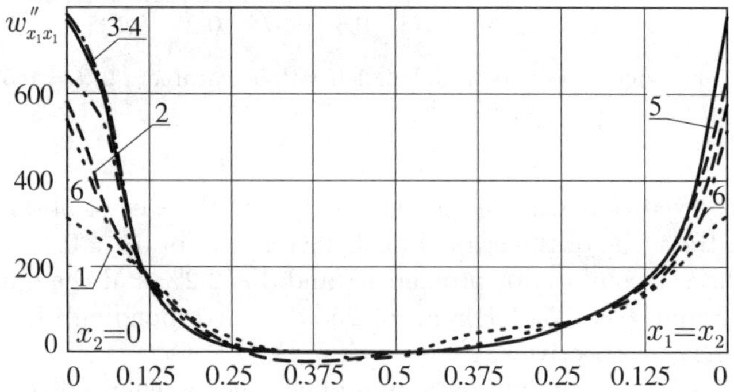

Fig. 7.43. Distribution of $w''_{x_1x_2}$ for problems 1–6

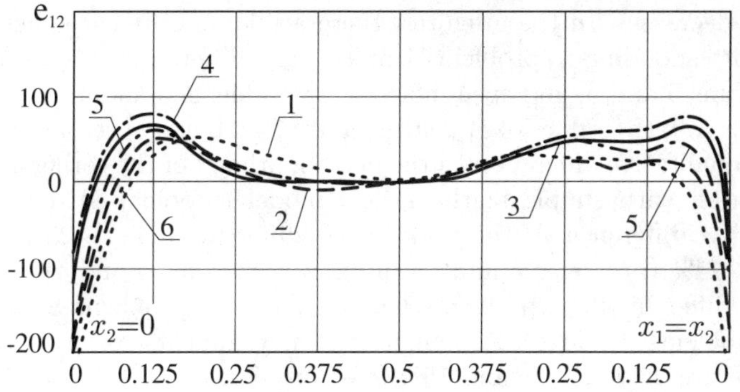

Fig. 7.44. Distribution of e''_{12} for problems 1–6

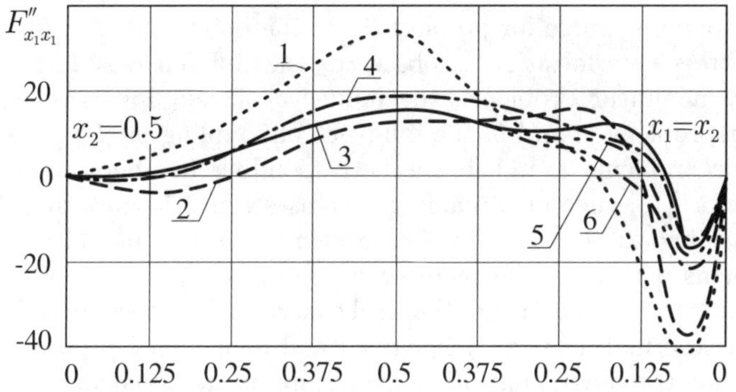

Fig. 7.45. Distribution of $F''_{x_1x_2}$ for problems 1–6

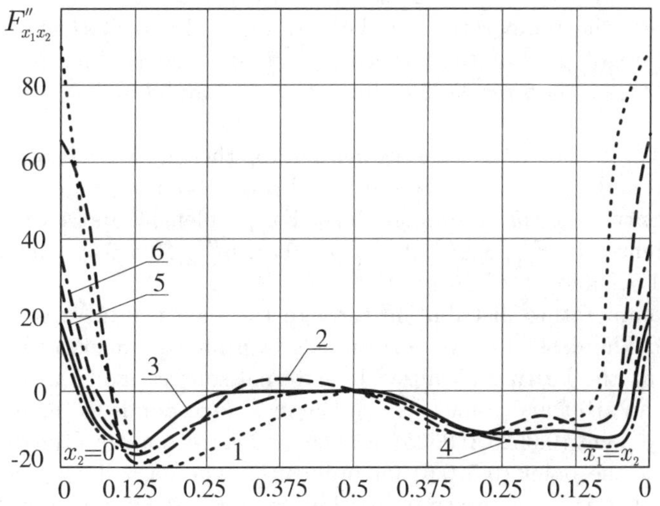

Fig. 7.46. Distribution of $F''_{x_1x_2}$ for problems 1–6

coupling decreases. In these figures, there are large differences between the curves corresponding to problems 1 and 2. The differences are much smaller for problems 5 and 6, and even more so for problems 3 and 4.

For $q_T = 0$, the values of ε_{12}, $w''_{x_1x_2}$ and $F''_{x_1x_2}$ have rather large quantitative and qualitative differences in the neighbourhood of the corners. However, computation with simple mathematical model (problem 2) does not give a qualitative difference. At the point $(0,0)$ a difference in ε_{12} is 25% (problems 1 and 2), 24% (problems 5 and 6) and 21% (problems 3 and 4). The values of $w''_{x_1x_2}$ differ by 46% (problems 1 and 2), 14.5% (problems 5 and 6), and 1.2% (problems 3 and 4). A similar tendency appears for $F''_{x_1x_2}(0,0)$: the difference are 25%, 21.5% and 15.9% for problems 1 and 2, 5 and 6, and 3 and 4, respectively.

For both $w''_{x_1x_1}$ and $F''_{x_1x_1}$, the effects of coupling are large in the physically linear problems. At the shell centre, $F''_{x_1x_2}$ for problem 1 is 2.7 times the corresponding value for problem 2. At the point $(0.0625, 0.0625)$, where a large stress concentration is observed, the difference is 11.9%. For the physically nonlinear problems, the influence of coupling is of less importance: for problems 5 and 6, the influence of coupling on $F''_{x_1x_2}$ is 29% and 10.5%; for problems 3 and 4, the corresponding numbers are 21.5% and 2.9%. For $w''_{x_1x_2}$, the corresponding increases are 2.7 times and 21.7% for problems 1 and 2; 30% and 8% for problems 5 and 6, and 11.8% and 1.9% for problems 3 and 4. The increase in of q_{cr} for $q_T \neq 0$ does not affect qualitatively the self-heating of the shell caused by the mechanical load. Comparing the thermal fields of the coupled and noncoupled problems, one can observe that the maximum difference occurs at the moment corresponding to maximum deflection, close to the corner of the surface $x_3 = 0.5$. For problems 3 and 4, the difference at the point $(0.0625, 0.0625)$ reaches 13.06°, whereas for problems 1 and 2 the difference reaches 10.7°. The differences observed in the temperature gradient through the shell thickness due to warming up of part of the surface $x_3 = -0.5$ and cooling down of part of the surface $x_3 = 5$ reach 22.5° for problems 3 and 4 and 15° for problems 1 and 2.

Inclusion of coupling causes changes in the thermal field. It also decreases the strain and stress values. However, it brings about an important increase of the exactness of the results obtained. For problems 3 and 4 with $q_T = 25$, the differences in $F''_{x_1x_2}(0,0)$, $F''_{x_1x_1}(0.5, 0.5)$, $w''_{x_1x_1}(0.5; 0.5)$ reach values of 15.9%, 21.5% and 11.8%, respectively.

The temperature distribution through the shell thickness is reported in Fig. 7.47. The dashed curve corresponds to θ for the uncoupled problem 4 at $t = 0.54$ the heating is caused by external sources, and this dependence holds for an arbitrary point (x_1, x_2). Curve 1 corresponds to the distribution of θ at the point $(0.0625, 0.0625)$, and curve 2 corresponds to the distribution of θ at the the point $(0.5, 0.5)$ for problem 3. The dash–dot curves 3 and 4 correspond to the temperature distribution at $t = 0.625$ for problems 3 and 4 at the same point as for curve 1.

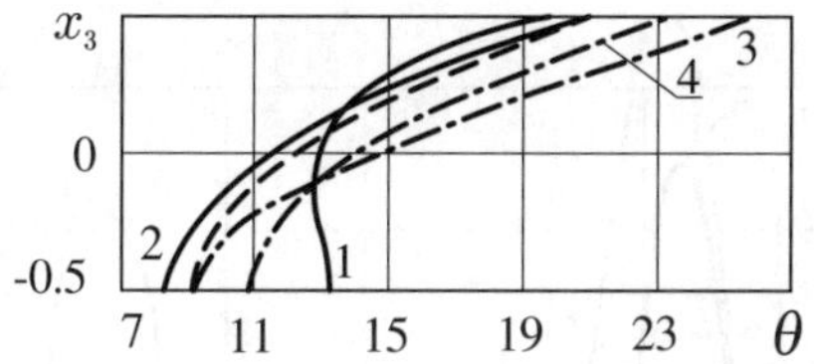

Fig. 7.47. Distribution of temperature θ through the shell thickness

The coupling effect significantly influences distribution of θ along x_3, especially at the moment when the maximum deflection is reached. For a shell with given physical and geometrical parameters, the influence of coupling during thermal shock is most evident in the neighbourhood of a corner point.

Now we consider a problem related to the dynamic reaction of a thin, shallow ($K_1 = K_2 = 48$) physically nonlinear shell to the load specified in Sect. 7.1, applied to the surface $x_3 = -0.5$. The boundary conditions (7.3) are used. The other geometrical and physical parameters are the same as earlier.

Also, as earlier we assume that the shell satisfies the thinness condition $f/a < 1/8$. According to the definition given in [229], a spherical shell is shallow if the angle subtended at the cetre by a chord which forms a side of a shell, square in plan, does not exceed 60° (in our case, for $K_1 = K_2 = 48$, this angle is 26°).

The deflections at the shell centre versus time for $q = 900$, for problems 1–4, are shown in Fig. 7.48. It can be seen from the figure that the use of physically linear and physically nonlinear models gives qualitatively different results. The maximum deflections at the shell centre for problems 3 and 4 in the first period are more than twice the deflections for problems 1 and 2; such deflections have not been observed for a shell with $K_1 = K_2 = 24$ under pre-critical loads.

It is clear that for $K_1 = K_2 = 48$ and loads close to the critical load, the use of a physically nonlinear model improves the accuracy of the results of computations. For $K_1 = K_2 = 24$, the influence of physical nonlinearity on the deflections is not significant.

In the physically nonlinear problem, coupling decreases the deflection at the shell centre by 10.5% during the first period of oscillation. In problem 3 (GPN(c)), the dynamic reaction of the shell has been traced up to $t = 1.3$ (in the figure, results for a shorter time are reported), and after the first period the deflection of the shell at its centre does not exceed 0.8 times the shell thickness. This means that the deflections are comparable to those obtained in problems 1 and 2. In the problem 4 (GPN), damping of oscillations does not occur and the values of $w(0.5; 0.5)$ in the subsequent periods are similar.

It turns out that the load $q = 900$ cannot be treated as critical in problems 3 and 4. During a small increase of the load from this value a small deflection increase was observed. The time needed to achieve the maximum deflection

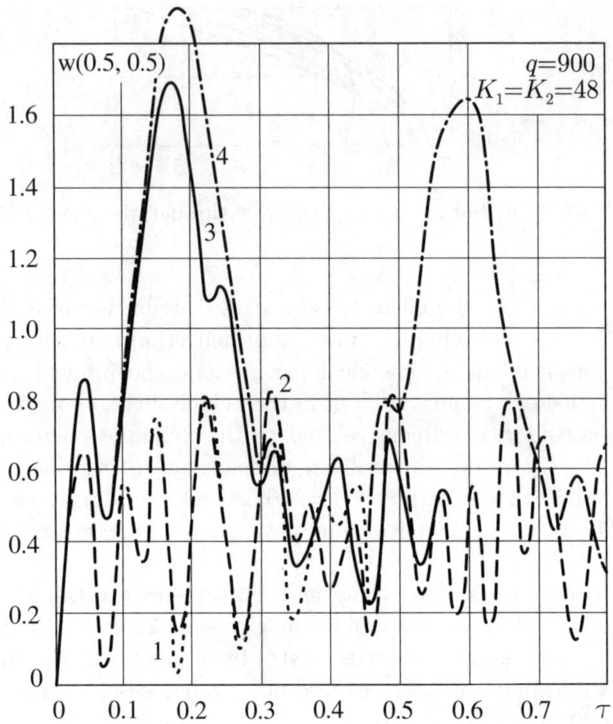

Fig. 7.48. Deflection of the shell centre for problems 1–4, $q = 900$, $K_1 = K_2 = 48$

increases and the deflection at the shell centre does not reach twice the value of the height of the shell over the plane.

Now we compare the stress–strain states of the shell in problems 3 and 4 at $t = 0.17$, when the deflections (and stresses) are maximum. The results are presented in Figs. 7.49–7.54.

In these figures (except for Figs. 7.53 and 7.54), the stress–strain state of the shell is reported for problems 1 and 2 at $t = 0.22$ and at $t = 0.17$ (a picture for $t = 0.22$ is omitted here). As Fig. 7.48 illustrates, at $t = 0.22$ the deflections in problems 1 and 2 are maximum, and the shell is working in extremely difficult circumstances; therefore this moment of time has been chosen in order to make comparisons with problems 3 and 4. The data for $t = 0.17$ illustrate the stress–strain state of the shell obtained from different models at the same moment of time. Additional information about the data is included in the figures. The results shown in the Figs. 7.49–7.52 indicate a large difference between the solutions obtained for problems 3 and 4 and for problems 1 and 2. The distributions of the various functions over the middle surface of the shell possess complex character. The values of $F_{x_1x_1}$ in problems 3 and 4 at the shell centre are more than four times smaller than in problems 1 and 2, for $t = 0.22$. At the point $(0.1875, 0.5)$, $F''_{x_1x_1}$

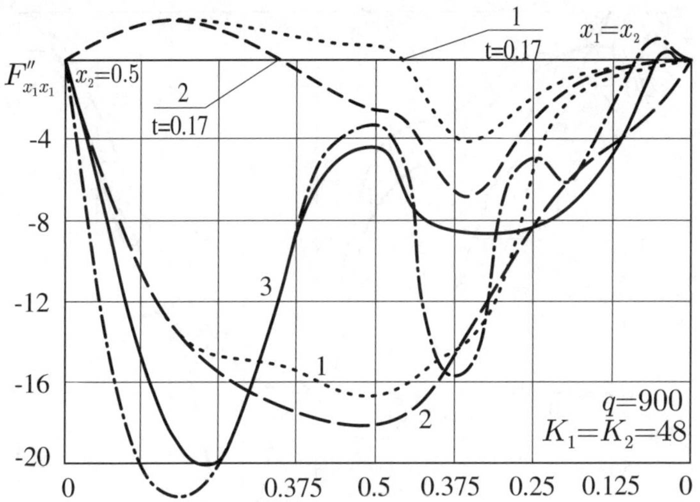

Fig. 7.49. Distribution of $F''_{x_1x_1}$, for $t = 0.17$, $q = 900$, $K_1 = K_2 = 48$

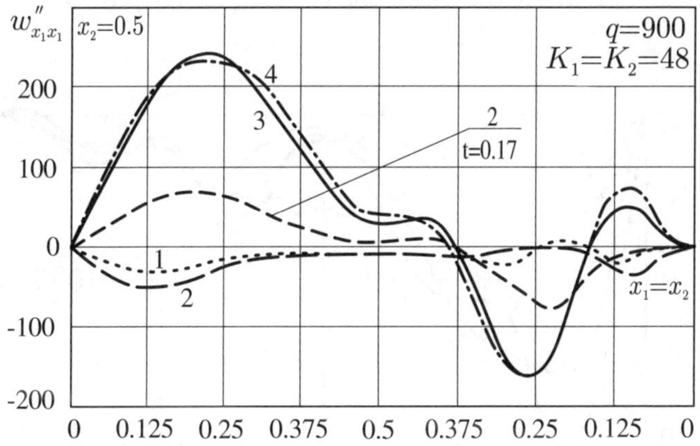

Fig. 7.50. Distribution of $W''_{x_1x_1}$, for $t = 0.17$, $q = 900$, $K_1 = K_2 = 48$

in problem 4 (GPN) is almost 36% higher than in problems 1 and 2. The difference between the value for problem 3 and those for problems 1 and 2 is 26%. A comparison of $F''_{x_1x_1}$ along the x_2 axis in problems 3 and 4 and problems 1 and 2 for $t = 0.17$ gives an even greater difference.

The distribution of $w''_{x_1x_1}$ along the axis $x_2 = 0.5$ in problems 3 and 4 also differs qualitatively from that obtained in problems 1 and 2 at the time moments chosen. Because of the complex variation of $F''_{x_1x_1}$ and $w''_{x_1x_1}$ along the diagonal $x_1 = x_2$, it is difficult to compare these functions (the difference varies from a few to a hundred percent). Also the distributions of

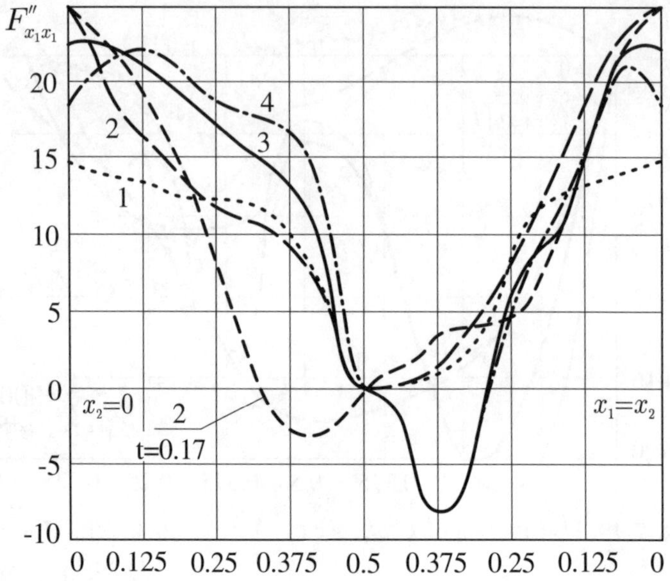

Fig. 7.51. Distribution of $F''_{x_1x_1}$, for $t = 0.17$

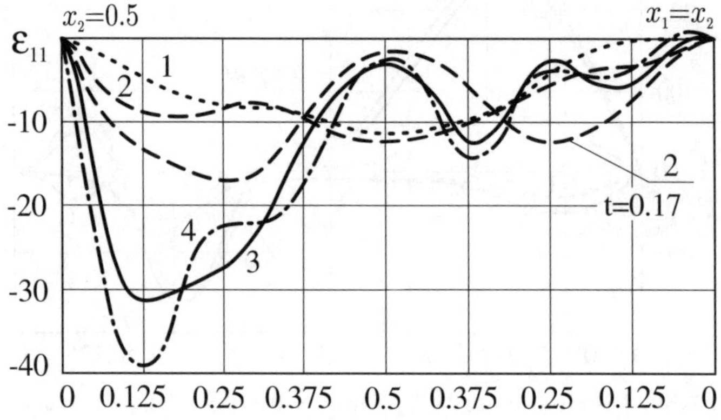

Fig. 7.52. Distribution of ε_{11}, for $t = 0.17$

$F''_{x_1x_2}$ and ε_{11} are complex and a large difference between problems 3 and 4 and problems 1 and 2 is observed.

Therefore, we draw the following conclusion. The physically linear model when used in the computation of the stress–strain state of a shell with the given curvature parameters under loads close to the critical load, leads to unacceptable errors.

The influence of coupling on the stress–strain state of the shell in problems 3 and 4 is presented in Figs. 7.49–7.54. It appears that coupling does not

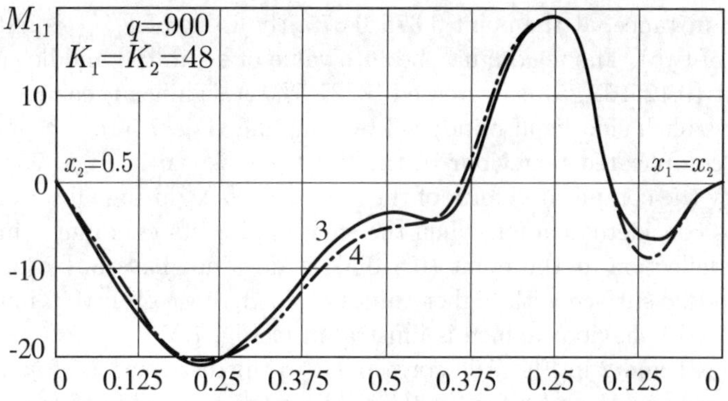

Fig. 7.53. Distribution of M_{11}, for $q = 900$, $K_1 = K_2 = 48$

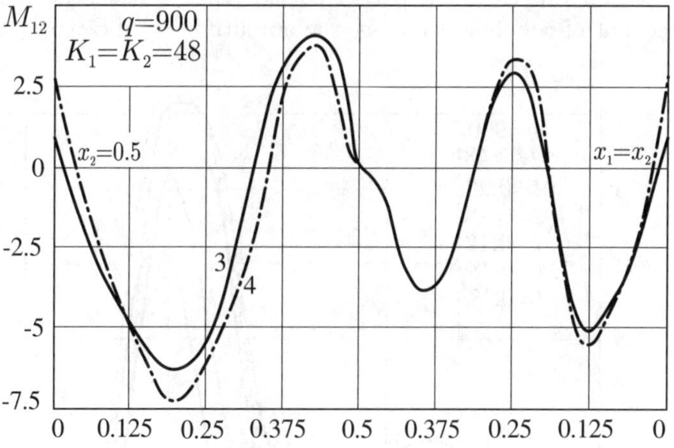

Fig. 7.54. Distribution of M_{12}, for $q = 900$, $K_1 = K_2 = 48$

change significantly the bending and rotational moments (Figs. 7.53 and 7.54), nor $w''_{x_1x_1}$ (Fig. 7.50). For instance, at the point $(0.25, 0.5)$ the difference between the values of $w''_{x_1x_1}$ for problems 3 and 4 is 4.3% (a similar observation can be made for M_{11} and M_{12}). More obvious differences occur at the points $(0.5, 0.5)$ and $(0.125, 0.125)$. The differences between the values of $w''_{x_1x_1}$ for problems 3 and 4 at these two points are 30% and 36%, respectively. The corresponding differences for m_{11} are 23% and 20%, respectively. However, at those points the values of $w''_{x_1x_1}$ and M_{11} are far from the maximum values achieved at that time moment on the middle surface of the shell.

More evident coupling effects are exhibited by the functions $F''_{x_1x_1}$ and ε_{11} (Figs. 7.49, 7.51 and 7.52). The behaviour reported here has been reported earlier. For a shell with the given parameters, when the maximum deflection is achieved the coupling decreases the maximum absolute stress $F''_{x_1x_1}$ by

14% (for instance, at the point 0.375, 0.375, the value of $F''_{x_1x_1}$ is reduced by a factor of two). The maximum absolute value of ε_{11} in the middle surface at the point (0.125, 0.125) is decreased by 21.5%. A significant change in $F''_{x_1x_2}$ of the two shell sides is also induced by coupling (Fig. 7.52).

The complicated behaviour of the functions shown in Figs. 7.49–7.54 is caused by the complicated form of the middle surface during vibration. In the problems considered earlier, when the shell reaches its maximum (in a given period) deflection at the point (0.5, 0.5), it does not have any other points of the middle surface with higher values of w. Here we observe a much more complicated behaviour, which is illustrated in Fig. 7.55.

Curves 1 and 2 in Fig. 7.55 correspond to the distribution of w along the axes in problems 1 and 2 for $t = 0.22$. The maximum value of w is achieved in the shell centre. Curves 3 and 4 correspond to the distribution of w in problems 3 and 4 when, at the shell centre, the maximum deflection has been achieved within the vibration period considered. As has been mentioned, taking account of coupling decreases the amplitude of vibration greatly not

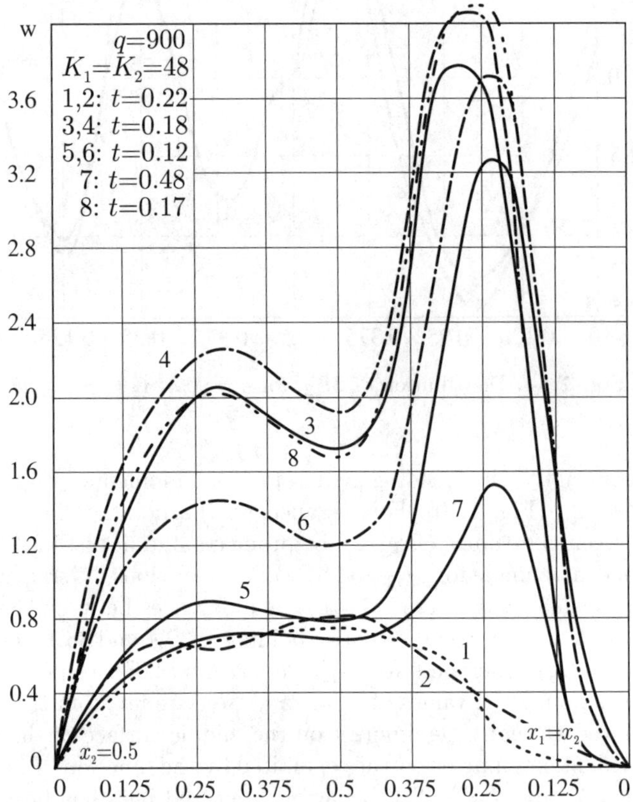

Fig. 7.55. Distribution of w for different problems and time instants, $q = 900$, $K_1 = K_2 = 48$

only in the centre but also over all the middle surface. Curves 5 and 6 correspond to the distributions for problems 3 and 4 at the time moment $t = 0.12$, and curves 7 and 8 correspond to the distributions at $t = 0.48$ and $t = 0.6$, respectively.

The deflections indicated by curves corresponding to uncoupled physically nonlinear problem practically overlap at the point $(0.25, 0.25)$, but in the centre (Fig. 7.49) they show a significant difference. In the coupled problem, the values of w in the first period of vibration (curve 3) and in the next period (curve 7) differ greatly. For $t = 0.48$, which corresponds to the highest deflection value in one of the periods, $w(0.5, 0.5)$ differs from the value related to the first period by 2.5 times, and $w(0.25, 0.25)$ differs by 2.16 times.

A significant increase of the deflection in the neighbourhood of the point $(0.25, 0.25)$ is caused by zones of plastic deformation. In order to illustrate the complicated shape of the shell deflection, contour lines linking the same deflection values for problem 3 at $t = 0.17$ have been shown in the Fig. 7.56 (a similar like behaviour has been observed for problem 4, but with slightly higher w values).

The temperature distribution is much more complicated than in the previously considered coupled problem. A few areas of heating and cooling occur on the external shell surfaces $x_3 = \pm 0.5$. As an example, the distribution of

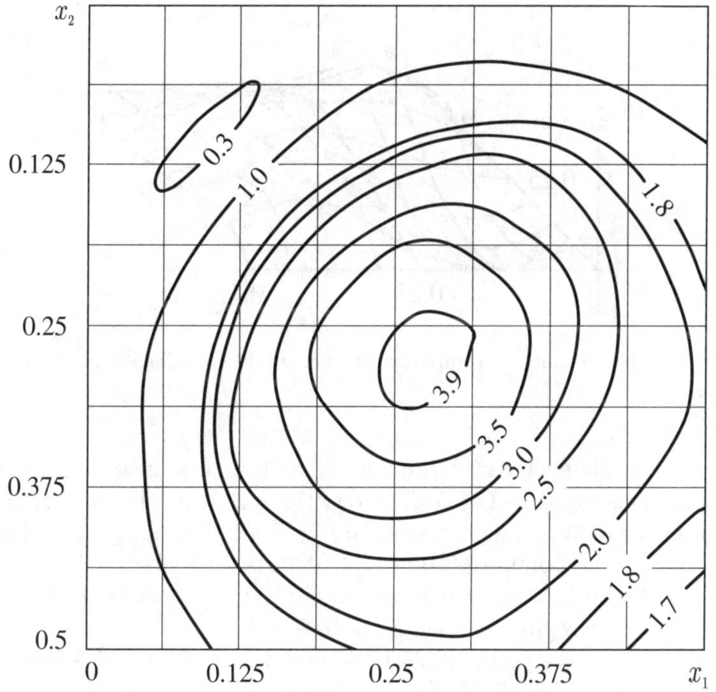

Fig. 7.56. Distribution of w for problem 3, $t = 0.17$

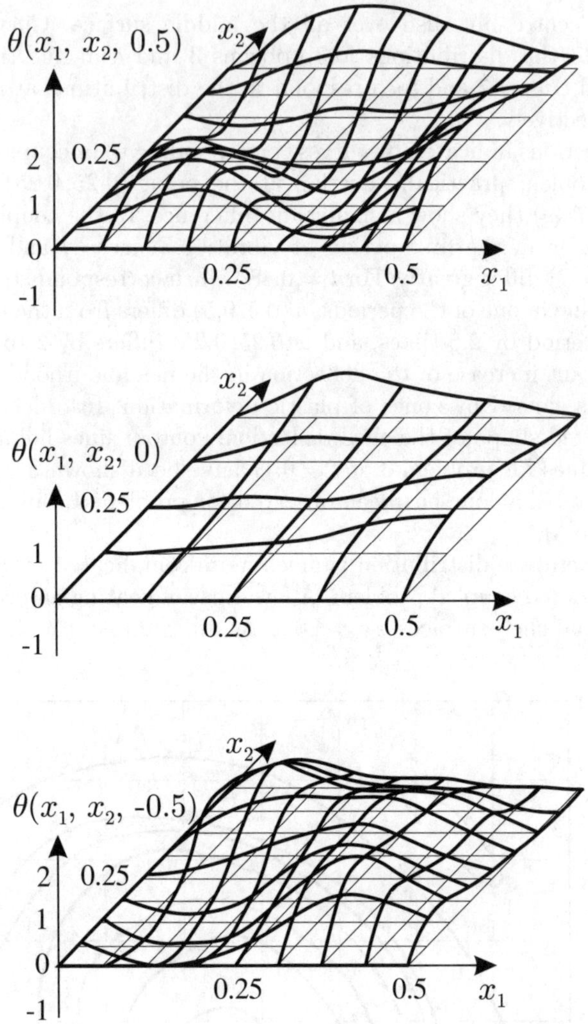

Fig. 7.57. Distribution of temperature over middle surface and through shell thickness

θ over the middle surface for problem 3, at the time instant of maximum deflection, is reported in Fig. 7.57. Now, the maximum heating of the shell does not occur at this time moment. In Fig. 7.58, the change of temperature θ with time at the points $(0.5, 0.5)$ and $(0.25, 0.25)$ (curves 1 and 2, respectively), on the surface $x_3 = 0.5$ is reported. The distribution of θ through the thickness at the time moments $t = 0.07, 0.017$ and 0.39 (curves 1, 2 and 3, respectively) at the points $(0.25; 0.25)$ (solid curves) and $(0.5, 0.5)$ (dashed curves) is shown in Fig. 7.59. The distribution of θ through thickness at the point $(0.5, 0.25)$ at the same times is presented in Fig. 7.60. The self-heating

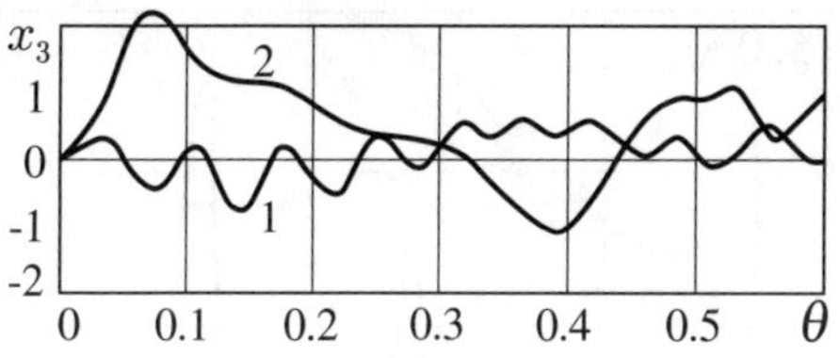

Fig. 7.58. Temperature θ on the surface $x_3 = 0.5$; curve 1, $(0.5, 0.5)$; curve 2, $(0.25, 0.25)$

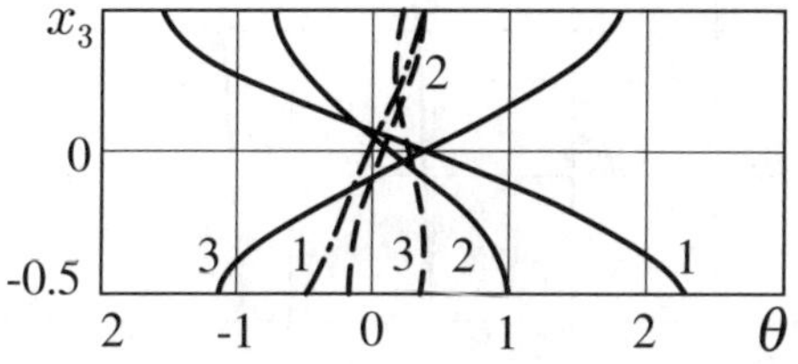

Fig. 7.59. Temperature distribution through the shell thickness for $t = 0.07,\ 0.17,\ 0.39$ (curves 1, 2, 3); solid curves $(0.25, 0.25)$; dashed curves $(0.5, 0.5)$

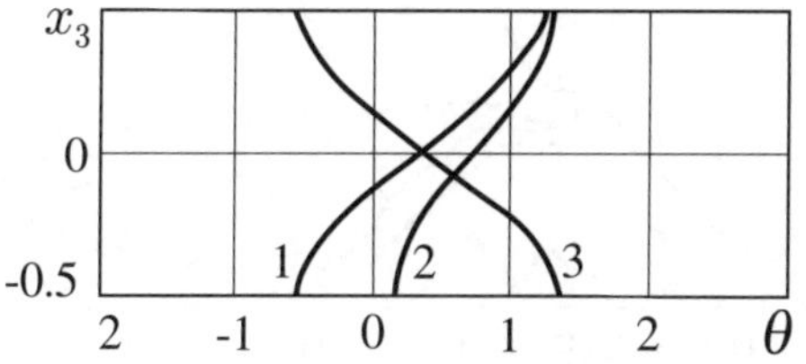

Fig. 7.60. Temperature distribution through the shell thickness for $t = 0.07,\ 0.17,\ 0.39$ (curves 1, 2, 3) at point $(0.5, 0.25)$

of the shell changes slightly. For $t = 0.07$ the self-heating reaches 12.8°, for $t = 0.17$ it reaches about 6° and for $t = 0.39$ it reaches little more than 6.5°. The maximum heating of the shell in the surface $x_3 = -0.5$ is observed at this point at $t = 0.07$ (it is equal to 7.5°).

Now we consider the influence of coupling on the stress–strain state and the self-heating of the shell with an increase of the load q up to its critical value. Our numerical experiments show that for problems 3 and 4, the critical loads q_{cr} are 945 and 925, respectively. The change of the deflection at the shell centre versus time is shown in Fig. 7.61 for problems 3 and 4 with loads equal to q_{cr} and for the pre-critical loads $q = 940$ and $q = 920$.

As earlier, coupling slightly increases the critical load for problems 3 and 4 by about 2.2% (an increase of 1.4% for problems 3 and 4 was reported in Sect. 7.3). Physically linear models (problems 1 and 2) give qualitatively different results and cannot be used to define the critical load.In Fig. 7.61, the

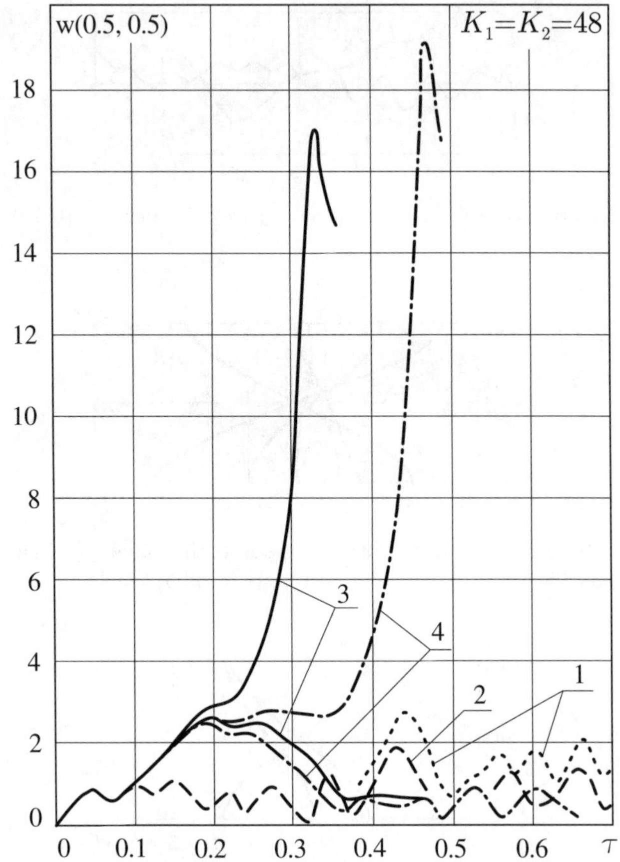

Fig. 7.61. Deflection at the shell centre for critical ($q_{cr} = 945$ and $q_{cr} = 925$) and pre-critical ($q = 940$ and $q = 920$) loads, for problems 3 and 4, and for $q = 1080$ for problems 1 and 2

shell centre deflection for problems 1 and 2 with $q = 1080$ are also presented. This means that the load is greater than q_{cr} by 14% and 17% for problems 3 and 4. The loads in problems 1 and 2 are 14% and 17% larger than those in problems 3 and 4. The dynamic reaction of the shell in problem 1 (GN(c)) has been traced until $t = 0.07$, and in problem 2 (GN) it has been traced up to $t = 1.4$, and a deflection value equal to three shell thickness has never been observed. For $q = 900$ unacceptable divergences of the results for problems 1 and 2 and problems 3 and 4 have been observed.

To conclude, it has been shown that computations in the framework of different mathematical models for a load close to the critical load lead to qualitatively different results because the load can be, depending on the model, either pre- or post-critical. Here it has been clearly shown that the use of linear models must be strictly avoided.

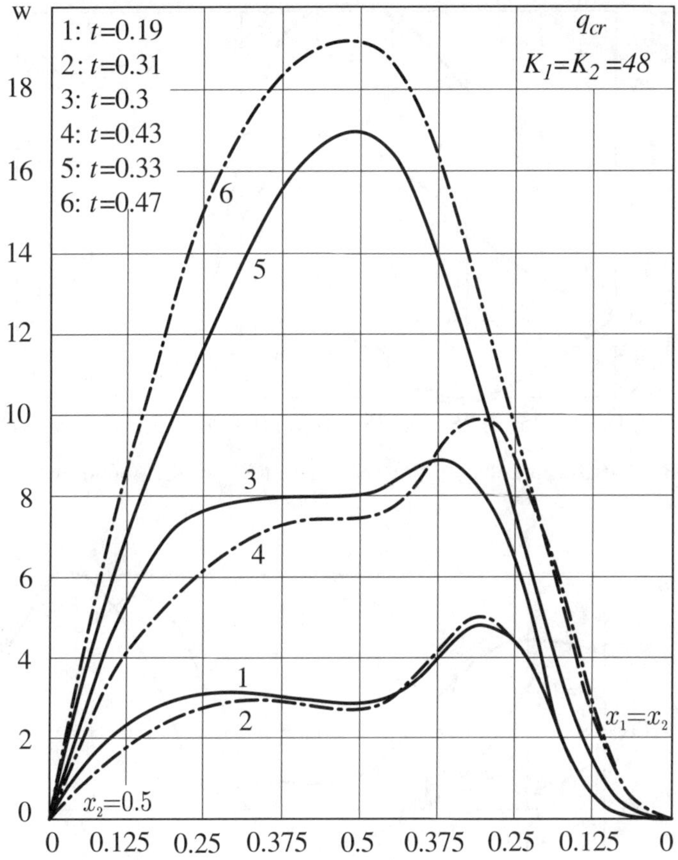

Fig. 7.62. Distribution of w during jumping

Let us compare the stress–strain states of the shell obtained from calculations for problems 3 and 4 at the critical load. The shell deflections during the jumping process at different time moments are presented in Fig. 7.62. Curves 1, 3 and 5 show the deflections for problem 3 at the moments $t = 0.19, 0.3$ and 0.33, respectively. Curves 2, 4 and 6 correspond to problem 4 at $t = 0.312, 0.43, 0.47$. The coupling decreases the maximum deflection by 11.5% in spite of the fact that q_{cr} for the coupled problem exceeds q_{cr} for the uncoupled one.

In Figs. 7.63 and 7.64, the distributions of $F''_{x_1x_1}$, $w''_{x_1x_1}$ along the symmetry axes of the shell for problems 3 and 4 at the moments corresponding to the maximum deflections are shown. In addition in the Fig. 7.63, the values of $F''_{x_1x_1}$ for problems 3 and 4 at $t = 0.14$ are reported. The latter results illustrate the fact that coupling, in the initial stages of jumping, practically no influence on the stress–strain state. However, when the shell reaches its maximum deflection, the coupling has an important effect.

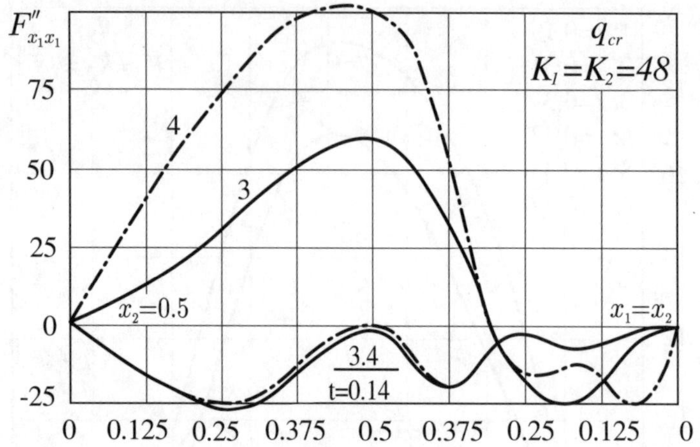

Fig. 7.63. Distribution of $F''_{x_1x_1}$ at the time instants corresponding to maximum deflection

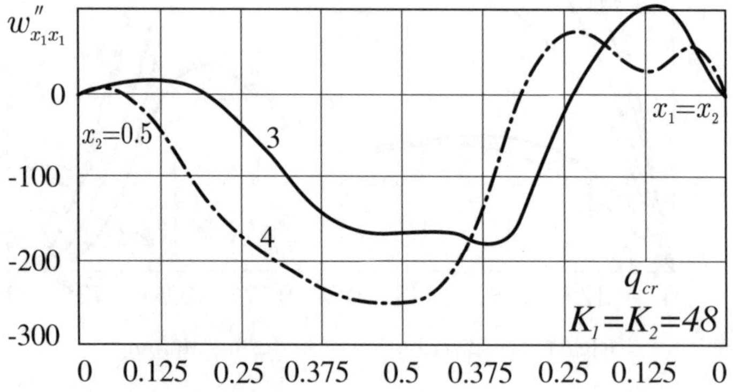

Fig. 7.64. Distribution of $w''_{x_1x_1}$ at the time instants corresponding to maximum deflection

Analysing, in addition, Figs. 7.65 and 7.66, where the distributions of $F''_{x_1x_2}$ and $w''_{x_1x_2}$ are reported, one can conclude that for the problem considered in this section, coupling greatly decreases the maximum values of $F''_{x_1x_1}$, $w''_{x_1x_1}$, $F''_{x_1x_2}$, $w''_{x_1x_2}$ at the time when the shell reaches its maximum deflection. The following decrease are obtained: for $F''_{x_1x_1}$ at the point $(0.5, 0.5)$, 14%; for $w''_{x_1x_1}$ at the point $(0.5, 0.5)$, 33%.

For $F''_{x_1x_2}$, at the point $(0, 0.25)$ the decrease is 70%, whereas at the point $(0.25, 0.25)$, it is 35%. For $w''_{x_1x_2}$, at the point $(0.2, 0.2)$, the difference is 14%. Even larger differences in $w''_{x_1x_2}$ and $F''_{x_1x_2}$ have been observed at other points of the shell. The value of $F''_{x_1x_2}$ reported in Fig. 7.65 for problems 3 and 4 at $t = 0.14$ prove again that, at the initial moment of shell jumping, coupling induces negligible changes.

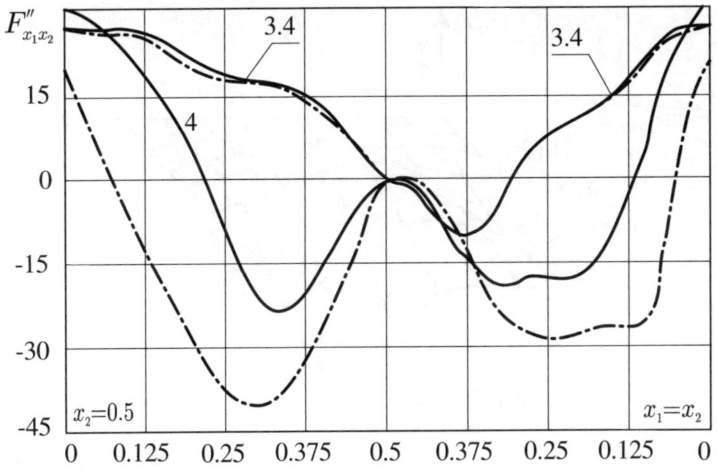

Fig. 7.65. Distribution of $F''_{x_1x_2}$ for time instants corresponding to maximal deflections

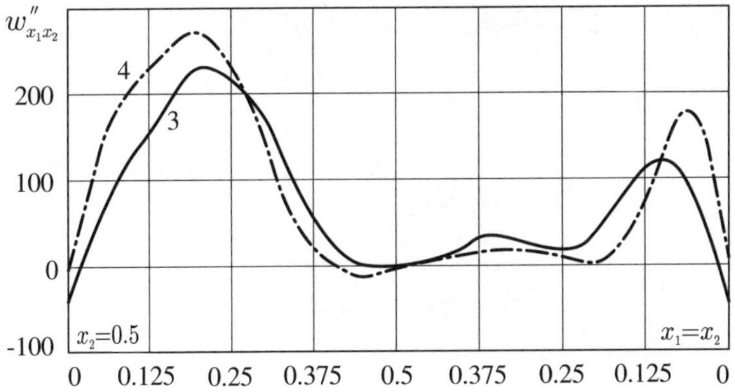

Fig. 7.66. Distribution of $w''_{x_1x_2}$ for time instants corresponding to maximal deflections

However, coupling greatly changes the temperature during jumping phenomena. In Fig. 7.67, the distribution of the temperature θ over the middle surface at the moment of maximum deflection is shown. In the centre of the shell, at this time, a relatively large area of cooling occurs owing to a sudden increase of the volume expansion e. The fall of θ at this time in the centre of the shell is almost 26°.

In Figs. 7.68–7.70, the dependence of θ on time on the surface $x_3 = -0.5$ (solid curves) and $x_3 = 0.5$ (dashed curves) at the points $(0.5, 0.5)$, $(0.5, 0.2)$ and $(0.1875, 0.0625)$ are reported.

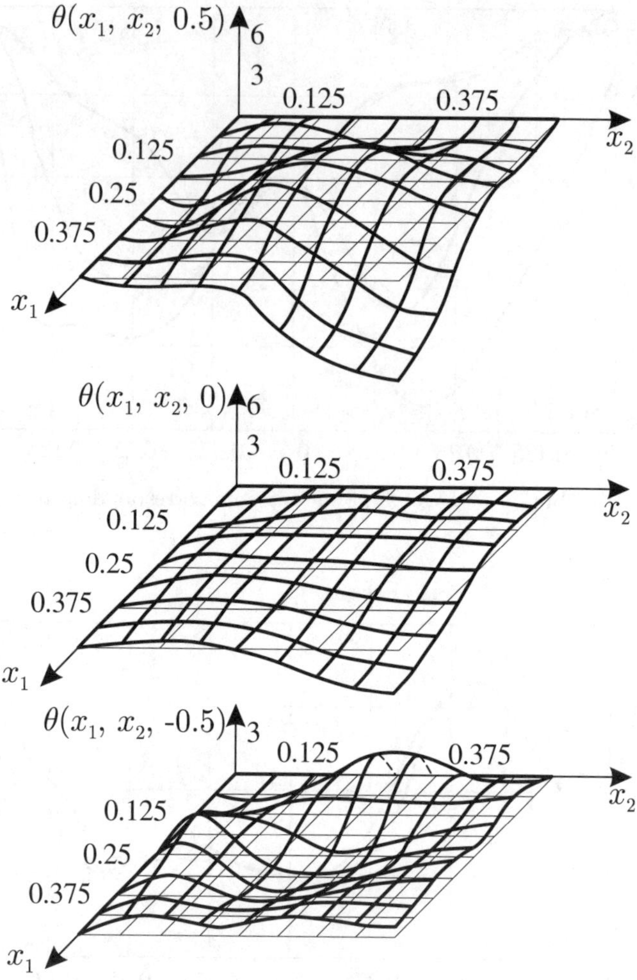

Fig. 7.67. Distribution of temperature θ over the shell for time instant corresponding to maximum deflection

The figures show the appearance of large temperature gradients at the moment of the sudden increase of shell deflection. For instance, at the point $(0.5, 0.25)$ at $t = 0.28$, the fall in θ is $28.6°$, and the fall in θ shown in Fig. 7.70 at $t = 0.35$ is even more larger and is over $35°$. Therefore, in the problem under consideration during shell jumping a zone of cooling appears in the central part, which extends from the surface $x_3 = 0.5$ to the surface $x_3 = -0.125$. Relatively large gradients of θ occur in the neighbourhood of the sides of the shell, as a result of self-heating on the surface $x_3 = -0.5$ and cooling on the surface $x_3 = 0.5$.

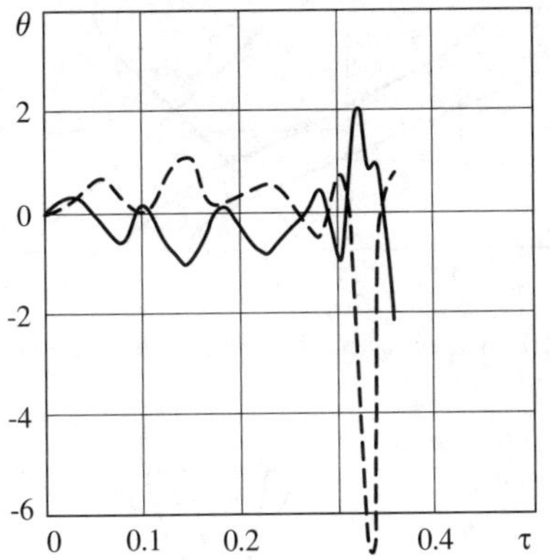

Fig. 7.68. Temperature θ versus time for the point $(0.5, 0.5)$: *solid curve*, $x_3 = -0.5$; *dashed curve*, $x_3 = 0.5$

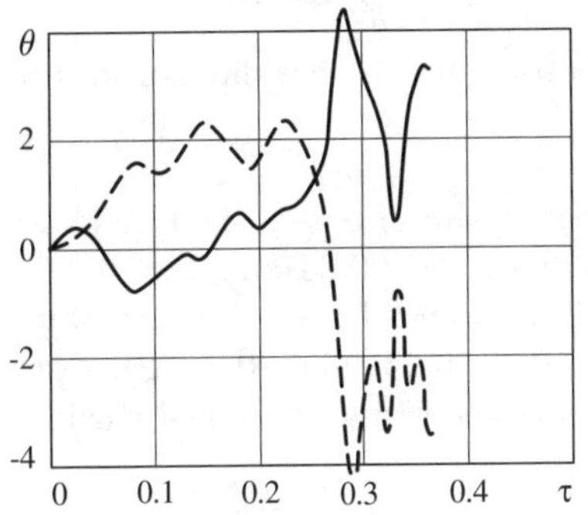

Fig. 7.69. Temperature θ versus time for the point $(0.5, 0.2)$: *solid curve*, $x_3 = -0.5$; *dashed curve*, $x_3 = 0.5$

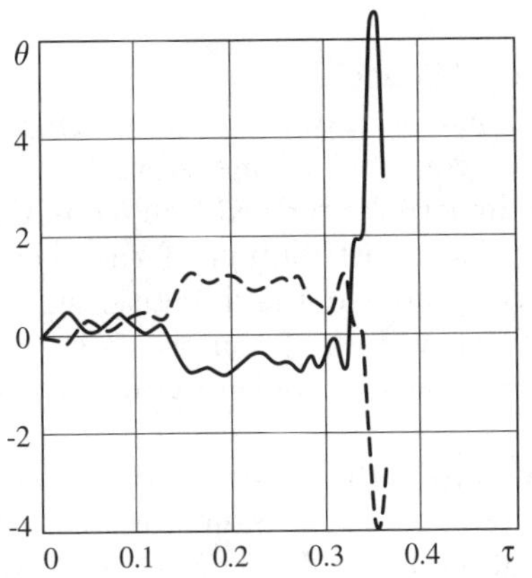

Fig. 7.70. Temperature θ versus time for the point $(0.1875, 0.0625)$: *solid curve*, $x_3 = -0.5$; *dashed curve*, $x_3 = 0.5$

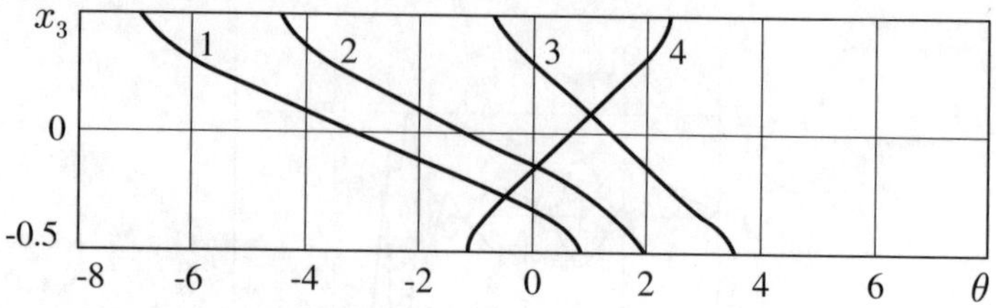

Fig. 7.71. Temperature distribution through shell thickness for $t = 0.33$

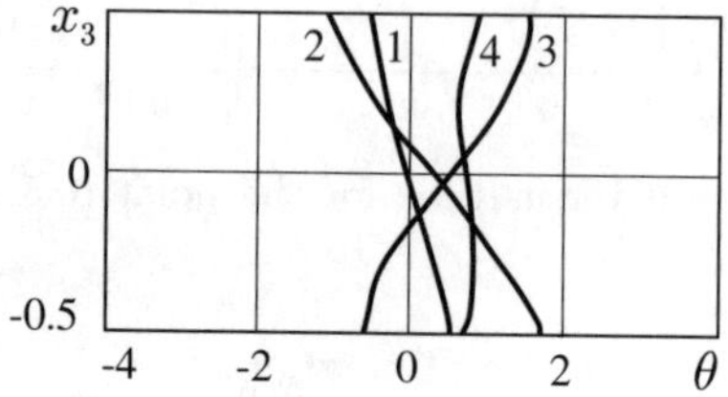

Fig. 7.72. Temperature distribution through shell thickness for $t = 0.28$

The distribution of θ through the shell thickness at some points are reported for time moments $t = 0.33$ and $t = 0.28$ in Figs. 7.71 and 7.72, respectively. Curves 1, 2, 3 and 4 show the distributions of θ at the points $(0.5, 0.5)$, $(0.5, 0.2)$, $(0.0625, 0.25)$ and $(0.2, 0.2)$, respectively. This distribution qualitatively repeats the results obtained earlier.

7.6 Influence of Coupling and Rotational Inertia on Stability

In previous chapters of this book, the problem of the stress–strain state of a shallow shell in a nonuniformly distributed temperature field has been stated. The problem is governed by a system of nonlinear, coupled equations of parabolic and hyperbolic type with different dimensions. It is caused essentially by the nonlinear distribution of the thermal field through the shell thickness. We have proposed and discussed methods of solution to this problem and their efficiency. In addition the existence of a solution to the nonlinear coupled thermoelastic problem within a Timoshenko-type model has been proved.

In this section, we prove the existence of a solution to the nonlinear coupled thermoelastic problem in the framework of the Kirchhoff–Love model including the rotational inertia of the shell elements. In addition, we present a numerical algorithm to solve problems of this class, which will be illustrated by numerical examples.

The physically linear equations of shell theory with coupling between the temperature and deformation fields read as follows (the equations are given in the general form for a physically nonlinear material in Sect. 6.3):

$$A(w) \equiv \rho h \frac{\partial^2 w}{\partial t^2} + D\Delta^2 w + \frac{E\alpha_T}{1-\nu}\Delta\left\{\int_{-\frac{h}{2}}^{\frac{h}{2}} x_3\theta\, dx_3\right\}$$

$$-\rho\frac{h^3}{12}\Delta\left(\frac{\partial^2 w}{\partial t^2}\right) - k_1\frac{Eh}{1-\nu^2}[\varepsilon_{11}+\nu\varepsilon_{22}]$$

$$-k_2\frac{Eh}{1-\nu^2}[\varepsilon_{22}+\nu\varepsilon_{11}] + (k_1+k_2)\frac{E\alpha_T}{1-\nu}\int_{-\frac{h}{2}}^{\frac{h}{2}}\theta\, dx_3$$

$$-\frac{\partial}{\partial x_1}\left(\frac{Eh}{1-\nu^2}[\varepsilon_{11}+\nu\varepsilon_{22}]\frac{\partial w}{\partial x_1} + \frac{Eh}{2(1+\nu)}\varepsilon_{12}\frac{\partial w}{\partial x_2}\right)$$

$$-\frac{\partial}{\partial x_2}\left(\frac{Eh}{1-\nu^2}[\varepsilon_{22}+\nu\varepsilon_{11}]\frac{\partial w}{\partial x_2} + \frac{Eh}{2(1+\nu)}\varepsilon_{12}\frac{\partial w}{\partial x_1}\right)$$

$$+\frac{E\alpha_T}{1-\nu}\left(\frac{\partial}{\partial x_1}\left\{\frac{\partial w}{\partial x_1}\int_{-\frac{h}{2}}^{\frac{h}{2}}\theta\, dx_3\right\} + \frac{\partial}{\partial x_2}\left\{\frac{\partial w}{\partial x_2}\int_{-\frac{h}{2}}^{\frac{h}{2}}\theta\, dx_3\right\}\right) = q_1,$$

$$B(u) \equiv \rho h\frac{\partial^2 u}{\partial t^2} - \frac{Eh}{1-\nu^2}\frac{\partial}{\partial x_1}[\varepsilon_{11}+\nu\varepsilon_{22}]$$

$$+\frac{E\alpha_T}{1-\nu}\frac{\partial}{\partial x_1}\left\{\int_{-\frac{h}{2}}^{\frac{h}{2}}\theta\, dx_3\right\} - \frac{Eh}{2(1+\nu)}\frac{\partial w}{\partial x_2}\varepsilon_{12} = p_1, \tag{7.46}$$

$$C(v) \equiv \rho h\frac{\partial^2 v}{\partial t^2} - \frac{Eh}{1-\nu^2}\frac{\partial}{\partial x_2}[\varepsilon_{22}+\nu\varepsilon_{11}]$$

$$+\frac{E\alpha_T}{1-\nu}\frac{\partial}{\partial x_2}\left\{\int_{-\frac{h}{2}}^{\frac{h}{2}}\theta\, dx_3\right\} - \frac{Eh}{2(1+\nu)}\frac{\partial w}{\partial x_1}\varepsilon_{12} = p_2,$$

$$D_1(\theta) \equiv C\frac{\partial\theta}{\partial t} - k_T\Delta\theta + \frac{E\alpha_T T_0}{1-\nu}[\varepsilon_{11}+\varepsilon_{22}-x_3\Delta w] = q_2,$$

where ε_{11}, ε_{22}, ε_{12} have the form (6.2) and (6.3), $k_1, k_2 = const \geq 0$.

The following initial and boundary conditions are applied:

$$
\begin{aligned}
&w\big|_{t=0} = \varphi_1(x_1,x_2),\ u\big|_{t=0} = \varphi_2(x_1,x_2),\ v\big|_{t=0} = \varphi_3(x_1,x_2),\\
&\theta\big|_{t=0} = \varphi_4(x_1,x_2,x_3),\ \left.\frac{\partial w}{\partial t}\right|_{t=0} = \Psi_1(x_1,x_2),\\
&\left.\frac{\partial u}{\partial t}\right|_{t=0} = \Psi_2(x_1,x_2),\ \left.\frac{\partial v}{\partial t}\right|_{t=0} = \Psi_3(x_1,x_2),\\
&w\big|_{\Gamma} = \left.\frac{\partial w}{\partial n}\right|_{\Gamma} = u\big|_{\Gamma} = \nu\big|_{\Gamma} = 0,\ \theta_S\big|_S = 0.
\end{aligned}
\tag{7.47}
$$

Here and in what follows $t \in [0;T]$; Ω_1 is the space in which x_1 and x_2 vary; $\Omega_1 \subset R^2$; $\overline{\Omega}_1 = \Omega_1 \cup \partial\Omega_1$; $\Omega_1 = \Omega_1 \times (0 \times T)$; $\Gamma = \partial\Omega_1 \times [0,T]$; $\Omega_2 = \Omega_1 \times (-\frac{h}{2}, \frac{h}{2})$; $\overline{\Omega}_2 = \Omega_2 \cup \partial\Omega_2$; $Q_2 = Q_2 \times (0 \times T)$; $S = \partial\Omega_2 \times [0,T]$; E, h, ρ, c, k_T, α_T, T_0, ν, D are positive constants (all much greater than 0); $0 < \nu \ll 1/2$; $w(x_1,x_2,t)$, $u(x_1,x_2,t)$, $v(x_1,x_2,t)$, $\theta(x_1,x_2,x_3,t)$ are the sought functions; and $g_1(x_1,x_2,t)$, $p_1(x_1,x_2,t)$, $p_2(x_1,x_2,t)$, $g_2(x_1,x_2,t)$ are given functions; $\Delta^2 = \Delta\Delta$ is the biharmonic operator; and Δ is the Laplace operator (two- or three-dimensional).

We introduce the space $L^P(0,T;X)$, which is the space of functions $f(t)$ with values in the space X measured through the Lebesgue measure dt in the interval $[0,T]$ and having a finite norm $\|f\| = [\int_0^T \|f\|_x^p\, dt]^{1/p}$, where $\|f\|_x$ is the norm of the elements in the space X (various spaces will be used as X) [167].

For the problem (7.46), (7.47) formulated above, the following theorem holds:

Theorem 7.1 *Let $\partial\Omega_1$ be a sufficiently smooth boundary of Ω_1; let $g_1, p_1, p_2 \in L^2(0,T;L_2(\Omega_1))$, $g_2 \in L^2(0,T;L_2(\Omega_1))$, $\varphi_1 \in \mathring{W}_2^2(\Omega_1)4$, $\varphi_2, \varphi_3 \in \mathring{W}_2^1(\Omega_1)$, $\varphi_4 \in \mathring{W}_2^1(\Omega_2)$, $\Psi_1 \in \mathring{W}_2^1(\Omega_2)$, and $\Psi_2, \Psi_3 \in L_2(\Omega_1)$. Then:*

1. *There exists at least one choice from the four functions $\{w,u,v,\theta\}$ that is a solution to the problem (7.46), (7.47), and the following conditions are satisfied:*

$$
w \in L^2(0,T;\mathring{W}_2^2(\Omega_1)),\quad \frac{\partial w}{\partial t} \in L^2(0,T;W_2^1(\Omega_1)),
\tag{7.48}
$$

$$
u,v \in L^2(0,T;\mathring{W}_2^1(\Omega_1)),\quad \frac{\partial u}{\partial t}, \frac{\partial v}{\partial t} \in L^2(0,T;L_2(\Omega_1)),
\tag{7.49}
$$

$$
\theta \in L^2(0,T;\mathring{W}_2^1(\Omega_2)).
\tag{7.50}
$$

2. *An approximate solution to the problem defined above can be found with the help of the Bubnov–Galerkin (Bubnov–Galerkin) method in the form*

$$\begin{aligned} w_n &= \sum_{i=1}^{n} g_{1i}(t)X_{1i}(x_1,x_2), \quad u_n = \sum_{i=1}^{n} g_{2i}(t)X_{2i}(x_1,x_2), \\ v_n &= \sum_{i=1}^{n} g_{3i}(t)X_{2i}(x_1,x_2), \quad \theta_n = \sum_{i=1}^{n} g_{4i}(t)X_{3i}(x_1,x_2), \end{aligned} \tag{7.51}$$

where $\{X_{1i}\}$ *is a basis system in* $\mathring{W}_2^2(\Omega_1)$*, orthonormalized through the norm* $W_2^1(\Omega_1)$*;* $\{X_{2i}\}$ *is a basis function in* $\mathring{W}_2^1(\Omega_1)$*, orthonormalized in* $L_2(\Omega_1)$*; and* $\{X_{3i}\}$ *is a basis function in* $\mathring{W}_2^1(\Omega_2)$*, orthonormalized in* $L_2(\Omega_2)$*. In addition, we assume that* $\forall i$*,* $X_{3i} = a_i(x_1,x_2)b_i(x_2)$ *(note that the existence of such a basis system implies result given in [166]). We can find the unknown* $g_{ji}(t)$*,* $j = 1,2,3,4$*, from the following system of ODEs given below, which is solvable for* $\forall n$ *in the whole interval* $[0,T]$*.*

3. *The whole set of approximate solutions* $\{w_n, u_n, v_n, \theta_n\}$ *is weakly compact in the spaces corresponding to* (7.48)–(7.50).

We present here only the fundamental steps of the proof, based on the method of compactness [147].

Step 1. Construction of the approximating solution using the Bubnov–Galerkin method. We substitute the functions (7.51) into the system (7.46), (7.47) and find the unknown $g_{ji}(t)$, $j = 1,2,3,4$, from the following system of ODEs:

$$\begin{gathered} (A(w_n) - g_1, X_{1k})_{L_2(\Omega_1)} = 0, \\ (B(u_n) - p_1, X_{2k})_{L_2(\Omega_1)} = 0, \\ (C(v_n) - p_2, X_{2k})_{L_2(\Omega_1)} = 0, \\ (D_1(\theta_n) - g_2, X_{3k})_{L_2(\Omega_2)} = 0 \\ (k = 1,2,\ldots,n), \end{gathered} \tag{7.52}$$

with the boundary conditions

$$\begin{aligned} g_{1k}(0) &= (\varphi_1, X_{1k})_{L_2(\Omega_1)}, \quad g_{2k}(0) = (\varphi_2, X_{2k})_{L_2(\Omega_1)}, \\ g_{3k}(0) &= (\varphi_3, X_{3k})_{L_2(\Omega_1)}, \quad g_{4k}(0) = (\varphi_4, X_{3k})_{L_2(\Omega_2)}, \\ \frac{dg_{1k}(0)}{dt} &= (\Psi_1, X_{1k})_{L_2(\Omega_1)}, \quad \frac{dg_{2k}(0)}{dt} = (\Psi_2, X_{2k})_{L_2(\Omega_1)}, \\ \frac{dg_{3k}(0)}{dt} &= (\Psi_3, X_{2k})_{L_2(\Omega_1)}, \quad k = 1,2,\ldots,n. \end{aligned} \tag{7.53}$$

Similarly to the Peano theorem ([98]) on the existence of a solution to the Cauchy problem, the existence of a solution to the problem (7.52), (7.53) on $[0,T_l], T_l \leq T$, in the space $W_2^2(0,T_l)$, can be proved.

Step 2. Prior estimates. In order to obtain prior estimates, we multiply each of the equations in (7.52) by $\dfrac{dg_{1k}(0)}{dt}$, $\dfrac{d2_{1k}(0)}{dt}$, $\dfrac{dg_{3k}(0)}{dt}$ and $g_{4k}(0)$, respectively, sum the resulting over k and add the four equations obtained. We then obtain

$$
\begin{aligned}
&\frac{1}{2}\frac{d}{dt}\left\{\rho h\left(\left|\frac{\partial w_n}{\partial t}\right|^2_{L_2(\Omega_1)}+\left|\frac{\partial u_n}{\partial t}\right|^2_{L_2(\Omega_1)}+\left|\frac{\partial v_n}{\partial t}\right|^2_{L_2(\Omega_1)}\right)\right.\\
&+\rho\frac{h^3}{12}\left|\frac{\partial w_n}{\partial t}\right|^2_{\overset{\circ}{W}{}_2^1(\Omega_1)}+D\left|w_n\right|^2_{\overset{\circ}{W}{}_2^2(\Omega_1)}+\frac{C}{T_0}\left|\theta_n\right|^2_{L_2(\Omega_2)}\\
&\left.+\int\limits_{\Omega_1}\left(\frac{Eh}{1-\nu^2}[\varepsilon_{11}^2+\varepsilon_{22}^2+2\nu\varepsilon_{11}\varepsilon_{22}]+\frac{Eh}{2(1+\nu)}\varepsilon_{12}^2\right)d\Omega_1\right\} \qquad (7.54)\\
&+\frac{k_T}{T_0}\left|\theta_n\right|^2_{\overset{\circ}{W}{}_2^1(\Omega_1)}=\int\limits_{\Omega_1}\left(g_1\frac{\partial w_n}{\partial t}+p_1\frac{\partial u_n}{\partial t}+p_2\frac{\partial v_n}{\partial t}\right)d\Omega_1\\
&+\int\limits_{\Omega_2} g_2\theta_n\,d\Omega_2.
\end{aligned}
$$

Denoting the right-hand side of (7.54) by R, we estimate it using the Cauchy inequality:

$$
\begin{aligned}
R\le\;&\frac{1}{2\rho h}\int\limits_{\Omega_1}\left|(g_1)^2+(p_1)^2+(p_2)^2\right|d\Omega_1+\frac{T_0}{2C}\int\limits_{\Omega_1}(g_2)^2\,d\Omega_2\\
&+\frac{1}{2}\rho h\left(\left|\frac{\partial w_n}{\partial t}\right|^2_{L_2(\Omega_1)}+\left|\frac{\partial u_n}{\partial t}\right|^2_{L_2(\Omega_1)}+\left|\frac{\partial v_n}{\partial t}\right|^2_{L_2(\Omega_1)}\right)\\
&+\frac{1}{2}\frac{C}{T}\left|\theta_n\right|^2_{L_2(\Omega_2)}+\frac{1}{2}\rho\frac{h^3}{12}\left|\frac{\partial w_n}{\partial t}\right|^2_{\overset{\circ}{W}{}_2^1(\Omega_1)}+\frac{1}{2}D\left|w_n\right|^2_{\overset{\circ}{W}{}_2^2(\Omega_1)} \qquad (7.55)\\
&+\int\limits_{\Omega_1}\left(\frac{Eh}{1-\nu^2}[\varepsilon_{11}^2+\varepsilon_{22}^2+2\nu\varepsilon_{11}\varepsilon_{22}]+\frac{Eh}{2(1+\nu)}\varepsilon_{12}^2\right)d\Omega_1.
\end{aligned}
$$

The last three terms in (7.55) have been added for convenience of later considerations. Integrating (7.54) with respect the variable t over an interval $(0,\tau)$, $\tau\le T_1$, taking (7.55) into account and using Gronwill's lemma [98] the following inequality is obtained:

$$\rho h\left(\left|\frac{\partial w_n}{\partial t}\right|^2_{L_2(\Omega_1)}+\left|\frac{\partial u_n}{\partial t}\right|^2_{L_2(\Omega_1)}+\left|\frac{\partial v_n}{\partial t}\right|^2_{L_2(\Omega_1)}\right)$$
$$+\rho\frac{h^3}{12}\left|\frac{\partial w_n}{\partial t}\right|^2_{\overset{\circ}{W}{}^1_2(\Omega_1)}+D\left|w_n\right|^2_{\overset{\circ}{W}{}^2_2(\Omega_1)}+\frac{C}{T}\left|\theta_n\right|^2_{L_2(\Omega_2)} \tag{7.56}$$
$$\int\limits_{\Omega_1}\left(\frac{Eh}{1-\nu^2}[\varepsilon_{11}^2+\varepsilon_{22}^2+2\nu\varepsilon_{11}\varepsilon_{22}]+\frac{Eh}{2(1+\nu)}\varepsilon_{12}^2\right)d\Omega_1<\beta,$$
$$\beta=const>0.$$

To obtain the constant β, Lemma 1 from [167] is used. Returning to (7.54) and taking into account (7.56), obtain

$$\frac{k_T}{T_0}\int\limits_0^T\left|\theta_n\right|^2_{\overset{\circ}{W}{}^1_2(\Omega_1)}dt\le\beta. \tag{7.57}$$

From the last term of the left-hand side of the inequality (7.56), the following estimates are obtained using the Cauchy inequality:

$$\int\limits_{\Omega_1}\left(\frac{\partial u_n}{\partial x_1}\right)^2 d\Omega_1\le\beta,\quad \int\limits_{\Omega_1}\left(\frac{\partial v_n}{\partial x_2}\right)^2 d\Omega_1\le\beta,$$
$$\int\limits_{\Omega_1}\left(\frac{\partial u_n}{\partial x_2}+\frac{\partial v_n}{\partial x_1}\right)d\Omega_1\le\beta.$$

After applying the Korn inequality, these estimates yield

$$\left|u_n\right|^2_{\overset{\circ}{W}{}^1_2(\Omega_1)}\le\beta,\quad \left|v_n\right|^2_{\overset{\circ}{W}{}^1_2(\Omega_1)}\le\beta. \tag{7.58}$$

The inequalities (7.56)–(7.58) are prior estimates of the solution to the problem (7.47), (7.48). Using these estimates, the solution to the system (7.52), (7.53) can be extended to the whole interval $[0,T]$.

Step 3. Limiting transition. As is known, an arbitrary bounded set in a Hilbert space is weakly compact. Consequently, using the prior estimates (7.56)–(7.58), one can extract the following subseries:

$$\frac{\partial u_{n_k}}{\partial t}\to\frac{\partial u}{\partial t},\ \frac{\partial v_{n_k}}{\partial t}\to\frac{\partial v}{\partial t}\quad \text{weak in } L^2(0,T;L_2(\Omega_1)),$$
$$\frac{\partial w_{n_k}}{\partial t}\to\frac{\partial w}{\partial t}\quad \text{weak in } L^2(0,T;\overset{\circ}{W}{}^1_2(\Omega_1)),$$
$$w_{n_k}\to w\quad \text{weak in } L^2(0,T;\overset{\circ}{W}{}^2_2(\Omega_1)),$$
$$\theta_{n_k}\to\theta\quad \text{weak in } L^2(0,T;\overset{\circ}{W}{}^1_2(\Omega_1)), \tag{7.59}$$
$$u_{n_k}\to u,\ v_{n_k}\to v\quad \text{weak in } L^2(0,T;\overset{\circ}{W}{}^1_2(\Omega_1)).$$

A limiting transition in the spaces related to (7.48)–(7.50) can be performed using (7.59) for the hyperbolic-type equations (7.46) similarly to the liming transition presented in [147]. Therefore, we shall examine only one term in the parabolic-type system. For an arbitrary function $\Phi = \sum_{i=1}^{n} \alpha_i(t) \otimes X_{3i}(x_1, x_2, x_3)\ \forall i\ \alpha_i(t) \in C'([0,T]),\ \alpha_i(t) = 0$, we have

$$\begin{aligned}
&\int_0^T \int_{\Omega_1} \int_{-\frac{h}{2}}^{\frac{h}{2}} \left(\frac{\partial^2 w_n}{\partial t\, \partial x_1} - \frac{\partial w_n}{\partial x_1}\Phi - \frac{\partial^2 w_n}{\partial t\, \partial x_1} - \frac{\partial w_n}{\partial x_1}\Phi \right) dx_3\, d\Omega_1\, dt \\
&= \int_0^T \int_{\Omega_1} \int_{-\frac{h}{2}}^{\frac{h}{2}} \left\{ \left(\frac{\partial w_n}{\partial x_1} - \frac{\partial w}{\partial x_1} \right) \frac{\partial^2 w_n}{\partial t\, \partial x_1} + \left(\frac{\partial^2 w_n}{\partial t\, \partial x_1} \right.\right. \\
&\left.\left. - \frac{\partial^2 w}{\partial t\, \partial x_1} \right) \frac{\partial w}{\partial x_1} \right\} \Phi\, dx_3\, d\Omega_1\, dt \leq \left| \frac{\partial^2 w}{\partial t v \partial x_1} \right|_{L_2(Q_1)} \left| \frac{\partial w_n}{\partial x_1} \right. \\
&\left. - \frac{\partial w}{\partial x_1} \right|_{L_2(Q_1)} \left| \Phi_1 \right|_{L_2(Q_1)} + \int_0^T \int_{\Omega_1} \left(\frac{\partial^2 w_n}{\partial t\, \partial x_1} - \frac{\partial^2 w}{\partial t\, \partial x_1} \right) \frac{\partial w}{\partial x_1} \Phi_1\, d\Omega_1\, dt.
\end{aligned} \tag{7.60}$$

Note that, owing to the special choice of the basis function $X_{3i}(x_1, x_2, x_3)$, we first integrate the components of Φ along x_3 (because there is no other function depending on x_3 in the expression under consideration). So, only the components that depend on two spatial coordinates x_1, x_2 and on time remain. Therefore $\Phi_1 \in W_2^{1,1}(Q_1)$, where $\Phi_1 = \sum_{i=1}^{n} \alpha_i(t) \otimes a_i(x_1, x_2) c_i$, $c_i = \int_{-\frac{h}{2}}^{\frac{h}{2}} b_i(x_3)\, dx_3$, $X_{3i} = a_i(x_1, x_2) b_i(x_3)$. In what follows, Lemma 10 from [164], on the compactness of the elements $W_2^{1,1}(Q_1)$ included in $L_4(Q_1)$, $\Omega_1 \subset R^2$, can be used.

The first term in (7.60) approaches zero, owing to Lemma 5.2 from [164]. The second term also approaches zero because of the weak convergence of $\frac{\partial w_n}{\partial t}$ in $L^2(0, T; \mathring{W}_2^1(\Omega_1))$, or of $\frac{\partial w}{\partial x_1} \times \Phi_1 \in L_2(Q_1)$. In the remaining nonlinear terms, the limiting transition can be carried out in a similar way.

The satisfaction of the initial conditions is understood in a generalized sense [123].

Remark 7.1 *A weak convergence in (7.59) takes place for practical purposes in the spaces $L^\infty(0, T; H)$; however, it is not important to prove this.*

Remark 7.2 *Note that in the linearized system of equations (7.46), (7.47) the uniqueness of the solution is apparent for initial data that are smoother than those required by the theorem. However, if the energetic equality holds* *in the conditions of the theorem related to the linearized problem then the uniqueness of the solutions to the problem (7.46), (7.47) is not difficult to prove using the methodology described in [164].*

In order to solve the ODEs (7.46) using the Bubnov–Galerkin method in its higher approximations, the system of equations is transformed to the following non-dimensional form:

$$\frac{1}{1-\nu^2}\frac{\partial}{\partial x_1}[\lambda^{-1}\varepsilon_{11}+\nu\lambda\varepsilon_{22}]-\frac{1}{1-\nu}\frac{\partial}{\partial x_1}\left\{\int_{-\frac{1}{2}}^{\frac{1}{2}}\theta\, dx_3\right\}$$

$$+\frac{\lambda}{2(1+\nu)}\frac{\partial}{\partial x_2}\varepsilon_{12}-\frac{\bar{c}}{\lambda_1\lambda_2}\frac{\partial^2 u}{\partial t^2}+P_{x_1}=0 \qquad (x_1\leftrightarrow x_2),\ (1\leftrightarrow 2),$$

$$\frac{1}{12(1-\nu^2)}\left\{-\frac{\partial^2}{\partial {x_1}^2}\left[\lambda^{-2}\frac{\partial^2 w}{\partial {x_1}^2}+\nu\frac{\partial^2 w}{\partial {x_2}^2}\right]-2(1-\nu)\right.$$

$$\left.\times\frac{\partial^2}{\partial x_1\partial x_2}\left[\frac{\partial^2 w}{\partial x_1\partial x_2}\right]-\frac{\partial^2}{\partial {x_2}^2}\left[\lambda^2\frac{\partial^2 w}{\partial {x_2}^2}+\nu\frac{\partial^2 w}{\partial {x_1}^2}\right]\right\}$$

$$-\frac{1}{1-\nu}\left(\lambda^{-1}\frac{\partial^2}{\partial {x_1}^2}\left\{\int_{-\frac{1}{2}}^{\frac{1}{2}}\theta x_3\, dx_3\right\}+\lambda\frac{\partial^2}{\partial {x_2}^2}\left\{\int_{-\frac{1}{2}}^{\frac{1}{2}}\theta x_3\, dx_3\right\}\right)$$

$$+\frac{\bar{c}}{12}\frac{\partial^2}{\partial t^2}\left[\lambda_2^{-2}\frac{\partial^2 w}{\partial {x_1}^2}+\lambda_1^{-2}\frac{\partial^2 w}{\partial {x_2}^2}\right]+k_1\left[\frac{1}{1-\nu^2}\{\lambda^{-2}\varepsilon_{11}\right.$$

$$\left.+\nu\varepsilon_{22}\}-\frac{1}{\lambda(1-\nu)}\int_{-\frac{1}{2}}^{\frac{1}{2}}\theta\, dx_3\right]+k_2\left[\frac{1}{1-\nu^2}\{\lambda^{-2}\varepsilon_{22}+\nu\varepsilon_{11}\}\right.$$

$$\left.+\frac{\lambda}{1-\nu}\int_{-\frac{1}{2}}^{\frac{1}{2}}\theta\, dx_3\right]+\frac{\partial}{\partial x_1}\left[\frac{\partial w}{\partial x_1}\left\{\frac{1}{1-\nu^2}\{\lambda^{-2}\varepsilon_{11}+\nu\varepsilon_{22}\}\right.\right.$$

$$\left.\left.-\frac{1}{\lambda(1-\nu)}\int_{-\frac{1}{2}}^{\frac{1}{2}}\theta\, dx_3\right\}+\frac{\partial w}{\partial x_2}\left\{\frac{1}{2(1+\nu)}\varepsilon_{12}\right\}\right]+\frac{\partial}{\partial x_2}\left[\frac{\partial w}{\partial x_2}\right.$$

$$\times\left\{\frac{1}{1-\nu^2}\{\lambda^{-2}\varepsilon_{22}+\nu\varepsilon_{11}\}-\frac{\lambda}{1-\nu}\int_{-\frac{1}{2}}^{\frac{1}{2}}\theta\, dx_3\right\}+\frac{\partial w}{\partial x_1}\left\{\frac{1}{2(1+\nu)}\right.$$

$$\left.\left.\times\varepsilon_{12}\right\}\right]-\frac{\bar{c}}{\lambda_1^2\lambda_2^2}\frac{\partial^2 w}{\partial t^2}+g_1=0, \qquad (7.61)$$

$$\frac{\partial\theta}{\partial t} - \left(\lambda_1^2 \frac{\partial^2\theta}{\partial {x_1}^2} + \lambda_2^2 \frac{\partial^2\theta}{\partial {x_2}^2} + \frac{\partial^2\theta}{\partial {x_3}^2}\right)$$
$$= -\gamma \frac{\partial}{\partial t} \times \left[\lambda^{-1}\varepsilon_{11} + \lambda\varepsilon_{22} - x_3\left(\lambda^{-1}\frac{\partial^2 w}{\partial {x_1}^2} + \lambda\frac{\partial^2 w}{\partial {x_2}^2}\right)\right] + g_2.$$

The equations (7.61) are in nondimensional form (the overbars are omitted). In what follows, we add the following additional nondimensional parameters (which have not been used earlier):

$$u = \frac{a}{h^2}\overline{u},\ v = \frac{b}{h^2}\overline{v},\ t = \frac{K_T}{ch^2},\ \theta = \frac{\alpha_T ab}{h^2}\overline{\theta},$$

$$\gamma = \frac{E\alpha_T^2 T_0}{(1-\nu)c}\overline{\theta},\ \overline{c} = \frac{\rho K_T^2}{c^2 Eh^2},\ g_1 = \frac{a^2b^2}{Eh}\overline{g_1},$$

$$g_2 = \frac{ab\alpha_t}{K_T}\overline{g_2},\ P_{x_1} = \frac{a^2 b}{Eh^3}\overline{P_{x_1}},\ P_{x_1} = \frac{ab^2}{Eh^3}\overline{P_{x_2}}.$$

The mechanical boundary conditions for the system (7.61) are the same as those for (7.46)–(7.47). For the heat transfer equation, besides the condition (7.47), we solve the problem for the condition

$$\left.\frac{\partial\theta}{\partial n}\right|_S = 0. \tag{7.62}$$

We have solved the systems (7.61), (7.46) and (7.61), (7.62) using the Bubnov–Galerkin method in its higher approximations. The series of approximating functions $\{w_n, u_n, v_n, \theta_n\}$ were approximated in the following way:

$$w_n = \sum_i \cos i\pi x_1 \cos^2 \pi x_1 \sum_j \cos j\pi x_2 \cos^2 \pi x_2 \times g_{1ij}(t)$$
$$(i, j = 1, 3, 5),$$

$$\begin{aligned} u_n &= \sum_i \cos \pi x_1 \sin i\pi x_1 \sum_j \cos \pi x_2 \cos j\pi x_2 \times g_{2ij}(t) \\ &(i, j = 1, 2, 3), \\ v_n &= \sum_{i=1} \cos \pi x_1 \cos i\pi x_1 \sum_j \cos \pi x_2 \cos j\pi x_2 \times g_{3ij}(t) \\ &(i, j = 1, 2, 3). \end{aligned} \tag{7.63}$$

For the first type of thermal part we have used

$$\theta_n = \sum_l \left(x_1^2 - \frac{1}{4}\right)x_1^l \sum_k \left(x_2^2 - \frac{1}{4}\right)x_2^k \sum_m \left(x_3^2 - \frac{1}{4}\right)x_3^m \times g_{4lkm}(t) \tag{7.64}$$
$$(l, k, m = 0, 1).$$

For the second type we have used

$$\theta_n = \left(1 + \sum_i \sin i\pi x_1\right)\left(1 + \sum_j \sin j\pi x_2\right)\left(1 + \sum_k \sin k\pi x_3\right) \times g_{4ijk}(t)$$
$$(i, j, k = 1, 2). \quad (7.65)$$

After application of the procedure of the Bubnov–Galerkin method to the systems (7.61), (7.46) and (7.61), (7.62), we obtained second-order ODEs for g_{1ij}, g_{2ij}, g_{3ij}, g_{4ijk}. The second-order equations were reduced by a change of variables to first-order ODEs, which were then solved using a Runge–Kutta method of the fourth order. The integrals related to S and $\Omega_1 = \left(-\frac{1}{2}, \frac{1}{2}\right) \times \left(-\frac{1}{2}, \frac{1}{2}\right)$ were computed analytically.

We have considered a spherical shell, rectangular in plan, clamped on its edge, made from an isotropic material, with boundary conditions for the thermal part of the problem of the first and the second type, $\Psi_1 = \Psi_2 = \Psi_3 = 0$, $\varphi_1 = \varphi_2 = \varphi_3 = \varphi_4 = 0$, $g_1 = const$, $g_2 = 0$. The following values of the dimensionless parameters were used: a_1 is a scaling multiplier, $\lambda = 1$ $\lambda_1 = 1/50$, $\lambda_2 = 1/50$, $\nu = 0.3$, $k_1 = k_2 = 9$ and 15, $a_1 = 300$, $\gamma = 0.03a_1$, $\bar{c} = 0.3 \times 10^{-9} \times a_1^2$.

The results of the calculations are presented in Figs. 7.73–7.76. In Fig. 7.73, the deflection function in the central point $w(0, 0, t)$ of the shell for $K_1 = K_2 = 9$ is shown. Curves 1–4 correspond to the uncoupled problem for $q = 50, 57, 70$ and 100, respectively. Curves 1–8 correspond to the first type of thermal boundary conditions, and curve 9 corresponds to the critical load $q_{cr} = 69$ for boundary conditions of the second type. Curves 2 and 6 correspond to the critical loads for the uncoupled and the coupled problem, respectively. In an analogous way, computations have been carried out for both the coupled and the uncoupled problem for $k_1 = k_2 = 15$. Two approaches have been used to detect the q_{cr} value:

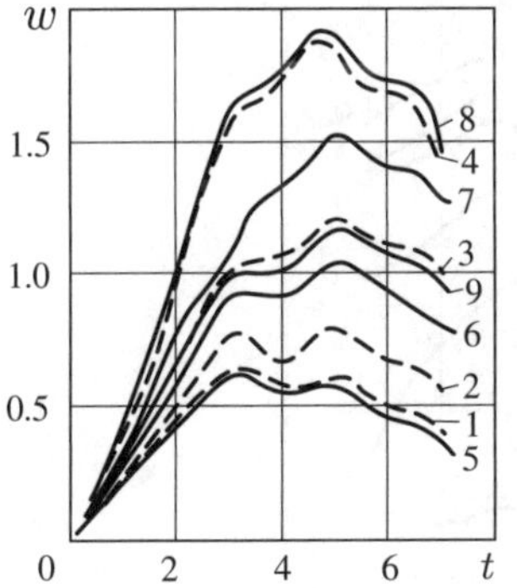

Fig. 7.73. Deflection of the shell centre $\theta(0, 0, 0, t)$ for different loads and boundary conditions ($K_1 = K_2 = 9$)

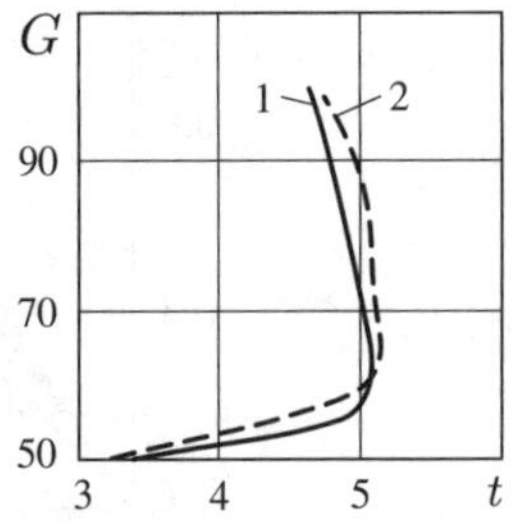

Fig. 7.74. Deflection of the shell centre $\theta(0, 0, 0, t)$ for $K_1 = K_2 = 15$: 1, uncoupled problem; 2, coupled problem

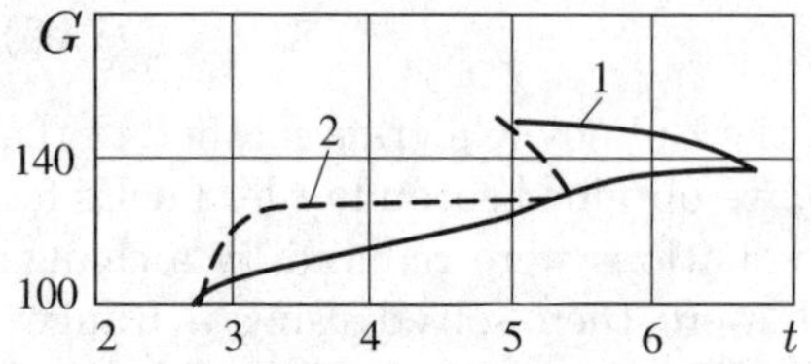

Fig. 7.75. Deflection of the shell centre $\theta(0, 0, 0, t)$ for $K_1 = K_2 = 9$: 1, uncoupled problem; 2, coupled problem

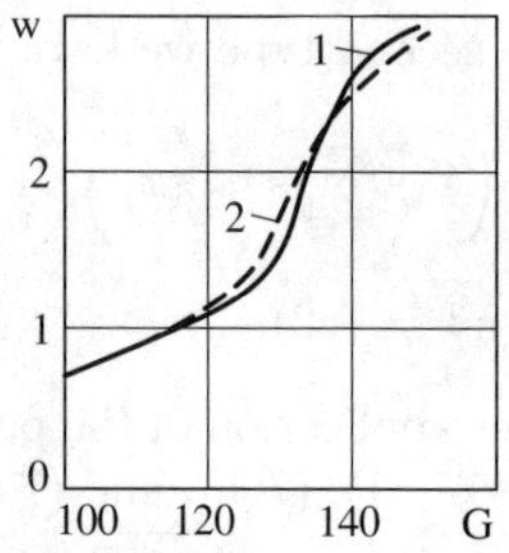

Fig. 7.76. Deflection of the shell centre $\theta(0, 0, 0, t)$ for $K_1 = K_2 = 15$: 1, uncoupled problem; 2, coupled problem; the Budiansky–Roth criterion has been applied

1. In Figs. 7.74 and 7.75, curve 1 corresponds to the uncoupled problem and curve 2 corresponds to the coupled problem. The Shian at al. [199] criterion has been applied for $k_1 = k_2 = 15$ and 9.
2. In Fig. 7.76, curve 1 corresponds to the uncoupled problem, and curve 2 corresponds to the coupled problem. The Budiansky–Roth criterion has been applied for $k_1 = k_2 = 15$ [55].

Both of these criteria give similar results. For the uncoupled problem with $k_1 = k_2 = 15$, $q_{cr} = 130$, and for $k_1 = k_2 = 9$, $q_{cr} = 57$. For the coupled problem with $k_1 = k_2 = 15$, $q_{cr} = 137$, and for $k_1 = k_2 = 9$, $q_{cr} = 65$. The graphs in Figs. 7.74–7.77 correspond to the first type of thermal boundary conditions.

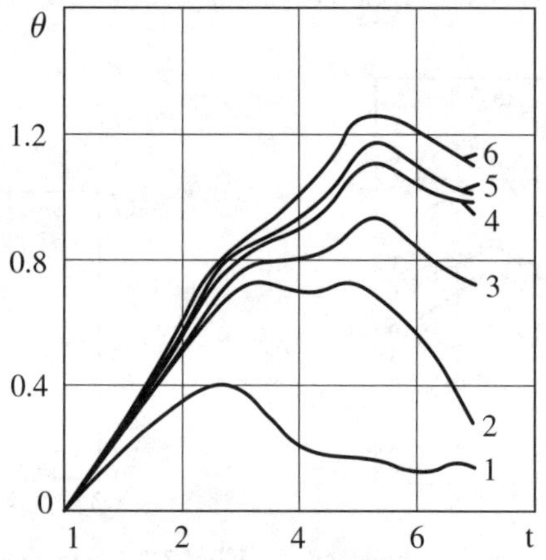

Fig. 7.77. Temperature field in the shell centre $\theta(0, 0, 0, t)$ for different loads and first type of the thermal boundary conditions ($K_1 = K_2 = 15$)

Table 7.6. Comparison of effects of rotational inertia

Uncoupled						Coupled					
Rotational inertia						Rotational inertia					
Yes			No			Yes			No		
g	w_{max}	t	g	w_{max}	t	g	w_{max}	t	g	w_{max}	t
100	0.7725	2.7	100	0.7728	2.7	100	0.7333	2.7	100	0.7337	2.7
126	1.3808	3.2	125	1.3462	3.2	130	1.5347	5.3	130	1.5367	5.3
129	1.666	5.3	130	1.7835	5.3	136	2.2369	5.4	136	2.2363	5.4
–	–	–	140	2.5324	5.2	137	2.3211	6.7	137	2.3236	6.7
150	2.8793	4.8	150	2.8778	4.3	150	2.9189	5.0	150	2.9210	5.0

In Fig. 7.77, the curves corresponding to the temperature field in the centre of the shell $\theta(0,0,0,t)$ versus time for $k_1 = k_2 = 15$ with the first type of thermal boundary conditions are shown. Curves 1–6 are related to the coupled problem without rotational inertia of the shell elements for $q = 100, 130, 133, 136, 137$ and 139, respectively. It should be noted, that in the computations, a dependence of the thermal field in the shell on the thermal boundary conditions is observed. In particular, for $k_1 = k_2 = 9$ and the first type of thermal boundary conditions with $q_{cr} = 65$, the maximum of the function $\theta(0,0,0,t)$ versus t is $\theta(0,0,0;4.9) = 0.333$, at $t = 4.9$, whereas for the second type of thermal boundary conditions $q_{cr} = 69$, and the maximum of $\theta(0,0,0,t)$ is achieved at $t = 5$, and is equal to $\theta(0,0,0;5) = 0.1618$.

Inclusion of the affect of rotational inertia has practically no influence on the value of the dynamic critical load. In Table 7.6, the maximum values of the deflection in the centre $w(0,0,t)$ versus time, and the corresponding loads with and without the effect of rotational inertia, are reported for $k_1 = k_2 = 15$.

7.7 Numerical Tests

In order to provide numerical tests, we have reduced some high-dimensional problems of plate theory to one-dimensional problems (ODEs with respect to time), using the finite-differences method (with respect to the spatial coordinates and with the help of two approximations, $O(h^4)$ and $O(h^2)$). Hence, a question concerning the convergence of the methods with respect to the spatial and time coordinates, as well as about the optimal choice of a solution to algebraic equations arises. We have solved the first of this problem using the Runge principle: solutions obtained with different integration steps with respect to the spatial coordinates are compared.

For the numerical computations, we have used the following two types of boundary conditions:

1. Clamped edge–ball support on ribs that are unstretchable in the tangiential plane:

$$w = \frac{\partial^2 w}{\partial x^2} = F = \frac{\partial^2 F}{\partial x^2} = 0 \quad \text{for} \quad x = 0 \text{ and } x = 1,$$
$$w = \frac{\partial^2 w}{\partial y^2} = F = \frac{\partial^2 F}{\partial y^2} = 0 \quad \text{for} \quad y = 0 \text{ and } y = 1. \tag{7.66}$$

2. Loosely clamped edge ribs that are unstretchable in the tangiential plane:

$$w = \frac{\partial w}{\partial x} = F = \frac{\partial^2 F}{\partial x^2} = 0 \quad \text{for} \quad x = 0 \text{ and } x = 1,$$
$$w = \frac{\partial w}{\partial y} = F = \frac{\partial^2 F}{\partial y^2} = 0 \quad \text{for} \quad y = 0 \text{ and } y = 1. \tag{7.67}$$

The initial conditions must satisfy the boundary conditions (7.66), and (7.67) and they are as follows:

1. For the boundary conditions (7.66),

$$w|_{t=0} = A \sin \pi x \sin \pi y, A = const, \quad \left.\frac{\partial w}{\partial t}\right|_{t=0} = 0. \tag{7.68}$$

2. For the boundary conditions (7.67),

$$w|_{t=0} = A \sin^2 \pi x \sin^2 \pi y, A = const, \quad \left.\frac{\partial w}{\partial t}\right|_{t=0} = 0. \tag{7.69}$$

A square plate ($\lambda = a/b = 1$) with the conditions (7.66) and (7.68) and with an initial excitation amplitude $A = 1 \times 10^{-4}$ was studied as an example. A longitudinal load $P_x = P_0 \sin \omega t$ (frequency $\omega = 5.72$, damping coefficient $\varepsilon = 1$) was applied. Only one quadrant of the plate was taken into account, owing to the two axes of symmetry along Ox and Oy, i.e. only symmetric solutions with respect to Ox and Oy were considered. The calculations were carried out using the algorithm, related an $O(h^4)$ approximation. The computational step was taken as $h = \{1/8,\ 1/10,\ 1/16\}$. The algebraic equations were solved using the Gauss method and the relaxation method. The load excitation amplitude P_0 was taken as 6 and 8. The fundamental characteristics $w(0.5, 0.5, t)$ and $w_t(w)$ for $x = y = 0.5$, fast Fourier transforms (FFTs) and power spectra are presented in Figs. 7.78 and 7.79.

An analysis of the results obtained leads to the conclusion that for further investigations it is sufficient to take $h = 1/8$ with respect to the spatial

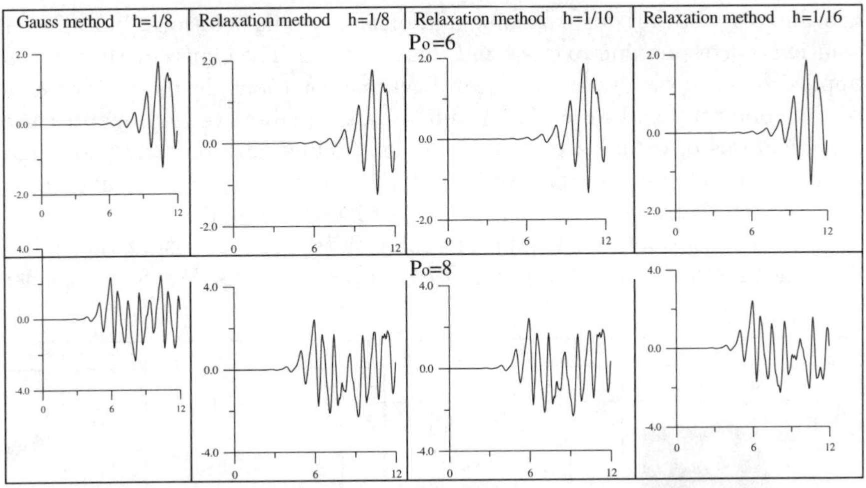

Fig. 7.78. Influence of different methods (Gauss or relaxation) and different spatial steps h on time history responses

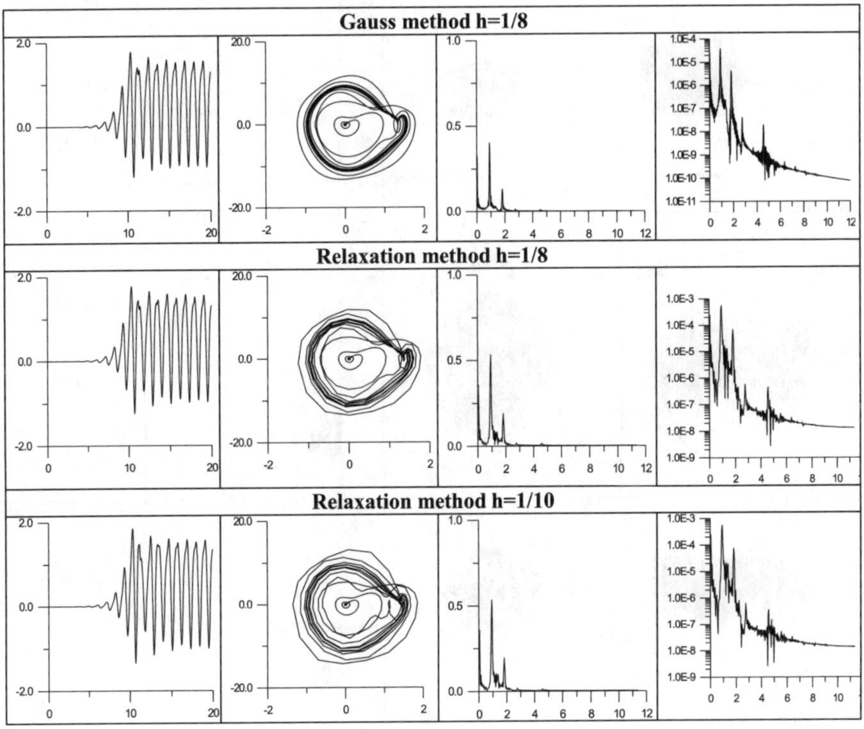

Fig. 7.79. Influence of different methods and different spatial step on phase portraits, FFTs and power spectra

coordinates, not only for harmonic and quasi-periodic solutions, but also for solutions corresponding to crisis and chaotic states. The Gauss method, when applied to solve the system of linear algebraic equations, leads to a decrease of the computational time. The result of both qualitative and quantitative investigations of complex plate oscillations, excited harmonically and longitudinally, practically overlap for this step h. According to the Runge rule, $\Delta t = 2 \times 10^{-4}$ has been used in the Runge–Kutta method.

Now we consider the computational results obtained using the finite-difference method with $O(h^4)$ and $O(h^2)$ approximations. We shall consider

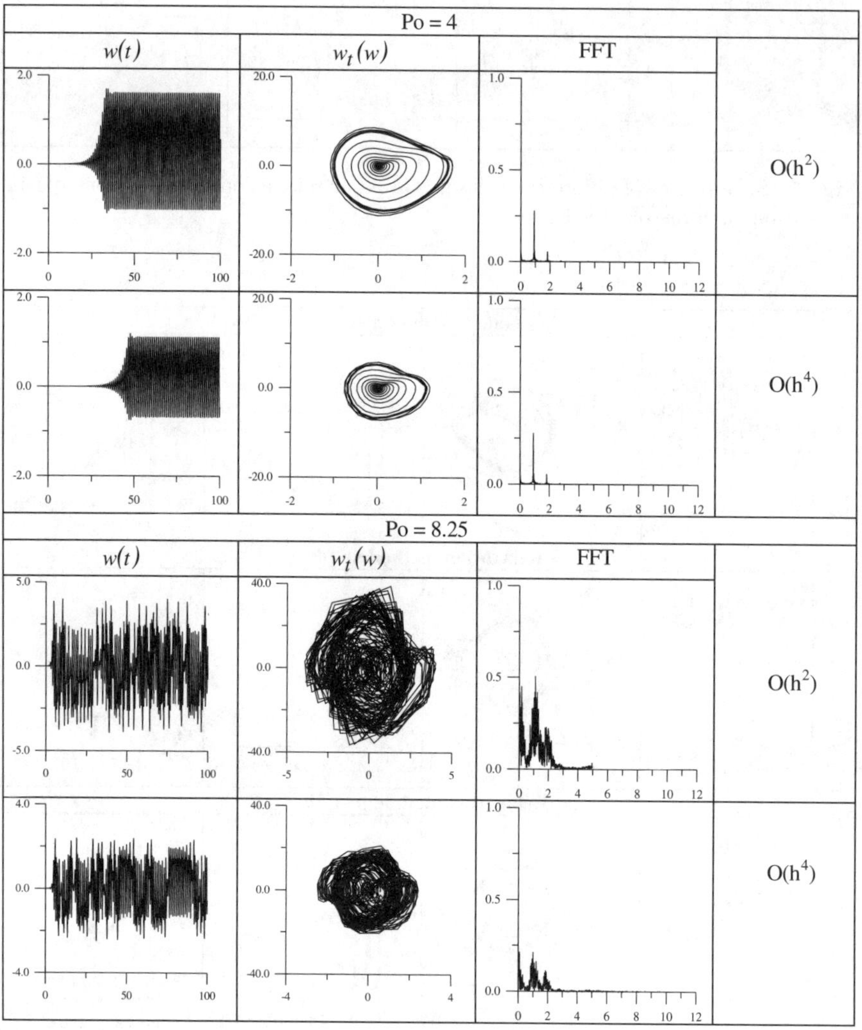

Fig. 7.80. Time histories, phase portraits and FFTs for different values of P_0 and two $O(h^k)$ approximations, $k = 2, 4$

the results for two areas, which we refer to as area II and area IV (crisis); these correspond to $P_0 = 4$ and $P_0 = 8.25$, respectively (an explanation of these areas will be given later). In Fig. 7.80, the time histories $w(0.5, 0.5; t)$, phase portraits and FFTs for the point $x = y = 0.5$ and these two values of P_0 are reported. An analysis of the characteristics shown here indicates that, qualitatively and quantitatively, the results are in very good agreement for both the $O(h^4)$ and $O(h^2)$ approximations; this agreement is clearly visible in the phase portraits and the FFTs.

7.8 Influence of Damping ε and Excitation Amplitude A

The influence of the control parameters A and ε has been investigated for $\lambda = 1$, $\nu = 0.3$, the boundary conditions (7.66) and the initial conditions (7.68). The analysis was carried out for $A = 1 \times 10^{-4}, 1.25$ and 1.375 for the fixed values $\varepsilon = 1$ and $P_0 = 6$. The time history $w(0.5, 0.5; t)$, the phase portrait, the FFT and the power spectrum for $A = 1 \times 10^{-4}$ are shown in Fig. 7.79. For $A = 1.25$ and $A = 1.375$, these characteristics are shown in Fig. 7.81.

In Fig. 7.81, the three-dimensional phase portraits (w_{tt}, w_t, w) and their projections $w_{tt}(w_t)$, $w_{tt}(w)$, $w_t(w)$ are reported, as well as the modal portraits $w_x(w)$ for the point $x = y = 0.25$, and the dependence of $w_{in} = \int_0^1 \int_0^1 w(x, y, t)\, dx\, dy$ on time.

In Fig. 7.82, the Poincaré pseudosections, the Poincaré sections, FFTs and the power spectra are shown for the three values of A considered. An analysis of the computational results shows that oscillations occur at two frequencies, and their values do not depend on the initial excitation amplitude. For each of the amplitudes considered, an internal synchronization at the frequency w_{17} (Table 7.7) occurs.

In comparing the computational results for $A = 1.25$ and $A = 1.375$, it should be emphasized that the phase portraits in a plane and in three-dimensional space are rotated relative to one another by about 180°. The indicator $w_{av} = \dfrac{w_{in}^{+} + w_{in}^{-}}{2}$, which depends on the initial excitation amplitude of the symmetric and nonsymmetric oscillation modes (Fig. 7.83) was con-

Table 7.7. Synchronization frequencies for ω_{17} ($\otimes$ - fundamental frequency, $\times$ - independent frequency, $N\,\omega_k$ - linear combinations)

No ω	ω	$A = 0.0001$	$A = 1.25$	$A = 1.375$
2	0.00555	$\times$	$\times$	$\times$
17	0.90998	$\otimes$	$\otimes$	$\otimes$
32	1.81996	$2\omega_{17}$	$2\omega_{17}$	$2\omega_{17}$

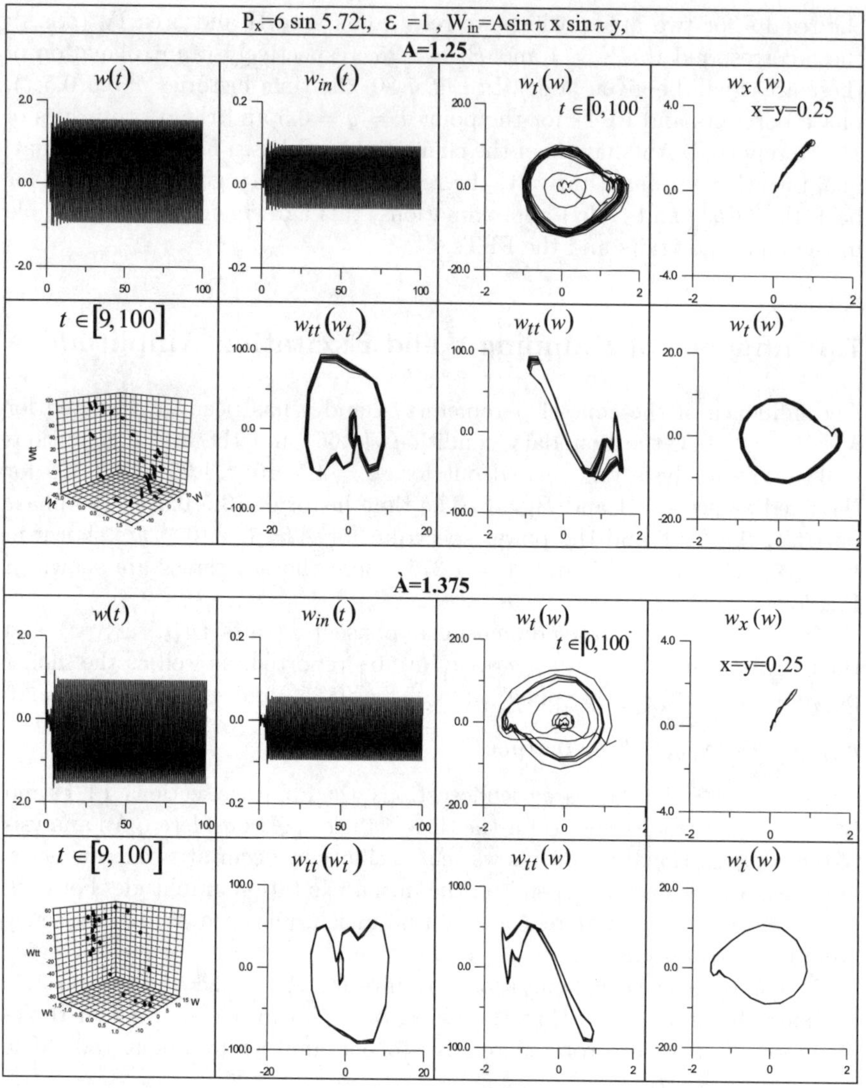

Fig. 7.81. Time histories, modal portraits, and three- and two-dimensional phase portraits, for different values of the amplitude A

structed (nonsymmetric oscillations will be discussed later). Here the critical values of A, at which oscillations of the plate around a new equilibrium state occur, i.e. a slight change of A leads to a change of the state of the plate. For nonsymmetric oscillations (Fig. 7.83), there are two such values of A. This corresponds to a rotation by 180° of the phase portraits, Poincaré pseudosections and Poincaré sections $w_t(w)$ for every period of the load that is causing excitation. The Poincaré pseudosection is defined as $w(t)$ plotted

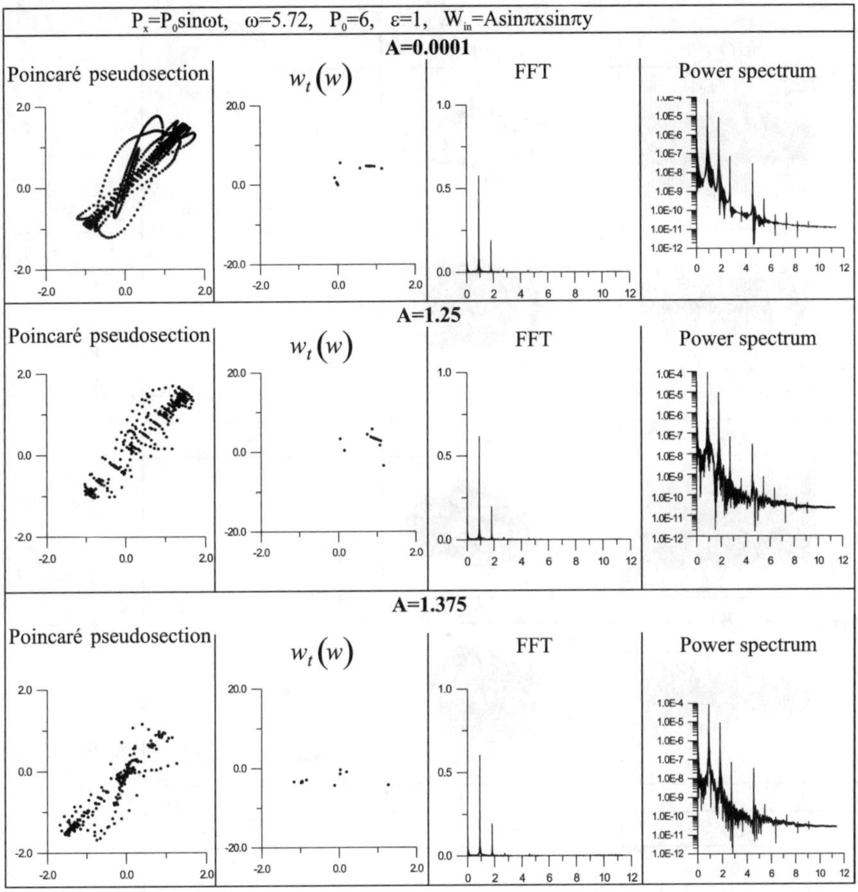

Fig. 7.82. Poincaré sections and pseudosections, $w_t(\omega)$, FFTs and power spectra for $A = 0.0001, 1.25, \text{and} 1.375$

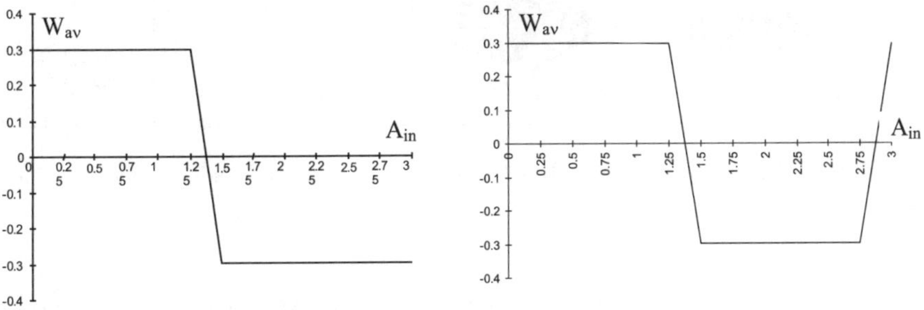

Fig. 7.83. Symmetric and nonsymmetric oscillations ($P_x = 6 \sin 5.72t$, $w_{in} = A \sin \pi x \sin \pi y$, $\varepsilon = 1$)

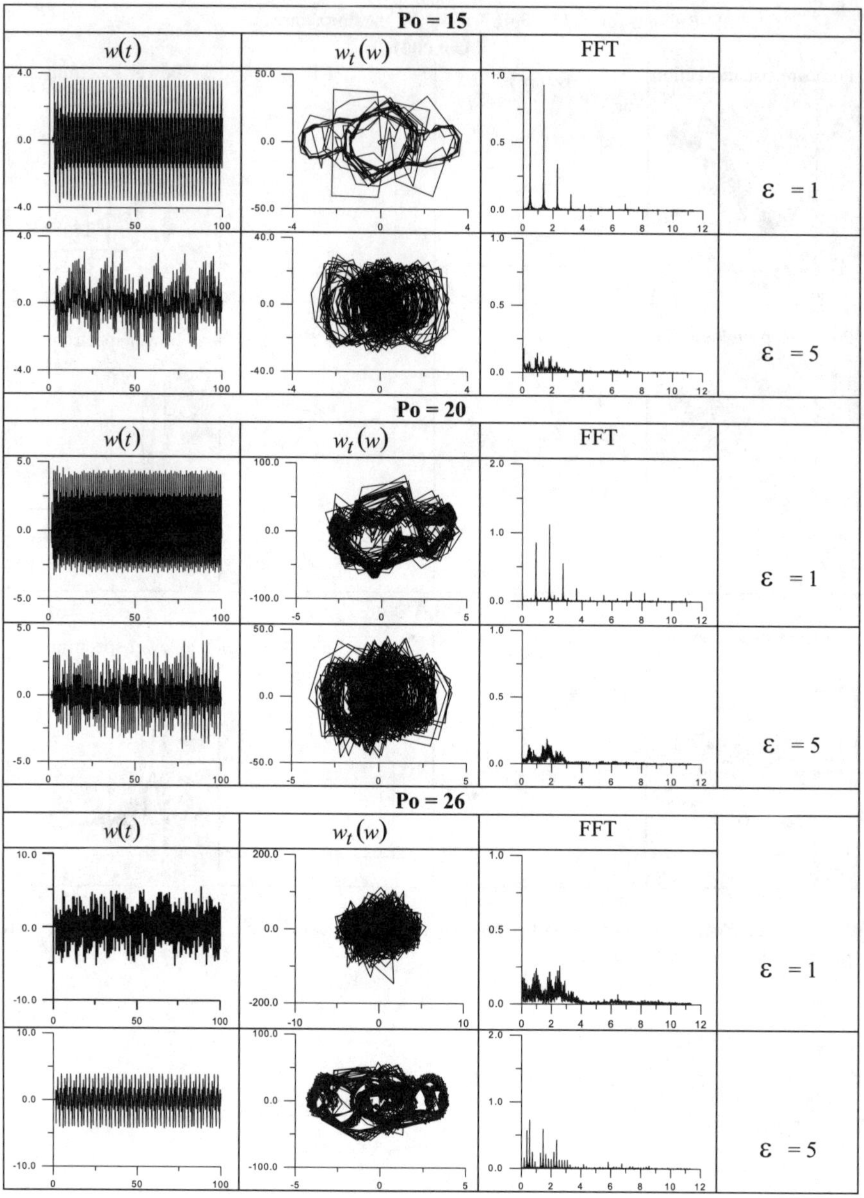

Fig. 7.84. Time histories $w(t)$, phase portraits $w_t(\omega)$ and FFTs for different values of P_0 and ε

against the deflection shifted in time, i.e. $[w(t), w(t+T)]$. The deflection $w(t+T)$ is related to the velocity $w_t(t)$ and, as a result, we obtain similar properties to those obtained when a full phase plane $[w(t),\ w_t(t)]$ is used.

We must emphasize that the above considerations are valid for fixed values of P_0 and the boundary and initial conditions applied here as well as for the period of the longitudinal load used here.

The nonlinear phase dynamics have been investigated for the fixed values $\varepsilon = 1$ and $\varepsilon = 5$, with the boundary and initial conditions given above, for $A = 1 \times 10^{-4}$ and $P_0 \in \{15, 20, 26\}$. In Fig. 7.84 the time histories $w(0.5, 0.5, t)$, phase portraits and FFTs for the plate centre ($x = y = 0.5$) are reported.

The numerical analysis leads to the conclusion that an increase of ε leads to an important change of the scenario leading to chaos using P_0 as the control parameter. For instance, for $\varepsilon = 1$ and $P_0 = 26$ we have a chaotic state, whereas for $\varepsilon = 5$ the plate undergoes post-crisis oscillations.

For $P_0 = 15$ and $\varepsilon = 5$ the plate is in a crisis state, and the number of quasi-periodic windows increases, but their durations in time greatly decrease, i.e. we observe intermittency with a higher number of oscillation modes in comparison with $\varepsilon = 1$.

7.9 Spatial–Temporal Symmetric Chaos

We have analysed the complex dynamics of symetric, flexible plates with harmonic longitudinal excitation. Although scenarios leading to chaotic dissipative oscillations have been reported in many publications, the problems of plates investigated here belong to a different class of dynamical systems: we have to work with high-dimensional systems possessing two spatial coordinates and a dependence of time, i.e. oscillations with infinitely many degrees of freedom must be taken into account.

In order to analyse multifrequency and chaotic oscillations, we need to use many dynamical characteristics, such as the oscillation processes at all points of the plate, phase portraits, FFTs, power spectra, Poincaré sections and pseudosections, and bifurcation diagrams.

In contrast to investigations in the fields of radiophysics, electronics, etc., where usually the transition to tempral chaos is studied, in problems related to plate theory a key point is the detection and analysis of spatial–temporal chaos.

A problem related to spatial chaos can be investigated relatively simply if the problem is homogeneous with respect to x and y. This situation occurs when x and y do not explicitly appear in the differential equations of the problem. If we consider one variable as time, then the system is autonomous. This means that from the point of view of the mathematics, the problem is analogous to that of temporal chaos. However, in the theory of plates with finite dimensions (in our case $\lambda = \frac{a}{b} = 1$), it is impossible to proceed in this way.

If we consider the deflection of the plate $w(x, y)$, then we have a physical interpretation up to the fourth derivative. The first derivatives $w_x(x, y)$ and

$w_y(x, y)$ correspond to the tangents of the angles between the slopes of the deflection and the corresponding axes at the points where the derivatives are calculated. The second-order derivatives $w_{xx}(x, y), w_{yy}(x, y)$ and $w_{xy}(x, y)$ correspond to the curvatures of the deflection functions at the points where the derivatives are calculated. If they are multiplied by the corresponding constants and summed, then the bending and torsion moments are obtained.

Therefore, by analysing those characteristics as a function of time, one can draw conclusions about the spatial variation of the plate surface. We refer to these characteristics later as the characteristics of the "modal portraits" when solving problems of plates using the method of finite differences. In a phase portrait the realtion between a deflection and its velocity is reported for each time moment, whereas in a modal portrait we have a relation between the deflection and the tangent of its slope relative to the corresponding coordinate axis in a plane. If we consider the phase portraits in space, then for each time moment we have a relation between the deflection, velocity and acceleration. If we consider the modal portraits in space, then each for each point of the plate we have information about the deflection, the tangents of its slope and the curvatures. As a result, we can obtain all required information about the character of the deflection of the plate surface. For the analysis of temporal chaos, one needs to consider the intersection of phase portraits with the hyperplanes at the required time intervals.

The intersections of the trajectories of modal portraits allow extraction of information about multifrequency oscillations and spatial chaos. The chaotic motions include regular and stochastic parts. In the modal portraits there exist saddle equilibrium states, saddle periodic orbits and other saddle-type invariant structures.

Different spatial structures are expected (equilibrium, periodic or chaotic behaviour) depending on whether we have a saddle-type equilibrium, saddle-type periodic orbit or saddle chaotic motion. The occurrence of periodic or chaotic spatial structures is analogous to the occurrence of periodic or stochastic self-excitations. A difference arises in investigations of the stability of regular motion and investigations of the saddle character of a solution.

The tools described above can be used to trace temporal and spatial order, as well as temporal and spatial chaos.

We must emphasize that it is rather difficult to distinguish between regular and chaotic motion, especially in cases of intermittency. In addition, an arbitrary chaotic motion is characterized by a certain regularity, certain time rules and certain spatial structures. One problem of investigations of chaos is to detect and quantify temporal and spatial behaviour possessing both deterministic and stochastic aspects.

The observations discussed above have been proved by tracing a scenario of development of spatial–temporal chaos by changing the parameters of the longitudinal load $\{P_0, \omega\}$ along the Ox axis. We fixed the value of $\omega = 5.72$ and varied $P_0 \in [0, 30]$ for $t \in [0, 100]$ (the plate oscillations were analysed only for this time interval). We considered the symmetric form of the plate

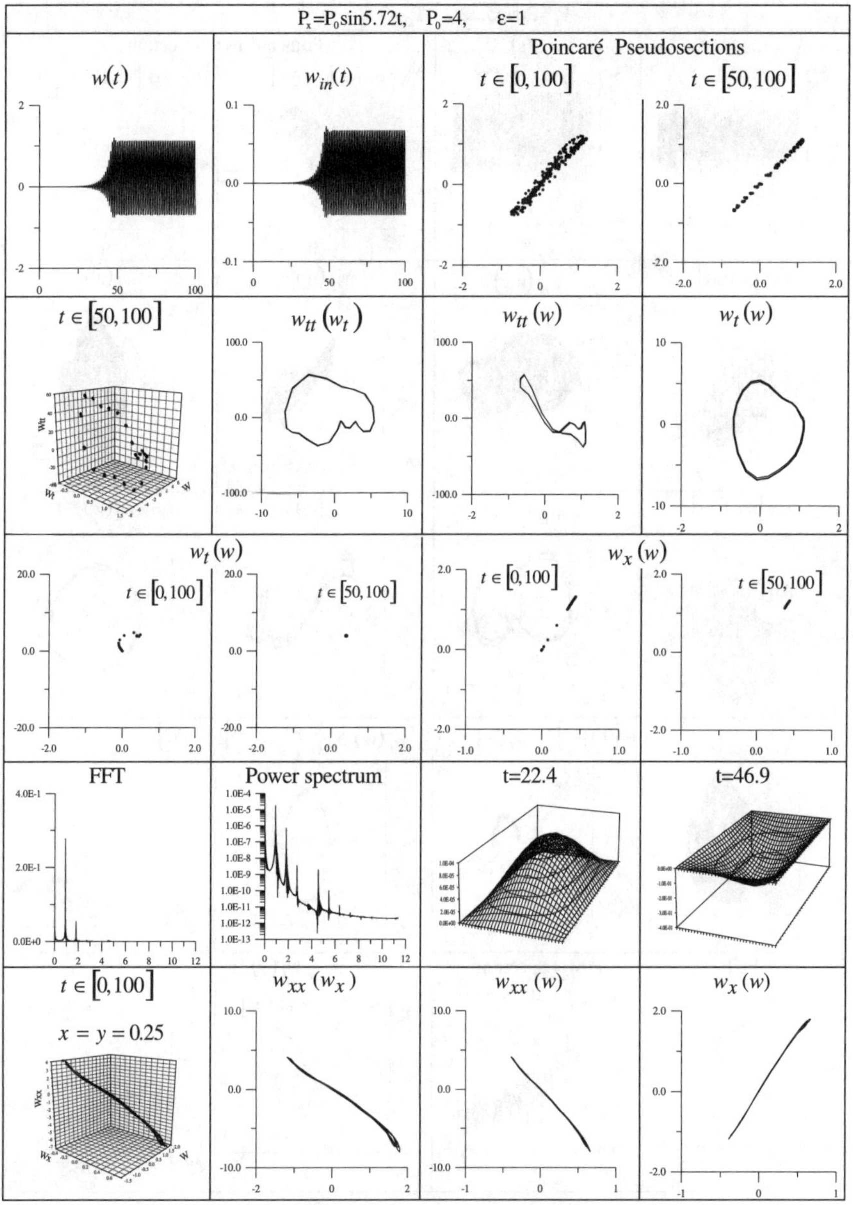

Fig. 7.85. Time histories, Poincaré sections and pseudosections, FFT, power spectrum, and modal portraits for the parameters and spatial configuration indicated ($P_0 = 4$)

oscillations, i.e. one-quarter of the plate was considered. The following indicators are reported in Figs. 7.85–7.91: the time history $w(t)$ for the centre of the plate; $w_{in}(t) = \int\int_S w(x,y,t)ds$; and Poincaré pseudosections for the

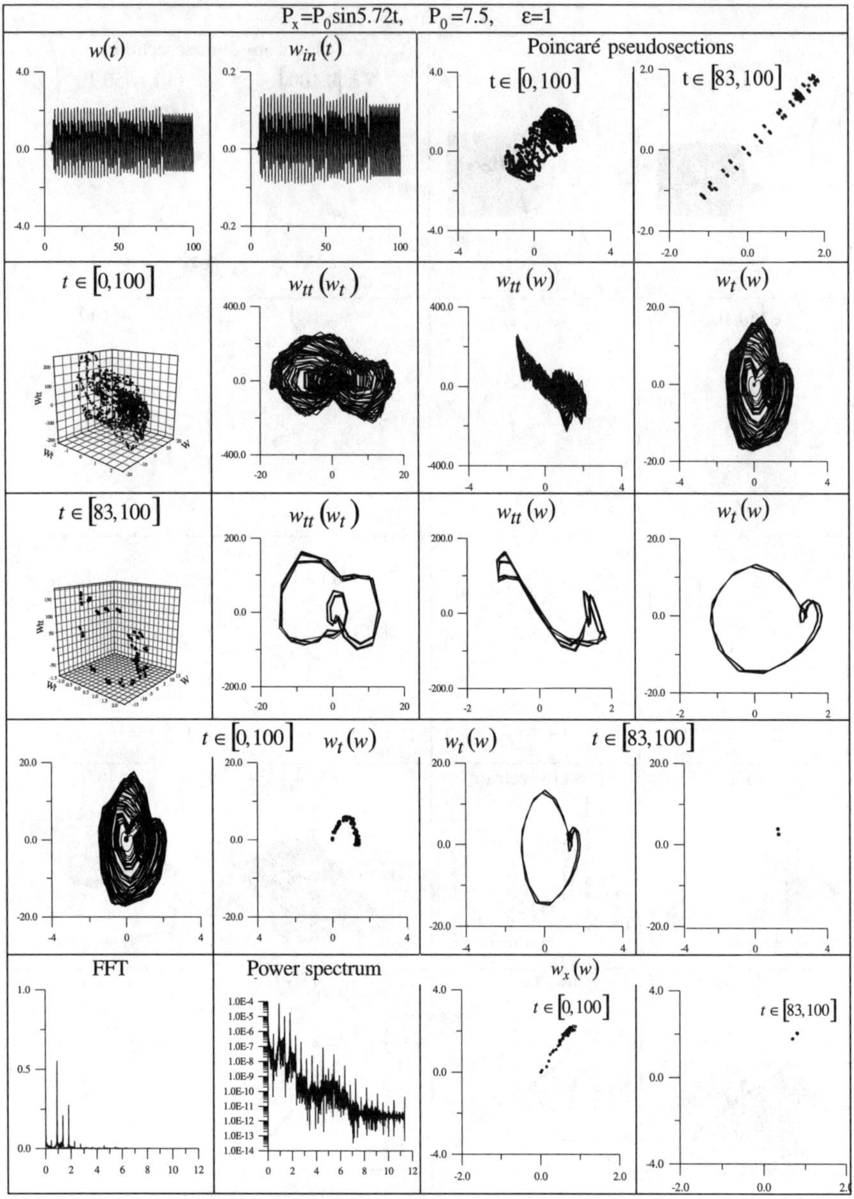

Fig. 7.86. Time histories, Poincaré sections and pseudosections, FFT, power spectrum, and modal portraits for the parameters indicated ($P_0 = 7.5$)

whole time interval, as well as for some time intervals where periodic and quasi-periodic orbits were observed.

The corresponding data for the modal spaces (w_{xx}, w_x, w) and their projections on the planes $w_{xx}(w_x), w_{xx}(w), w_x(w)$, the mapping of the phase

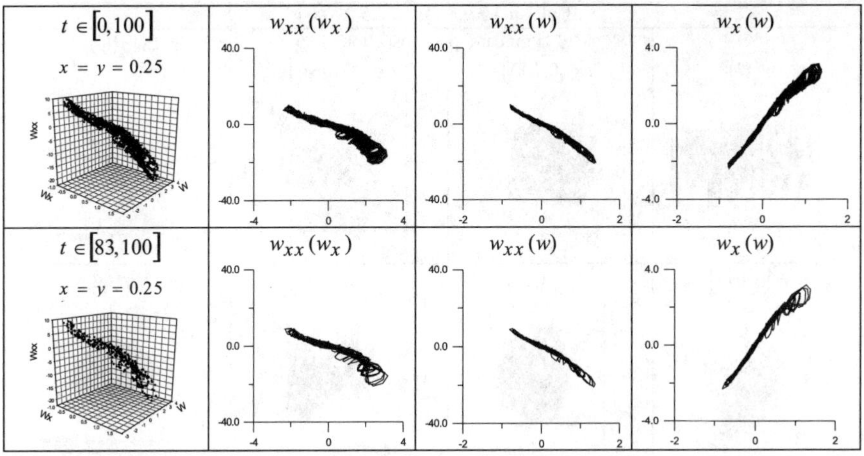

Fig. 7.86. (continued)

and modal portraits with respect to the period of the excitation, the FFT and power spectrum, the deflection of the plate surface at the characteristic time moments, and the values of the fundamental frequencies ω_i versus P_0 (i corresponds to the frequency number) are reported in Table 7.8. Graphs of $t_{in}(P_0)$ and $N\omega_i(P_0)$ are shown in Fig. 7.92, where t_{in} is the time interval for which the system is initially driven and $N\omega_i$ corresponds to the number of independent frequencies which govern the plate oscillations.

The one-parameter scenario leading to chaos, the corresponding values of $w_{av}(P_0)$, and various two-parameter portraits of the scenario leading to chaos, which will be discussed later, are shown in Figs. 7.93–7.95.

The following parameters are used here: $w_{av}(P_0)$, $w_{av}(t)$, $w^*_{av} = \int_0^T w_{av}(t)\,dt$, $w_{in} = \int\int_S w(x,y,t)\,ds$ and $w_{av} = \frac{1}{2}(w^+_{in,n} + w^-_{in,n})$. The dependence $w_{av}(t)$ allows identification of the character of the oscillation after a bifurcation, i.e. whether the oscillation of the plate takes place in the neighbourhood of an initially well-defined equilibrium, or the oscillation takes place in a regime of jumps between two equilibria. The characteristics $w_{in}(t)$ and $w_{av}(t)$ are reported in Fig. 7.96 for a series of P_0 values. Analysis of the dependence $w_{av}(t)$ leads to the following conclusions: after a bifurcation and until a crisis, the oscillations take place around the initial equilibrium state, whereas in a state of crisis a series of jumps (the plate turns inside out) are observed, ending in the inside-out state, with oscillations around the positive or negative equilibrium state. After the crisis, up to the pre-chaotic state, the oscillations are accompanied by continuous turning inside out. In the pre-chaotic state, only oscillations around the initial equilibrium state take place.

The interval $P_0 \in [0, 30]$ has been divided into a series of subintervals, whithin each of which the oscillations have similar general properties, as listed below.

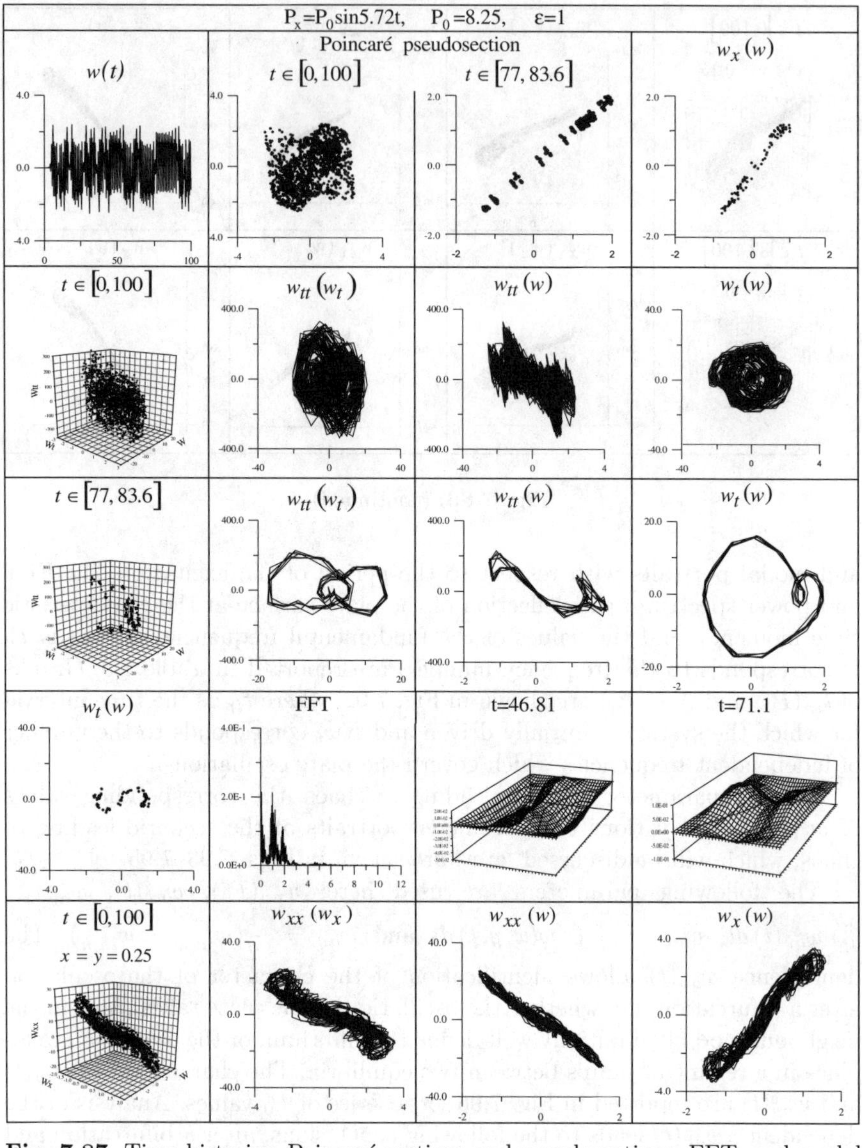

Fig. 7.87. Time histories, Poincaré sections and pseudosections, FFT, and modal portraits for the parameters and spatial configuration indicated ($P_0 = 8.25$)

I. Stable fixed point for $P_0 \in [0, 35]$. In this subinterval, the solution of the equations does not undergo qualitative changes, and the origin of $w_t(w)$ remains a singular point (for $P_0 < 3.5$ it is an attractor).

II. Bifurcation, and harmonic or quasi-periodic oscillations, $P_0 \in [3.5, 7.25]$ *(see Fig. 7.85.)* The main features of the oscilations in this interval are as follows:

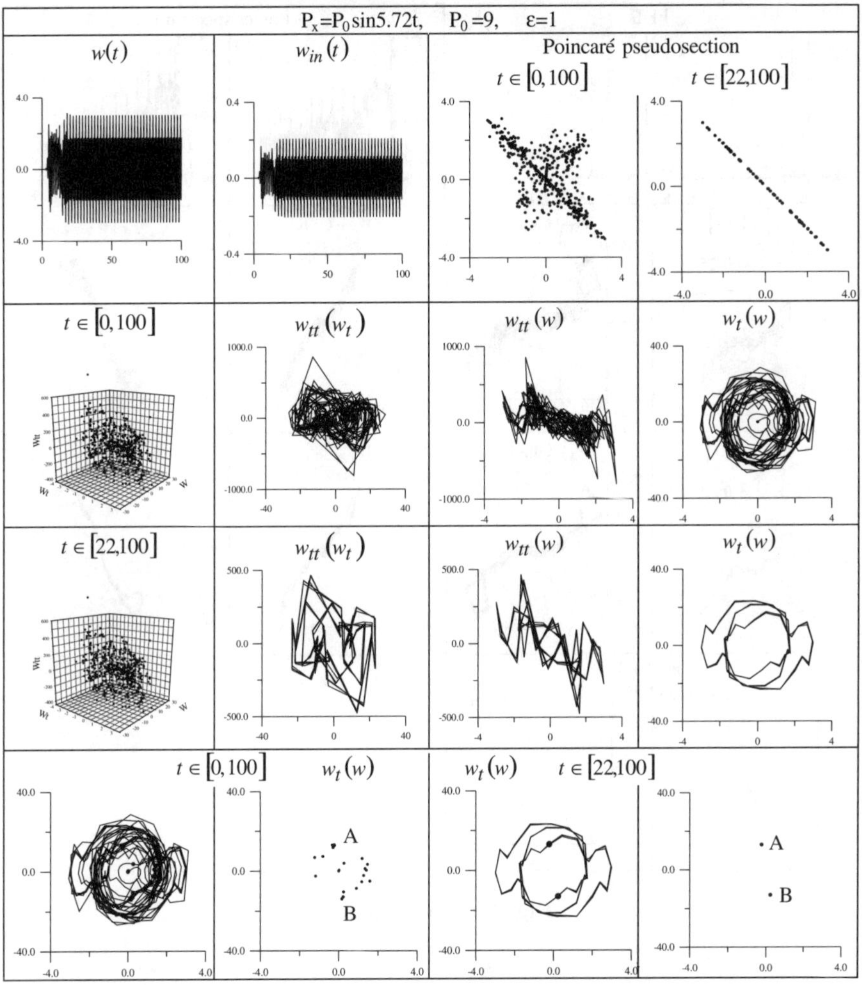

Fig. 7.88. Time histories, Poincaré sections and pseudosections, FFTs, power spectra, and modal portraits for the parameters indicated ($P_0 = 9$) (continued on page 316)

1. For $P_0 \geq 3.5$, the first bifurcation occurs, and the plate oscillates around a new equilibrium state. The time needed for bifurcation to occur is largest for $P_0 = 3.5$, and this time is referred to as the critical time t_{cr}. With increase of P_0, i.e. $P_0 \in [3.5, 7.25]$, t_{cr} significantly decreases. When the static problem is solved using the relaxation method, we obtain $P_{0_{cr}} = 3.61$.
2. In this interval of P_0, oscillations take place in the time interval $t \in [0, 100]$ at two frequencies ω_{17}, and internal synchronization occurs, because the third frequency is $\omega_{32} = 2\omega_{17}$ (see Table 7.9).

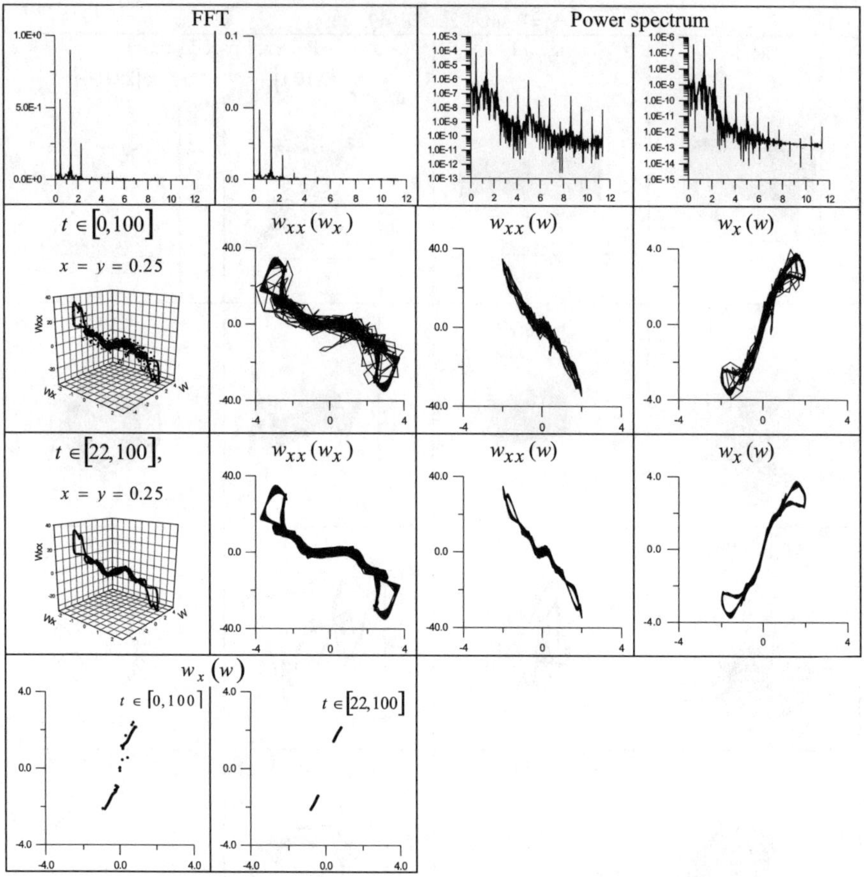

Fig. 7.88. (continued)

3. In the interval $t \in [50, 100]$ the steady state is characterized by one frequency ω_{17}, because $\omega_{32} = 2\omega_{17}$, and the amplitude corresponding to ω_2 is much less than those for ω_{17} and $\omega_{32} = 2\omega_{17}$ (as shown by the FFT), and we obtain one point in the Poincaré map (see Fig. 7.85 for $P_0 = 4$).
4. The modal portrait $w_x(w)$ for the plate point $x = y = 0.25$, as well as for other points, forms an angle of 45° with the coordinates. Deflected plate surfaces are given for the two time moments $t = 22.4$ and $t = 46.9$. The projections of the space modal portrait (w_{xx}, w_x, w) are almost straight lines, which implies almost periodic oscillations in the steady state (see Fig. 7.85).
5. After bifurcation, the plate oscillates in an excited regime, the duration of which decreases with increase of P_0, and for $P_0 = 7$ this duration is at its minimum (see Fig. 7.92). The phase portrait trajectories and the

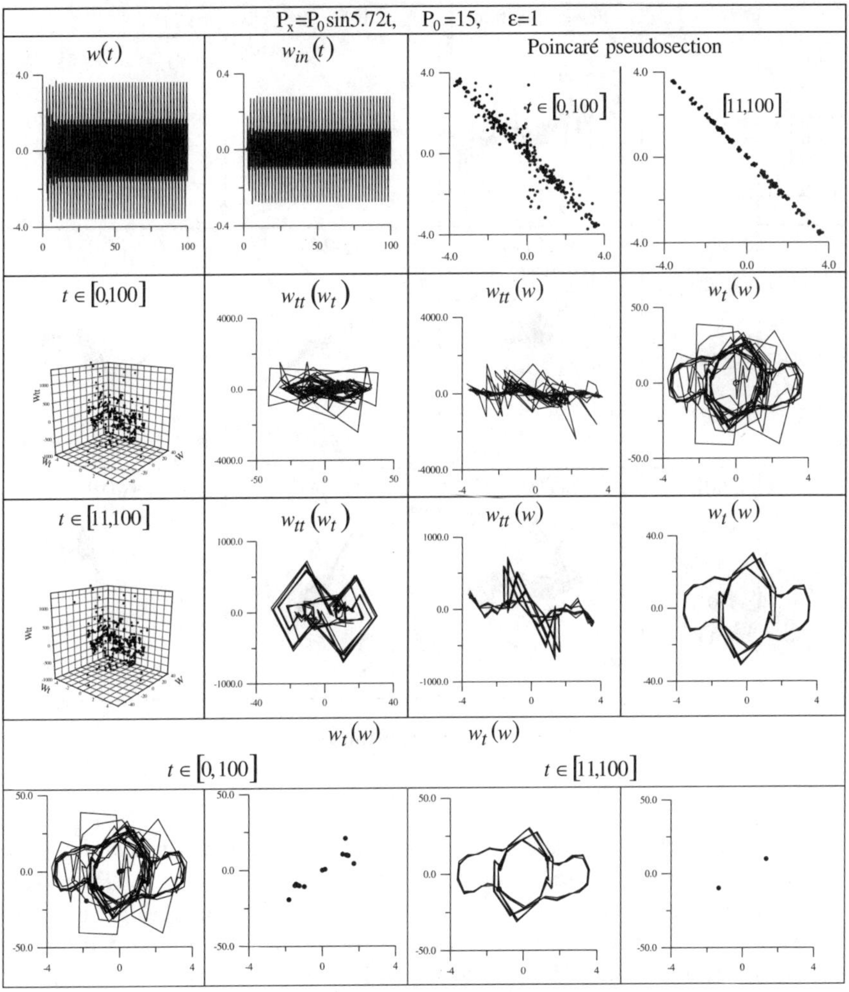

Fig. 7.89. Time histories, Poincaré sections and pseudosections, FFT, power spectrum, and modal portraits for the parameters indicated ($P_0 = 15$) (continued on page 318)

sequence of points of the Poincaré series create a sequence of curves and points, which are attached to an ellipse and one point.

6. The Poincaré pseudomaps $w^{t+T}(w^{(t)})$, which are defined by the deflection values at the times t and $t+T$, where T is the excitation period, also characterize the oscillations of the system.

III. Pre-crisis - $P_0 \in [7.25; 7.95]$. This dynamical state is described by the example of $P_0 = 7.5$ (see Fig. 7.86). The dynamical behaviour for other values of $P_0 \in [7.25, 7.95]$ is qualitatively similar. The main features of the oscillations in this interval are as follows:

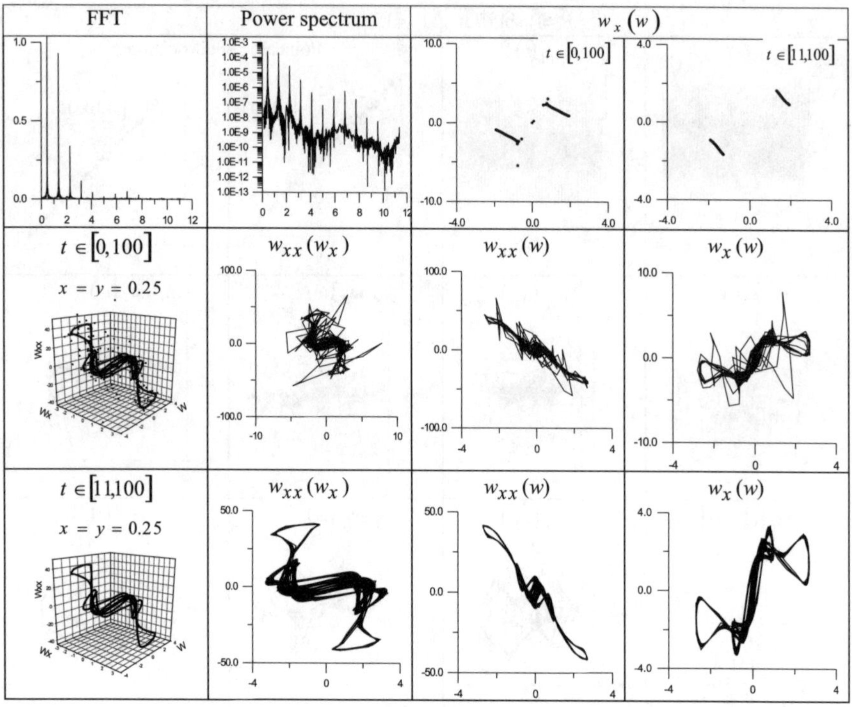

Fig. 7.89. (continued)

1. Here we observe oscillations at three frequencies, and a transition from two frequencies of oscillation to three frequencies occurs at $P_0 = 7.25$ via a jump, where the character of the oscillations changes greatly (see Table 7.9). It is known that at least three frequencies are needed for the occurrence of chaos. Here we have the following frequencies: ω_2, ω_{17} and ω_{28}; ω_2 and ω_{17} appeared also in the previous interval of P_0, which implies an intermittent character of the oscillations. Similarly to the previous scenario, an interval synchronization of the form $\omega_{32} = 2\omega_{17}$ occurs here. The additional frequency ω_{28} also appears, which causes another occurrence of internal synchronization. In addition, two new frequencies, dependent on ω_{28} and ω_{17}, appear: $\omega_{11} = \omega_{28} - \omega_{17}$ and $\omega_{41} = \omega_{28} + \omega_{17}$. To conclude, the number of frequencies increases, but the three frequencies ω_2, ω_{17} and ω_{28} play the fundamental role.
2. The excited oscillations are transformed to chaotic oscillations, and their duration greatly increases. The occurrence of chaos is characterized by the volume and plane phase portraits, the power spectrum (which is broadband and continuous), and the structure of the Poincaré pseudomaps. Analysis of the modal portraits and bent plate surfaces shows the complex form of the surface of the bending plate. The modal portraits $w_x(w)$

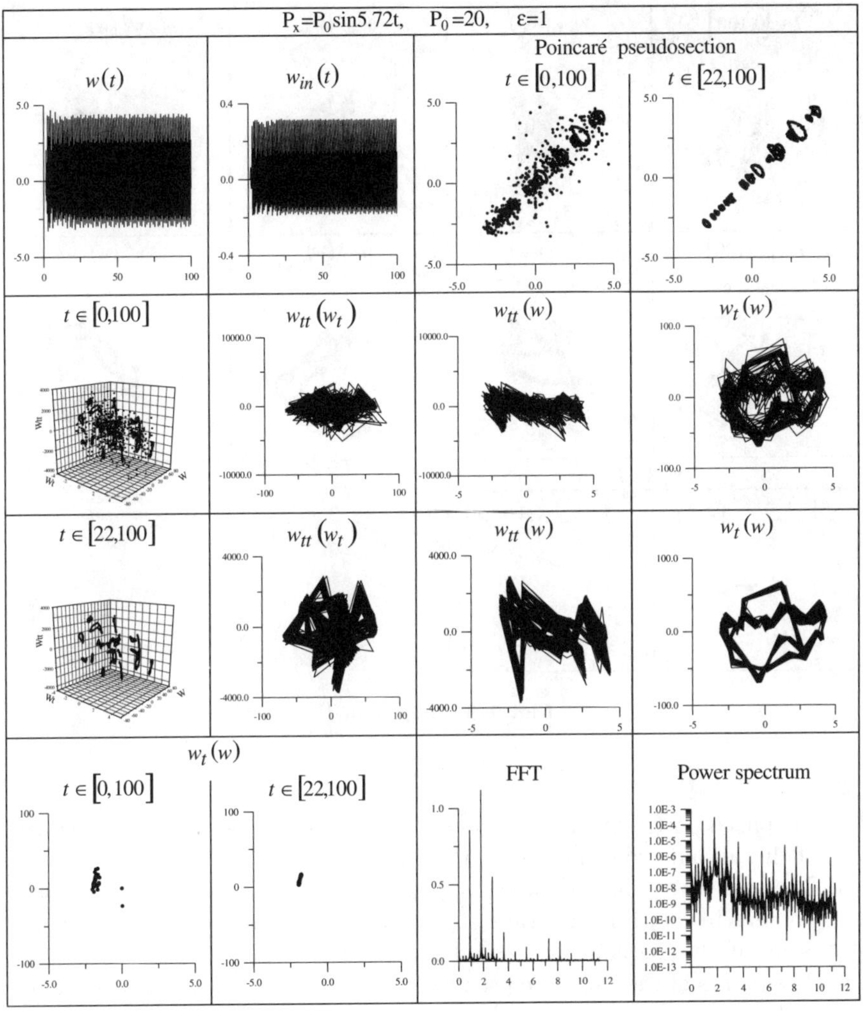

Fig. 7.90. Time histories, Poincaré sections and pseudosections, FFT, power spectrum, and modal portraits for the parameters indicated ($P_0 = 20$) (continued on page 320)

for $x = y = 0.25$ are broadbanded, and at the boundary, steady state oscillating loops appear. The oscillations are unsymmetric with respect to the equilibrium position, and they decrease with increase of P_0. Also, the angle of the tangent line to the deflection at the point $x = y = 0.25$ is changed.

3. In the centre of the interval $[0, 83]$, a strange attractor is visible in the Poincaré map, which proves the existence of chaotic motion.
4. In the interval $t \in [83, 100]$, quasi-periodic orbits occur, which is testified by the space phase portrait and its projection onto the three planes. In

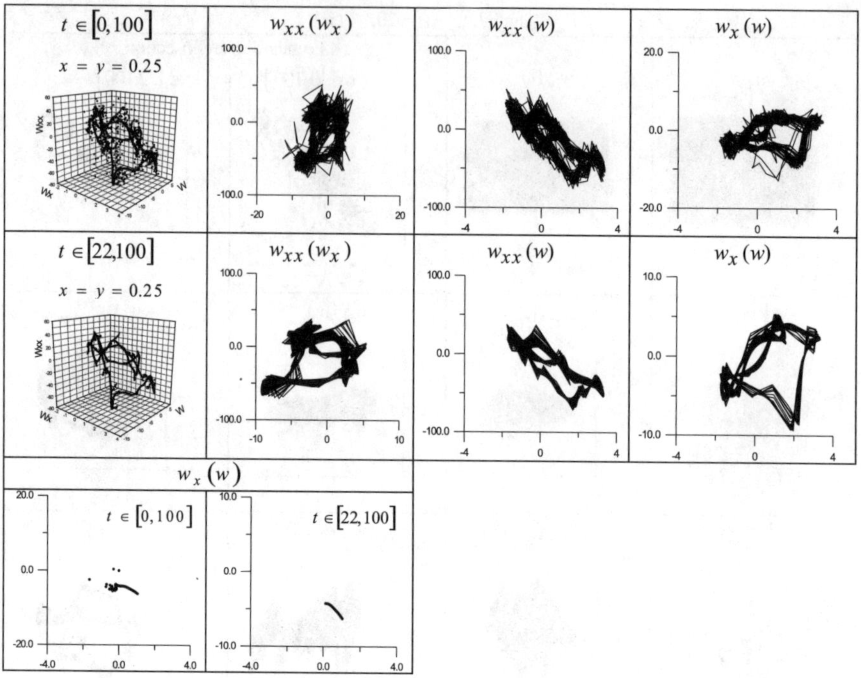

Fig. 7.90. (continued)

the phase portraits there are loops, which correspond to a large increase of the second-order harmonics (for instance ω_{28}).

5. There are two points of the phase portrait in the cross-section of the phase portrait in the time interval $t \in [83, 100]$, which indicate that oscillations occur at the frequencies ω_{17} and ω_{28}.
6. The points of the Poincaré pseudosections are close to a line with a slope of 45°, as in the previously discussed case.
7. A sudden decrease is observed in the dependence $w^*_{av}(P_0)$, indicating that the system is approaching a new bifurcation point. The summed deflection $w^*_{av} = \int_1^{100}\int_0^1\int_0^1 w(x,y,t)\,dx\,dy\,dt$ over the interval $t \in [0, 100]$ changes its sign (Fig. 7.94).

IV. Crisis $P_0 \in [7.95; 8.5]$ *(see Fig. 7.87).* The main features of the oscillations in this interval are as follows:

1. Two quasi-periodic windows are observed within the ocean of chaos for $t \in [0, 100]$, and intermittency behaviour is also observed (see Fig. 7.92).
2. The phase portraits of the quasi-periodic windows are similar to those of the pre-crisis state (see Figs. 7.86 and 7.87).

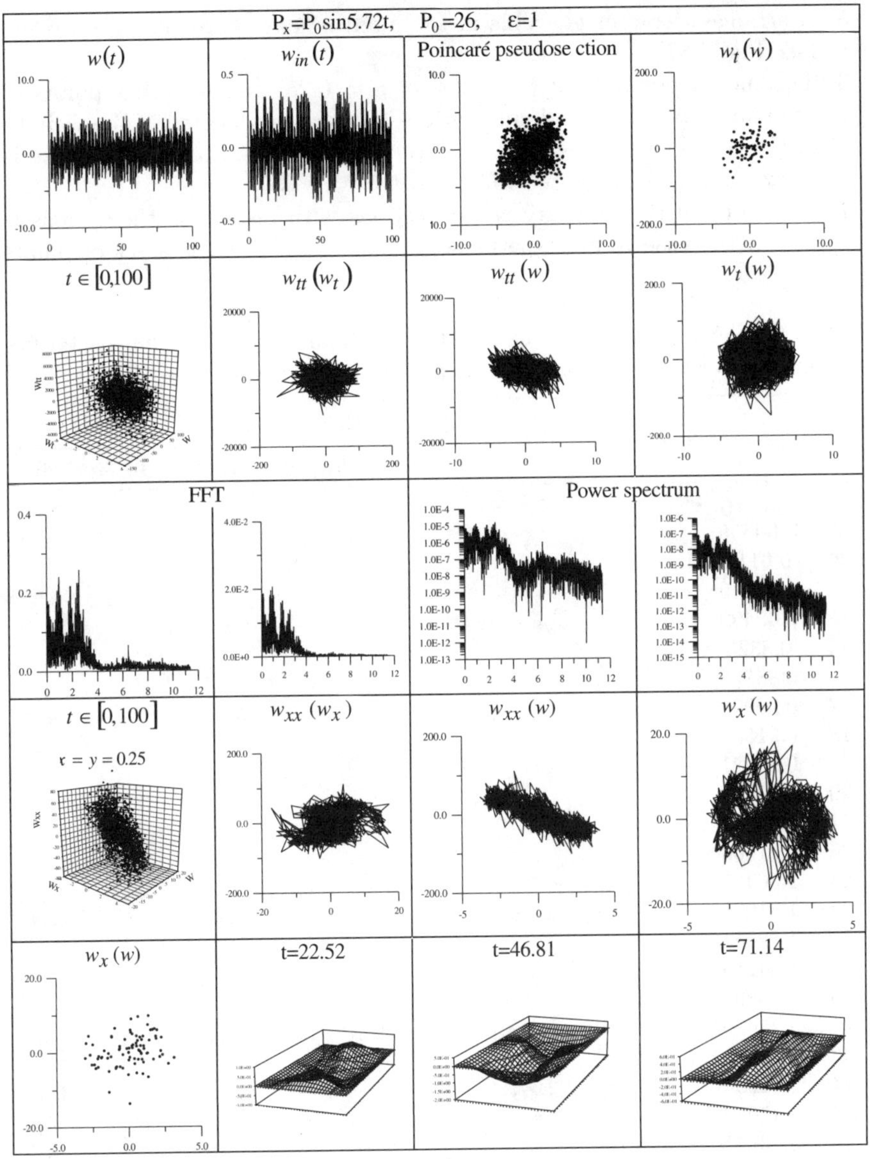

Fig. 7.91. Time histories, Poincaré sections and pseudosections, FFTs, power spectra, and modal portraits for the parameters and spatial configurations ($P_0 = 26$)

3. The oscillations in the crisis state occur at three or four independent frequencies (see Fig. 7.92).
4. In the cross-sections of the phase portrait $w_t(w)$, a structure corresponding to a strange attractor is present.

5. A strange attractor is visible in the intersections of the modal portraits (see Fig. 7.87).
6. The modal portraits $w_x(w)$ for $x = y = 0.25$, and for other points of the plate, are broadband, and various loops occur, located in the vicinity of the diagonal of the graph of $w_x(w)$. The complexity of the surface is shown in Fig. 7.87 for $t = 46.81$ and $t = 71.1$.
7. The slope of the Poincaré pseudosections with respect to the w_t axis is 45° in the case of quasi-periodicity. Otherwise, there is a set of points

Table 7.8. Frequencies of oscillations ($P_x = P_0 \sin 5.72t$) (⊗ - fundamental frequency, × - independent frequency, $N\,\omega_k$ - linear combinations)

No	ω	P_0									
ω		4	7	7.5	8	8.25	8.5	9	10	15	20
1	9.45×10^{-14}										×
2	0.00555	×	×	×	×		×				
3	0.0111							×			
4	0.02219										
11	0.45499			$\omega_{28}-\omega_{17}$				⊗	⊗	⊗	
14	0.8323					×					
15	0.88778										
17	0.90998	⊗	⊗	⊗					$2\omega_{11}$		⊗
18	0.91553					×					
19	0.99321				×						
20	0.99876										
21	1.04869			×							
26	1.31503					×					
28	1.36497			⊗				$3\omega_{11}$	$3\omega_{11}$	$3\omega_{11}$	
32	1.81996	$2\omega_{17}$	$2\omega_{17}$	$2\omega_{17}$					$4\omega_{11}$		$2\omega_{17}$
33	1.82551					×					
34	1.84215				×						
36	1.95867			×							
41	2.27495			$\omega_{28}+\omega_{17}$				$5\omega_{11}$	$5\omega_{11}$	$5\omega_{11}$	
43	2.72495										
44	2.72994			$2\omega_{28}$					$6\omega_{11}$		$3\omega_{17}$
45	3.17938									⊗	
46	3.63437										×
47	4.08936									$\omega_{45}+2\omega_{11}$	
48	4.54434										×
51	4.99933							×			
52	5.00488								×		
54	5.45432										×
55	6.81929									×	
56	7.27428										×
57	7.72927									×	
58	8.18426										×

Table 7.9. Frequencies of oscillations for various values of P_0 (see Table 7.8 for more details)

No	ω	P_0									
ω		4	7	7.5	8	8.25	8.5	9	10	10.25	10.5
2	0.00555	$\times$	$\times$	$\times$	$\otimes$	$\otimes$	$\otimes$		$\times$	$\otimes$	
3	0.0111										
5	0.02774										$\times$
6	0.03329										
7	0.04439							$\times$			
8	0.05549										
9	0.32737					$\times$					
10	0.4217										
11	0.45499		$\omega_{28}-\omega_{17}$				$\otimes$	$\otimes$	$\otimes$		
12	0.46609									$\otimes$	
13	0.47164										$\times$
15	0.88778							$\times$			
16	0.89333										
17	0.90998	$\otimes$	$\otimes$	$\otimes$	$\otimes$						
18	0.91553					$\otimes$					
22	1.13747		$\times$								
23	1.21515				$\times$						
24	1.23735					$\otimes$					
25	1.24845			$\times$							
27	1.33168										
28	1.36497		$\otimes$				$3\omega_{11}$	$3\omega_{11}$	$3\omega_{11}$		
29	1.37607									$\otimes$	
30	1.38161										$\times$
31	1.44265									$\times$	
32	1.81996	$2\omega_{17}$	$2\omega_{17}$	$2\omega_{17}$	$2\omega_{17}$			$4\omega_{11}$			
33	1.82551					$2\omega_{18}-\omega_2$					
35	1.85325										
37	2.15228					$\omega_{24}+\omega_{18}$					
38	2.24165										
39	2.2583									$\omega_{29}-\omega_{12}-\omega_2$	
40	2.26694								$\times$		
41	2.27495		$\omega_{28}+\omega_{17}$				$5\omega_{11}$	$5\omega_{11}$			
42	2.28604										$\times$
43	2.72439				$3\omega_{17}-\omega_2$						
44	2.72994		$2\omega_{28}$	$3\omega_{17}$			$7\omega_{11}-\omega_2$				
45	3.17938										
48	4.54434		$\times$								
49	4.82178					$\times$					
50	4.85507				$\times$						
51	4.99933								$\times$		
52	5.00488						$\times$	$\times$			
53	5.02708										
54	5.45432		$\times$								

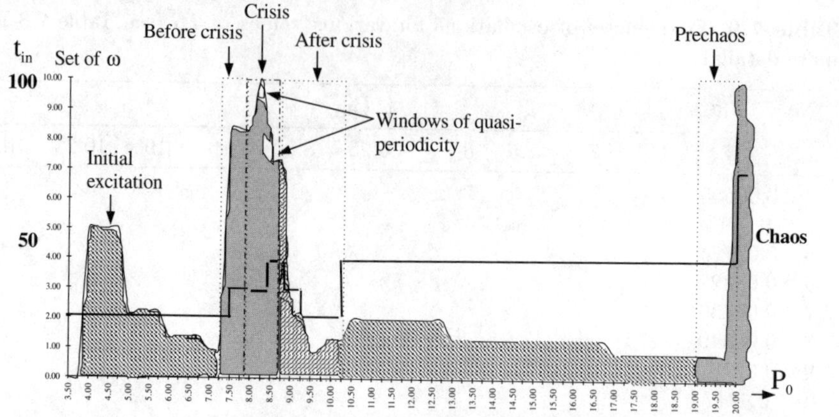

Fig. 7.92. Scenario to chaos versus P_0 (thick line characterizes the dependence $N\omega_i(P_0)$)

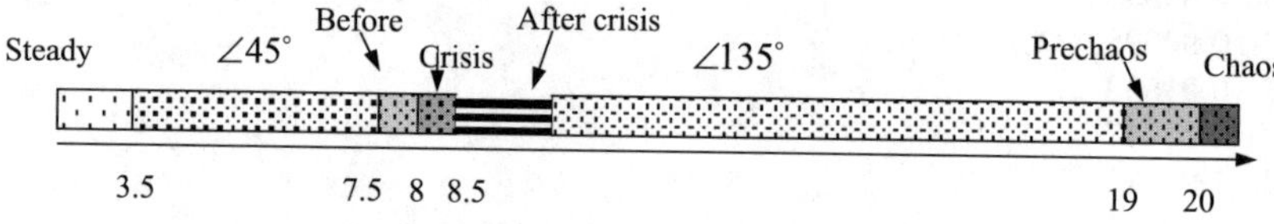

Fig. 7.93. Dynamics versus P_0 ($\omega = 5.72$)

with a slope of 135° with respect to the w_t axis for $t \in [0, 100]$ (see Fig. 7.93).

8. Crisis is a necessary behaviour if a chaotic state is to be achieved, and it depends on many parameters.
9. In the Fig. 7.94 ($w_{av}(t)$ against P_0), a series of bifurcations can be observed i.e. a continuous change of the sign of the deflection occurs in the time interval considered. Only in the case of quasi-periodic oscillations does w_{av} become constant, with either a positive or a negative sign.

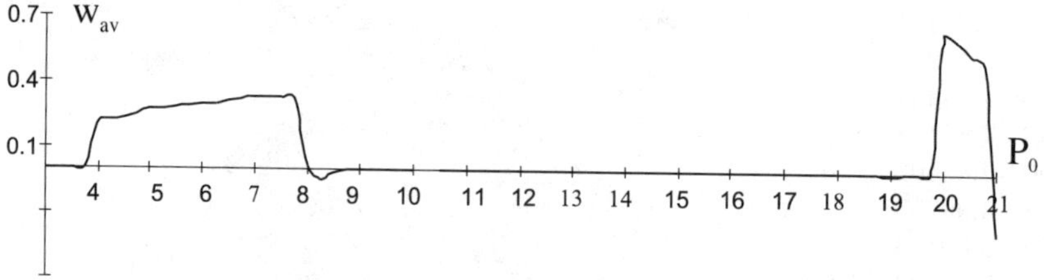

Fig. 7.94. w_{av} versus P_0 (see text)

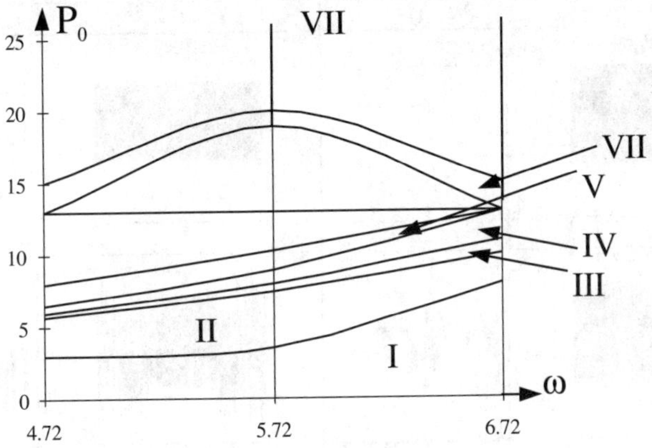

Fig. 7.95. Various dynamical states within the two-parameter space (see text for a detailed description)

V. Post-crisis, $P_0 \in [8.5, 10.25]$ *(see Fig. 7.88 and Figs. 7.92–7.95).* The "post-crisis state" describes a narrow zone next to the crisis state of the mechanical system. The characterisitc properties of the state of the plate state are given in Figs. 7.88 (for $P_0 = 9$). The main features of the oscillations in this post-crisis interval are as follows:

1. The number of independent frequencies at which oscillation occurs decreases (see Table 7.8). Here we have three frequencies, internal synchronization occurs on ω_{11}: $\omega_{28} = 3\omega_{11}, \omega_{41} = 5\omega_{11}$. In a steady state, two frequencies of oscillation are observed (see the cross-section $w_t(w)$ for $t \in [22, 100]$).
2. The time needed to achieve quasi-periodic solutions decreases and becomes equal to that observed in the pre-crisis state.
3. The phase portrait is changed qualitatively. Higher-order harmonics occur, as demonstrated by the FFT and power spectrum.
4. The oscillations at $x = y = 0.5$ are symmetric with respect to the zero equilibrium position, i.e. jumps appear (Fig. 7.96). At the maximum displacement, only one loop occurs in the modal portrait (Fig. 7.88). Also, symmetric loops appear in the $w_{xx}(w)$ dependence.
5. The Poincaré pseudosection has a slope of 135° with respect to the w axis, but in the excited state there is a set of points that have a slope with respect to w_t of 45° (see Fig. 7.93).
6. The pre-crisis and post-crisis states can be treated as intervals of crisis, where the plate oscillations are transformed to the crisis state and then into a steady state with two or three frequencies, respectively.

VI. Steady post-crisis state, $P_0 \in [10.25;, 18.9]$. In this case the oscillations are qualitatively similar to those of the post-crisis state (see Fig. 7.89). The

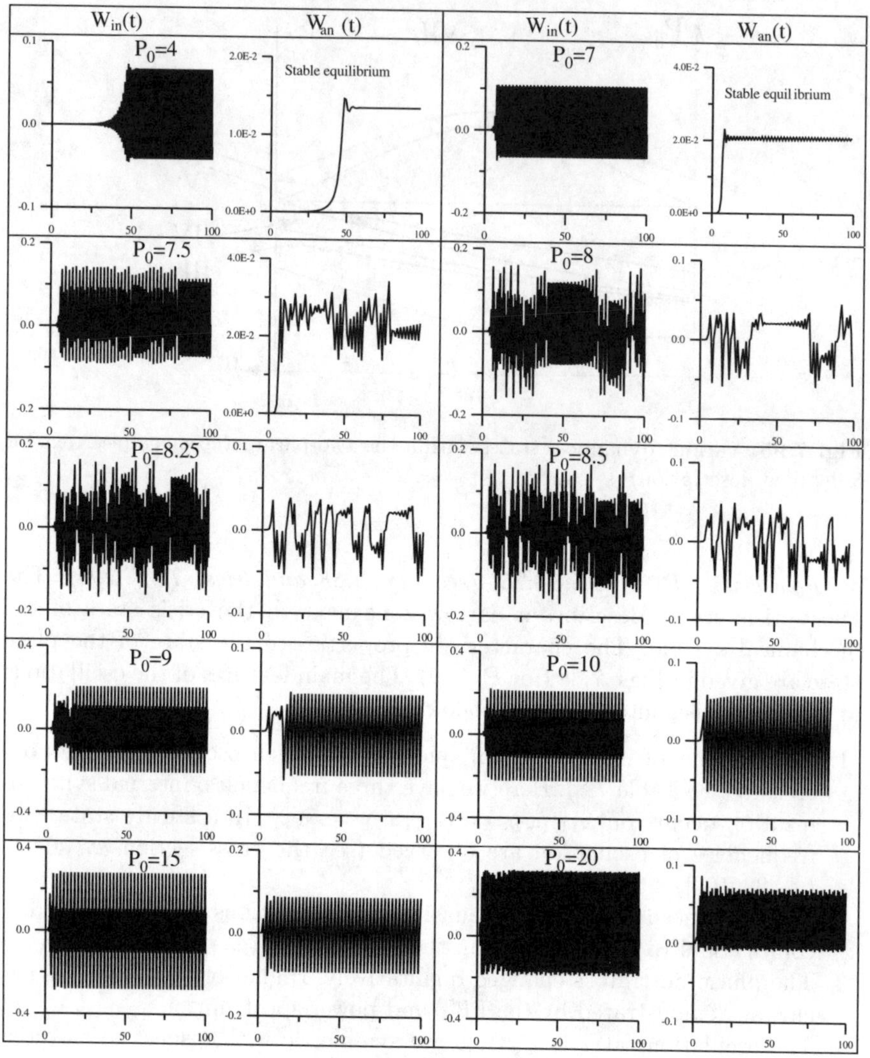

Fig. 7.96. w_{in} and w_{av} (see text) for different values of P_0

phase and modal portraits are similar to that of the previous state, but the loops in their limiting states in the modal portraits are larger.

VII. Pre-chaotic state, $P_0 \in [18.9, 20]$ (see Fig. 7.90). The main features of the oscillations in this interval are as follows:

1. A number of independent frequencies occurs (see Table 7.8, and the FFT and power spectrum in Fig. 7.90). An internal synchronization on ω_{17}, i.e. $\omega_{32} = 2\omega_{17}$, $\omega_{44} = 3\omega_{17}$, occurs.

2. The points of the Poincaré pseudo-section are close to a line with a slope of 45° with respect to the w_t axis. The phase and modal portraits have closed subspaces.
3. A series of bifurcations occurs and the equilibrium states change (Fig. 7.96).
4. The deflection surfaces have a chaotic character.

VIII. Chaos, $P_0 > 20$ (see Fig. 7.91). The term "chaos" is applied in deterministic problems, where there are no unpredictable or stochastic forces or parameters. For $P_0 > 20$, a slight change of P_0 leads to large changes of all characteristics of the system analysed.

7.10 Dissipative Nonsymmetric Oscillations

We have considered nonsymmetric oscillations of a square, isotropic plate ($\lambda = 1$) subjected to the action of a longitudinal one-sided periodic load. The scenarios leading to chaotic state with a change of $\{P_0, \omega\}$ is investigated. The problem was solved using the algorithm described earlier. The fourth-order $0(h^4)$ approximation was used, and a condition of axial symmetry in the mesh space $\bar{G}_h$ was not been applied. The whole mesh space $\bar{G}_h$ was used, and the number of nodes was increased by a factor of two in comparison with the symmetric case. Therefore, we were able to find all possible symmetric and nonsymmetric solutions to the differential equations considered. The frequencies of the exciting force $P_x = P_0 \sin \omega t$ were taken from the interval $\{4.72, 672\}$. Figures 7.97–7.107 correspond to the symmetric oscillations considered earlier.

To illustrate the scenarios leading to chaos in a flexible plate we have taken $\omega = 5.72$, $\varepsilon = 1$, $A = 1 \times 10^{-4}$. In Table 7.9, the frequencies of oscillation of the plate for a series of values of P_0 are given. The dependences $t_*(P_0)$ and $N\omega_i(P_0)$ are shown in Fig. 7.103. Fig. 7.106 shows a classification of the oscillations, which is similar to that in the case of symmetric oscillations (Fig. 7.95). We shall describe only some of the differences. States I and II are qualitatively similar in the symmetric and nonsymmetric cases, but in state II the plate oscillates about a symmetric form. The corresponding intervals of P_0 in the nonsymetric case are reported in Figs. 7.103 and 7.105. In state II we have a pre-crisis, and the oscillations exhibit 3–6 frequencies. The fundamental characteristics of this state are presented in Fig. 7.97.

In the phase portrait, three loops and three points of the Poincaré map correspond to quasi-periodic orbits. A strange attractor is visible for $t \in [0, 100]$. The mode portrait (Fig. 7.97) includes a series of closed orbits with complicated shapes.

One of the peculiarities of the crisis (see Figs. 7.98 and 7.99) is an exchange of symmetric and nonsymmetric forms. At the initial time moment, the oscillations begin symmetrically and then approach a nonsymmetric state. This

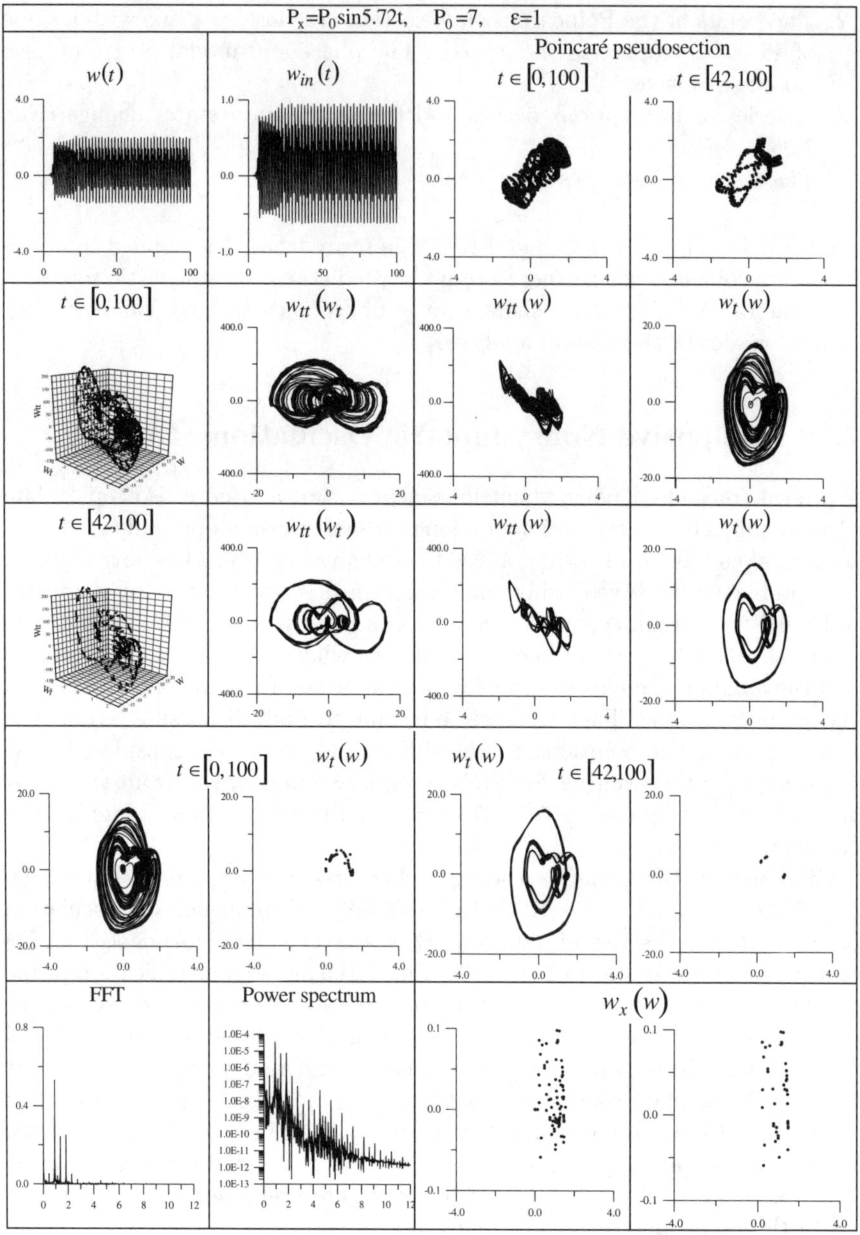

Fig. 7.97. Time histories, Poincaré sections and pseudosections, modal portraits, FFT and power spectrum ($P_0 = 7$) (continued on page 330)

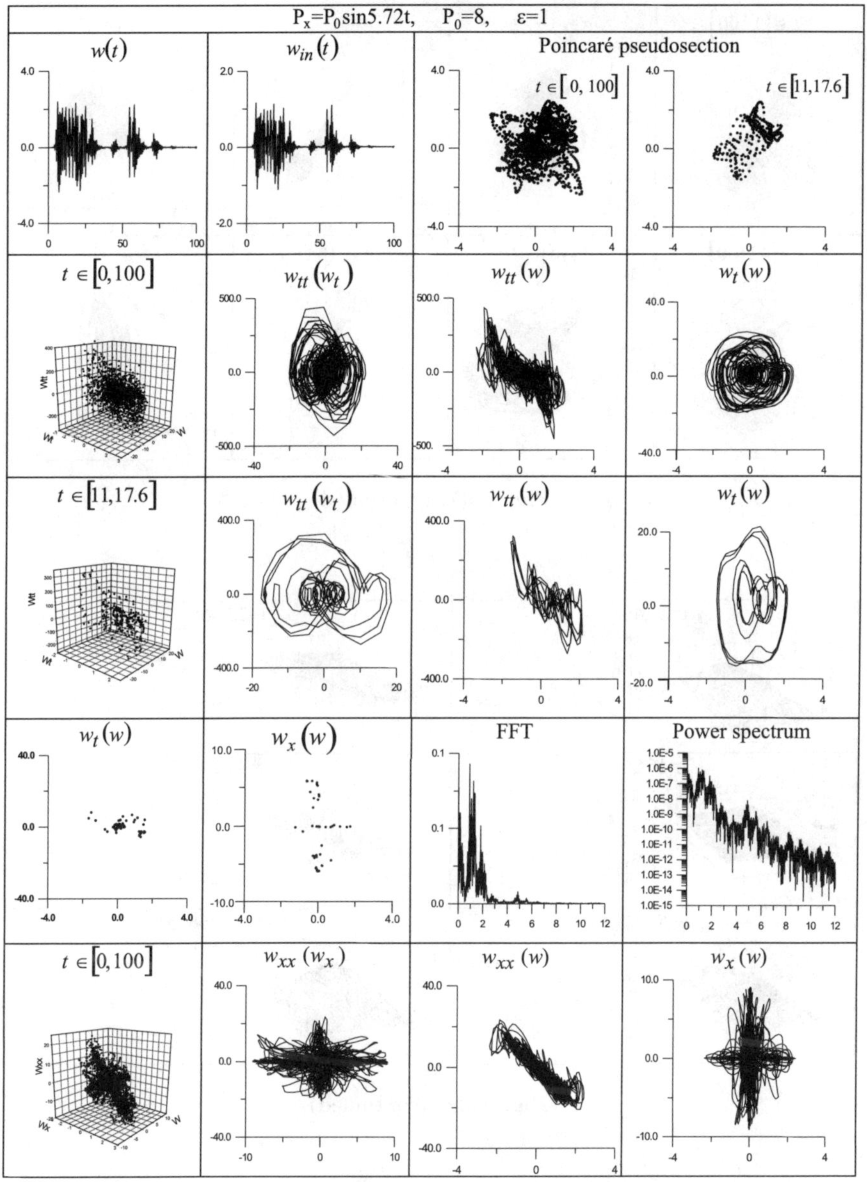

Fig. 7.98. Time histories, Poincaré sections and pseudosections, modal portraits, FFT and power spectrum, and spatial configurations for specified time moments ($P_0 = 8$) (continued on page 330)

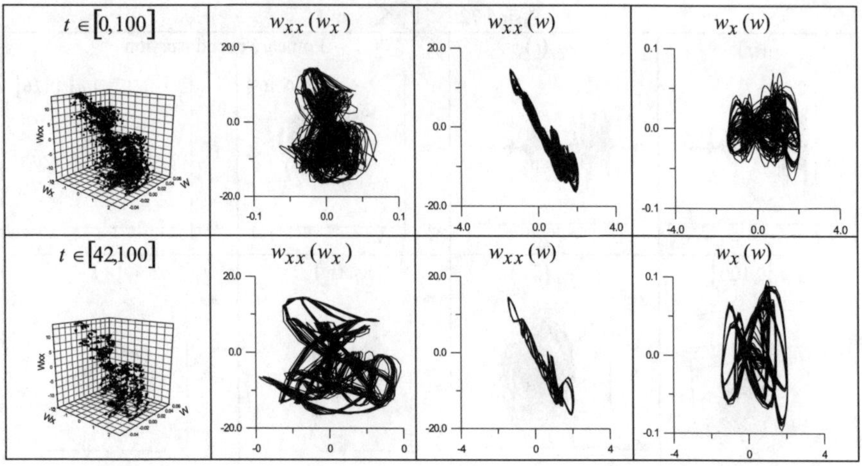

Fig. 7.97. (continued)

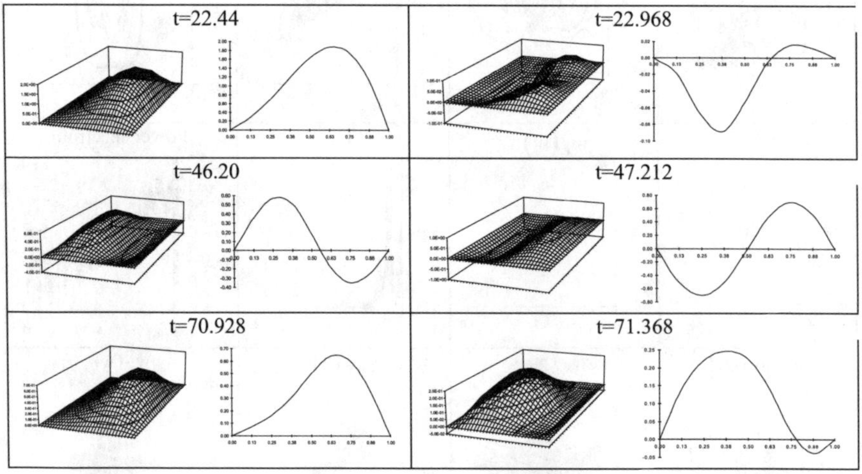

Fig. 7.98. (continued)

is clearly manifested when the dependence $w(0.5, 0.5, t)$ and the deflection surface are analysed. In the crisis state, the deflection surfaces have extremely complicated shapes for both the symmetric and the nonsymmetric oscillations (Fig. 7.98).

In the quasi-periodic windows, the oscillations of the plate occur in the neighbourhood of various equilibrium states, and a series of space bifurcations appears (Fig. 7.107). This means that very often the magnitude and sign

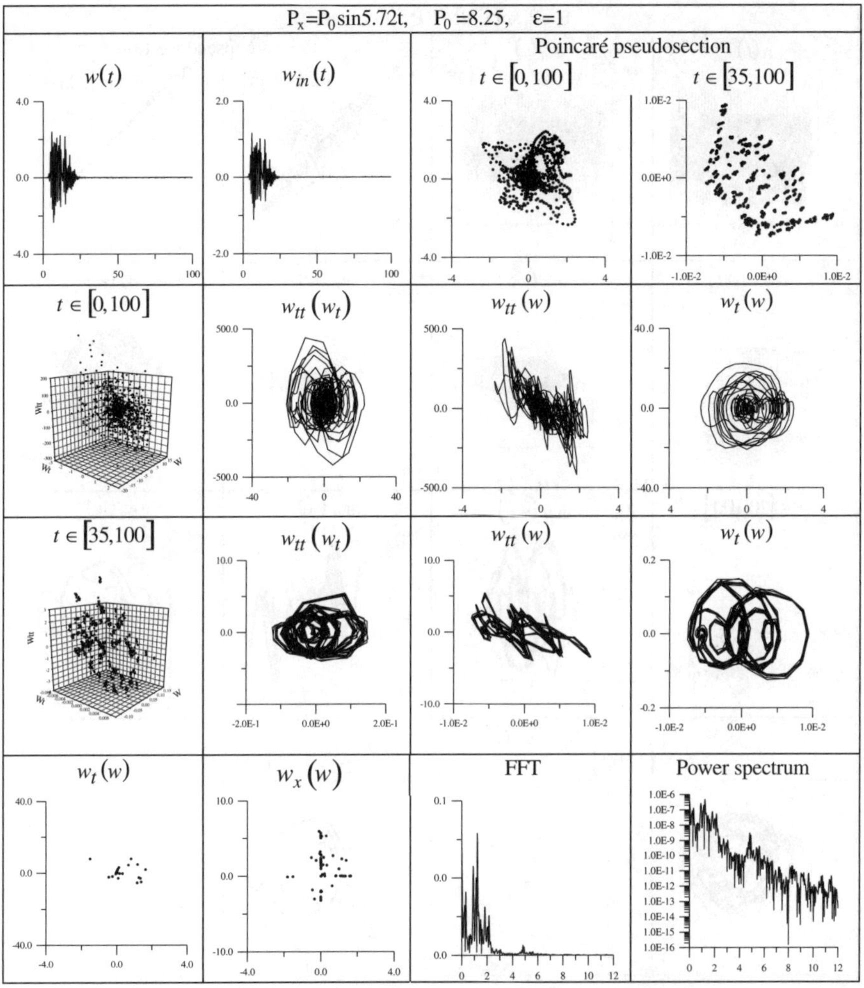

Fig. 7.99. Time histories, Poincaré sections and pseudosections, modal portraits, FFT and power spectrum ($P_0 = 8.25$) (continued on page 333)

of the equilibrium deflection change, and the oscillations take place in the vicinity of this equilibrium deflection.

In state V, a post-crisis state is observed (Figs. 7.100 and 7.101), and oscillations occur in a regime of concavity and convexity. The strange attractors collapse, and there are two or four points in the intersections of the phase and modal portraits.

The corresponding characteristics for the chaotic state are reported in Fig. 7.102. The modal portrait is composed of a set of arbitrarily distributed curves, which create cross, which is interpreted as the creation of two mutually

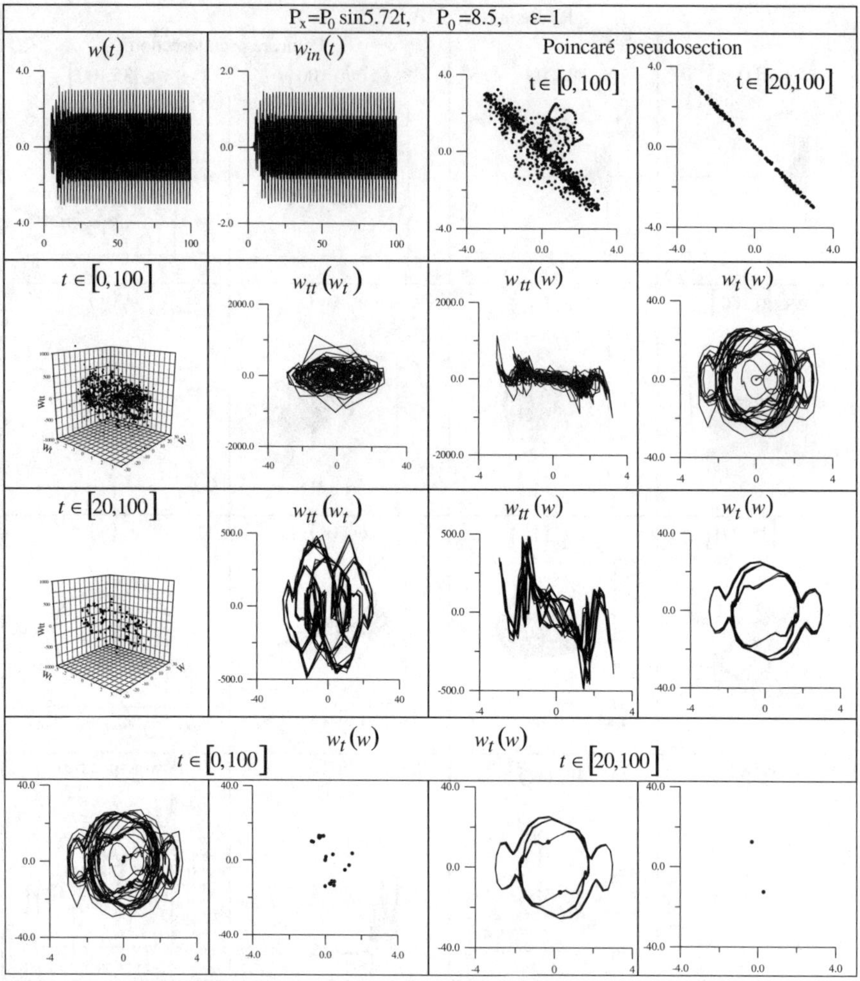

Fig. 7.100. Time histories, Poincaré sections and pseudosections, modal portraits, FFT and power spectrum ($P_0 = 8.5$) (continued on page 333)

perpendicular waves at different times (see Fig. 7.102, surfaces for $t = 22.44$ and $t = 46.772$). The oscillations occur in the longitudinal direction with the creation of two or four waves, and in the transverse direction with the creation of one, three or five waves.

7.11 Solitary Waves

In this section we describe some results of applying a method described earlier to reduce the dimension of the PDEs by projection to ODEs, i.e.

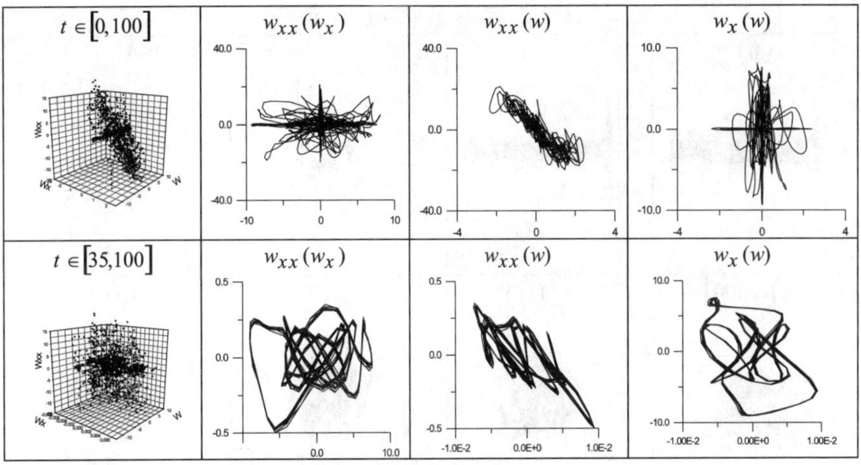

Fig. 7.99. (continued)

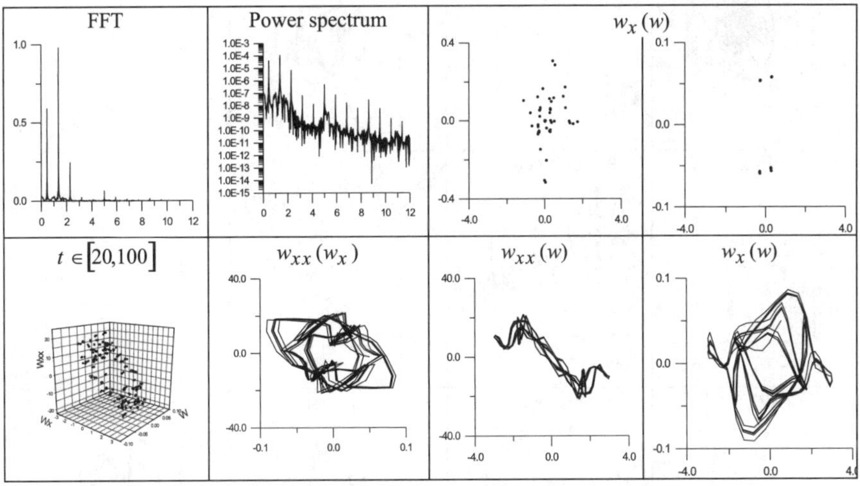

Fig. 7.100. (continued)

the method of finite differences with a spatial approximation of order $O(h^4)$. This approach has been used to study standing and travelling waves in a space occupied by a thin plate subjected to the action of a one-sided periodic longitudinal load.

Infinite gradients can appear in a wave profile in various physical situations. For instance, a wave on a fluid surface, may collapse by spurting.

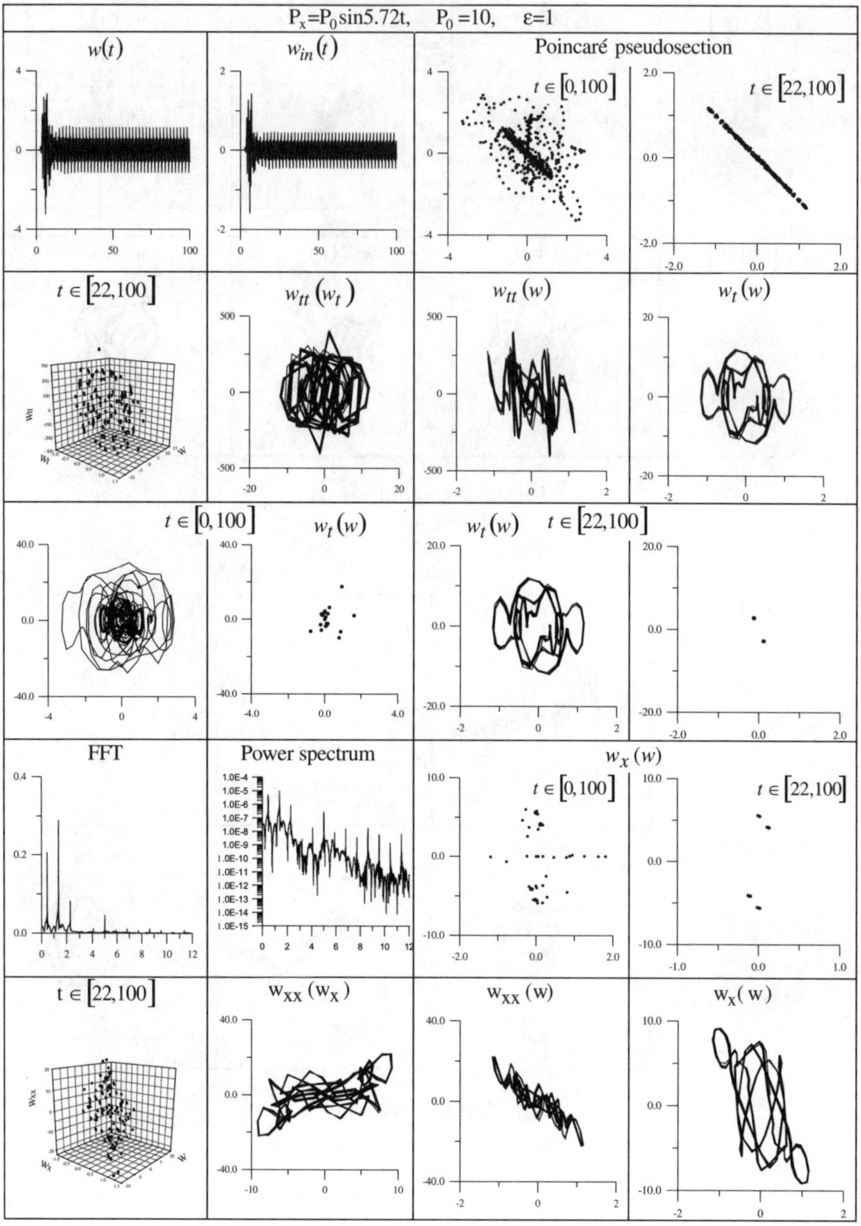

Fig. 7.101. Time histories, Poincaré sections and pseudosections, modal portraits, FFT and power spectrum ($P_0 = 10$)

If the flow is composed of independent particles, then nonuniqueness appears in the wave profile. After the occurrence of a discontinuity in the

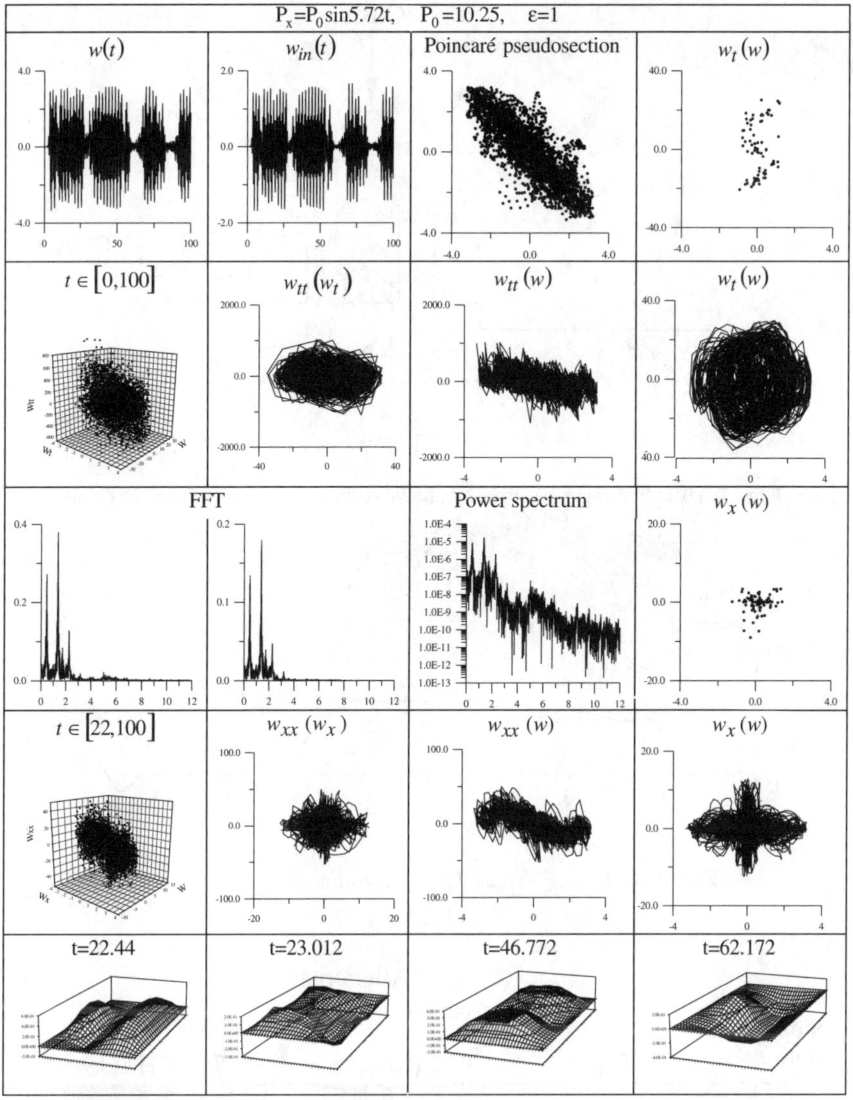

Fig. 7.102. Time histories, phase portraits, Poincaré sections and pseudosections, modal portraits, FFTs, power spectrum and spatial configurations ($P_0 = 10.25$)

fundamental flow, several different flows appear, moving with vary different velocities (multiflow). For a sound or electromagnetic field, where nonuniqueness is not allowed, the further development of the nonlinear wave depends on whether dissipative or dispersive effects will dominate in the neighbourhood of a rapid field change. The following one-wave PDE has been analysed:

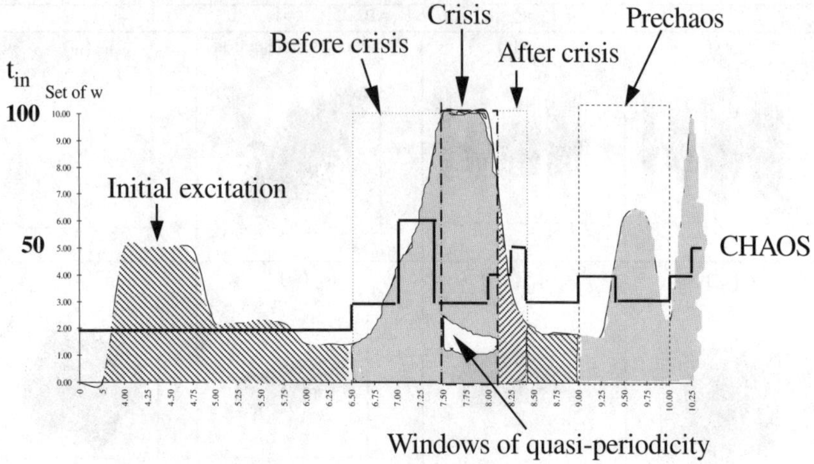

Fig. 7.103. Scenario leading to chaos versus P_0 (nonsymmetric case)

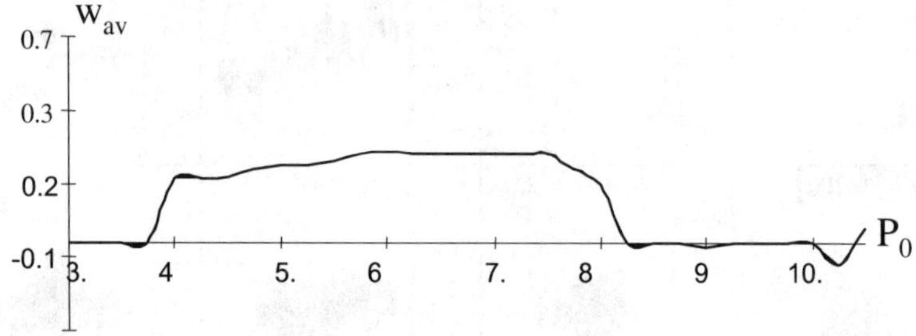

Fig. 7.104. w_{av} versus P_0 (nonsymmetric case); see text

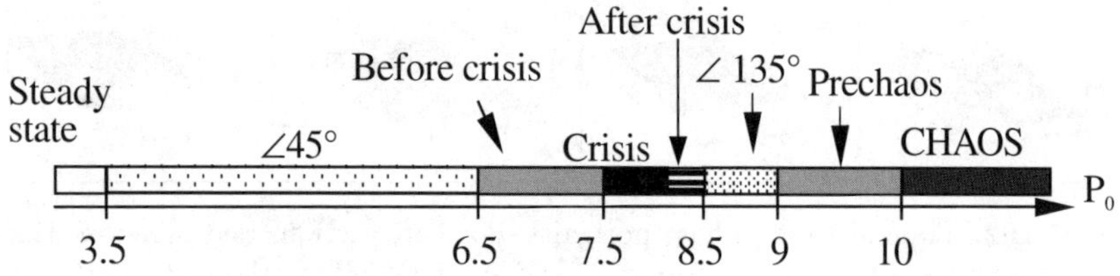

Fig. 7.105. Dynamics versus P_0 ($\omega = 5.72$) (nonsymmetric case)

$$\frac{\partial u}{\partial t} + \nu(u)\frac{\partial u}{\partial x} + \beta\frac{\partial^3 u}{\partial x^3} - \alpha\frac{\partial^2 u}{\partial x^2} = 0,$$

where u denotes the average velocity.

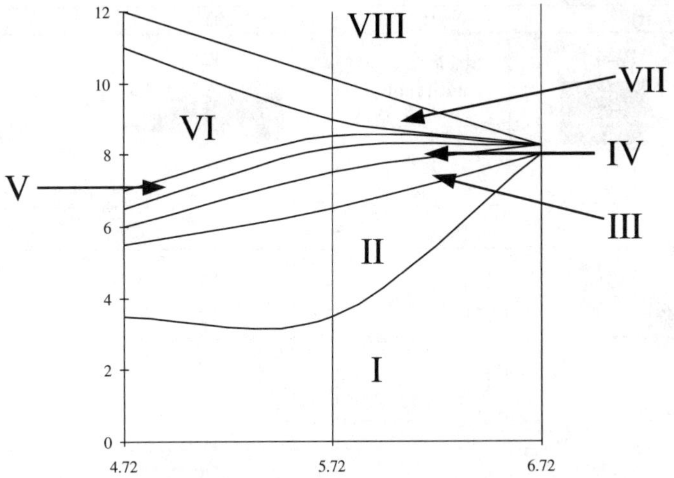

Fig. 7.106. Various dynamical states within the two-parameter space (nonsymmetric case); see text for a detailed description

This equation can be reduced either to the Kortweg–de Vries equation ($\alpha = 0$) or to Burger's equation ($\beta = 0$). Owing to high-frequency dissipation, a discontinuity will not lead to stability loss. In addition, the dispersion bounds the width of the discontinuity, and as a result a "stationary" travelling wave with a constant profile can appear. The stationary waves of the Kortweg–de Vries equation correspond to a conservative nonlinear oscillator.

Periodic motions in the neighbourhood of separatrices are waves. A spatially solution that can be described by one soliton is associated with a separatrix.

The one-dimensional stationary waves that appear in one-dimensional continuous systems (e.g. transmission lines) and in the case of plane waves (for instance, solitons on water governed by the Kortweg–de Vries equation) are well studied. On the other hand, it is clear that solitons in a falling water layer and ionic-sound solitons in a plasma must depend on two spatial coordinates. A simple model, which is a generalization of the Kortweg–de Vries equation of the form

$$\frac{\partial}{\partial x} + (u_t + uu_x - \delta u_{xxx}) = \gamma u_{yy}.$$

is given in [183].

As has been mentioned in [174], a particular feature of the atmosphere of Jupiter is the Great Red Spot, which is a two-dimensional Rossby soliton. Rossby waves, in a linear approximation, correspond to waves in a rotating atmosphere, and the main reason for their existence is related to the change with a latitude of the horizontal projection of the Coriolis force.

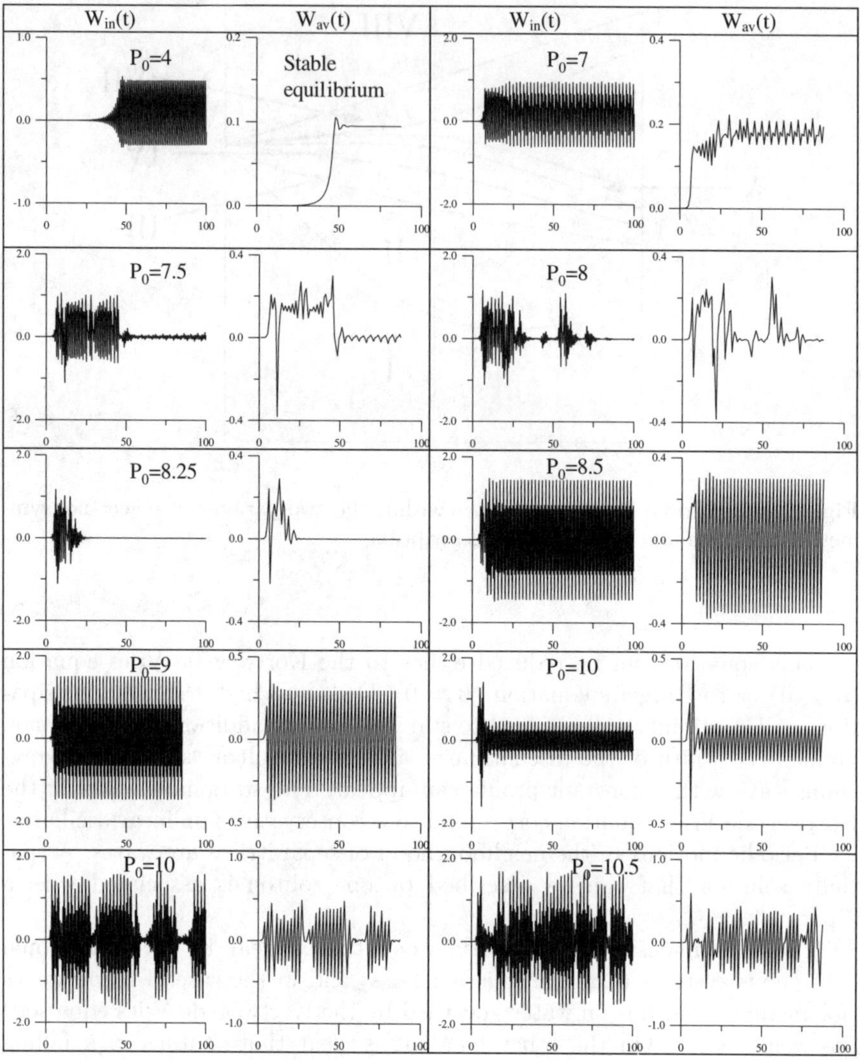

Fig. 7.107. Time histories of w_{in} and w_{av} for different values of P_0 (compare with Fig. 7.96)

For a medium with dissipation, the Chochleva–Zabolotskij equation plays a representative role in describing the waves that occur [189]. However, in the case of thin flexible plates, solitary waves have not previously been detected within the kinematic Kirchhoff–Love model. Here, we illustrate and analyse this behaviour.

The boundary and initial conditions (7.67) and (7.69) were applied, and the full mesh space $\bar{G}_h$ was used without symmetry conditions. A spatial step $h = 1/16$ and a time step $\Delta t = 2 \times 10^{-4}$ were used.

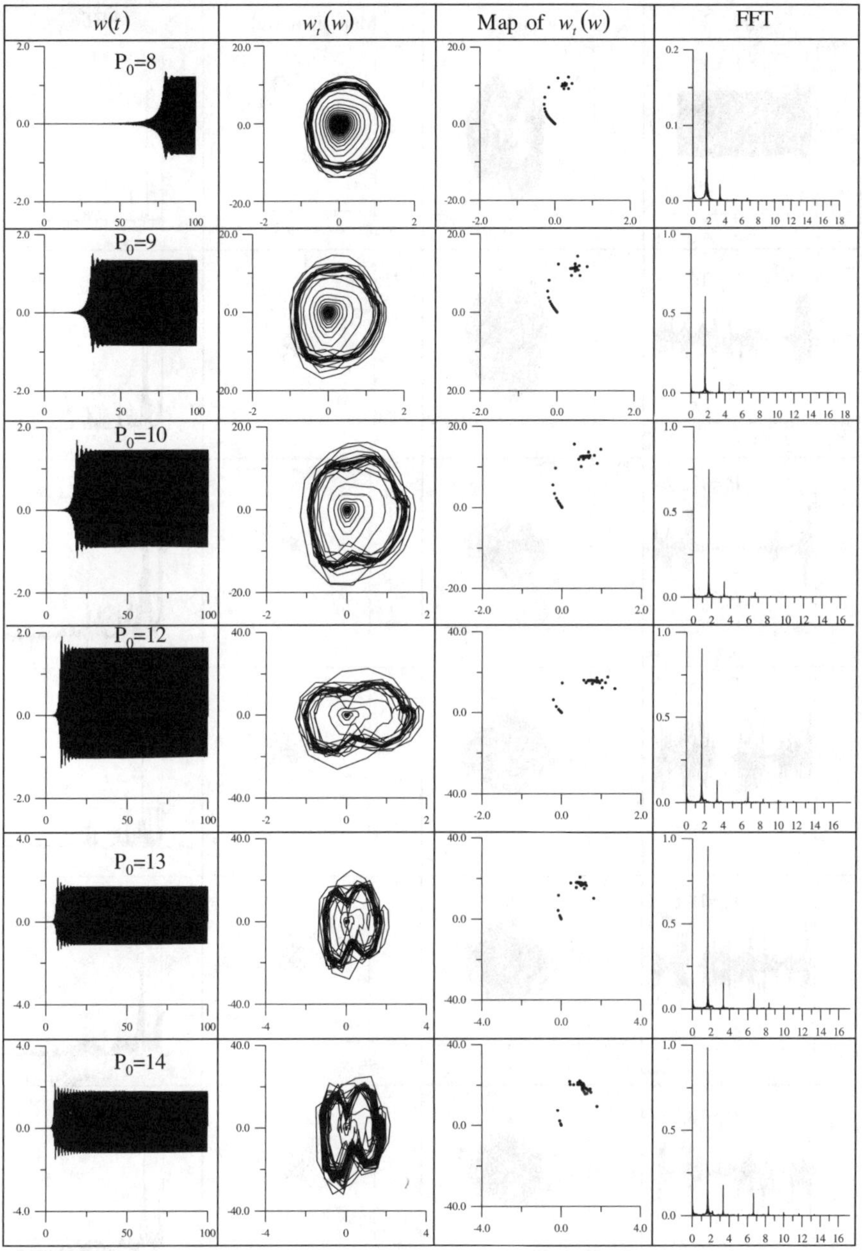

Fig. 7.108. Time histories of $w(t)$, phase portraits and Poincaré maps of $w_t(w)$, and FFTs for different P_0 values (continued on page 340 and 341)

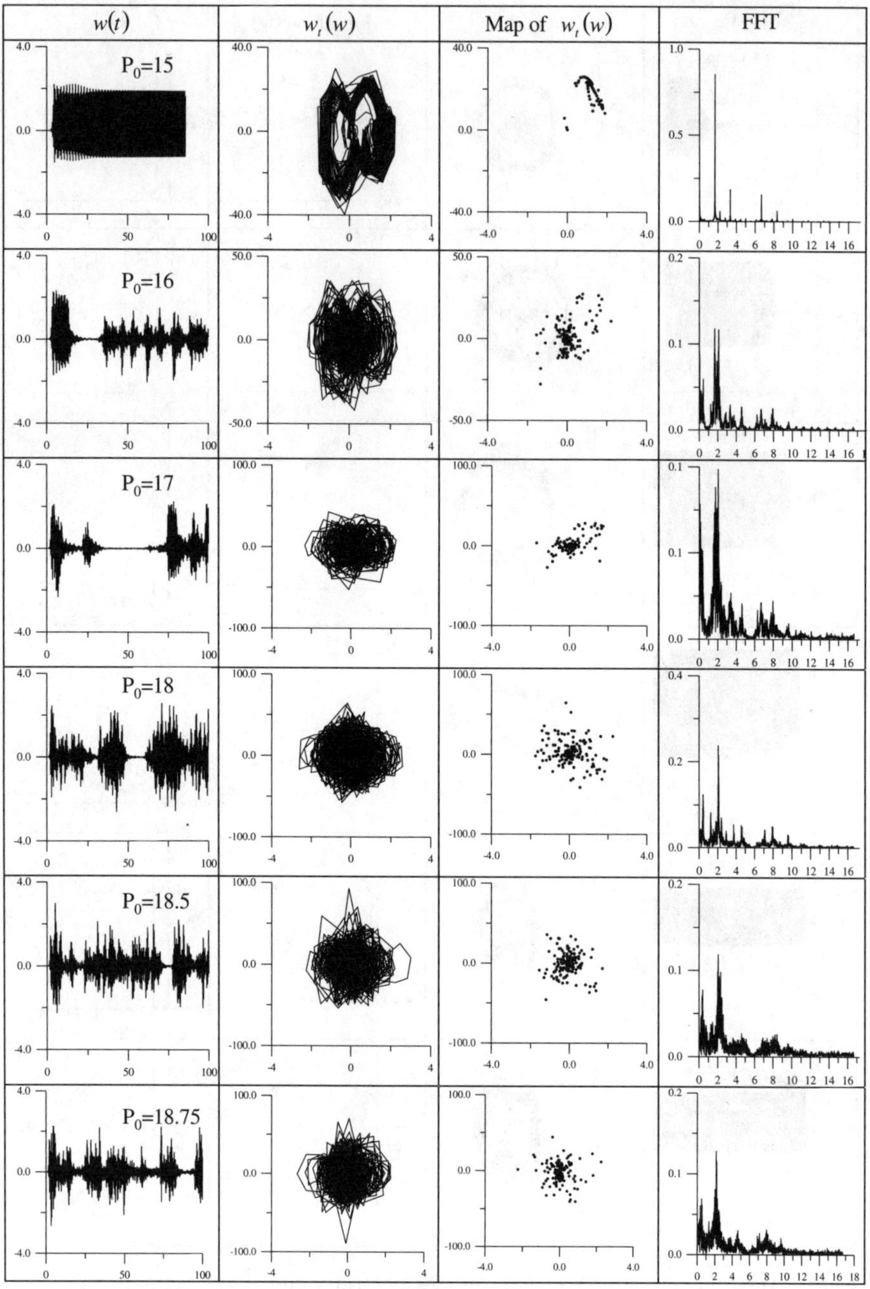

Fig. 7.108 (continued)

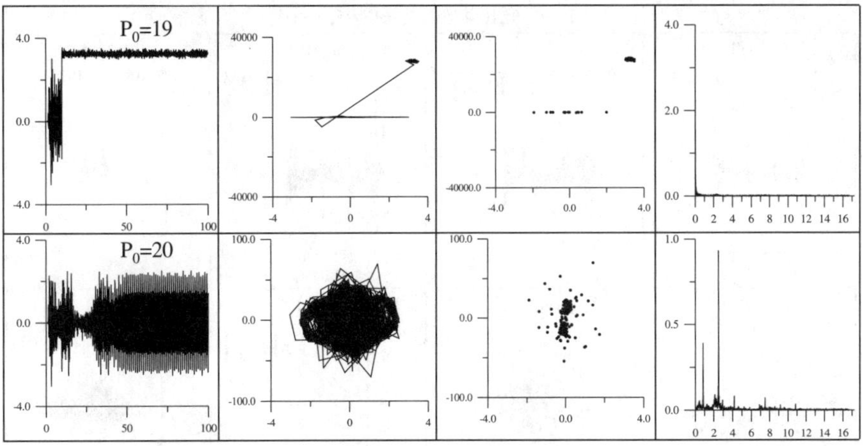

Fig. 7.108 (continued)

To analyse the very sensitive structure of multifrequency and stochastic oscillations, and to study transitions between different oscillation regimes, we have used the same characteristics as used earlier.

We took P_0 as the control parameter, and the other parameters were fixed: $\omega = 10.47$, $\lambda = \varepsilon = 1$, $\nu = 0.3$. In Figs. 7.107 and 7.108, the deflections $w(0.5, 0.5, t)$, phase portraits, Poincaré maps and FFTs are reported for various values of P_0. Beginning with $P_0 \geq 16$, the mechanical system analysed here is in a chaotic state, and a series of transitions between symmetric and nonsymmetric oscillation forms is observed.

For $P_0 > 19.25$, neither solitons nor jump phenomena appear. Instead, travelling waves and regular oscillations are observed for $t \in [48, 100]$ (see Fig. 7.112).

We shall now focus our attention on the interval $P_0 \in [18.95; 19.25]$. In this interval, an oscillation jump is observed, which results in a change of the spatial–temporal configuration of the dynamical state, and in the occurrence of standing and travelling waves. For $P_0 = 18.95$ and $P_0 = 19$, the fundamental characteristics are reported in Figs. 7.109 and 7.110. The time before a jump occurs is different in the two cases. The oscillations before the jump are chaotic, as is indicated by the characteristics presented. Travelling bending waves are observed, shown in the Figs. 7.109 and 7.110 for selected time moments, and the maximum deflections occur in the direction of two mutually pendicular symmetry axes. The travelling waves become a standing wave (Fig. 7.111, $t = 10.04, P_0 = 19$), and then a jump to another deflection level occurs, as indicated by the characteristics shown in Fig. 7.110. At the centre of the plate, the effect of a discontinuity mentioned above is observed in the phase and modal portraits. The space phase and modal portraits and their projections into the planes show that chaos occurs (as indicated by the

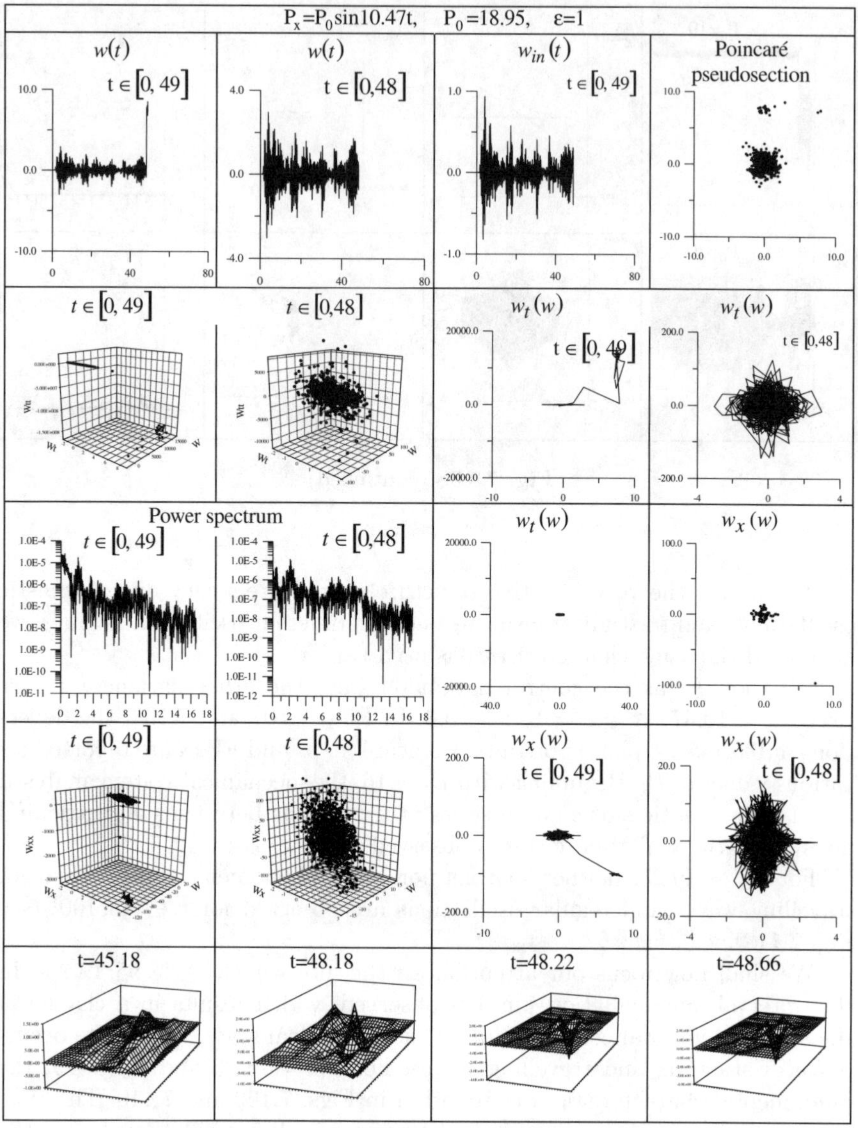

Fig. 7.109. Time histories of $w(t)$ and $w_{in}(t)$, phase portraits and Poincaré maps of $w_t(w)$, power spectra, modal portraits, and spatial configurations for the time moments and intervals indicated

broadband character of the power spectrum). But a collapse of the standing waves starting from $t = 10.04, P_0 = 9$ does not appear. A new type of wave self-organization occurs in the standing wave shown in Fig. 7.110 at $t = 20$. This type of wave does not change with time and is practically stable over

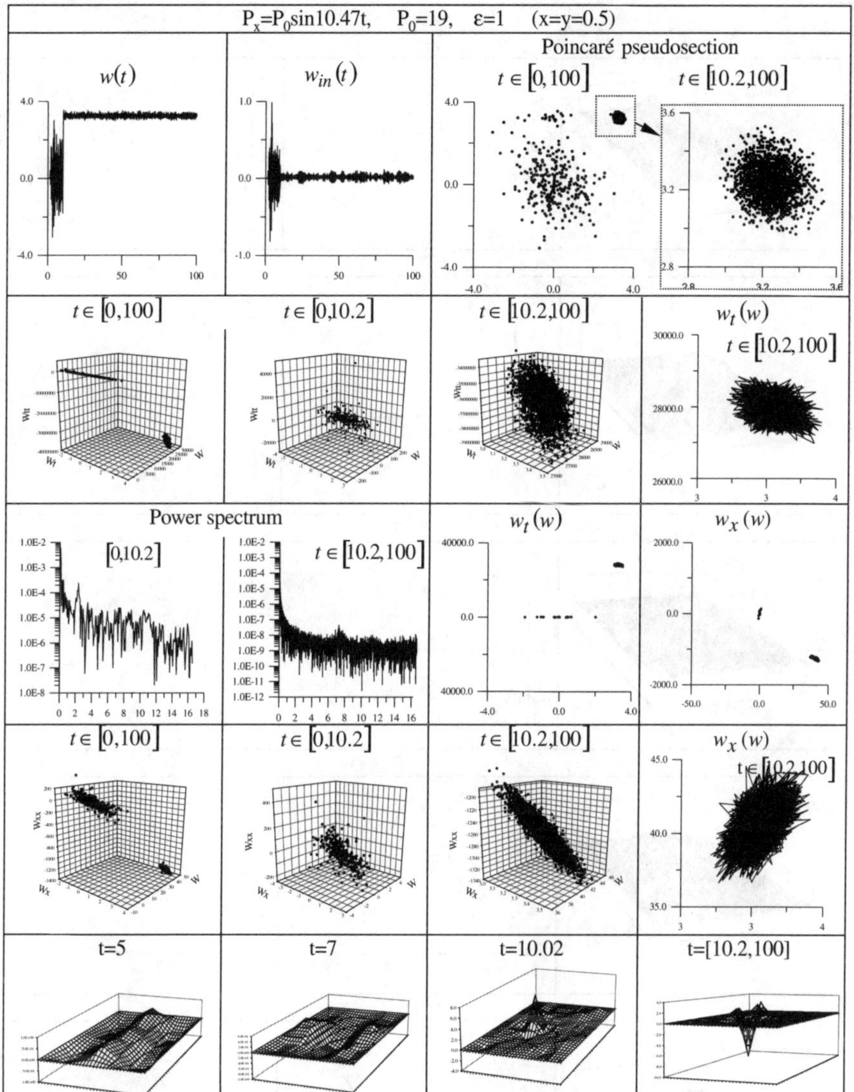

Fig. 7.110. Time histories of $w(t)$ and $w_{in}(t)$, Poincaré pseudosections, modal portraits, power spectra, and spatial configurations for the time moments and intervals indicated

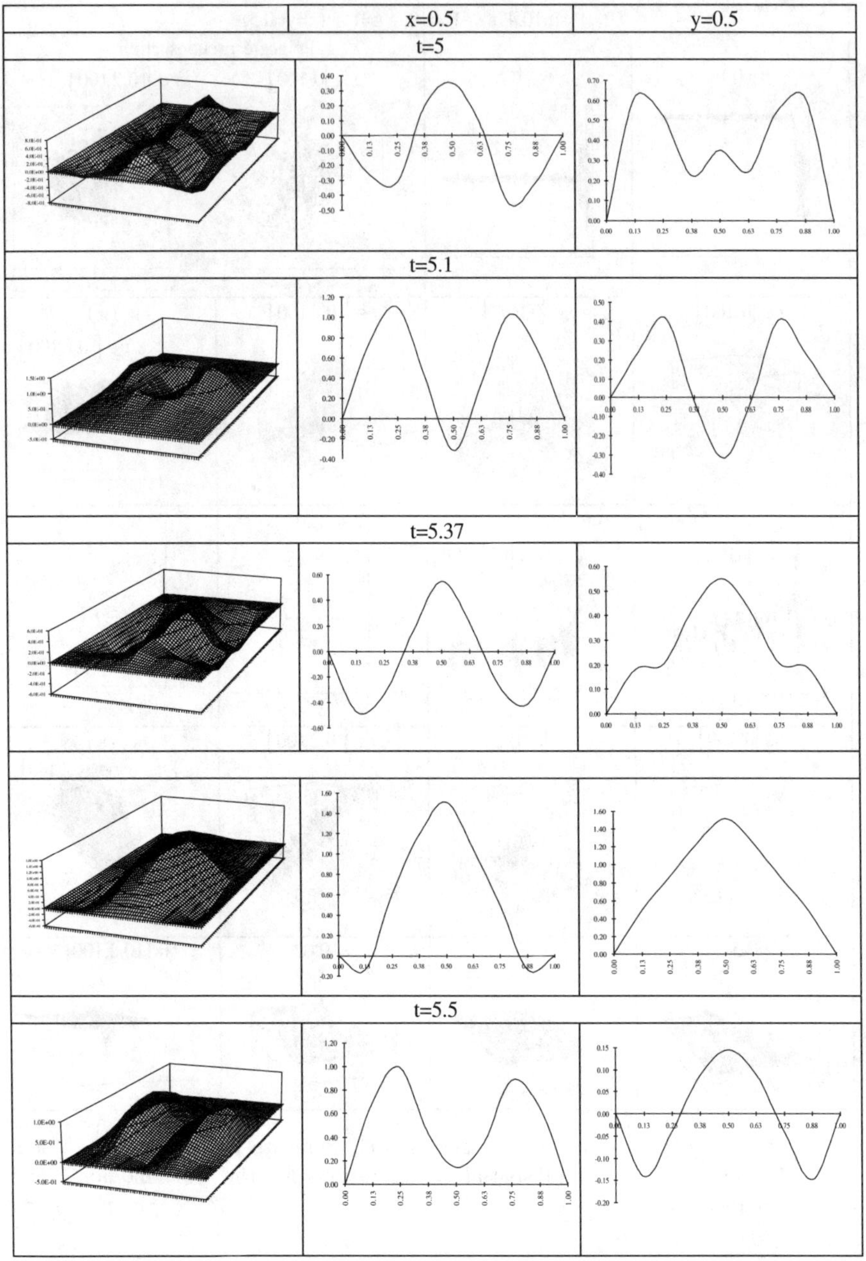

Fig. 7.111. Spatial configurations and cross-sections for the times and values of x and y indicated (see text)

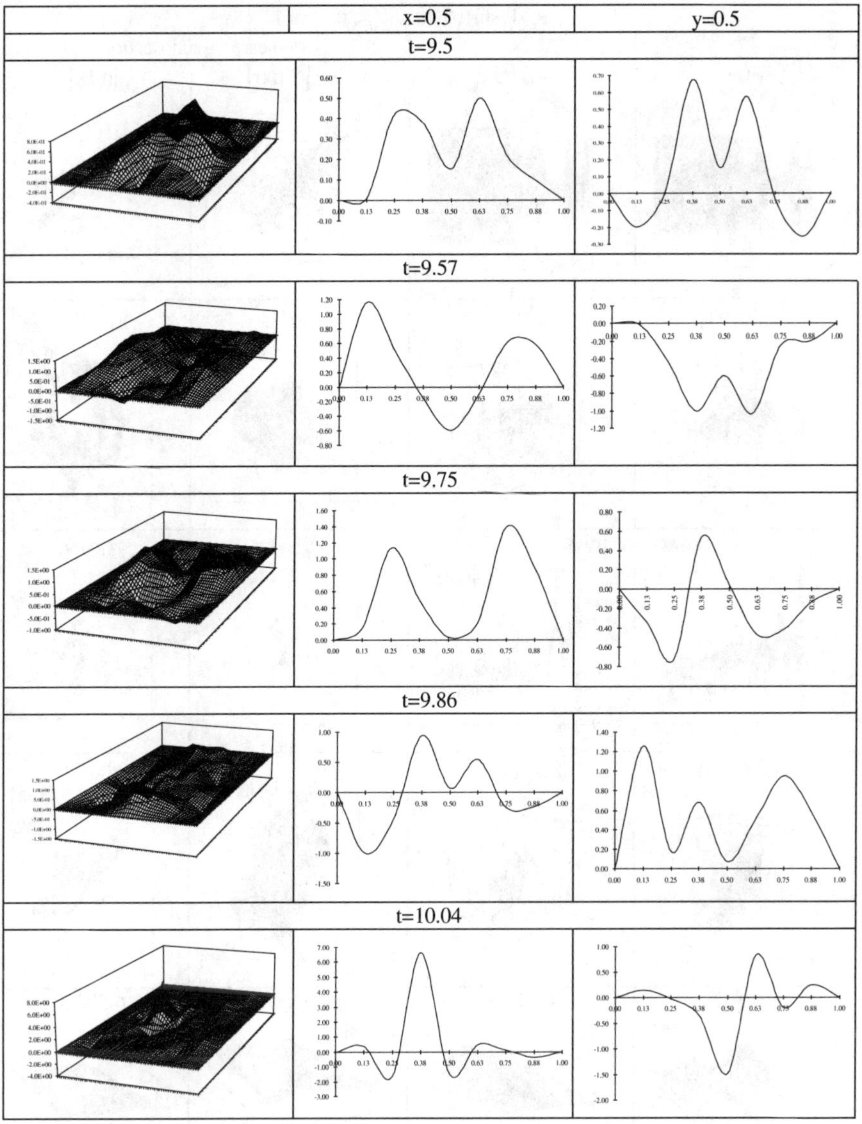

Fig. 7.111 (continued)

$t \in [10.04; 100]$. This observation leads to the conclusion that, within a chaotic state, a new self-organized behaviour, which we call two-dimensional standing solitary waves, can appear.

A periodically excited flexible plate is a complex dynamical system which can exhibit various dynamical behaviours. The vibrational process can be characterized by complex resonance structures, a collapse of a vibrational

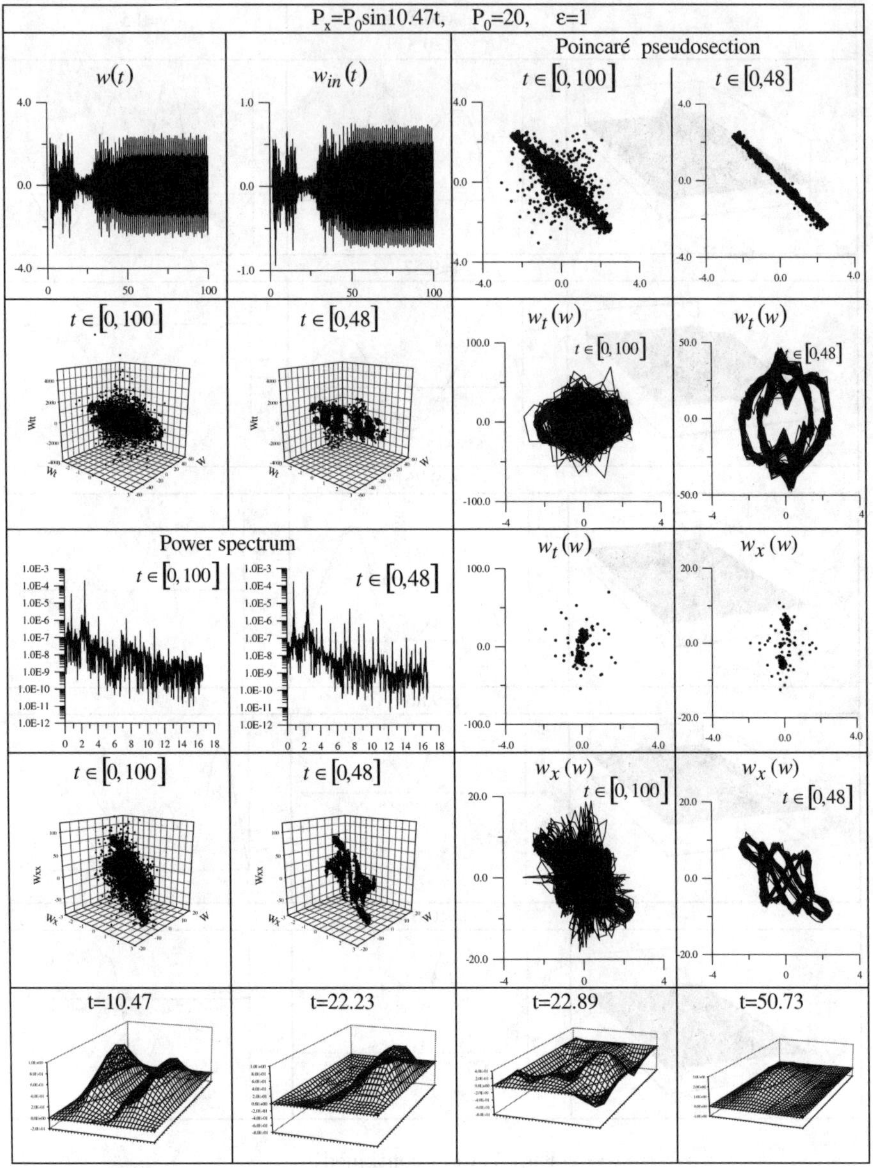

Fig. 7.112. Time histories of $w(t)$ and $w_{in}(t)$, Poincaré pseudosections, phase portraits and Poincaré sections of $w_t(w)$, power spectra, modal portraits and maps, and spatial configurations for the time moments indicated

regime leading to a change of the spatial–temporal state, the occurrence of standing or travelling waves, stability loss with respect to either symmetric or nonsymmetric modes, and other features. This various features have been discussed and illustrated in this section.

8 Dynamics of Thin Elasto-Plastic Shells

In Chap. 6, a mathematical model governing the oscillations of a flexible shell with physical nonlinearity and coupling between the thermal and deformation fields has been presented. However, the loading and unoading processes overlap in the $\sigma_i(e_i)$ diagram, and the remaining elasto-plastic deformation has not been taken into account. In a theoretical treatment of complicated shell oscillations, in order to analyse the stress–strain state properly, the elasto-plastic deformation should be considered, as well as the fatigue behaviour of the material. Only in this case can the mathematical model be close to the real behaviour of a structure. The aim of this chapter is to describe the development of complex mathematical models of structures, including elasto-plastic deformation and cyclic loading.

The fundamental relations of the theory of small elastic–plastic deformations with a cyclic load are formulated in Sect. 8.1. In Sect. 8.2, a method of solution of the equations governing the elasto–plastic deformation of flexible shells is described. The oscillations and stability of elasto-plastic shells are considered in Sect. 8.3, where many computational examples are reported.

8.1 Fundamental Relations

We present below the fundamental relations of the theory of small elastic–plastic deformations when a cyclic load is applied, using the the Maugin principle as generalized by Moskvitin [166, 167]:

$$\begin{aligned} \sigma_{11} &= 3Ke_0 + \frac{2}{3}\frac{\sigma_i}{e_i}(e_i - e_0), \quad (\overleftrightarrow{1,2,3}) \\ \sigma_{12} &= \frac{2}{3}\frac{\sigma_i}{e_i}e_{12}. \end{aligned} \tag{8.1}$$

Here (as earlier), K, e_0 and ν correspond to the modulus of volume compression, the average deformation and Poisson's ratio;

$$e_i = \left\{\frac{2}{3}\left[(e_{11} - e_0)^2 + (e_{22} - e_0)^2 + (e_{33} - e_0)^2 + e_{12}^2 + e_{23}^2 + e_{31}^2\right]\right\}^{1/2}$$

characterizes the intensity of the deformation. We use also the relation $\sigma_i = f(e_i) = 3G(1-\overline{w})e_i$, introduced earlier, where $\overline{w}$ is the Iliushin function. The relation above can be generalized in the case of a cycle load to

$$\overline{\sigma}_i^{(n)} = \alpha_n \sigma_T \Phi' \left(\frac{\overline{e}^{(n)}}{\alpha_n e_T} \right), \tag{8.2}$$

for the nth cycle of deformation. The elasto-plastic properties for an arbitrary number of load cycles are defined by the initial strain diagram. In (8.2),

$$\alpha_n = \alpha_2 (n-1)^\gamma, \tag{8.3}$$

where $\gamma < 1$ and α_n is a constant associated with the cycle. If we take $\alpha_2 = 2$ and $\gamma = 0$, then (8.3) describes the properties of a material that is cyclically ideal with respect to the Maugin principle. In this case the formula (8.2) has the following form in the first loading cycle:

$$\sigma_i^{(n)} = \sigma_T \Phi' \left(\frac{e^{(1)}}{e_T} \right). \tag{8.4}$$

For variable loads, we obtain

$$\overline{\sigma}_i^{(n)} = 2\sigma_T \Phi' \left(\frac{\overline{e}^{(n)}}{2e_T} \right). \tag{8.5}$$

Comparing these two relations, one can conclude that the stress intensity in the system of coordinates for the nth cycle is equal to twice the value of the stress intensity during the first loading. The latter corresponds to half the value of the deformation intensity in the initial coordinate system. Using Moskvitin's generalized model related to elastic–plastic unloading, for the nth loading cycle Iliushin's relation can be rewritten in the following form [104]:

$$\begin{aligned} \overline{\sigma}_{11}^{(n)} &= 3K\overline{e}_0^{(n)} + \frac{2}{3}\left(\frac{\overline{\sigma}^{(n)}}{\overline{e}_i^{(n)}} \right)(\overline{e}_{11}^{(n)} - \overline{e}_0^{(n)}), \\ \overline{\sigma}_{12}^{(n)} &= \frac{2}{3}\frac{\overline{\sigma}_i^{(n)}}{\overline{e}_i^{(n)}} \quad (\overleftrightarrow{1,2,3}), \end{aligned} \tag{8.6}$$

where

$$\begin{aligned} \overline{e}_{11}^{(n)} &= e_{11}^{(n-1)} - e_{11}^{(n)}, \quad \overline{\sigma}_{11}^{(n)} = \sigma_{11}^{(n-1)} - \sigma_{11}^{(n)} \quad (\overleftrightarrow{1,2,3}), \\ \overline{e}_{12}^{(n)} &= e_{12}^{(n-1)} - e_{12}^{(n)}, \quad \overline{\sigma}_{12}^{(n)} = \sigma_{12}^{(n-1)} - \sigma_{12}^{(n)}. \end{aligned} \tag{8.7}$$

Now $\sigma_{11}^{(n)}$, $e_{11}^{(n)}$, $\sigma_{12}^{(n)}$, $e_{12}^{(n)}$ $(\overleftrightarrow{1,2,1})$ are components of the stress and deformation tensors in the nth loading cycle in the initial coordinate system; $\sigma_{11}^{(n-1)}$, $e_{11}^{(n-1)}$, $\sigma_{12}^{(n-1)}$, $e_{12}^{(n-1)}$ $(\overleftrightarrow{1,2,3})$ are the stress tensor components at the end of the previous, $(n-1)$th loading cycle in the initial coordinate system; and

$\overline{\sigma}_{11}^{(n)}, \overline{e}_{11}^{(n)}, \overline{\sigma}_{12}^{(n)}, \overline{e}_{12}^{(n)}$ $(\overleftrightarrow{1,2,3})$ are the components of the stress and deformation tensors in the nth cycle in the moving coordinate system where the origin coincides with the origin at the begining of the unloading process and where the axes are parallel to those of the initial system of coordinates.

The expressions for the deformation intensity $\overline{e}_i^{(n)}$, the average stress values $\overline{\sigma}_i^{(n)}$ and the average deformation $\overline{e}_0^{(n)}$ have the following forms for a cyclic load:

$$\overline{e}_i^{(n)} = \left\{ \frac{2}{3} \left[(\overline{e}_{11}^{(n)} - \overline{e}_0^{(n)})^2 + (\overline{e}_{22}^{(n)} - \overline{e}_0^{(n)})^2 + (\overline{e}_{33}^{(n)} - \overline{e}_0^{(n)})^2 \right.\right.$$
$$\left.\left. + (\overline{e}_{12}^{(n)})^2 + (\overline{e}_{23}^{(n)})^2 + (\overline{e}_{31}^{(n)})^2 \right] \right\}^{1/2}, \tag{8.8}$$

$$\overline{\sigma}_0^{(n)} = \frac{1}{3}(\overline{\sigma}_{11}^{(n)} + \overline{\sigma}_{22}^{(n)} + \overline{\sigma}_{33}^{(n)}), \tag{8.9}$$

$$\overline{e}_0^{(n)} = \frac{1}{3}(\overline{e}_{11}^{(n)} + \overline{e}_{22}^{(n)} + \overline{e}_{33}^{(n)}). \tag{8.10}$$

The coupling between the spherical stress and deformation tensors for the nth loading cycle has the form

$$\overline{\sigma}_0^{(n)} = 3K\overline{e}_0^{(n)}. \tag{8.11}$$

When the theory of small elasto-plastic deformations is used, the relation between the stress and the deformation intensity reads:

$$\overline{\sigma}_0^{(n)} = 3G(1 - \overline{w}^{(n)})\overline{e}_i^{(n)}, \tag{8.12}$$

where $\overline{w}^{(n)} = f(\overline{e}_i^{(n)})$ is the Iliushin function [104] for the nth loading cycle, which does not change on moving from one cycle to another. This function depends on the material and on the type of approximation used. The various $\sigma_i(e_i)$ relations are described and discussed in the monograph [133]. We present only the form for $\overline{w}^{(n)}(\overline{e}_i^{(n)})$ related to the bilinear approximation of $\sigma_i(e_i)$ in the moving coordinate system, which will be used later:

$$\overline{w}^{(n)} = \begin{cases} 0 \quad \text{for} \quad \overline{e}_i^{(n)} \le \alpha_n e_T, \\ \left(1 - \dfrac{G_2}{G_1}\right)\left(1 - \dfrac{\alpha_n e_T}{\overline{e}_i^{(n)}}\right) \quad \text{for} \quad \overline{e}_i^{(n)} > \alpha_n e_T, \end{cases} \tag{8.13}$$

where G_1 and G_2 are the shear and tangent moduli of elasticity.

The Kirchhoff–Love hypothesis will be used to describe the shell structure in what follows. This hypothesis assumes that $\sigma_{33} = 0$ and $e_{23} = e_{31} = 0$ (it is also possible to obtain equations using the Timoshenko and Reussner kinematic hypotheses, and others).

For the Kirchhoff–Love kinematic model, the relations (8.6) and (8.7) for the nth loading cycle read

$$e_{33}^{(n)} = e_{33}^{(n-1)} + \frac{1+\nu-(1-2\nu)(1-\overline{w}^{(n)})}{1+\nu+(1-2\nu)(1-\overline{w}^{(n)})}(e_{11}^{(n-1)} + e_{22}^{(n-1)} + e_{11}^{(n)} + e_{22}^{(n)}), \tag{8.14}$$

where $e_{33}^{(n)}$ is obtained by assuming a plane stress–strain state ($\sigma_{33} = 0$). This quantity appears in the expression for the deformation intensity, together with other components of the deformation tensor:

$$\begin{aligned}
\sigma_{11}^{(n)} &= \frac{E(1-\overline{w}^{(n)})}{(1+\nu)(1-\overline{\beta}^{(n)})}(e_{11}^{(n)} + \overline{\beta}^{(n)} e_{22}^{(n)}) + \sigma_{11}^{(n-1)} \\
&- \frac{E(1-\overline{w}^{(n)})}{(1+\nu)(1-\overline{\beta}^{(n)})}(e_{11}^{(n-1)} + \overline{\beta}^{(n)} e_{22}^{(n-1)}) \quad (1 \leftrightarrow 2), \\
\sigma_{12} &= \frac{E(1-\overline{w}^{(n)})}{2(1+\nu)} e_{12}^{(n)} + \overline{\sigma}_{12}^{(n-1)} - \frac{E(1-\overline{w}^{(n)})}{2(1+\nu)} e_{11}^{(n-1)},
\end{aligned} \tag{8.15}$$

where

$$\overline{\beta}^{(n)} = \frac{1}{2}\left(\frac{3-2\alpha(1-\overline{w}^{(n)})}{3+2\alpha(1+\overline{w}^{(n)})}\right), \qquad \alpha = \frac{3}{2}\frac{1-2\nu}{1+\nu}. \tag{8.16}$$

By integrating the stresses (8.15) with respect to $z \left(-\frac{h}{2} \le z \le \frac{h}{2}\right)$, we obtain the following stresses in the middle surface:

$$\begin{aligned}
T_{11}^{(n)} &= \frac{Eh}{1-\nu^2}(\varepsilon_{11}^{(n)} + \nu\varepsilon_{22}^{(n)}) + \Delta T_{11}^{(n)} \quad (1 \leftrightarrow 2), \\
T_{12}^{(n)} &= \frac{Eh}{2(1+\nu)}\varepsilon_{12}^{(n)} + \Delta T_{12}^{(n)},
\end{aligned} \tag{8.17}$$

where

$$\begin{aligned}
\Delta T_{11}^{(n)} &= \frac{Eh}{1+\nu}(\overline{A}_1^{(n)} \varepsilon_{11}^{(n)} + \overline{B}_1^{(n)} \varepsilon_{22}^{(n)}) \\
&+ \frac{Eh^2}{1+\nu}(\overline{A}_2^{(n)} æ_{11} + \overline{B}_2^{(n)} æ_{22}) + \int_{-\frac{h}{2}}^{\frac{h}{2}} \sigma_{11}^{(n-1)}\, dx_3 \\
&- \frac{E}{1+\nu}\int_{-\frac{h}{2}}^{\frac{h}{2}} \frac{1-\overline{w}^{(n)}}{1-\overline{\beta}^{(n)}}(e_{11}^{(n-1)} + \overline{\beta}^{(n)} e_{22}^{(n-1)})\, dx_3 \quad (1 \leftrightarrow 2), \\
\Delta T_{12}^{(n)} &= \frac{Eh}{2(1+\nu)}\overline{C}_1^{(n)} \varepsilon_{12}^{(n)} + \frac{Eh^3}{1+\nu}\overline{C}_2^{(n)} æ_{12}^{(n)} \\
&+ \int_{-\frac{h}{2}}^{\frac{h}{2}} \sigma_{12}^{(n-1)}\, dx_3 - \frac{E}{2(1+\nu)}\int_{-\frac{h}{2}}^{\frac{h}{2}} (1-\overline{w}^{(n)})e_{12}^{(n-1)}\, dx_3.
\end{aligned} \tag{8.18}$$

In (8.18), the following notation has been used:

$$\overline{F}_1^{(n)} = \frac{1-\overline{w}^{(n)}}{1-\overline{\beta}^{(n)}} - \frac{1}{1-\nu}, \quad \overline{F}_2^{(n)} = \frac{1-\overline{w}^{(n)}}{1-\overline{\beta}^{(n)}}\overline{\beta}^{(n)} - \frac{1}{1-\nu},$$

$$\overline{A}_m^{(n)} = \frac{1}{n^j}\int_{-\frac{h}{2}}^{\frac{h}{2}} \overline{F}_1^{(n)} x_3^{(j-1)}\, dx_3, \quad \overline{B}_m^{(n)} = \frac{1}{n^j}\int_{-\frac{h}{2}}^{\frac{h}{2}} \overline{F}_2^{(n)} x_3^{(j-1)}\, dx_3, \tag{8.19}$$

$$\overline{C}_m^{(n)} = \frac{1}{n^j}\int_{-\frac{h}{2}}^{\frac{h}{2}} (-\overline{w}^{(n)}) x_3^{(j-1)}\, dx_3 \quad (m = 1,2,3).$$

Multiplying $\sigma_{11}^{(n)}$, $\sigma_{22}^{(n)}$, $\sigma_{12}^{(n)}$ in (8.15) by x_3 and integrating over the thickness, we obtain the following moments:

$$\begin{aligned} M_{11}^{(n)} &= D(æ_{11}^{(n)} + \nu æ_{22}^{(n)}) + \Delta M_{11}^{(n)} \quad (1 \leftrightarrow 2),\\ M_{12}^{(n)} &= D æ_{12}^{(n)} + \Delta M_{12}^{(n)}, \end{aligned} \tag{8.20}$$

where

$$D = \frac{Eh^3}{12(1-\nu^2)}$$

is the stiffness of the plate,

$$\begin{aligned} \Delta M_{11}^{(n)} &= \frac{Eh^2}{1+\nu}(\overline{A}_2^{(n)}\varepsilon_{11}^{(n)} + \overline{B}_2^{(n)}\varepsilon_{22}^{(n)})\\ &\quad + 12(1-\nu)D(\overline{A}_3^{(n)} æ_{11}^{(n)} + \overline{B}_3^{(n)} æ_{22}^{(n)}) + \int_{-\frac{h}{2}}^{\frac{h}{2}} \sigma_{11}^{(n-1)} x_3\, dx_3\\ &\quad - \frac{E}{1+\nu}\int_{-\frac{h}{2}}^{\frac{h}{2}} \frac{1-\overline{w}^{(n)}}{1-\overline{\beta}^{(n)}}(e_{11}^{(n-1)} + \overline{\beta}^{(n)} e_{22}^{(n-1)}) x_3\, dx_3 \quad (1 \leftrightarrow 2), \end{aligned} \tag{8.21}$$

$$\begin{aligned} \Delta M_{12}^{(n)} &= \frac{Eh^2}{2(1+\nu)}\overline{C}_2^{(n)}\varepsilon_{12}^{(n)} + 12(1-\nu)D\overline{C}_3^{(n)} æ_{12}^{(n)}\\ &\quad + \frac{E}{2(1+\nu)}\int_{-\frac{h}{2}}^{\frac{h}{2}} (1-\overline{w}^{(n)}) e_{12}^{(n-1)} x_3\, dx_3. \end{aligned}$$

The terms ΔN_{ij} and ΔM_{ij} $(i,j = 1,2)$ include the final plastic deformations.

Similarly to the approach presented in Sect. 6.1, we solve the equations (8.17) for the deformations $\varepsilon_{ij}^{(n)}$ $(i,j = 1,2)$ of the middle surface, and we obtain

$$
\begin{aligned}
\varepsilon_{11}^{(n)} &= \frac{1}{Eh}\left[(T_{11}^{(n)} - \nu T_{22}^{(n)}) - (\Delta T_{11}^{(n)} - \nu \Delta T_{22}^{(n)})\right] \quad (1 \leftrightarrow 2),\\
\varepsilon_{12}^{(n)} &= \frac{2(1+\nu)}{Eh}(T_{12}^{(n)} - \Delta T_{12}^{(n)}).
\end{aligned}
\tag{8.22}
$$

Using the variational equations introduced in Sect. 6.2 and the earlier described methodology of derivation of equations, we obtain (analogously to the differential equations (6.51)) the following equations governing the dynamics of an elasto-plastic flexible shell:

$$
\begin{aligned}
\frac{1}{Eh}\nabla^2\nabla^2 F &= -\nabla_k^2 w - \frac{1}{2}L(w,w) + \frac{1}{Eh}\left[(\Delta T_{11}^{(n)} - \nu \Delta T_{22}^{(n)})_{x_1x_2}\right.\\
&\quad \left. + (\Delta T_{22}^{(n)} - \nu \Delta T_{11}^{(n)})_{x_1x_1} - 2(1+\nu)(\Delta T_{12}^{(n)})_{x_1x_2}\right],\\
D\nabla^2\nabla^2 w &- L(w,F) - \nabla_k^2 F - (\Delta M_{11}^{(n)})_{x_1x_1} - (\Delta M_{22}^{(n)})_{x_2x_2}\\
&\quad - 2(\Delta M_{12}^{(n)})_{x_1x_2} - q + \rho h(\ddot{w} + \varepsilon\dot{w}) = 0.
\end{aligned}
\tag{8.23}
$$

The Airy function is introduced in the following way:

$$
N_{11}^{(n)} = \frac{\partial^2 F}{\partial {x_2}^2}, \quad N_{22}^{(n)} = \frac{\partial^2 F}{\partial {x_1}^2}, \quad N_{12}^{(n)} = \frac{\partial^2 F}{\partial x_1 \partial x_2}.
\tag{8.24}
$$

Analogously to Sect. 6.3, we obtain the following differential equations for the displacements when we take account of elastic–plastic deformation:

$$
\begin{aligned}
\rho_0\frac{h}{g}\ddot{u}_1 &= \Omega_1 + \Delta_1^{(n)},\\
\rho_0\frac{h}{g}\ddot{u}_2 &= \Omega_2 + \Delta_2^{(n)},\\
\rho\frac{h}{g}(\ddot{w} + \varepsilon\dot{w}) &= q - D\nabla^2\nabla^2 w + \Omega_3 + \Delta_3^{(n)},
\end{aligned}
\tag{8.25}
$$

where the following notation has been used:

$$
\begin{aligned}
\Omega_1 &= \frac{Eh}{1-\nu^2}\left[\frac{\partial^2 u_1}{\partial {x_1}^2} + \frac{1-\nu}{2}\frac{\partial^2 u_1}{\partial {x_2}^2} + \frac{1+\nu}{2}\left(\frac{\partial^2 u_2}{\partial x_1 \partial x_2}\right.\right.\\
&\quad \left. + \frac{\partial w}{\partial x_2}\frac{\partial^2 w}{\partial x_1 \partial x_2}\right) + \frac{\partial w}{\partial x_1}\left(\frac{\partial^2 w}{\partial {x_1}^2} + \frac{1-\nu}{2}\frac{\partial^2 w}{\partial {x_2}^2}\right)\\
&\quad \left. - (k_1 + \nu k_2)\frac{\partial^2 w}{\partial {x_1}^2}\right] \quad (1 \leftrightarrow 2),
\end{aligned}
\tag{8.26}
$$

$$
\begin{aligned}
\Omega_3 &= \frac{Eh}{1-\nu^2}\left\{\left[\frac{\partial u_1}{\partial x_1} + \nu\frac{\partial u_1}{\partial x_2} + \frac{1}{2}\left[\left(\frac{\partial w}{\partial x_1}\right)^2\right.\right.\right.\\
&\quad \left.\left. + \nu\left(\frac{\partial w}{\partial x_2}\right)^2\right] - w(k_1 + \nu k_2)\right]\left(k_1 + \frac{\partial^2 w}{\partial x^2}\right)
\end{aligned}
$$

$$+\left[\frac{\partial u_2}{\partial x_2}+\nu\frac{\partial u_1}{\partial x_1}+\frac{1}{2}\left[\left(\frac{\partial w}{\partial x_2}\right)^2+\nu\left(\frac{\partial w}{\partial x_1}\right)^2\right]\right.$$
$$\left.-w(k_2+\nu k_1)\right]\left(k_2+\frac{\partial^2 w}{\partial {x_2}^2}\right)+(1-\nu)\left(\frac{\partial u_2}{\partial x_1}\right.$$
$$\left.\left.+\frac{\partial u_1}{\partial x_2}+\frac{\partial w}{\partial x_1}\frac{\partial u_2}{\partial x_2}\right)\frac{\partial^2 w}{\partial x_1 \partial x_2}\right\}, \tag{8.27}$$

$$\Delta_1^{(n)}=(\Delta T_{11}^{(n)})_{x_1}+(\Delta T_{12}^{(n)})_{x_2} \quad (1\leftrightarrow 2), \tag{8.28}$$

$$\Delta_3^{(n)}=\left(k_1+\frac{\partial^2 w}{\partial {x_1}^2}\right)\Delta T_{11}^{(n)}+\left(k_2\frac{\partial^2 w}{\partial {x_2}^2}\right)\Delta T_{22}^{(n)}$$
$$+\frac{\partial^2 w}{\partial x_1 \partial x_2}\Delta T_{12}^{(n)}+(\Delta M_{11}^{(n)})_{x_1x_2}+2(\Delta M_{12}^{(n)})_{x_1x_2}+(\Delta M_{22}^{(n)})_{x_2x_2}. \tag{8.29}$$

The boundary conditions are formulated in a way similar to that to the described in Sect. 6.3. We need only to replace the deformations ε_{ij} by $\varepsilon_{ij}^{(n)}$, replace the moments M_{ij} by $M_{ij}^{(n)}$ and replace the loads on the middle surface T_{ij} by $T_{ij}^{(n)}$. The initial conditions used in Chap. 6 can now be applied.

8.2 Method of Solution

Similarly to the methodology used in Chap. 6, the hybrid partial differential equations obtained above are reduced to ODEs with respect to time, with the help of a finite-difference method with respect to the spatial coordinates x_1, x_2 that uses an $O(h^2)$ approximation.

The system of equations (8.23), after application of the difference operators (7.6) and the substitution of variables $\frac{dw}{dt}=\dot{w}$, is reduced to a system of first-order ODEs and a system of linear algebraic equations similar to (7.8) and (7.9):

$$\frac{dw_{ij}}{dt}=\dot{w}_{ij},$$

$$\frac{d\dot{w}_{ij}}{dt}+\varepsilon\dot{w}_{ij}=-\frac{1}{æ}\Big[A(w)+B(w,F)$$
$$+C(\Delta M_{11}^{(n)},\Delta M_{22}^{(n)},\Delta M_{12}^{(n)})\Big], \tag{8.30}$$

$$D(F_{ij})=E(w_{ij})+G(\Delta T_{11}^{(n)},\Delta T_{22}^{(n)},\Delta T_{12}^{(n)}). \tag{8.31}$$

The finite-difference operators appearing in (8.30) and (8.31) are similar to those in (7.10). One must simply replace N_T, $\Delta_T T$, M_T, $\Delta_T M$ by zero and replace the operators ΔT_{11}, ΔT_{22}, ΔT_{12}, ΔM_{11}, ΔM_{22}, ΔM_{12} by $\Delta T_{11}^{(n)}$,

$\Delta T_{22}^{(n)}$, $\Delta T_{12}^{(n)}$, $\Delta M_{11}^{(n)}$, $\Delta M_{22}^{(n)}$, $\Delta M_{12}^{(n)}$, respectively. Equations (8.30) and (8.31) are reduced to nondimensional form as in Chap. 7.

The algorithm that we have used for calculation of the elastic–plastic deformations of a flexible shell is similar to that used for geometrically and physically nonlinear problems described in Chap. 6. But this algorithm has some peculiarities and therefore will be considered in some detail.

The stress–strain state of the shell structure is computed for a series of successive steps in time Δt, the magnitude of which is chosen to satisfy requirements on the stability and convergence of the numerical process. In the present case, the conditions of dynamic equilibrium are considered at separate time moments corresponding to the beginning and end of each step.

At the initial time moment ($t_0 = 0$), the initial conditions $w_{ij}|_{t_0=0} = f_1$, $\dot{w}_{ij}|_{t_0=0} = f_2$ are applied. The stress field F_{ij} is then defined by the solution to the linear system of algebraic equations (8.31). The plastic supplements to the stresses ΔT_{11}, ΔT_{22}, ΔT_{12} and to the moments ΔM_{11}, ΔM_{22}, ΔM_{12} are taken to be zero. At the points (i, j, k) of the volume mesh, the deformation values e_{11}, e_{22}, e_{12} and the deformation intensities are equal to zero. We then consider a time step: by solving the system of algebraic equations (8.31) for F_{ij} and solving the ODEs (8.30), we find the values of F_{ij} at the end of the step Δt. Using w_{ij} and F_{ij}, we define the deformations e_{11}, e_{22}, e_{12} and also æ$_{11}$, æ$_{22}$, æ$_{12}$ at the points (i, j) of the middle surface. Using the Kirchhoff–Love hypothesis, we obtain the deformations in the shell volume, and also the deformation intensity $e_i|_{t+\Delta t}$ corresponding to the end of the interval Δt. This deformation intensity is compared with the value of e_i^t corresponding to the beginning of the interval. If $e_i|_{t+\Delta t} \leq e_i^t$, i.e. unloading has taken place, then a comparison of $e_i|_{t+\Delta t}$ with $\overline{e}^{(n)}\big|_t = \alpha_n e_T$ is made. Here two cases are possible: (1) if $e_i|_{t+\Delta t} \leq e_T^{(n)}$, then the shell point (i, j, k) is on an elastic branch of the deflection diagram (this means that the Iliushin function is equal to zero at this point); (2) if $e_i|_{t+\Delta t} > e_T^{(n)}$, then the point (i, j, k) is on the cyclic deformation diagram and the Iliushin function $\overline{w}^{(n)}$ is defined via (8.13), taking into account the number of the deformation cycle.

If $e_i|_{t+\Delta t} > e_i|_t$, then we have also two different cases: (1) for $e_i|_{t+\Delta t} \leq e_T^{(n)}$, the deformation proceeds on an elastic branch of the deformation diagram, i.e. the Iliushin function $\overline{w}^{(n)} = 0$; (2) for $e_i|_{t+\Delta t} > e_T^{(n)}$, active loading on the strengthening branch takes place and the Iliushin function $\overline{w}^{(n)}$ is defined by (8.13), taking into account the number of the deformation cycle.

When all $\overline{w}_{ijk}^{(n)}$ have been found, the values of the function $\overline{F}_1^{(n)}$ are computed for all points in the volume of the shell. The values of $\overline{F}_2^{(n)}$ at the points (i, j, k) of the shell volume and the functions $\overline{A}_m^{(n)}$, $\overline{B}_m^{(n)}$, $\overline{C}_m^{(n)}$ on the middle surface (i, j) are also computed. Then, through (8.18) and (8.21), the plastic supplements $\Delta T_{11}^{(n)}$, $\Delta T_{22}^{(n)}$, $\Delta T_{12}^{(n)}$, $\Delta M_{11}^{(n)}$, $\Delta M_{22}^{(n)}$, $\Delta M_{12}^{(n)}$ to the stresses and moments are found. The stresses and moments themselves are calculated from (8.24) and (8.20). The stresses in the volume (i, j, k) are defined by (8.15). Finally the next step in time is considered out and the procedure is repeated beginning with computation of the stress functions F_{ij} and the deflections w_{ij}.

The method and algorithm described above can be used to perform computations for shell structures with various mathematical models, including models of flexible elastic–plastic constructions (FEPC models), physically and geometrically nonlinear (PGN) models, physically nonlinear (PN) models, geometrically nonlinear (GN) models and linear (L) models. In addition, various $\sigma_i(e_i)$ relations and various boundary and initial conditions can be used. Static and dynamic problems can be solved, and also structures in either stationary or nonstationary fields can be analysed.

As an testing example, we have considered a square, uniformly loaded plate. The $\sigma_i(e_i)$ relation used has been defined earlier. We have taken $e_T = 0.98 \times 10^{-3}$, $\frac{E_2}{E_1} = 0.57735$, $a = b = 0.1$ m, $\frac{a}{h} = 50$. Free support was taken as the boundary condition. Three different mathematical models were analysed: elastic–plastic body (EPB), nonhomogeneous body (NB) and nonlinear elastic body (NEB). Computations for the EPB and NB models were carried out using the method and algorithm described earlier. The results for the NB model were obtained using the method of variable elastic parameters and of variational iterations, in the first approximation [133]. Some results of the computations are given in Table 8.1.

One can conclude, taking into account the results shown in Table 8.1, that all three mathematical models give reliable results.

Now we shall describe an investigation of the oscillations of this plate but with a suddenly applied load of infinite duration ($q \in \{23.4; 60.0; 70.3\}$), with the same geometrical and physical parameters. Damping ($\varepsilon = 30$) has been taken into account.

We have analysed the influence of choice of mathematical model on the forces $T_{22} = \frac{\partial^2 F}{\partial {x_1}^2}$ and the curvatures $w''_{x_1 x_1}$ along the lines $x_2 = 0.5$, $x_1 \in [0, 0.5]$ and $x_1 = x_2 \in [0.5, 1.0]$. Graphs are shown in Figs. 8.1 and 8.2 for

Table 8.1. Results of computations of deflection in the centre of a square plate for different mathematical models

q (uniformly distributed)	Mathematical model $w(0.5, 0.5)$ EPB	NEB [133]	NB	Differences 2 from 3 (%)	2 from 4 (%)
1	2	3	4	5	6
23.4	0.98	0.96	0.95	+2.0	+3.0
60.0	1.97	2.03	2.00	–3.0	–2.0
70.3	2.26	2.16	2.23	+4.0	+1.0

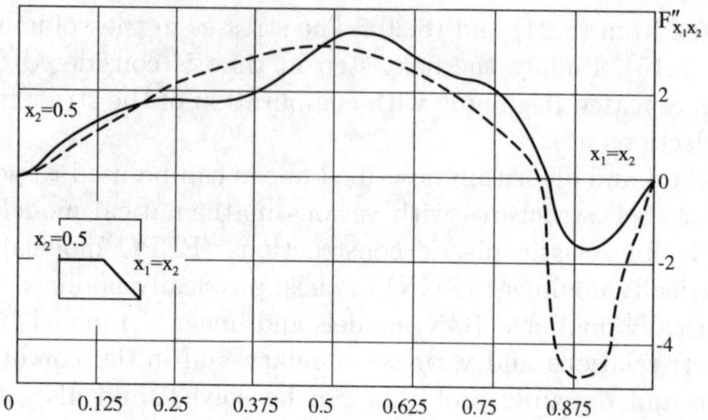

Fig. 8.1. Distribution of $F''_{x_1x_2}$ for $q = 60, t = 0.5$ taking into account the unloading cycle (*solid curve*) and without taking into account the unloading cycle (*dashed curve*)

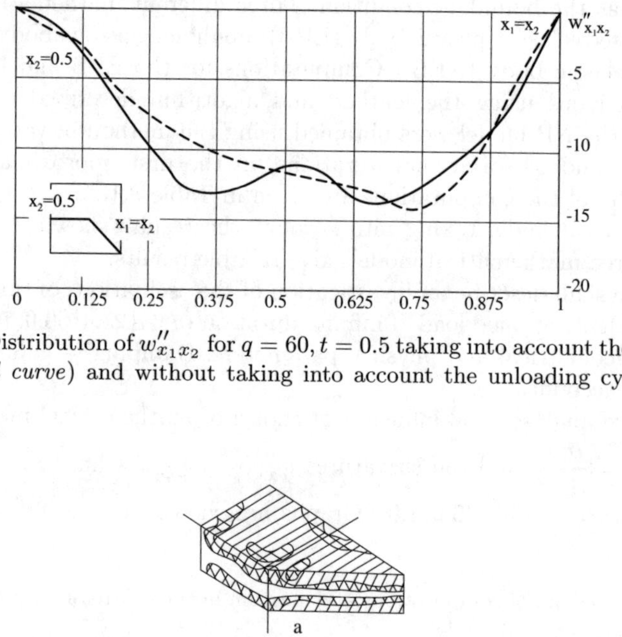

Fig. 8.2. Distribution of $w''_{x_1x_2}$ for $q = 60, t = 0.5$ taking into account the unloading cycle (*solid curve*) and without taking into account the unloading cycle (*dashed curve*)

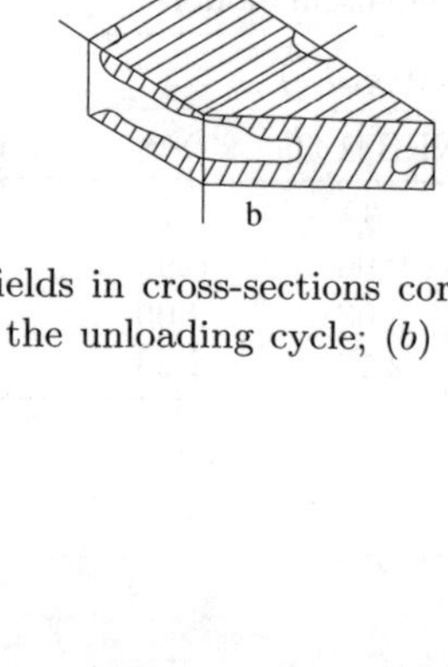

Fig. 8.3a,b. Deformation fields in cross-sections corresponding to Figs. 8.1 and 8.2: (*a*) taking into account the unloading cycle; (*b*) without taking into account the unloading cycle

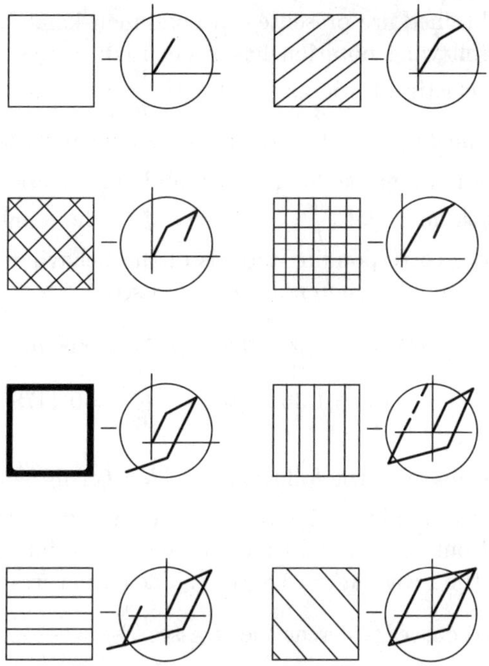

Fig. 8.4. Types of hatching used in Fig. 8.3

$q = 60.0$ and time $t = 0.5$, and the data listed above. The deformation fields in the corresponding cross-sections are given in Fig. 8.3 both with (Fig. 8.3a) or without (Fig. 8.3b) taking into account the unloading cycle. The various types of hatching used in Fig. 8.3 are explained in Fig. 8.4. An increase of the transverse load greatly changes the stress–strain state of the structure. In addition, the differences between the stresses at different times in the middle surface increase greatly, for separate time moments, which causes mixing of elastic and plastic zones.

8.3 Oscillations and Stability of Elasto-Plastic Shells

We consider the oscillations of a simply supported plate (with the boundary condition (6.60)), with the initial conditions $w|_{t=0} = \dot{w}|_{t=0} = 0$. Two different types of load have been considered:

$$q = const, \tag{8.32}$$

$$q = \begin{cases} q = const, & t < t_1, \\ 2q = const, & t \geq t_1. \end{cases} \tag{8.33}$$

The results obtained are presented in nondimensional form in Figs. 8.5–8.32, where the following notation has been used: curves labelled 1 and 2 correspond to an elastic–plastic material with $\frac{a}{h} = 111$ and 50, respectively; curves labelled 3 and 4 ($\frac{a}{h} = 111$) correspond to a linear elastic and an elastic–plastic material, for a step-like load (8.33); and curves labelled 5 correspond to an elastic–plastic material ($\frac{a}{h} = 111$, $\varepsilon = 35$). The numbers 1–5 are also used to refer to the corresponding models of the behaviour of the material. The following material parameters have been used:

$$E_1 = 69 \text{ GPa}, \quad \nu = 0.3, \quad \rho_0 = 28 \text{ kN/m}^3,$$
$$e_T = 1.35 \times 10^{-3}, \quad \frac{E_2}{E_1} = 0.4478. \tag{8.34}$$

The plate dimensions were the following: $a = b = 0.1$ m, $\frac{a}{h} = \{50, 111\}$. The shell volume was covered by a $16 \times 16 \times 8$ mesh. Analysis of the stress–strain state was carried out for various time moments. The integration step used in the Adams method was $\Delta t^k = 5 \times 10^{-4}$. The curvatures $w''_{x_1x_1}$, $w''_{x_1x_2}$ are denoted by dashed curves, whereas the stresses $T_{22} = \frac{\partial^2 F}{\partial {x_1}^2}$, $T_{12} = -\frac{\partial^2 F}{\partial x_1 \partial x_2}$ are denoted by solid curves. Graphs are presented for the cross-sections shown in Fig. 8.3.

The dependences $w(0.5, 0.5, t)$ for an elastic–plastic material (solid curve, labelled 1) and a linear elastic material (dashed curve, labelled 3), for $q = 503.7$ (load model (8.32)) are shown in Fig. 8.5. The use of model 1 causes a decrease of the static deflection by a factor of 1.4 and an increase of the amplitude by 10% (by the static deflection, we mean the deflection caused by a static load).

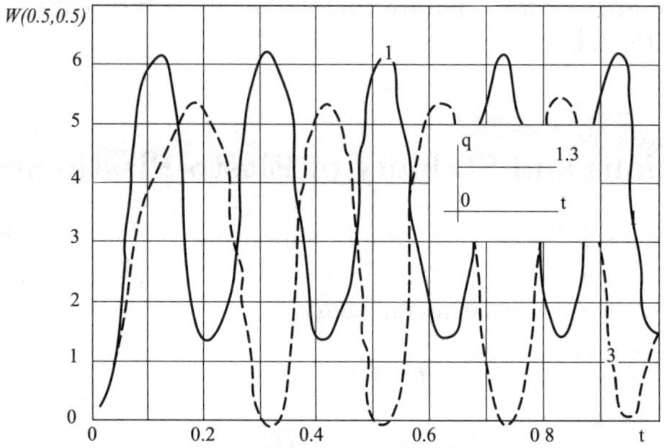

Fig. 8.5. Deflection of the plate centre $w(0.5, 0.5)$ as a function of time for different materials: 1, elastic–plastic material ; 3, linear elastic material

For model 1, the plate oscillates around the static-equilibrium position ($w_{st} = 3.6$) with a period of 0.23 and an amplitude of 5.3. It should be noted that the period of free oscillations of the elastic plate is 16% higher than that of the elastic–plastic. The stress–strain states of the plate differ greatly between the mathematical models considered here. The maximum stress for model 1 is 12% larger than for model 3. In one quarter of the plate, the stresses obtained with the elastic–plastic model are a factor of two smaller than those obtained with model 3. The values of the curvature $w''_{x_1x_1}$ values obtained with different models differ by 50% in the corner zones. The difference in $F''_{x_1x_1}$ reaches 65% on the shell edge. The graphs of the curvatures $w''_{x_1x_2}$ obtained with different models differ not only quantitatively but also qualitatively (Figs. 8.6 and 8.7). We obtain a negative curvature using the elastic–plastic model, whereas for the elastic model the curvature is positive

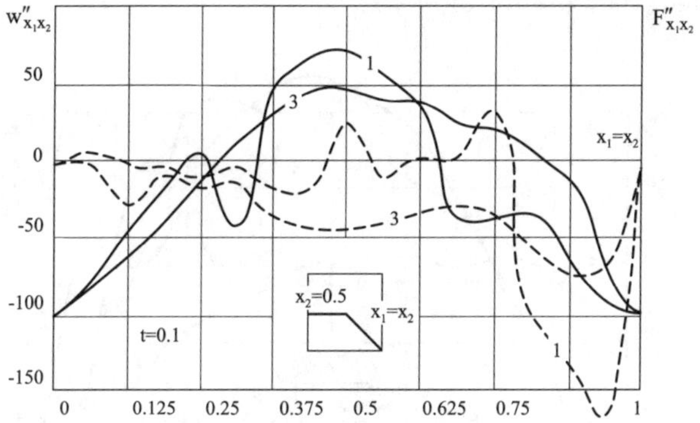

Fig. 8.6. Distribution of $F''_{x_1x_1}$ (*solid lines*) and $w''_{x_1x_1}$ (*dashed lines*) for $q = 503.7, t = 0.1$; 1, elastic–plastic material ; 3, linear elastic material

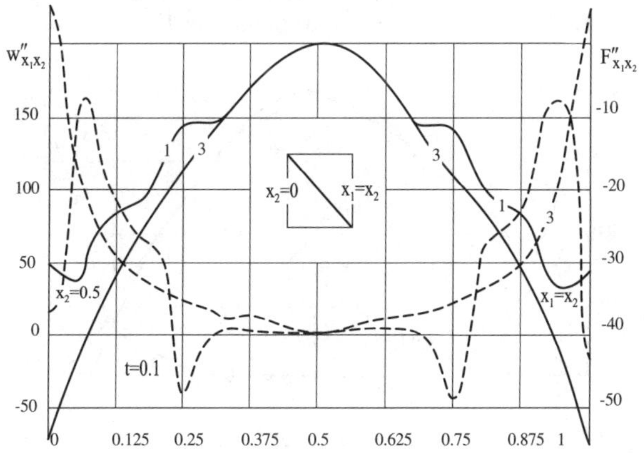

Fig. 8.7. Distribution of $F''_{x_1x_2}$ (*solid lines*) and $w''_{x_1x_2}$ (*dashed lines*) for $q = 503.7, t = 0.1$; 1, elastic–plastic material; 3, linear elastic material

(on the shell edge). The difference can be explained by the fact that for $t = 0.1$, $w = 6.0$ for the elastic–plastic model, and $w = 3.8$ for the elastic model (Fig. 8.5).

Diagrams of $F''_{x_1x_1}$ and $w''_{x_1x_1}$ are shown in the Figs. 8.8 and 8.9 for the time instant $t = 0.3$. At this time, the two models give opposite phases for the position of the plate. There is no qualitative difference in the plots of $F''_{x_1x_1}$, but there are qualitative and quantitative differences between the plots of the curvatures in Fig. 8.8. The values of $w''_{x_1x_2}$ obtained with the different models differ in sign. The difference between the stresses $F''_{x_1x_2}$ reaches 17% at the edge.

Graphs for $t = 0.42$ are presented in Figs. 8.10 and 8.11. At this time, the two models again give opposite phases. These results show that graphs of $F''_{x_1x_1}$ can differ even qualitatively for different models. Note that when the linear elastic model is used, the graphs of both stress and curvature are smoother than they are for model 1.

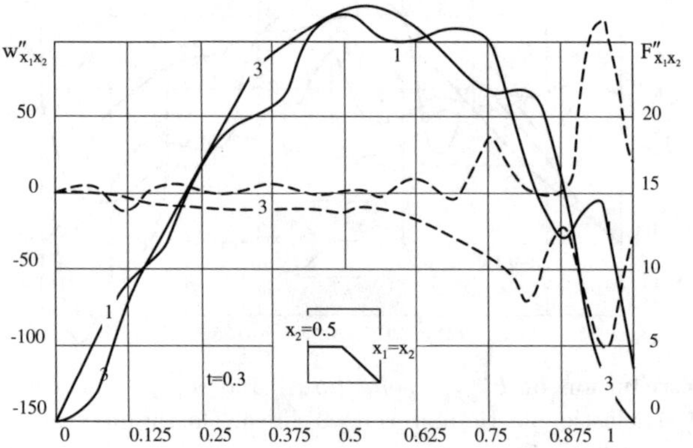

Fig. 8.8. Distribution of $F''_{x_1x_1}$ (*solid lines*) and $w''_{x_1x_1}$ (*dashed lines*) for $q = 503.7, t = 0.3$; 1, elastic–plastic material; 3, linear elastic material

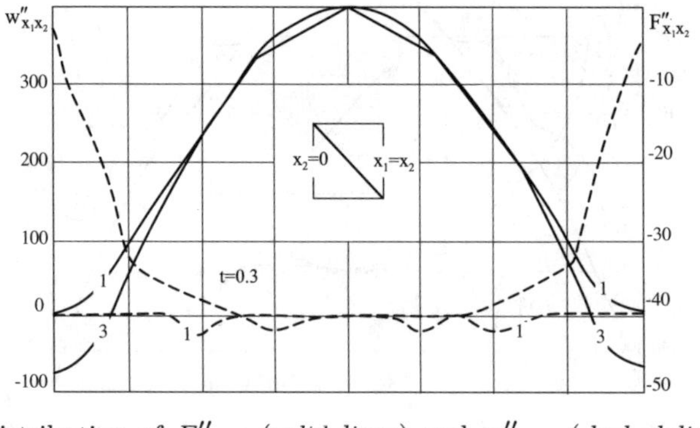

Fig. 8.9. Distribution of $F''_{x_1x_2}$ (*solid lines*) and $w''_{x_1x_2}$ (*dashed lines*) for $q = 503.7, t = 0.3$; 1, elastic–plastic material; 3, linear elastic material

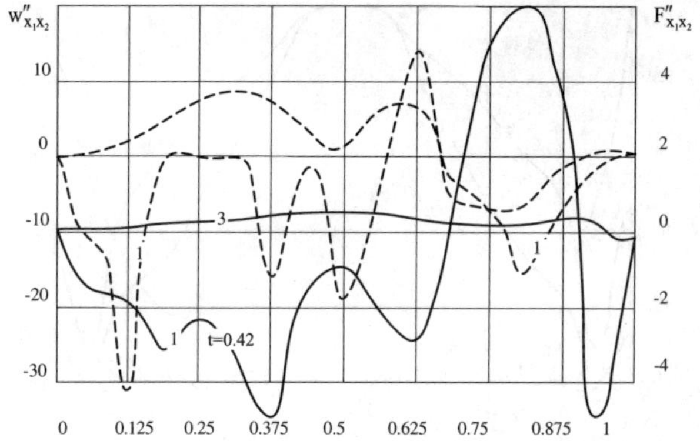

Fig. 8.10. Distribution of $F''_{x_1x_1}$ (*solid lines*) and $w''_{x_1x_1}$ (*dashed lines*) for $q = 503.7, t = 0.42$; 1, elastic–plastic material; 3, linear elastic material

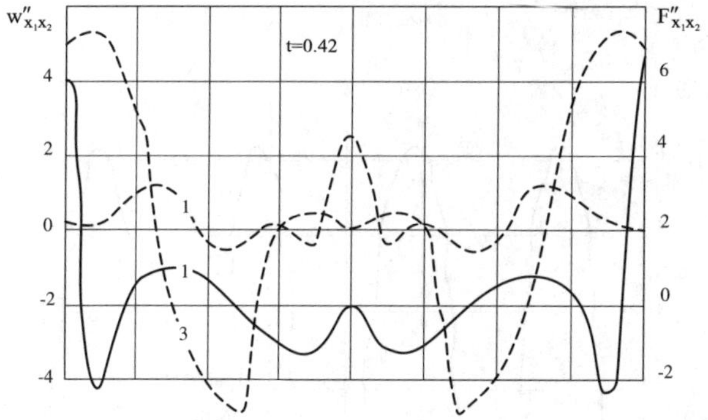

Fig. 8.11. Distribution of $F''_{x_1x_2}$ (*solid lines*) and $w''_{x_1x_2}$ (*dashed lines*) for $q = 503.7, t = 0.42$; 1, elastic–plastic material; 3, linear elastic material

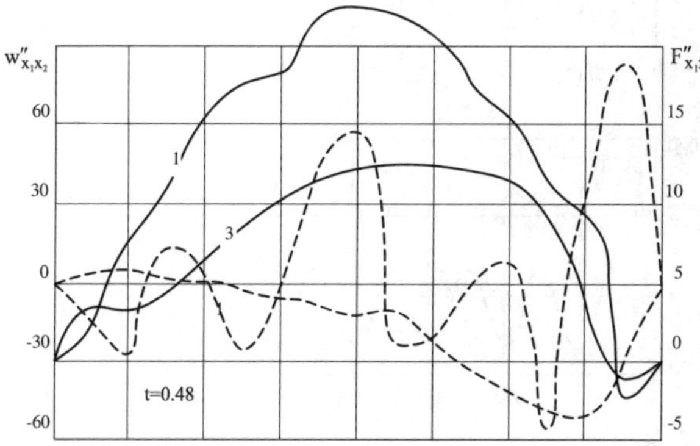

Fig. 8.12. Distribution of $F''_{x_1x_1}$ (*solid lines*) and $w''_{x_1x_1}$ (*dashed lines*) for $q = 503.7, t = 0.48$; 1, elastic–plastic material; 3, linear elastic material

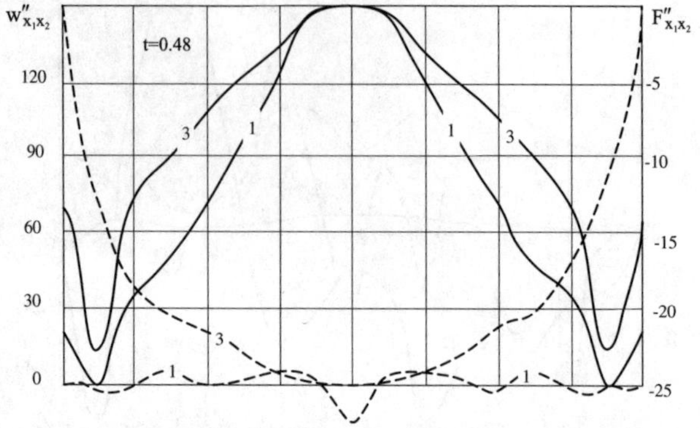

Fig. 8.13. Distribution of $F''_{x_1x_2}$ (*solid lines*) and $w''_{x_1x_2}$ (*dashed lines*) for $q = 503.7, t = 0.48$; 1, elastic–plastic material; 3, linear elastic material

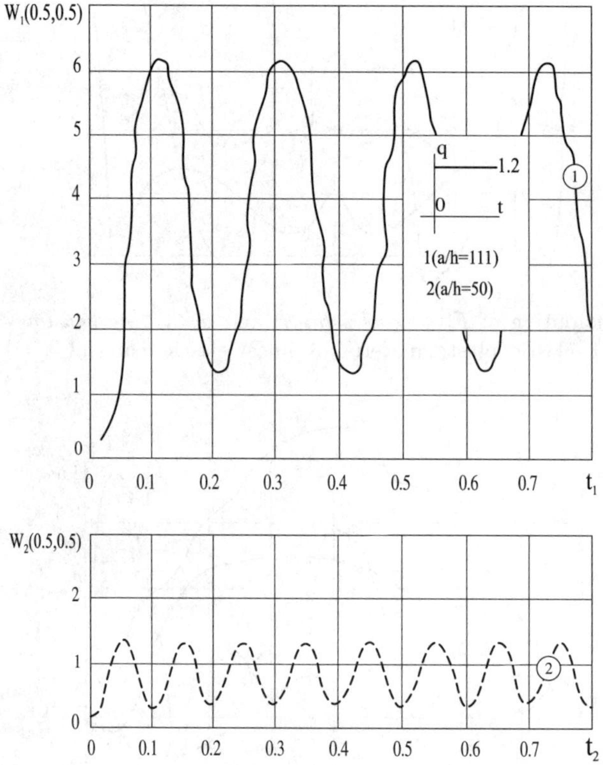

Fig. 8.14. Deflection of the plate centre $w(0.5, 0.5)$ as a function of time for $\frac{a}{h} = 111$ (*solid curve*) and for $\frac{a}{h} = 50$ (*dashed curve*)

Graphs for $t = 0.48$ are presented in Figs. 8.12 and 8.13. A qualitative agreement of the diagrams is observable, but the stresses $F''_{x_1x_1}$ in the centre of the plate obtained with the elastic–plastic model are 39% larger than the stresses obtained with the elastic model. The difference in the stresses reaches 38% on the edge of the plate, although in one quarter it reaches only 8%.

From the results presented in Fig. 8.5, one can conclude that the plates oscillate around their equilibrium positions. The amplitude of oscillation of the plate is smaller with the elastic–plastic model, and the value of the deflection from the undeformed state w_{st} is larger. Taking account of elastic–plastic material properties leads to remarkable mixing of different types of stresses with increase of time, caused by plastic deformation due to the impulse-type load.

The influence of the geometrical parameters on the oscillation of the plate and on the stress distribution is demonstrated in Figs. 8.14–8.23. In Fig. 8.14, the deflection at the centre of the shell is shown for the control parameter $\frac{a}{h} = 111$ (solid curve) and $\frac{a}{h} = 50$ (dashed curve). The static deflection in

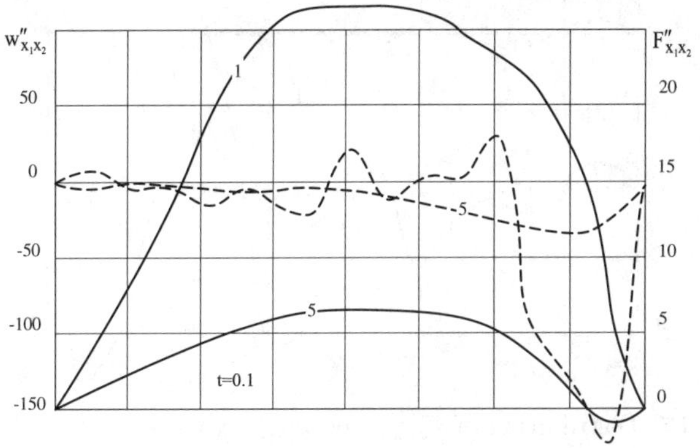

Fig. 8.15. Distribution of $F''_{x_1x_1}$ and $w''_{x_1x_1}$ for $\frac{a}{h} = 111$

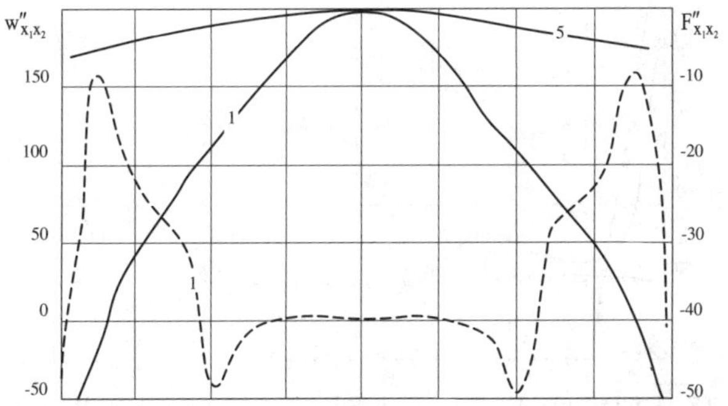

Fig. 8.16. Distribution of $F''_{x_1x_2}$ and $w''_{x_1x_2}$ for $\frac{a}{h} = 50$

the first case is three times bigger than in the second case, and the amplitude of oscillation is 2.3 times larger.

In Figs. 8.15 and 8.16, graphs of the stress and curvature at the time, when both plates are in the neighbourhood of their maximum deflection are shown. The stress $F''_{x_1x_1}$ at the plate centre for $\frac{a}{h} = 111$ is more than 30 times the stress for $\frac{a}{h} = 50$. The difference in curvature in one quarter of the plate reaches 95%. The difference in the stress $F''_{x_1x_2}$ at the edge reaches 96%.

In Figures 8.17 and 8.18, graphs are shown for the time when the deflection $w(0.5, 0.5)$ is at its minimum. Here we observe a difference in the characters of the graphs of stress and curvatures for plates with different thickness. However, for plates with the same thickness, the qualitative similarities of the graphs of $F''_{x_1x_1}$ and $w''_{x_1x_1}$, as well as those of $F''_{x_1x_2}$ and $w''_{x_1x_2}$ are remarkable. We note that larger changes of the stress–strain state have been observed for the more slender plate.

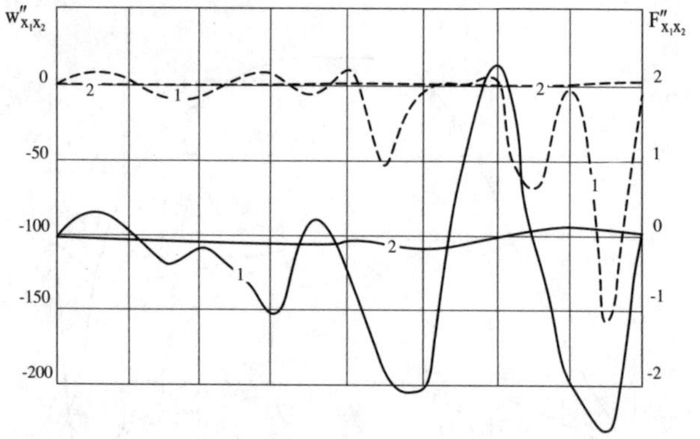

Fig. 8.17. Distribution of $F''_{x_1x_1}$ and $w''_{x_1x_1}$ for the smallest deflection

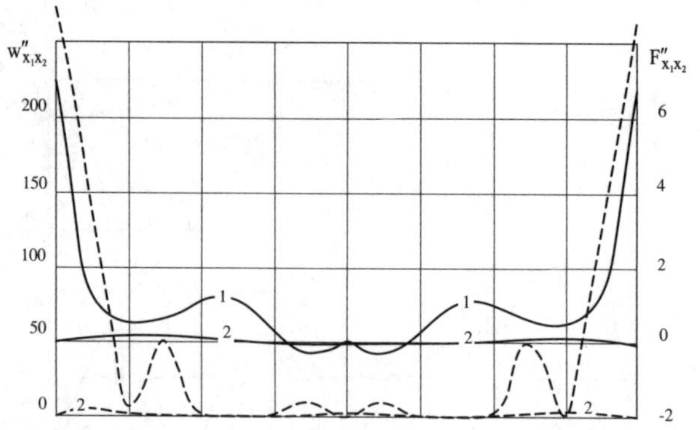

Fig. 8.18. Distribution of $F''_{x_1x_2}$ and $w''_{x_1x_2}$ for the smallest deflection

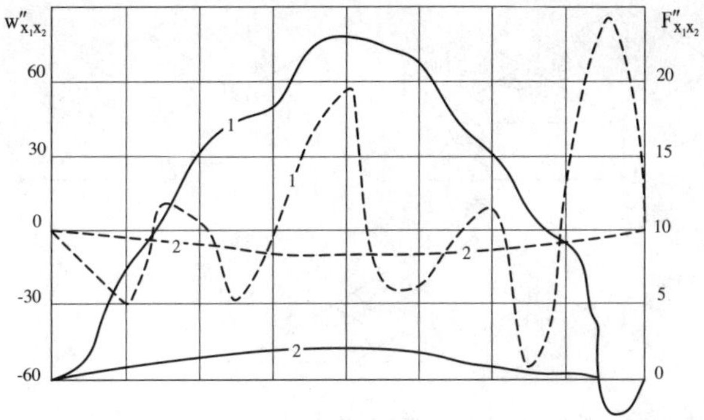

Fig. 8.19. Distribution of $F''_{x_1x_1}$ and $w''_{x_1x_1}$ for the greatest deflection

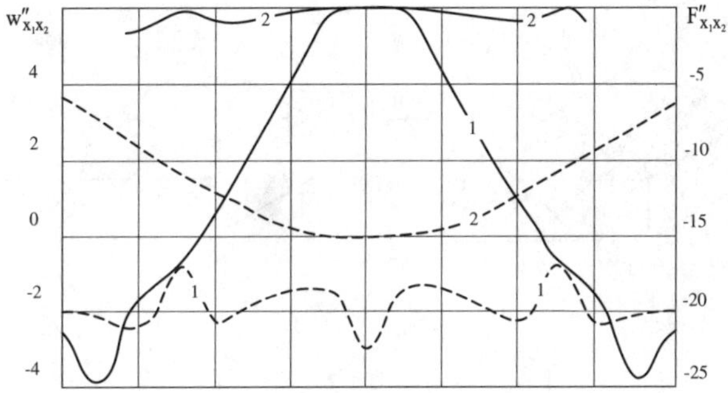

Fig. 8.20. Distribution of $F''_{x_1x_2}$ and $w''_{x_1x_2}$ for the greatest deflection

In Figs. 8.19 and 8.20, graphs of the stress and curvature at the time corresponding to the next maximum are shown. Here the difference in the stresses in the plate centre is 80%, and the difference in the curvature is 60%. It can be seen that a larger change in the stress–strain state occurs in the more slender plate at this time moment.

Figs. 8.21–8.23 show how the plastic zones develop in the shell volume with time (Figs. 8.21 and 8.22 correspond to $\frac{a}{h} = 111$, and Fig. 8.23 to $\frac{a}{h} = 50$). It can be seen how plastic zones develop in an initial period (up to $t = 0.12$), and then a stable distribution of plastic zones occurs, which for practical purposes does not change with time. Therefore, one can conclude that after achieving maximum deflection, the plate starts to oscillate like a linearly elastic one but with the plastic deformation that has occurred earlier. For the plate with $\frac{a}{h} = 50$ a stabilized distribution of plastic zones is observed from $t = 0.1$.

Comparing the character of the distribution of plastic zones in the volume of plates with different thickness, one can conclude that in relatively

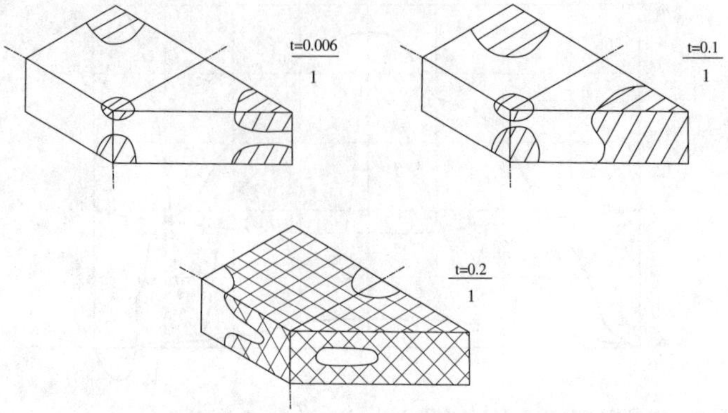

Fig. 8.21. Distribution of plastic zones at $t = 0.006, 0.1$ and 0.2 for a plate with $\frac{a}{h} = 111$

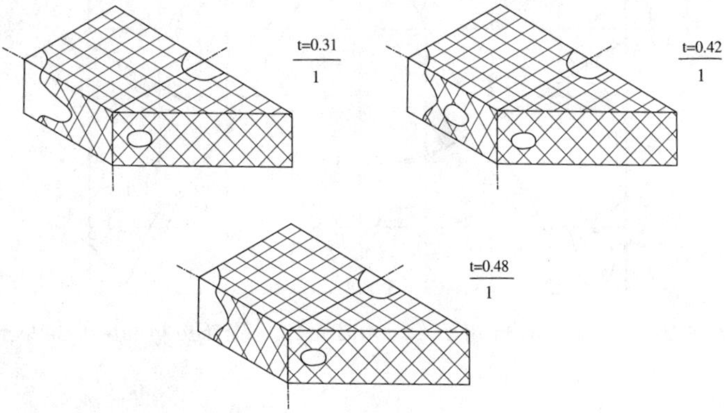

Fig. 8.22. Distribution of plastic zones at $t = 0.31, 0.42$ and 0.48 for a plate with $\frac{a}{h} = 111$

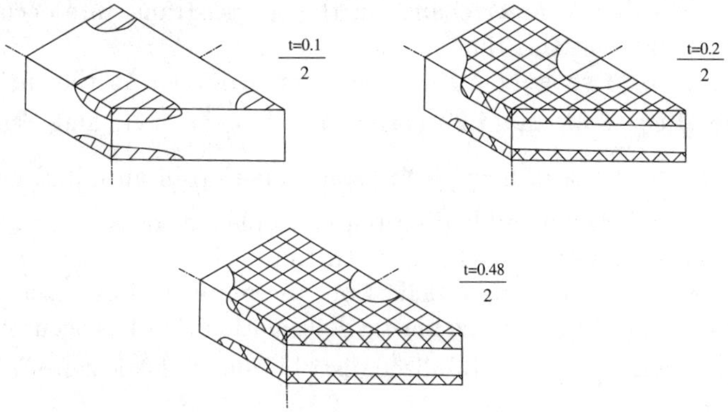

Fig. 8.23. Distribution of plastic zones at $t = 0.1, 0.2$ and 0.48 for a plate with $\frac{a}{h} = 50$

thin plates the plastic zones are distributed throughout the whole thickness, whereas for thicker plates the plastic zones are located close to the surfaces. In addition, in thinner plates nonbending stresses occur, which lead to a more uniform stress distribution through the thickness (Fig. 8.22). For thicker plates, bending stresses dominate and cause a nonuniform stress distribution through the thickness (Fig. 8.23), which leads to the occurrence of plastic zones but not through the whole thickness.

Now we consider the variation with time of the deflection of the shell centre for three cases: model 1, with an impulse load $q = 503.7$ of type (8.32) with successive oscillations; model 4, with an impulse load of type (8.33); and model 5, with an impulse load of type (8.32) with damping. In all cases $\frac{a}{h} = 111$ and the material is elastic–plastic (8.34). The curves shown in Fig. 8.24 illustrate that the static deflection corresponding to the load $2q$ of type (8.33) is 1.6 times larger than the static deflection corresponding to the load of type (8.32). After the extra load has been applied both the amplitude and the frequency of the oscillations change greatly. The amplitude decreases by a factor of 1.3 whereas the frequency increased 5 times. Analysis of the curve for model 5 (with damping) shows that intense damping of the oscillations

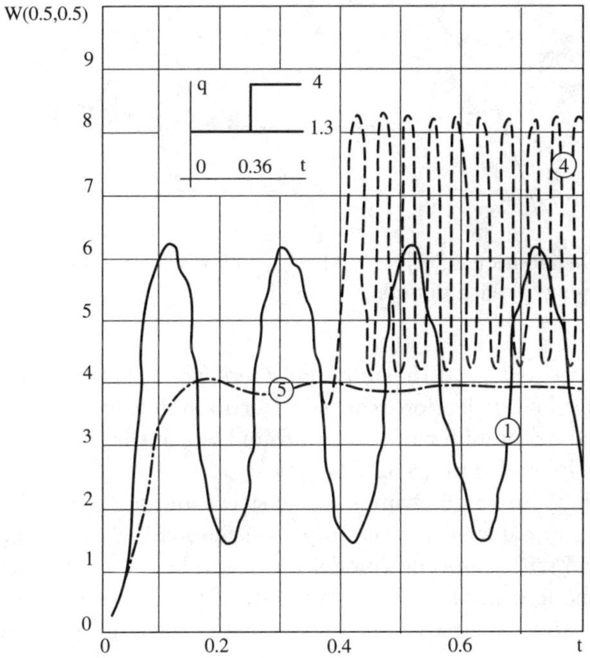

Fig. 8.24. Deflection of the plate centre $w(0.5, 0.5)$ for $\frac{a}{h} = 111$ under different loads: 1, impulse load $q = 503.7$ of type (8.32), 4, impulse load of type (8.33), 5, impulse load of type (8.32) with damping

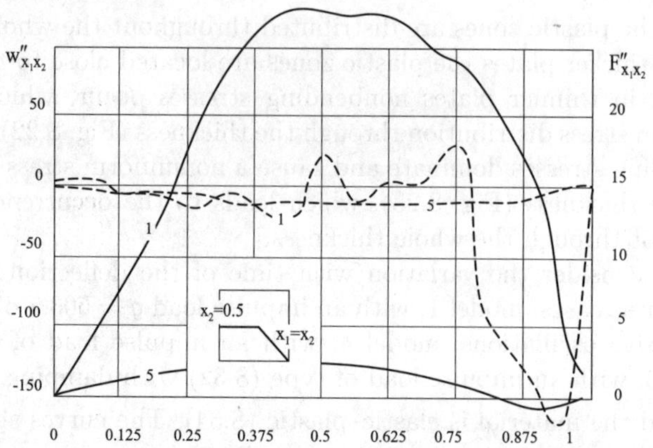

Fig. 8.25. Distribution of $F''_{x_1x_1}$ and $w''_{x_1x_1}$ at $t = 0.18$ for type (8.32) load: 1, without damping; 5, with damping

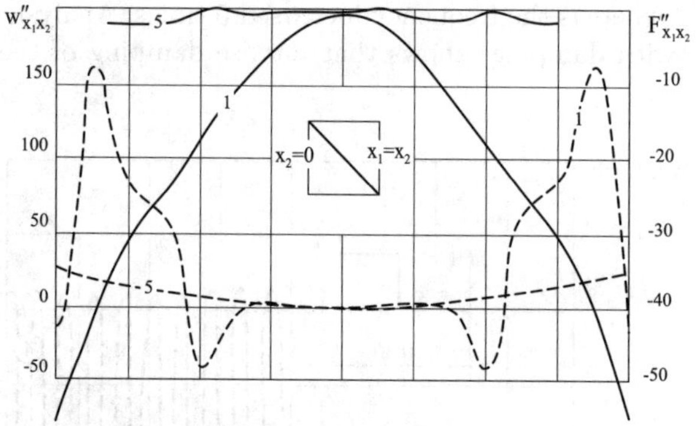

Fig. 8.26. Distribution of $F''_{x_1x_2}$ and $w''_{x_1x_2}$ at $t = 0.18$ for type (8.32) load: 1, without damping; 5, with damping

takes place up to $t = 0.15$, and that there are practically no oscillations after $t = 0.5$. The deflection reached after damping has occurred (curve 5) corresponds to the equilibrium state, around which the plate oscillates (curve 1) under the load of type (8.32).

In Figs. 8.25 and 8.26, graphs of the stress and curvature for the elastic–plastic model are shown, for the impulse-type load (8.32) without damping (model 1) and with damping (model 5). It can be seen that damping causes important changes in the stress–strain state (the deflection decreases by a factor of 1.8, and the stress $F''_{x_1x_1}$ in the plate centre decreases by a factor of 5). Damping smooths the graphs of the curvature and decreases the value of the curvature by a factor of about 4.8. Graphs of the stress and curvature at the time instant $t = 0.2$ for the same two cases are shown in Figs. 8.27 and 8.28. The divergence mentioned above has increased.

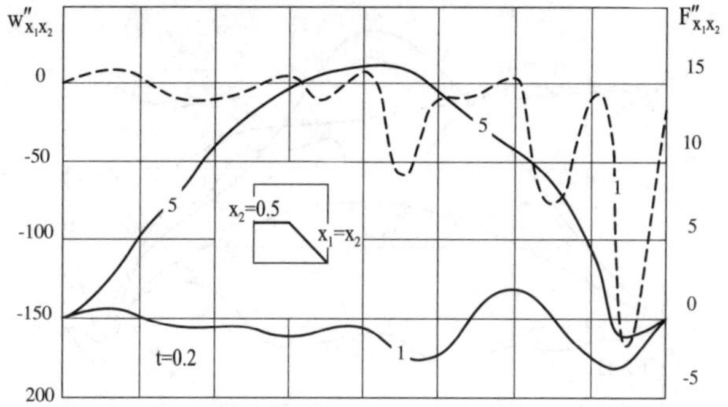

Fig. 8.27. Distribution of $F''_{x_1x_1}$ and $w''_{x_1x_1}$ at $t = 0.2$ for type (8.32) load: 1, without damping; 5, with damping

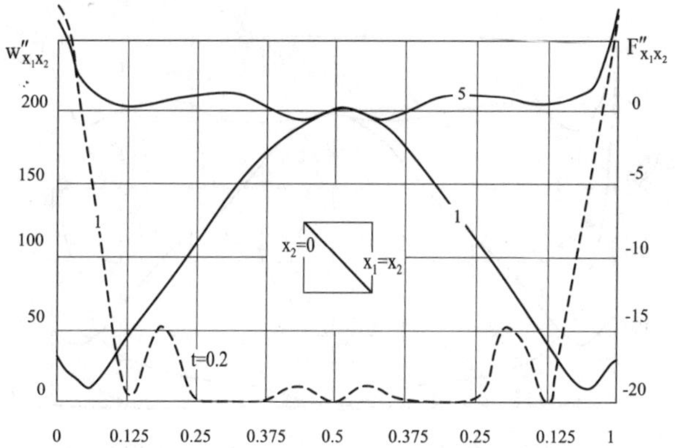

Fig. 8.28. Distribution of $F''_{x_1x_2}$ and $w''_{x_1x_2}$ at $t = 0.2$ for type (8.32) load: 1, without damping; 5, with damping

In Figs. 8.29 and 8.30, graphs of the stress and curvature at $t = 0.42$ for models 1 and 4, corresponding to a load of type (8.33), are presented. Results for model 5 are also shown in Fig. 8.29. A comparison of the graphs shows that the stress–strain state of the plate after the second step of the loading process differs significantly from the stress–strain state of the plate subjected only to the extended oscillations caused by an initial impulse load of type (8.32). A difference is visible not only in the values but also of the stresses in their sign. For model 5, the graphs of stress and curvature are located between those for models 1 and 4.

At $t = 0.48$, the positions of the plates for models 1 and 4 approach each other more closely than they do at $t = 0.42$, and the difference in character of the graphs becomes smaller. For instance, in the centre of the loaded plate, the stress $F''_{x_1x_1}$ is 17% higher than the stress for the load reported in Fig. 8.31. The curvatures of the shell for the two models differ in sign, because

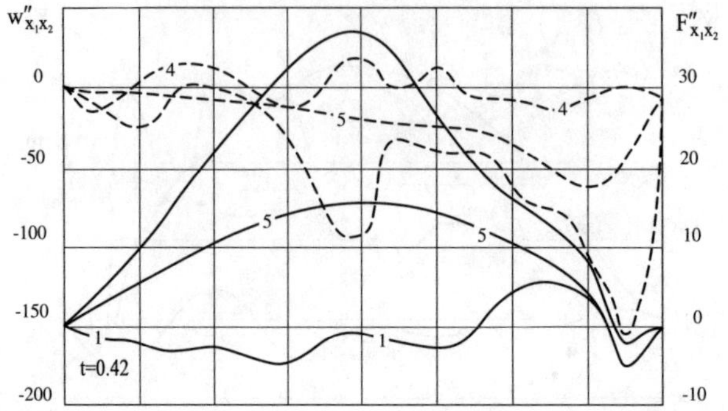

Fig. 8.29. Distribution of $F''_{x_1x_1}$ and $w''_{x_1x_1}$ at $t = 0.42$ for type (8.33) load

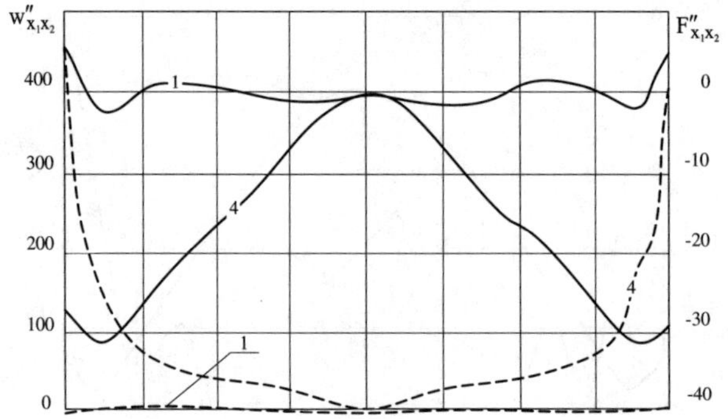

Fig. 8.30. Distribution of $F''_{x_1x_2}$ and $w''_{x_1x_2}$ at $t = 0.42$ for type (8.33) load

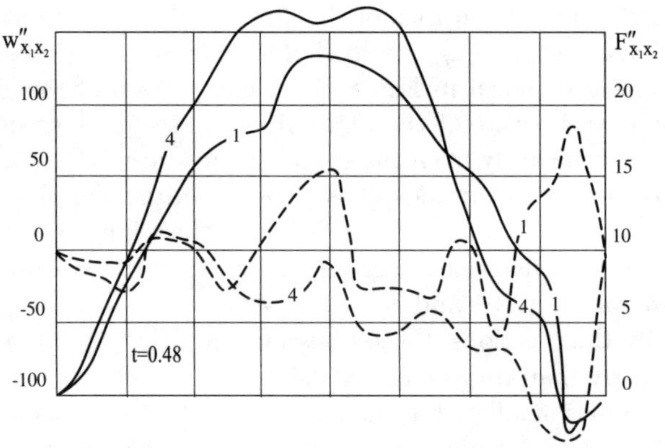

Fig. 8.31. Distribution of $F''_{x_1x_1}$ and $w''_{x_1x_1}$ at $t = 0.48$ for type (8.33) load

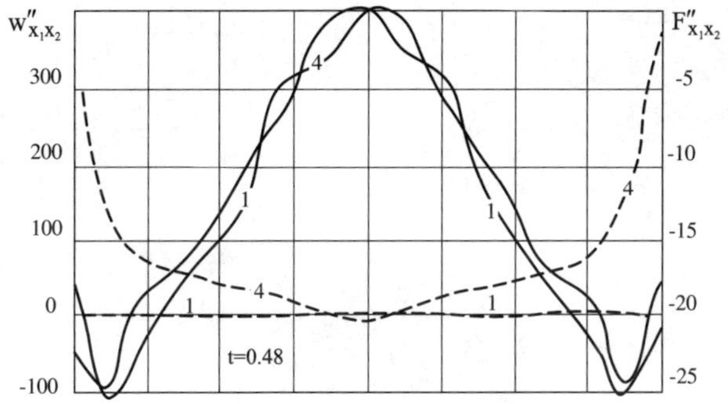

Fig. 8.32. Distribution of $F''_{x_1x_2}$ and $w''_{x_1x_2}$ at $t = 0.48$ for type (8.33) load

at $t = 0.42$ the plates are on different sides of their equilibrium states (Fig. 8.24). The difference in the values of the shear stress can be neglected (Fig. 8.32).

In Fig. 8.33, the distribution of elastic and plastic zones in the plate, in the case with damping, is presented for three time moments. Because the intense oscillations have already been damped at $t = 0.1$, difference in the distribution of plastic zones is not observed. Comparison of the distribution of plastic zones at $t = 0.1$ for models 1 and 5 leads to the conclusion that plastic deformation occurs only near corners and in zones lying close to the plate surfaces. In the centre of the plate and also close to the middle surface, plastic deformation is not observed.

In Fig. 8.34, the distributions of plastic zones in the plate volume at the time instants $t = 0.42$, $t = 0.46$ and $t = 0.48$ for model 4 are presented. The latter two distributions differ from each other. This is evidence of the stabilization of the distribution of plastic zones, which has occurred after

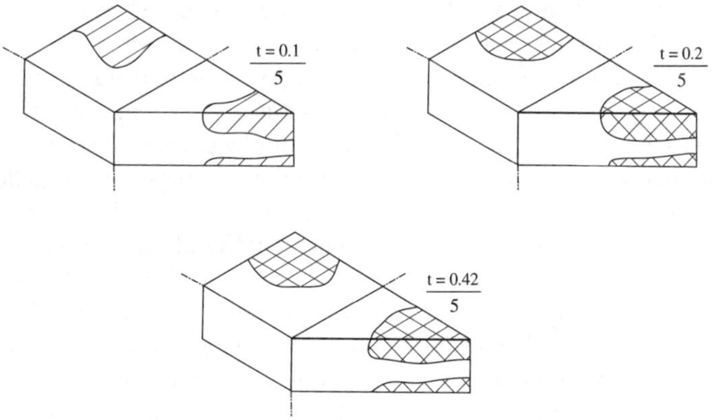

Fig. 8.33. Distribution of elastic and plastic zones at $t = 0.1, 0.2, 0.42$ (with damping)

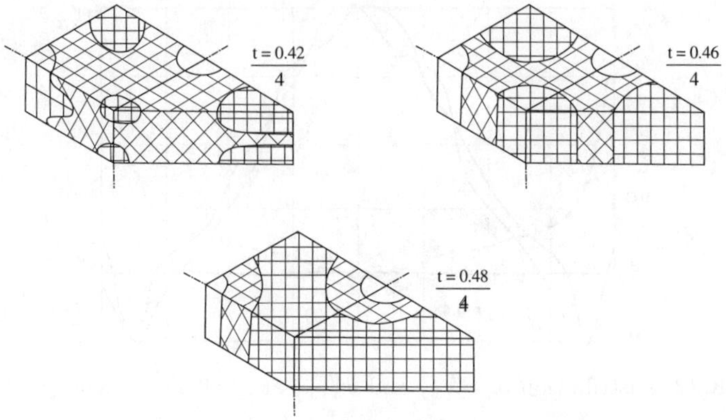

Fig. 8.34. Distribution of elastic and plastic zones at $t = 0.42, 0.46, 0.48$

the maximum deflection has occurred under the extra load. Comparing the character of the distribution of plastic zones at $t = 0.48$ for models 1 and 4, we discovered that after the additional loading process the elastic zones remain only around the centres of the sides and vanish completely in the rest of the plate volume. Also, additional plastic deformation occurs in the corner zones and around the plate centre.

We now investigate the dynamic stability and the stress–strain state of a flexible elastic–plastic shell under an impulse load of type (8.32). The initial equations were solved in the hybrid form (8.23). The computations were carried out for the boundary conditions (2.116) and the initial conditions (4.1). The program developed to do this, however, allows us to investigate other boundary conditions such as (7.33). Shells with the following geometric parameters have been considered: $a = 0.1$ m, $\lambda = 1$, $\frac{h}{a} = \frac{1}{111}$; the physical characteristics (8.34) were used. The nondimensional curvature parameters were $k_x = k_y = 24$ and 48. A mesh of size $16 \times 16 \times 8$ was used, and the integration step used in the Adams method was $\Delta t_k = 5 \times 10^{-4}$. The mesh size and integration step were determined by numerical experiments so as to satisfy the condition of stability of the solution. Dynamic stability criteria described earlier were applied. At the moment of a stability loss, an increase of 1% in the load above its critical value caused an increase in the deflection by more than three times. The maximum deflection exceeded twice the camber of the shell and the time required to reach the maximum deflection greatly decreased.

Analysis of the stress–strain state of a flexible rectangular plate showed that the results depend mainly on the material model used. In this section we investigate the influence of three models of material deformation (linear elastic, nonlinear elastic and elastic–plastic) on the stress–strain state, the distribution of plastic zones and the dynamic stability of a flexible, shallow, rectangular shell. In the figures, the data are presented in nondimensional form using the following notation: 1, results obtained using the linear elastic model; 2, results obtained using the nonlinear elastic model ; 3, results obtained using the elastic–plastic model.

A shell with $K_{x_1} = K_{x_2} = 24$ in a pre-critical state was investigated. An analysis of plots of $w(t)$ for the central point of a shallow shell under a pre-critical load $q = 198$ (Fig. 8.35) showed that the point describes almost the same trajectory for all three models. We examined the changes of the stresses $F''_{x_1x_1}$, $F''_{x_1x_2}$ and of the curvatures $w''_{x_1x_1}$, $w''_{x_1x_2}$ along the axis $x_2 = 0.5$ and along the diagonal using the "stop frame" method for the time moments $t = 0.14, 0.5, 0.58$ and 0.74.

In Fig. 8.36, the variation with time of the deflection at the shell centre is presented for all three models, for a pre-critical load (dashed line) and the

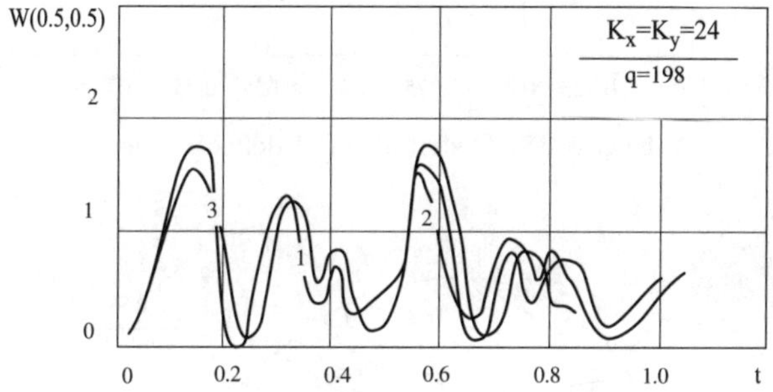

Fig. 8.35. Deflection of the plate centre $w(0.5, 0.5)$ for $K_1 = K_2 = 24$, $q = 198$

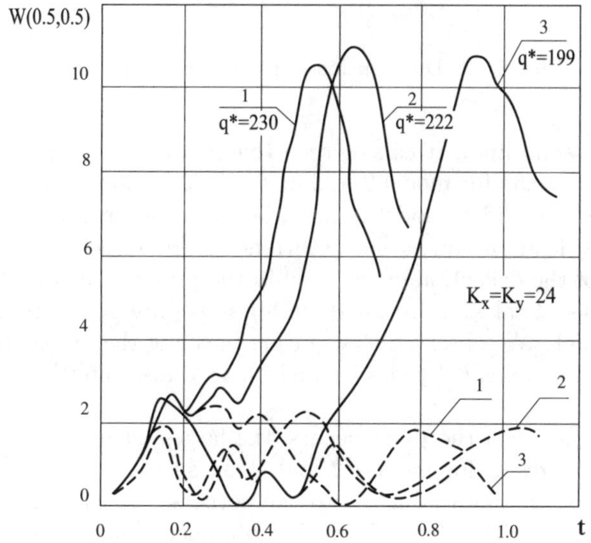

Fig. 8.36. Deflection of the plate centre $w(0.5, 0.5)$ for $K_1 = K_2 = 24$ and different loads

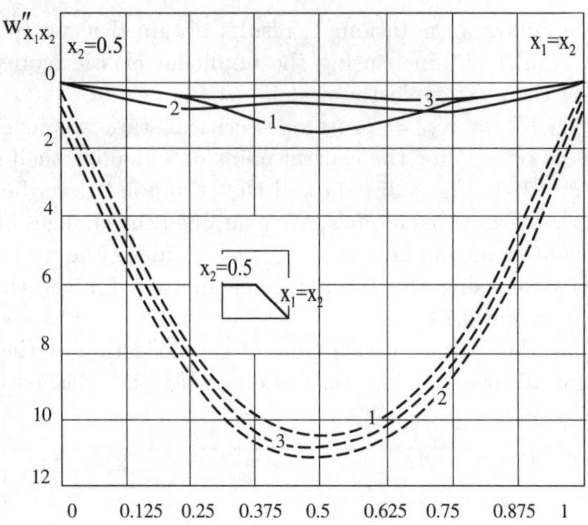

Fig. 8.37. Distribution of deflection w

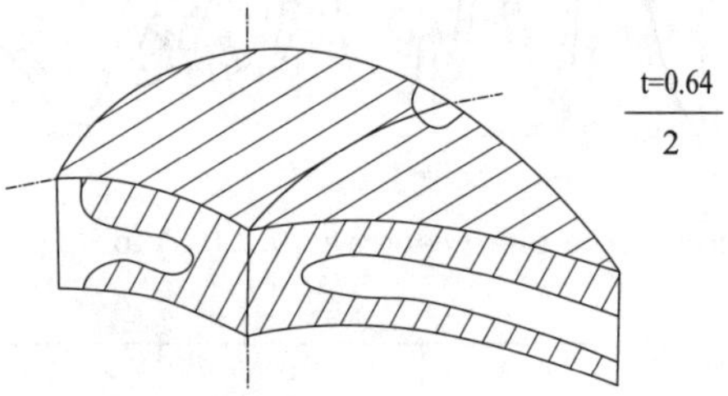

Fig. 8.38. Distribution of plastic zones at $t = 0.64$

critical load (solid line). It can be seen that buckling occurs at $t = 0.54$ for model 1, at $t = 0.64$ for model 2 and at $t = 0.92$ for model 3. Therefore, the use of a more simplified model of material deformation gives a higher value of the critical load and an earlier occurrence of buckling.

Graphs of the deflection of the shell in the pre-critical state (solid curve) and the post-critical state (dashed curve) states are shown in Fig. 8.37 for all three models. We observed that during buckling the maximum deflection increases by factors of 6.4 times (model 1), 18 times (model 2) and 26 times (model 3).

At the time when the jump occurs, the largest stresses are observed for model 3; the stresses for models 2 and 1 are smaller by factors of 1.6 and 2, respectively. In zones adjacent to the corners, the sign of the curvature obtained using model 3 differs from that obtained using models 1 and 2.

The distribution of plastic zones in the shell volume at $t = 0.64$ for model 2 is reported in Fig. 8.38.

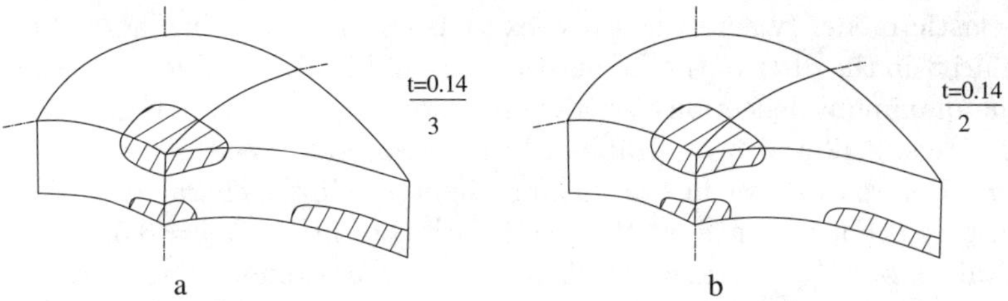

Fig. 8.39a,b. Distribution of plastic zones at $t = 0.14$

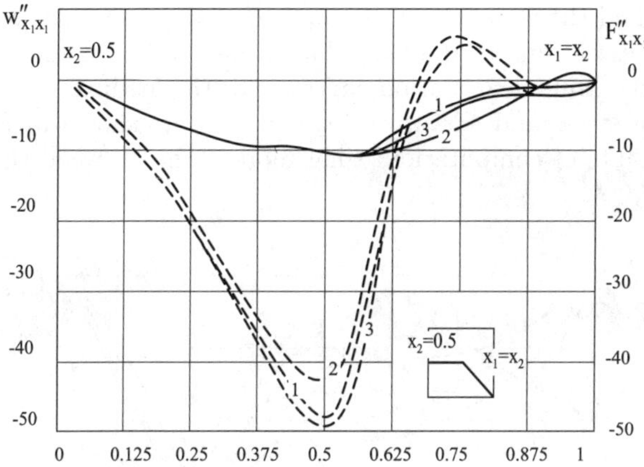

Fig. 8.40. Distribution of $F''_{x_1x_1}$ and $w''_{x_1x_1}$ at $t = 0.14$

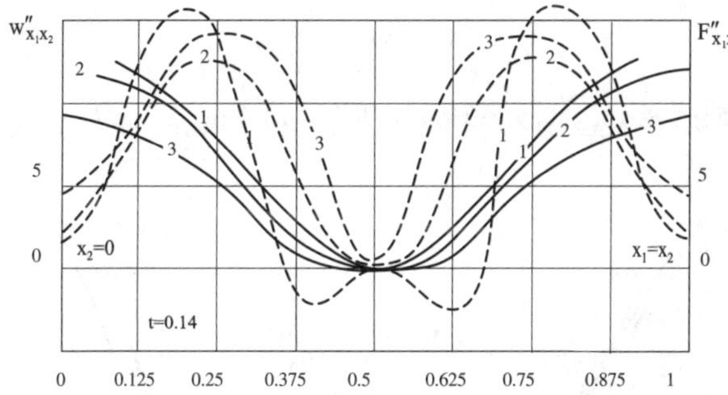

Fig. 8.41. Distribution of $F''_{x_1x_2}$ and $w''_{x_1x_2}$ at $t = 0.14$

The difference between the stress–strain states for the different models is slight; the results are shown in Figs. 8.40 and 8.41. In these figures, plots of $F''_{x_1x_1}$ (solid curves) and the curvatures $w''_{x_1x_2}$ (dashed curves) for $t = 0.14$ are shown.

The distribution of plastic zones obtained by use of the elastic–plastic model is shown in Fig. 8.39a, and the distribution obtained with the nonlinear elastic model (where the stress exceeds σ_T) is shown in Fig. 8.39b. The similarity in the distribution of plastic zones in Fig. 8.39 is due to overlapping of the maximum deflections for all three models at $t = 0.14$ (Fig. 8.35).

In Figs. 8.42 and 8.43, graphs of the stress and curvature at $t = 0.5$ are shown. Before this time instant, a large difference in the character of the plots is observed. For instance, at the point with coordinates $x_2 = 0.5$, $x_1 = 0.25$, the values of $F''_{x_1x_1}$ obtained with model 1 have a different sign from those calculated with model 3. The curvatures obtained with model 2 are 2–2.5 times smaller than those obtained with model 3 (Fig. 8.43).

Fig. 8.44 illustrates the distribution of plastic zones according to model 3 (Fig. 8.44a) and the zone where $\sigma_i \geq \sigma_T$ according to model 2 (Fig. 8.44b). The differences are due to the fact, that in model 2, a loading release takes place along the same deformation curve as for the loading.

Graphs of stress and curvature for $t = 0.58$ are shown in Fig. 8.45 and 8.46. The results of computation using model 3 lie between those obtained

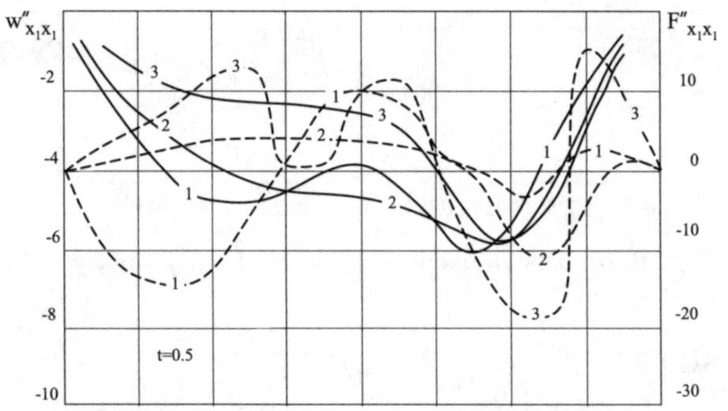

Fig. 8.42. Distribution of $F''_{x_1x_1}$ and $w''_{x_1x_1}$ at $t = 0.5$

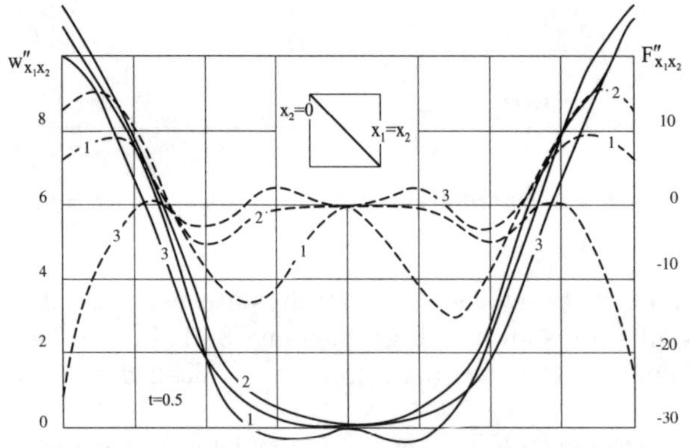

Fig. 8.43. Distribution of $F''_{x_1x_2}$ and $w''_{x_1x_2}$ at $t = 0.5$

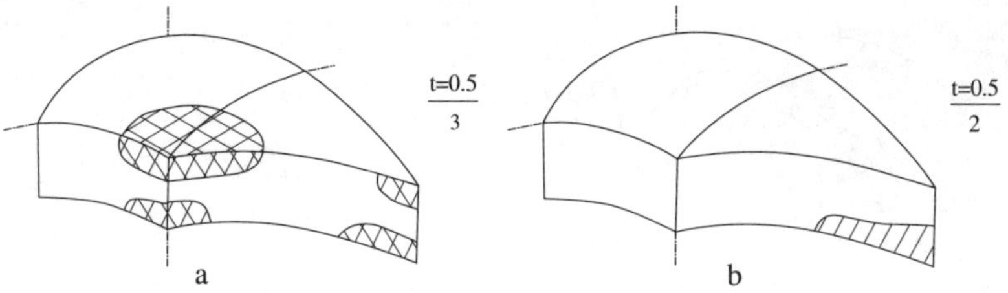

Fig. 8.44a,b. Distribution of plastic zones at $t = 0.5$ for models 3 and 2

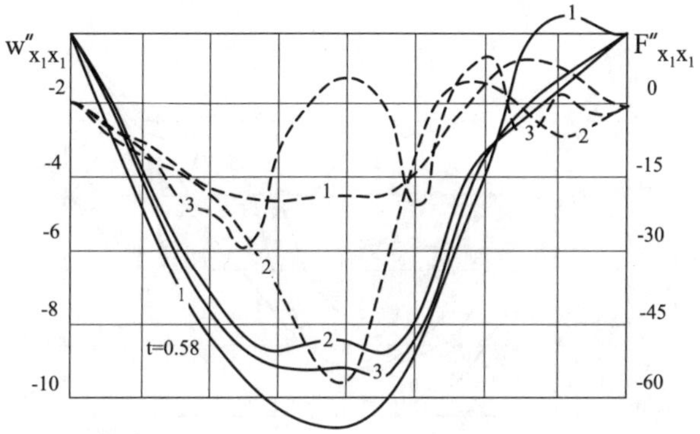

Fig. 8.45. Distribution of $F''_{x_1x_1}$ and $w''_{x_1x_1}$ at $t = 0.58$

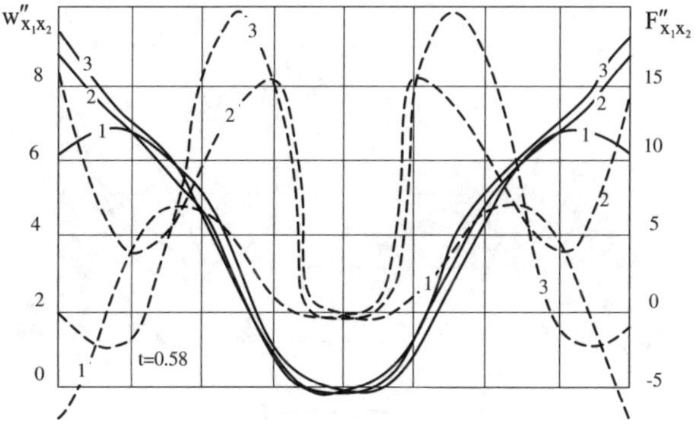

Fig. 8.46. Distribution of $F''_{x_1x_2}$ and $w''_{x_1x_2}$ at $t = 0.58$

using model 1 and model 2, and the differences are 12% and 9%, respectively. The difference between the deformations for models 3 and 2 is shown in Fig. 8.47 for the case of a loading release.

In Figs. 8.48 and 8.49, graphs of stress and curvature for $t = 0.74$ are shown. As Fig. 8.48 shows, the difference between the stresses computed with models 1 and 2 is insignificant, but these computed stresses differ strongly

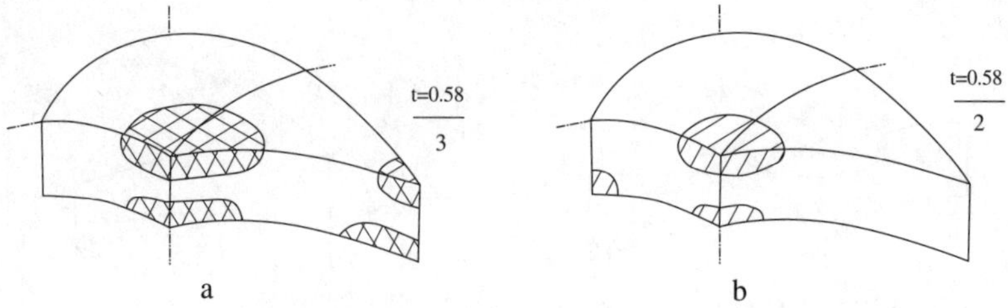

Fig. 8.47a,b. Deformations for (*a*) the model 3 and (*b*) model 2

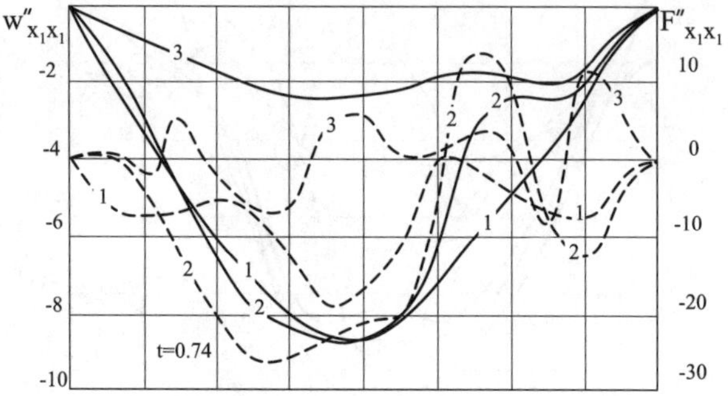

Fig. 8.48. Distribution of $F''_{x_1x_1}$ and $w''_{x_1x_1}$ at $t = 0.74$

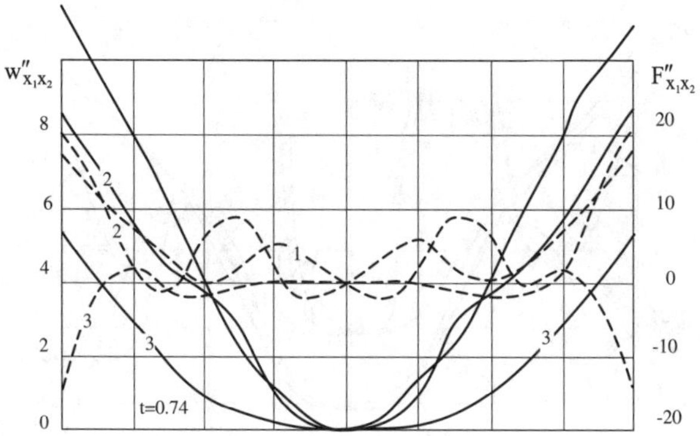

Fig. 8.49. Distribution of $F''_{x_1x_2}$ and $w''_{x_1x_2}$ at $t = 0.74$

from the results obtained with model 3. In a large part of the shell, the stresses obtained with model 3 differ in sign from those obtained with models 1 and 2. The character of the distribution of plastic zones for models 3 and 2, shown in Fig. 8.50, supports this observation.

In general, when we analyse the results for the pre-critical state, although the trajectories of the shell centre are close to each other for all three models,

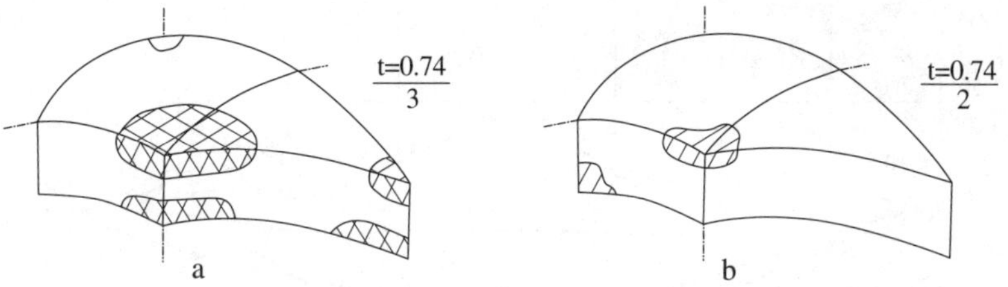

Fig. 8.50a,b. Distribution of plastic zones at $t = 0.58$ for models 3 and 2

a large difference in the stress–strain state can be observed at any given time moment. Evidently, however, the differences between the results obtained with models 2 and 3 are smaller than with models 1 and 3.

Now we shall describe the results of applying all three computational models to the problem of the dynamic stability of a shallow shell. According to models 1 and 2, the shell starts to exhibit buckling almost as soon as the critical load q^* is applied, in contrast to model 3, where the first "jump" is observed after oscillations around an equilibrium position. The values of the critical load obtained with the various models for a shell with geometric parameters $K_{x_1} = K_{x_2} = 24$ are given in Table 8.2. The mathematical models of elastic–plastic deformations plays the role of the reference model.

Graphs of stress and curvature at $t = 0.4$ and for the critical load are shown in Figs. 8.51–8.53. The results agree qualitatively for all models. However, $F''_{x_1x_1}$ is larger by a factor of 1.27 for model 2 and smaller by a factor of 5 times for model 1 than in the elastic–plastic model. The boundaries and locations of the plastic zones obtained with models 2 and 3 at $t = 0.4$ are shown in Fig. 8.54. The results are similar, although in the case of model 3 we observe that the material has already experienced load release.

The results of computation of the stresses and curvatures at $t = 0.5$ for the critical load are illustrated in Figs. 8.55–8.57. As earlier, one can observe a qualitative similarity here also. However, a qualitative difference can be observed in the graphs of $F''_{x_1x_1}$, where the stresses for model 3 lie between those for models 1 and 2. The stress in the corner is larger by a factor of 1.22 for model 1 and smaller by a factor of 6 times smaller for model 2 than that for model 3. In the case of the curvature $w''_{x_1x_1}$ in the corner zone, the curvature for model 2 is smaller by a factor of 1.6 than that for model 3.

Table 8.2. Critical loads q^* for a shell with $K_{x_1} = K_{x_2} = 24$, calculated for different models

No	Model	Critical load q^*	%
1	Linear-elastic	230	15%
2	Non linear elastic	222	11%
3	Elastic–plastic	199	–

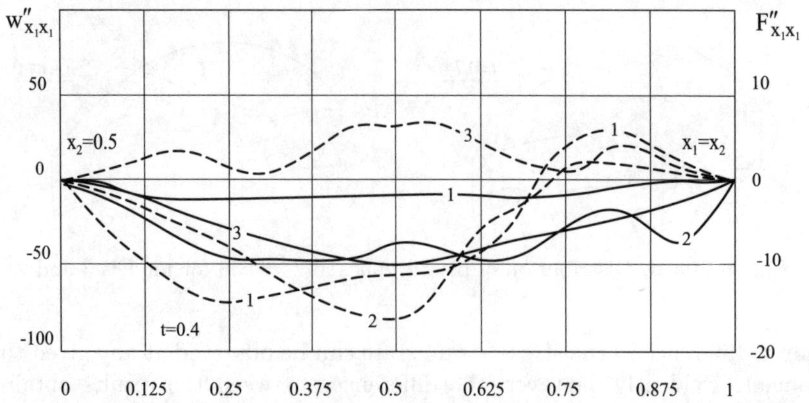

Fig. 8.51. Distribution of $F''_{x_1x_1}$ and $w''_{x_1x_1}$ for critical load at $t = 0.4$

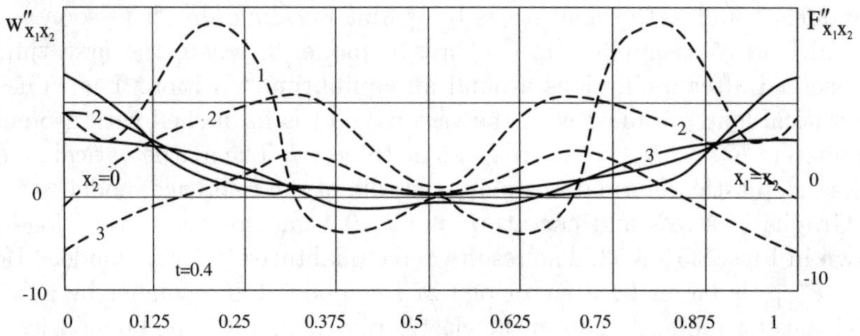

Fig. 8.52. Distribution of $F''_{x_1x_2}$ and $w''_{x_1x_2}$ for critical load at $t = 0.4$

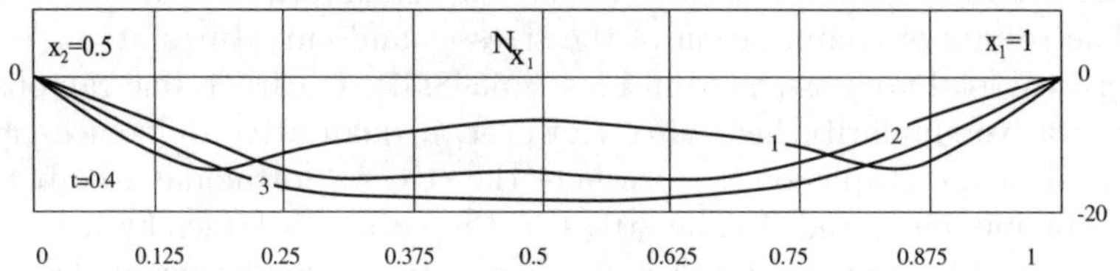

Fig. 8.53. Distribution of N_x for critical load at $t = 0.4$

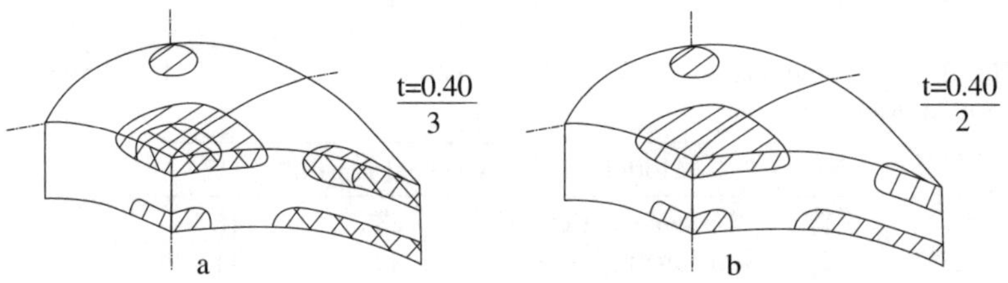

Fig. 8.54a,b. Distribution of plastic zones at $t = 0.4$ for models 3 and 2

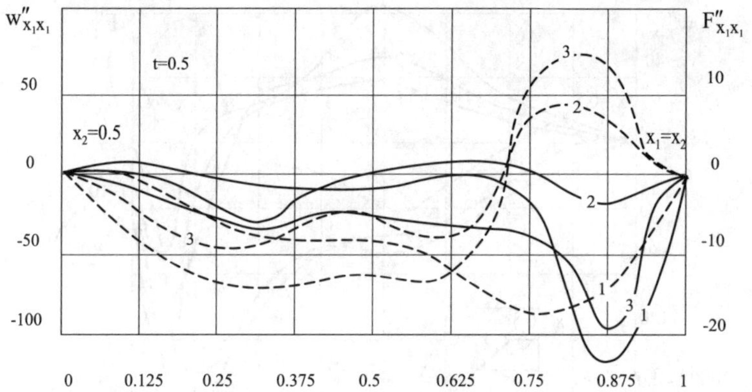

Fig. 8.55. Distribution of $F''_{x_1x_1}$ and $w''_{x_1x_1}$ for critical load at $t = 0.5$

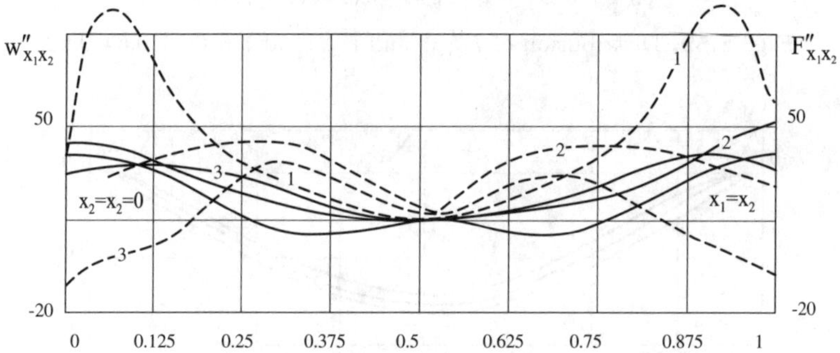

Fig. 8.56. Distribution of $F''_{x_1x_2}$ and $w''_{x_1x_2}$ for critical load at $t = 0.5$

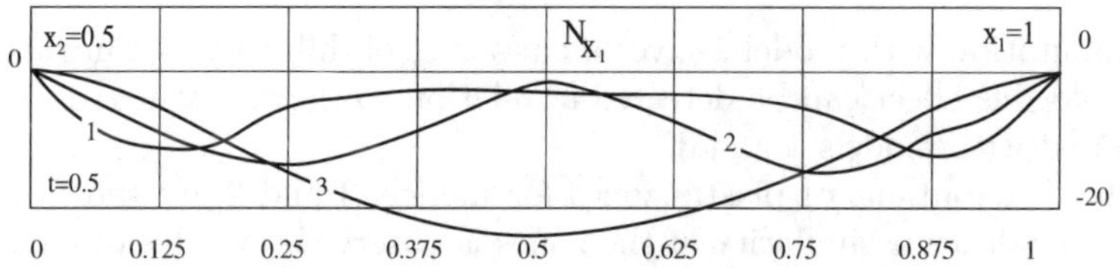

Fig. 8.57. Distribution of N_x for critical load at $t = 0.5$

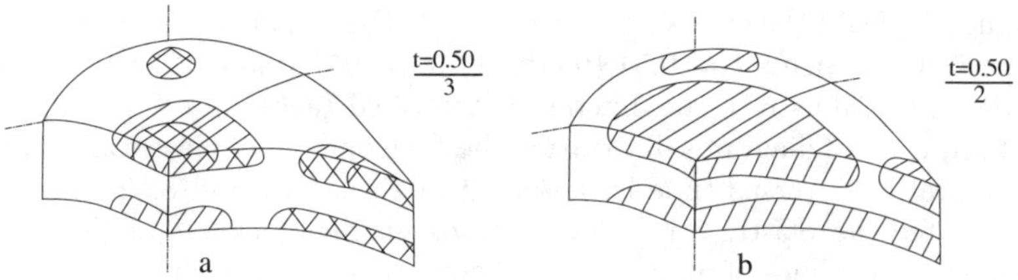

Fig. 8.58a,b. Distribution of plastic zones at $t = 0.5$ for models 3 and 2

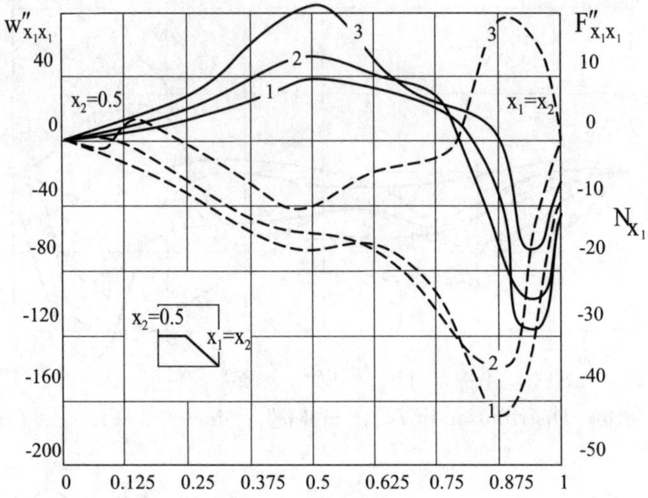

Fig. 8.59. Distribution of $F''_{x_1x_1}$ and $w''_{x_1x_1}$ at $t = 0.54, 0.64, 0.92$

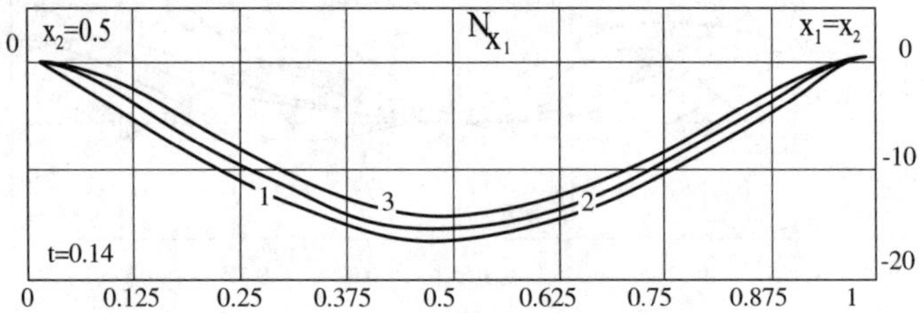

Fig. 8.60. Distribution of N_{x_1} for critical load at $t = 0.54, 0.64, 0.92$

A computation with model 1 gives a curvature of different sign in this zone. An analogous effect can be detected in relation to the curvatures of rotation computed with models 1 and 3.

The distributions of plastic zones for models 3 and 2 are shown in Fig. 8.58. A qualitative similarity of the zones is observed, but the volume occupied by the plastic zone for model 2 is larger than that for model 3. This is because, at the time considered, the deflection for model 2 exceeds the deflection for model 3 by a factor of 14 (Fig. 8.36).

The graphs of the stress $F''_{x_1x_1}$ and N_{x_1} in Figs. 8.59 and 8.60 correspond to the time moments $t = 0.54$ (model 1), $t = 0.64$ (model 2) and $t = 0.92$ (model 3) and illustrate the differences described earlier (see Fig. 8.37).

A sequence of diagrams illustrating the development of plastic zones under the critical load versus time is presented in Figs. 8.61 and 8.62. Beginning at $t = 0.85$, the plastic zones develop through the thickness of the shell in its centre. The zones of elastic deformation are located at the centres of the sides of the shell before and after the jump.

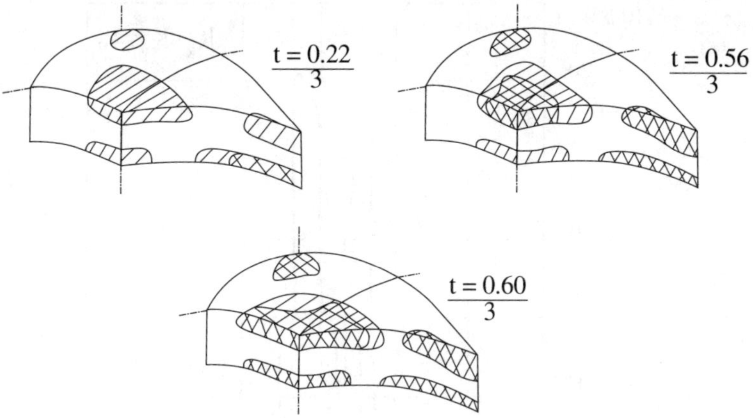

Fig. 8.61. Distribution of plastic zones for model 3 at $t = 0.22; 0.56; 0.60$

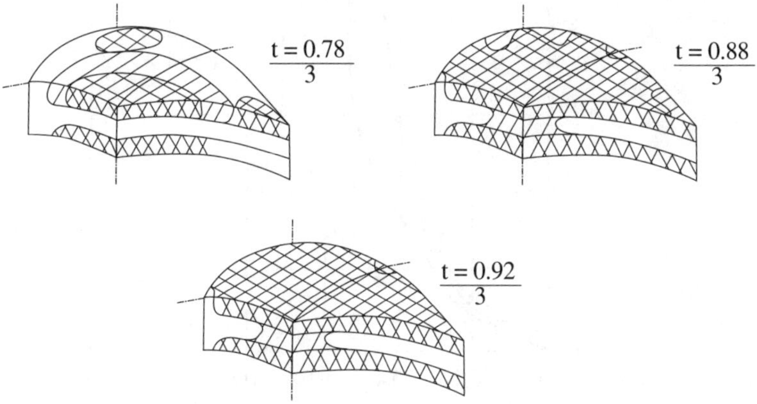

Fig. 8.62. Distribution of plastic zones for model 3 at $t = 0.78; 0.88; 0.92$

In Fig. 8.38, the distribution of plastic zones in the shell volume at $t = 0.64$ for model 2 is shown. Comparing this distribution with that shown in Fig. 8.62 for $t = 0.92$ (this time corresponds to the jump of the shell for model 3), we observe a good overlap of the boundaries of the plastic zones. This may be explained by the small difference ($< 2\%$) between the deflections of the shell at these times.

In conclusion, when elastic–plastic deformation can occur, one needs to use more sophisticated models, because simple models give results that are both qualitatively and quantitatively wrong.

We now analyse the dynamic behaviour of a shallow rectangular shell with the same material parameters (8.34) and the same dimensions, but with curvatures $k_{x_1} = k_{x_2} = 48$. Curves characterizing the variation with time of

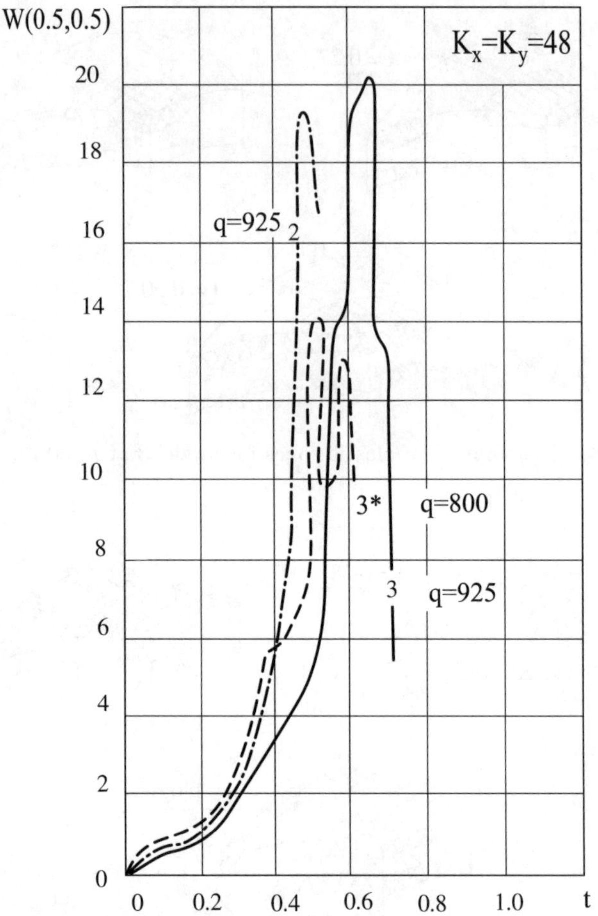

Fig. 8.63. Deflection of the centre of the shell $w(0.5, 0.5)$ as a function of time for different loads and materials: 2, nonlinear elastic material; 3, elastic–plastic material

the deflection at the centre of the shell given by the nonlinear elastic model (model 2) for one loading level and by the elastic–plastic model (model 3) for two loading levels are shown in Fig. 8.63. Graphs of stresses and curvatures for the shell with a load $q = 925$ given by models 2 and 3 are presented in Figs. 8.64–8.71. Analysis of those figures indicates that the stress–strain state of the shell for $k_{x_1} = k_{x_2} = 48$ is much more variable than that for $k_{x_1} = k_{x_2} = 24$.

In Figs. 8.72–8.75, the distributions of plastic zones for models 2 and 3 at different time moments are presented. The development of plastic zones shown here for the elastic–plastic shell (model 3) implies that strong plastic deformation occurs with the passage of time, although elastic properties

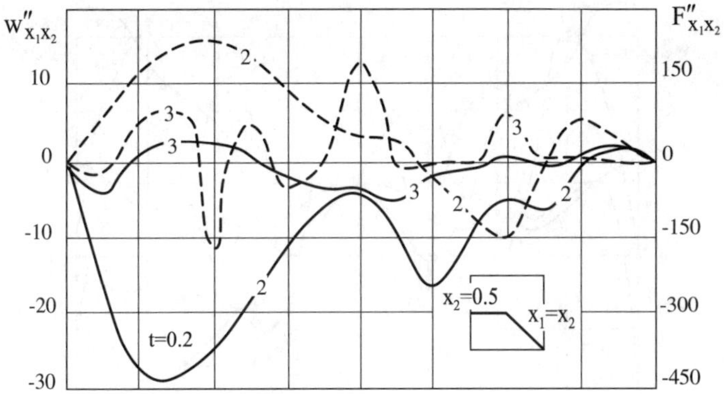

Fig. 8.64. Distribution of $F''_{x_1x_1}$ and $w''_{x_1x_1}$ for load $q = 925$ at $t = 0.2$

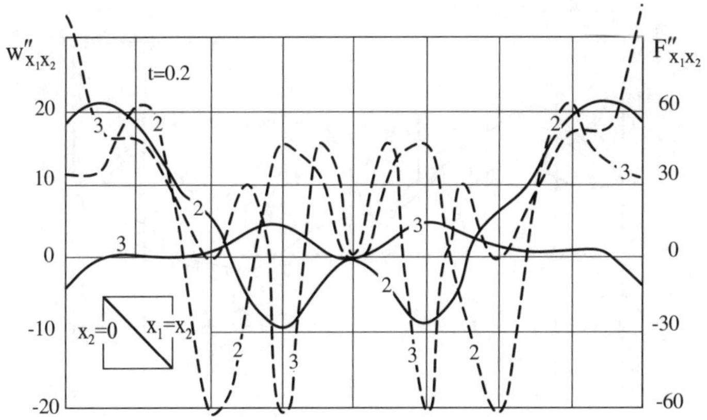

Fig. 8.65. Distribution of $F''_{x_1x_2}$ and $w''_{x_1x_2}$ for load $q = 925$ at $t = 0.2$

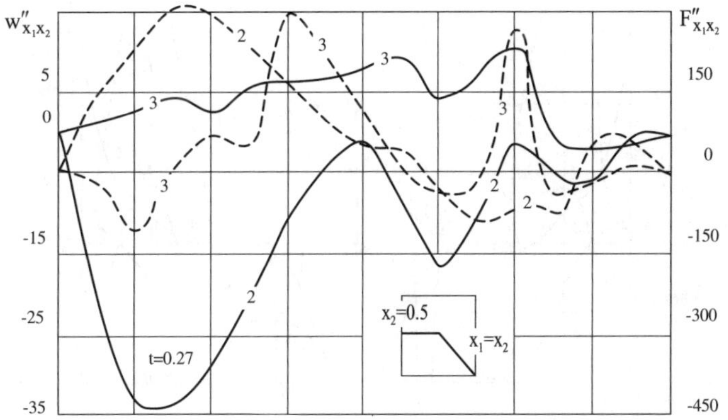

Fig. 8.66. Distribution of $F''_{x_1x_1}$ and $w''_{x_1x_1}$ for load $q = 925$ at $t = 0.27$

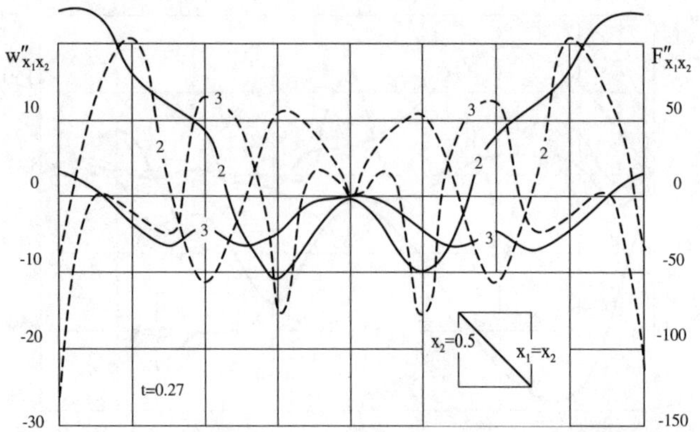

Fig. 8.67. Distribution of $F''_{x_1x_2}$ and $w''_{x_1x_2}$ for load $q = 925$ at $t = 0.27$

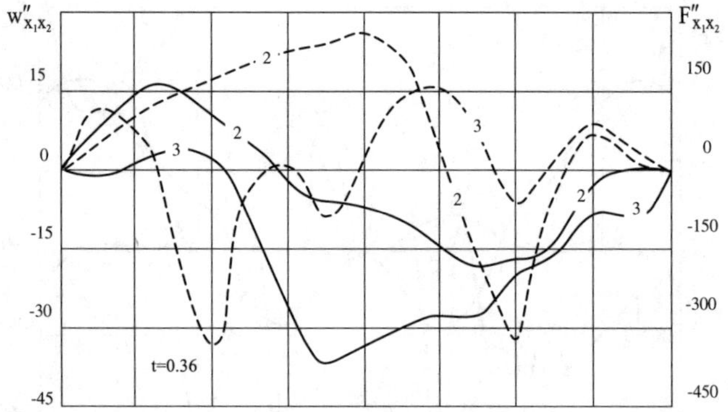

Fig. 8.68. Distribution of $F''_{x_1x_1}$ and $w''_{x_1x_1}$ for load $q = 925$ at $t = 0.36$

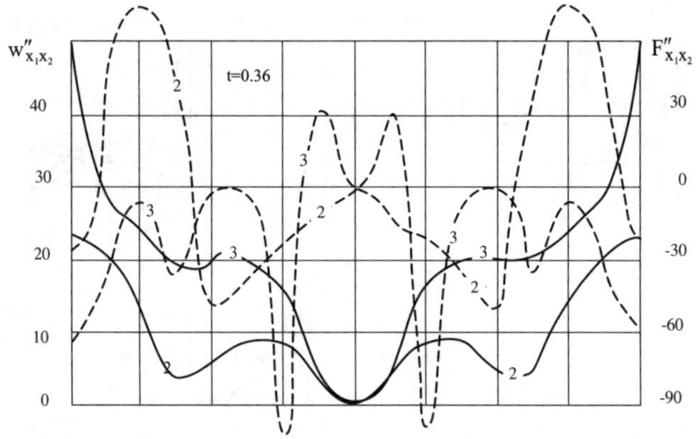

Fig. 8.69. Distribution of $F''_{x_1x_2}$ and $w''_{x_1x_2}$ for load $q = 925$ at $t = 0.36$

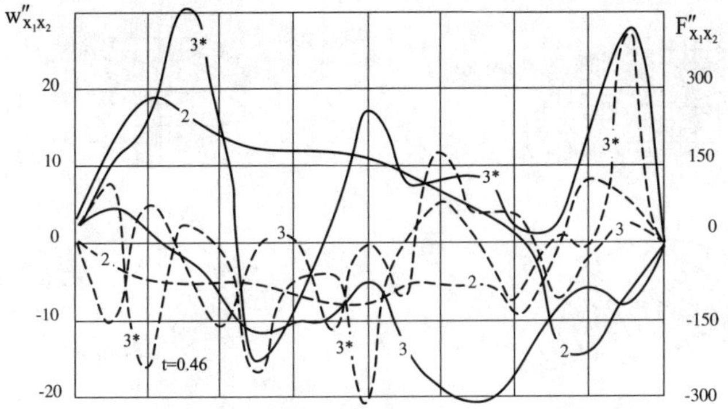

Fig. 8.70. Distribution of $F''_{x_1x_1}$ and $w''_{x_1x_1}$ for load $q = 925$ and $q = 800$ (asterisk) at $t = 0.46$

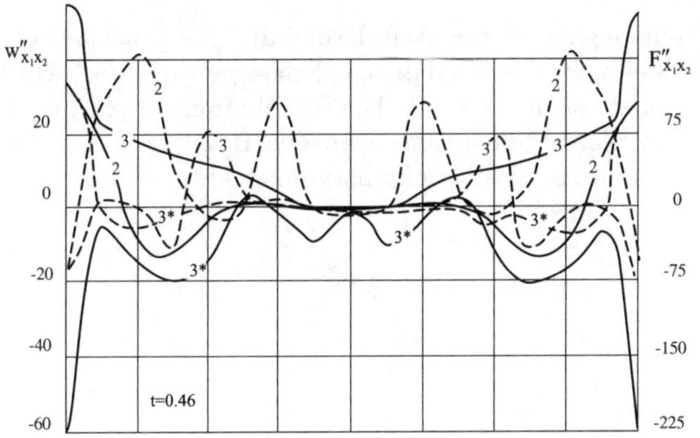

Fig. 8.71. Distribution of $F''_{x_1x_2}$ and $w''_{x_1x_2}$ for load $q = 925$ and $q = 800$ (asterisk) at $t = 0.46$

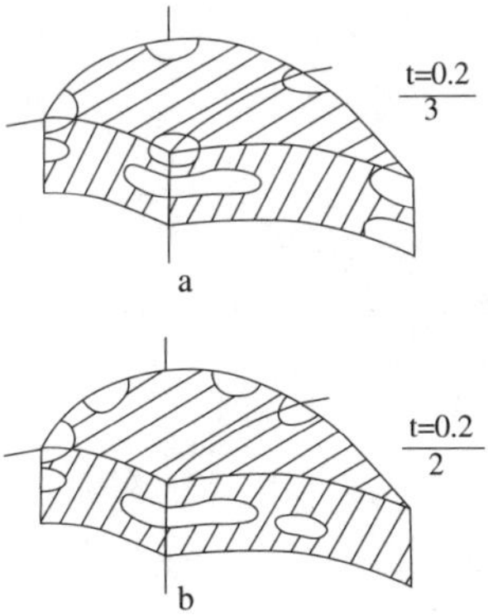

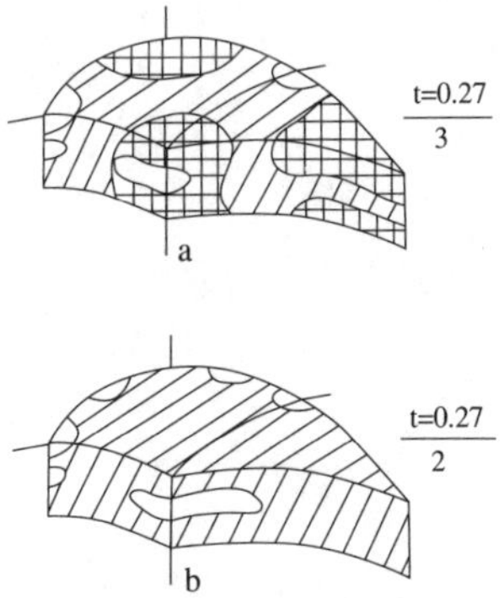

Fig. 8.72a,b. Distribution of plastic zones for models 2 and 3 at $t = 0.2$

Fig. 8.73a,b. Distribution of plastic zones for models 2 and 3 at $t = 0.27$

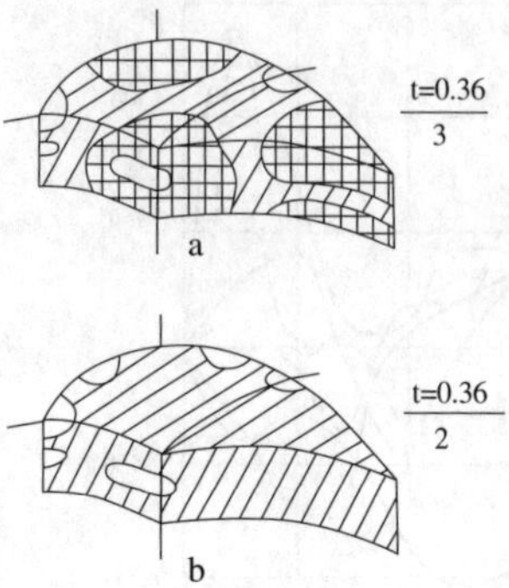

Fig. 8.74a,b. Distribution of plastic zones for models 2 and 3 at $t = 0.36$

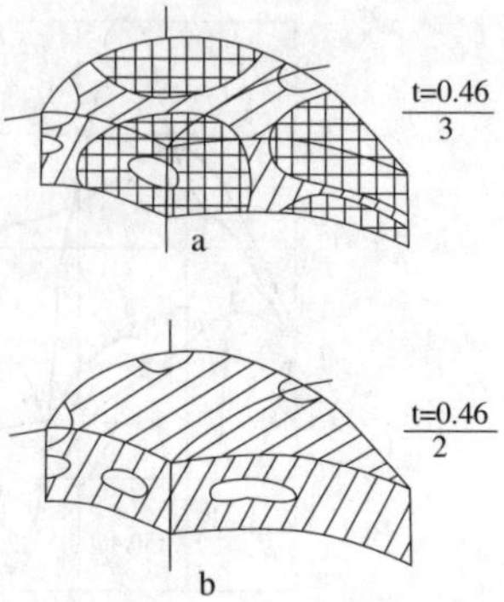

Fig. 8.75a,b. Distribution of plastic zones for models 2 and 3 at $t = 0.46$

persist in some regions of the shell. Examining the behaviour of the plastic zones for model 2, we observe only small changes with time, which indicates a high level of stress in the shell. The distributions of plastic zones models 2 and 3 are similar at some time moments. However, for model 3, a more complicated picture of plastic deformation is seen.

9 Unsolved Problems in Nonlinear Dynamics of Shells

The problems in the field of nonlinear dynamics of shells that remain to be solved can be summarized as follows:

1. Formulating the important boundary problems for the nonlinear dynamics of shells without assuming slenderness and central bending (i.e. for arbitrary rotations).
2. Formulating the important boundary problems for the coupled theory of thermoelasticity of shells without assuming slenderness and central bending, which means formulating the boundary problem for an initial set of hyperbolic and parabolic equations or of hyperbolic equations.
3. Obtaining numerical results for coupled problems of thermoelasticity for a wider class of boundary conditions than that investigated in this book.
4. Formulating a mathematical theory of the initial–boundary conditions within the Timoshenko theory of shells, taking into account different types of nonlinearity (geometric and physical nonlinearities, nonlinear elastic–plastic properties, and structural nonlinearities).
5. Formulating the important initial–boundary conditions for the coupled theory of thermoelasticity for polymer shells.
6. Investigation of the coupling of deflection and temperature fields for polymers in the framework of nonlinear geometry.
7. Research on obtaining an energy solution for nonlinear initial–boundary conditions in the vicinity of corners and of points or curves where the type of boundary conditions changes.
8. Detailed analysis of the regions of applicability of various nonlinear initial–boundary conditions in the theory of shallow shells rectangular in plan.
9. Investigation of the effect of local thermal shock in a three-dimensional theory, without the hypothesis of a linear distribution of the thermal field through the shell thickness.
10. Detailed investigation of the initial–boundary conditions in the theory of multilayer shells with coupled temperature and strain fields, in geometrically linear and nonlinear cases.
11. Investigation of the influence of local nonideal thermal contact on the stress–strain state in the theory of multilayer plates and shells for different kinematic models (Kirchhoff–Love, Timoshenko, etc.).

12. Derivation of the important initial–boundary problems in the nonlinear theory of shallow shells, including coupling of strain and temperature fields, from initial–boundary conditions of the three-dimensional coupled theory of thermoelasticity (for hyperbolic–parabolic and hyperbolic sets of equations).
13. Research on the initial–boundary conditions for nonlinear coupled thermoelastic problems of shells applying the quantitative theory of differential equations.
14. Finding a class of nonlinear initial–boundary problems of mathematical physics for which prior estimates of the solution can be obtained using the methods presented in this book.
15. Investigation of spatial–temporal chaos in the theory of plates and shells including coupling of strain and temperature fields with geometric, physical and structural nonlinearities.
16. Problems of aeroelasticity of shells interacting with transonic and supersonic flows of an ideal gas, including coupling of the strain, temperature and gas fields.
17. Research on nonsymmetric buckling of shells subjected to longitudinal periodic loads.

Some of the mentioned items have been analysed in this book. To conclude this book, a prior estimate of the error of the Bubnov–Galerkin method for an abstract hyperbolic equation with variable operator coefficients will be presented.

In Sect. 3.1, prior estimates of the error of the Bubnov–Galerkin method in the case of the Cauchy problems (3.59), for an abstract equation with constant operator coefficients, were presented. This problem generalizes class of initial–boundary problems in the theory of elasticity (for three-dimensional bodies and thin-walled structures). Here the Cauchy problem for an abstract, hyperbolic equation more general than (3.59) will be discussed. This equation has variable operator coefficients, containing non-adjoint operators of small modulus. Prior estimations of the error of the Bubnov–Galerkin method of type (3.60) and (3.62) will be presented for this case. An application of the results to a Cauchy–Dirichlet problem of partial differential equations of the second order, of hyperbolic type with variable coefficients and small elements, is also presented.

Let H be a Hilbert space, with a scalar product $(\cdot,\cdot)_H$ and a norm $\|\cdot\|_H$. Let us investigate the Cauchy problem

$$u''(t) + A(t)u(t) + B(t)u(t) = f(t), \tag{9.1}$$

$$u(0) = u'(0) = 0, \tag{9.2}$$

where u is a sought function that projects a segment $[0,T]$ of the number line into the space H, and $\{A(t)|t \in [0,T]\}$ is a family of unbounded self-adjoint, positive definite operators acting in H and determined on any linear set $D(A)$

included in H. $\{B(t)|t \in [0,T]\}$ is a family of linear (possibly unbounded) operators, defined on the same linear set $D(B)$, $D(A^{1/2}) \subset D(B) \subset H$; f is a given rational function that maps the segment $[0,T]$ into the space H. It is assumed that for all $t \in [0,T]$, the generalized derivatives $A'(t)$, $A''(t)$, $B'(t)$ are linear operators determined on $D(A)$, $D(A)$, $D(B)$, respectively. Furthermore, A_0 is a self-adjoint positive definite operator, having a completely continuous reverse operator A_0^{-1}. We assume also that the operator A_0 is convergent and forms an acute angle with all of the operators $A(t)$. The norms $\|A(t)A_0^{-1}\|_{H\to H}$ and $\|A_0^{1/2}(A(t))^{1/2}\|_{H\to H}$ are bounded uniformly from above in $t \in [0,T]$, and

$$\gamma[A] \equiv \inf_{0\le t\le T} \gamma(A(t), A_0) > 0, \tag{9.3}$$

where $\gamma(A(t), A_0)$ is the positive constant used in the acute-angle inequality (2.107) for the operators $A(t)$ and A_0. The following constrains are imposed on the operator coefficients in (9.1) and the operator A_0:

$$|(A'(t)u_1u_2)_H| \le C_1 \left\|A_0^{1/2}u_1\right\|_H \left\|A_0^{1/2}u_2\right\|_H, \tag{9.4}$$

$$|(A''(t)u_1u_2)_H| \le C_2 \left\|A_0^{1/2}u_1\right\|_H \left\|A_0^{1/2}u_2\right\|_H, \tag{9.5}$$

$$\|B(t)u_2\|_H \le C_3 \left\|A_0^{1/2}u_2\right\|_H, \tag{9.6}$$

$$\|B'(t)u_2\|_H \le C_4 \left\|A_0^{1/2}u_2\right\|_H, \tag{9.7}$$

where u_1 is any element from $D(A)$, u_2 is any element from $D(A^{1/2})$, and t is any number from $[0,T]$. Here and in what follows, the elements c_i $(i = 1,2,\dots,13)$ are various positive constants which can depend only on the operators $A(t)$, $B(t)$, A_0 and T. Conditions for boundness and differentiability are formulated in the inequalities (3.4)–(3.7). These conditions are used in the theorems given below about the uniqueness of the solution of the problem (9.1), (9.2), and also in obtaining the prior estimates of the error of the Bubnov–Galerkin method. The inequalities (9.5), (9.7) are used only for proving the existence of a solution of the problem (9.1), (9.2) for $f \in W_1^1(0,T;H)$. They are not used in obtaining the estimates of the error.

Definition 9.1 *The function $u : [0,T] \to H$ is called the generalized solution of the Cauchy problem (9.1), (9.2) if $u \in W_\infty^1(0,T;H) \cap L_\infty(0,T;D(A_0^{1/2}))$, $u(0) = 0$ and, for any function $v \in W_\infty^1(0,T;H) \cap L_1(0,T;D(A_0^{1/2}))$ satisfying conditions $v(T) = 0$, the following condition is satisfied:*

$$\int_0^T \Big[-(u'(t), v'(t))_H + \big((A(t))^{1/2}u(t), (A(t))^{1/2}v(t)\big)_H$$

$$+\big(B(t)u(t), v(t)\big)_H\Big]\, dt = \int_0^T (f(t), v(t))_H\, dt. \tag{9.8}$$

Definition 9.2 *The function* $u : [0,T] \to H$ *is called the "classical" solution of the Cauchy problem (9.1), (9.2) if* $u \in W^2_\infty(0,T;H) \cap W^1_\infty(0,T;D(A_0^{1/2})) \cap L_\infty(0,T;D(A_0))$, *and the condition (9.1) (almost always* $(t \in [0,T])$*) and conditions (9.2) are satisfied.*

Remark 9.1 *In the Definitions 9.1 and 9.2 we use a function with maximum smoothness* $u : [0,T] \to H$, *which can be estimated from the conditions on the operators* $A(t)$, $B(t)$ *and* A_0 *presented above for* $f \in L_2(0,T;H)$ *or for* $f \in W^1_2(0,T;H)$. *The integral condition (9.8) was obtained, following common assumptions used in defining generalized solutions in mathematical physics, by formally multiplying the differential equation (9.1) by a test function* $v(t)$ *and following it by integrating with respect to time, taking into account the initial condition* $u'(0) = 0$. *Hence, one can conclude that the "classical" solution (if it exists) of the problem (9.1), (9.2) is also its generalized solution.*

Let $\{v_i\}_{i=1}^\infty$ be a basis in H, constructed from eigenelements of the operator A_0, and let $\{\lambda\}_{i=1}^\infty$ be the corresponding series of eigenvalues of the operator $A_0 : A_0 v_i = \lambda_i v_i$, $i = 1, 2, \ldots$. The eigenelements and eigenvalues of the operator A_0 are treated as known.

Let H_n be a linear environment of the bounded system of basis elements $\{v_i\}_{i=1}^\infty$, let P_n be an orthogonal transformation H on H_n, and let R_n be an operator $E - P_H$, where E is an identity operator on H. The system of (9.1) and (9.2) is to be solved by means of the following equations given by the Bubnov–Galerkin method:

$$\begin{cases} P_n(u_n''(t) + A(t)u_n(t) + B(t)u_n(t)) = P_n f(t), \\ u_n(t) \in H_n, \ \ \forall t \in [0,T], \\ u_n(0) = u_n'(0) = 0. \end{cases} \tag{9.9}$$

The exact solution of the problem (9.9) is called the approximate solution of the problem (9.1), (9.2) obtained by the Bubnov–Galerkin method. For $f \in L_2(0,T;H)$ the problem (9.9) is equivalent to the Cauchy problem for a linear system of ordinary differential equations and consequently has an equivalent solution for any natural number n such that $u_n \in W^2_2(0,T;D(A_0))$, for $f \in L_\infty(0,T;H)$ (and for $f \in W^1_2(0,T;H)$, $u_n \in W^2_\infty(0,T;D(A_0))$). Other necessary conditions for the existence and uniqueness of a solution to the problem (9.1), (9.2) can be obtained from the following considerations: for $f \in L_2(0,T;H)$, a generalized solution of the problem (9.1) and (9.2) exists and is unique, and for $f \in W_2^1(0,T;H)$ a "classical" solution of the problem (9.1), (9.2) exists and is unique. The proof of the relevant theorems is based on the compactness method [147], and is analogous to that of Theorems 2.1 and 2.2 (Sect. 2.1.5) on the existence and uniqueness of the solution of the abstract problem (2.54)–(2.56). The solutions of the problems (9.1), (9.2), (9.9) satisfy the following prior estimates for $f \in L_2(0,T;H)$:

$$\left[\|u_n'(t)\|_H^2 + \|A_0^{1/2} u_2\|_H^2\right]^{1/2} \leq C_5 \|f\|_{L_2(0,T;H)}, \tag{9.10}$$

$$\left[\|u'(t)\|_H^2 + \|A_0^{1/2} u\|_H^2\right]^{1/2} \leq C_5 \|f\|_{L_2(0,T;H)} \tag{9.11}$$

(almost always ($t \in [0,T]$)). If $f \in W_2^1(0,T;H)$ then, in addition to (9.10) and (9.11),

$$\|u_n''(t)\|_H^2 + \|A_0^{1/2} u_n'\|_H^2 \leq C_6 \|f\|_{L_2(0,T;H)}^2$$

$$+C_7 \|f'\|_{L_2(0,T;H)}^2 + C_8 \|f(0)\|_H^2, \tag{9.12}$$

$$\|u''(t)\|_H^2 + \|A_0^{1/2} u'\|_H^2 \leq C_6 \|f\|_{L_2(0,T;H)}^2$$

$$+C_7 \|f'\|_{L_2(0,T;H)}^2 + C_8 \|f(0)\|_H^2 \tag{9.13}$$

(almost always ($t \in [0,T]$)). In the inequalities (9.10)–(9.13), $u(t)$ is an exact (generalized or "classical") solution of the problem (9.1), (9.2), and $u_n(t)$ is an approximate solution obtained with the Bubnov–Galerkin method from (9.9). The estimates (9.10)–(9.13) are obtained in the process of proving the theorems about the existence of the generalized and "classical" solutions of the problem (9.1), and (9.2) with the compactness method.

We shall now obtain prior estimates of the error of the Bubnov–Galerkin method for the problem (9.1), (9.2). We introduce

$$\Delta u_n(t) = u_n(t) - u(t),$$

where $u(t)$ is the exact solution and $u_n(t)$ is the approximate solution of the problem (9.1), (9.2).

Theorem 9.1 *For $f \in W_2^1(0,T;H)$, the accuracy of the Bubnov–Galerkin method using (9.9) to solve the problem (9.1), (9.2) can be estimated as follows:*

$$\begin{aligned}
\max_{0\le t\le T}&\left[\|\Delta u_n'(t)\|_H^2+\|A_0^{1/2}\Delta u_n\|_H^2\right]\\
&\le C_9\Big(C_6\|f\|^2_{L_2(0,T;H)}+C_7\|f'\|^2_{L_2(0,T;H)}\\
&+C_8\|f(0)\|_H^2\Big)^{1/2}\Big[\|f\|_{L_2(0,T;H)}\\
&+C_{11}\big(C_6\|f\|^2_{L_2(0,T;H)}+C_7\|f'\|^2_{L_2(0,T;H)}+\\
&+C_8\|f(0)\|_H^2\big)^{1/2}\Big]\tilde\lambda_{n+1}^{-1/2}\equiv 0(\tilde\lambda_{n+1}^{-1/2}),
\end{aligned} \tag{9.14}$$

where

$$\tilde\lambda_R=\min_{i\ge R}\lambda_i. \tag{9.15}$$

Proof. We rewrite the first equation of (9.9) in the following form:

$$P_nA(t)u_n(t)=P_n(f(t)-u_n''(t)-B(t)u_n(t)).$$

We then multiply both sides scalarly by $A_0u_n(t)$. From the equation obtained and the definition (9.3) of the constant $\gamma[A]$, we obtain

$$\begin{aligned}
\gamma[A]\|A_0u_n(t)\|_H^2&\le(A(t)u_n(t)A_0u_n(t))_H\\
=(f(t)-u_n''(t)-B(t)u_n(t),A_0u_n(t))_H&\\
\le\|f(t)-u_n''(t)-B(t)u_n(t)\|_H\times\|A_0u_n(t)\|_H.&
\end{aligned}$$

Hence, from the inequality (9.3), we obtain that

$$\|A_0u_n(t)\|_H\le\frac{1}{\gamma[A]}\|f(t)-u_n''(t)-B(t)u_n(t)\|_H.$$

Applying the above inequality and the condition of uniform boundedness of the norm $\|A(t)A_0^{-1}\|_{H\to H}$, we obtain

$$\begin{aligned}
\|A(t)u_n(t)\|_H&\le\frac{\|A(t)A_0^{-1}\|_{H\to H}}{\gamma[A]}\|f(t)-u_n''(t)-B(t)u_n(t)\|_H\\
&\le C_{12}\|f(t)-u_n''(t)-B(t)u_n(t)\|_H.
\end{aligned} \tag{9.16}$$

We introduce the notation

$$\delta_n(t)=u_n''(t)+A(t)u_n(t)+B(t)u_n(t)-f(t).$$

Applying the inequalities (9.16), (9.6), (9.10) and (9.12), we obtain

$$\begin{aligned}
\|\delta_n(t)\|_H&\le(1+C_{12})\,(\|u_n''(t)\|_H+\|B(t)u_n(t)\|_H+\|f(t)\|_H)\\
&\le(1+C_{12})\left(\|u_n''(t)\|_H+C_3\left\|A_0^{1/2}u_n(t)\right\|_H+\|f(t)\|_H\right)\\
&\le(1+C_{12})\Big[\left(C_6\|f\|^2_{L_2(0,T;H)}+C_7\|f'\|^2_{L_2(0,T;H)}+C_8\|f(0)\|_H^2\right)^{1/2}\\
&+C_3C_5\|f\|_{L_2(0,T;H)}+\|f(t)\|_H\Big].
\end{aligned}$$

Transforming to the norm of the space $L_2(0,T;H)$, we obtain the estimate

$$\|\delta_n(t)\|_{L_2(0,T;H)} \le C_{10}\|f\|_{L_2(0,T;H)} + C_{11}\Big(C_6\|f\|^2_{L_2(0,T;H)} + C_7\|f'\|^2_{L_2(0,T;H)} + C_8\|f(0)\|^2_H\Big)^{1/2}, \tag{9.17}$$

where

$$C_{10} = (1+C_{12})(1+C_3C_5T^{1/2}), \quad C_{11} = (1+C_{12})T^{1/2}.$$

Hence, from (9.1) and the first equation of (9.9), we obtain

$$\Delta u_n''(t) + A(t)\Delta u_n(t) + B(t)\Delta u_n(t) = R_n\delta_n(t).$$

We multiply this equation scalarly by $2\Delta u_n'(t)$ and then integrate by parts over the interval $[0,t]$, where $t \le T$. Taking into account the initial conditions $\Delta u_n(0) = \Delta u_n'(0) = 0$, we obtain

$$\|\Delta u_n'(t)\|^2_H + \|(A(t))^{1/2}\Delta u_n(t)\|^2_H] = 2\int_0^t \big(R_n\delta_n(\tau), \Delta u_n'(\tau)\big)_H\, d\tau - 2\int_0^t \big(B(\tau)\Delta u_n(\tau), \Delta u_n'(\tau)\big)_H\, d\tau + \int_0^t \big(A'(\tau)\Delta u_n(\tau), \Delta u_n(\tau)\big)_H\, d\tau. \tag{9.18}$$

Because of the special choice of the basis $\{v_i\}_{i=1}^{\infty}$, we obtain

$$\left\|A_0^{-1/2}R_n\right\|_{H\to G} = \tilde{\lambda}_{n+1}^{-1/2},$$

and consequently the first term of the right-hand side of (9.18) can be estimated in the following way:

$$2\int_0^t \big(R_n\delta_n(\tau), \Delta u_n'(\tau)\big)_H\, d\tau = 2\int_0^t \big(A_0^{-1/2}R_n\delta_n(\tau), A_0^{1/2}\Delta u_n'(\tau)\big)_H\, d\tau$$

$$\le 2\left\|A_0^{-1/2}R_n\right\|_{H\to H}\|\delta_n\|_{L_2(0,T;H)}\left\|A_0^{-1/2}\Delta u_n'\right\|_{L_2(0,T;H)}$$

$$\le 2\,\|\delta_n\|_{L_2(0,T;H)} \times \left(\left\|A_0^{-1/2}u_n'\right\|_{L_2(0,T;H)} + \left\|A_0^{-1/2}u\right\|_{L_2(0,T;H)}\right)\tilde{\lambda}_{n+1}^{-1/2}.$$

Hence (9.12), (9.13), (9.17) give

$$2\int_0^t \left(R_n\delta_n(\tau), \Delta u_n'(\tau)\right)_H d\tau \le 4\Big(C_6\|f\|^2_{L_2(0,T;H)} + C_7\|f'\|^2_{L_2(0,T;H)}$$
$$+ C_8\|f(0)\|^2_H\Big)^{1/2}\Big[C_{10}\|f\|_{L_2(0,T;H)} + C_{11}\big(C_6\|f\|^2_{L_2(0,T;H)} \tag{9.19}$$
$$+ C_7\|f'\|^2_{L_2(0,T;H)} + C_8\|f(0)\|^2_H\big)^{1/2}\Big]\tilde{\lambda}_{n+1}^{-1/2}.$$

The sums of the second and third terms of the right-hand side of (9.18) can be estimated using the conditions (9.4), (9.6) and the Cauchy inequality. We obtain

$$-2\int_0^t \left(B(\tau)\Delta u_n(\tau), \Delta u_n'(\tau)\right)_H d\tau + \int_0^t \left(A'(\tau)\Delta u_n(\tau), \Delta u_n(\tau)\right)_H d\tau$$
$$\le C_{13}\int_0^t \left(\|\Delta u_n'(\tau)\|^2_H + \|A_0^{1/2}\Delta u_n(\tau)\|^2_H\right) d\tau. \tag{9.20}$$

We now estimate the right-hand sides of (9.19), (9.20) using the inequalities (9.18). We obtain the following inequality:

$$\|\Delta u_n'(t)\|^2_H + \|(A(t))^{1/2}\Delta u_n(t)\|^2_H \le 4\Big(C_6\|f\|^2_{L_2(0,T;H)}$$
$$+ C_7\|f'\|^2_{L_2(0,T;H)} + C_8\|f(0)\|^2_H\Big)^{1/2}\Big[C_{10}\|f\|_{L_2(0,T;H)}$$
$$+ C_{11}\big(C_6\|f\|^2_{L_2(0,T;H)} + C_7\|f'\|^2_{L_2(0,T;H)} + C_8\|f(0)\|^2_H\big)^{1/2}\Big]\tilde{\lambda}_{n+1}^{-1/2}$$
$$+ C_{13}\int_0^t \left(\|\Delta u_n'(\tau)\|^2_H + \|A_0^{1/2}\Delta u_n(\tau)\|^2_H\right) d\tau.$$

Hence, applying the condition of uniform boundness of the norm

$$\left\|A_0^{-1/2}(A(t))^{-1/2}\right\|_{H\to H}$$

and the Gronwill lemma, we obtain the estimate (9.14).

Therefore Theorem 9.1 has been proved.

We now use the notation (3.21), (3.22), where H is the Hilbert space under consideration, f is the free element of (9.1).

Theorem 9.2 *For $f \in L_2(0,T;H)$, the accuracy of the Bubnov–Galerkin method using (9.9) to solve the problem (9.1), (9.2) can be estimated as follows:*

$$\begin{aligned}
&\sup_{0\le t\le T} ess \left(\|\Delta u_n'(t)\|_H^2 + \|A_0^{1/2}\Delta u_n(t)\|_H^2\right)^{1/2} \\
&\le D_1\omega(f,\tilde{\lambda}_{n+1}^{-1/4}) + C_9^{1/2}\Big[C_6(e(f,\tilde{\lambda}_{n+1}^{-1/4}))^2\tilde{\lambda}_{n+1}^{-1/2} \\
&+ D_2(\omega(f,\tilde{\lambda}_{n+1}^{-1/4}))^2\Big]^{1/4}\Big\{C_{10}e(f,\tilde{\lambda}_{n+1}^{-1/4})\tilde{\lambda}_{n+1}^{-1/4} \\
&+ C_{11}\Big[C_6(e(f,\tilde{\lambda}_{n+1}^{-1/4}))^2\tilde{\lambda}_{n+1}^{-1/2} + D_2(\omega(f,\tilde{\lambda}_{n+1}^{-1/4}))^2\Big]^{1/2}\Big\}^{1/2} \\
&\equiv 0(\max(\tilde{\lambda}_{n+1}^{-1/4},\omega(f,\tilde{\lambda}_{n+1}^{-1/4}))),
\end{aligned} \tag{9.21}$$

where $\tilde{\lambda}_R$ is the series defined by (9.15), D_1 and D_2 are positive constants, defined below by (9.29), and $e(f,\delta)$ is a quantity bounded for $\delta \to 0$, defined below by (9.30).

Proof. Let $\varphi(t)$ be any averaging kernel of class W_2^1 on the segment $[-1,1]$, i.e. an even function from the Sobolev space $W_2^1([-1,1])$, equal to zero on the ends of the segment $[-1,1]$ and satisfying (3.25). Let $\varphi_\delta(t) = \frac{1}{\delta}\varphi\left(\frac{t}{\delta}\right)$, $\delta > 0$, $-\delta \le t \le \delta$. We introduce the notation

$$f_\delta(t) = \int_{t-\delta}^{t+\delta} \tilde{f}(\tau)\varphi_\delta(t-\tau)\,d\tau,\ \ \Delta f_\delta(t) = f(t) - f_\delta(t),$$

where $\delta > 0$, $0 \le t \le T$, and $\tilde{f}$ is the extension of the function f from $[0,T]$ to the whole number line, defined by (3.21). Analogous, (3.28)–(3.30) lead to the inequalities:

$$\|\Delta f_\delta\|_{L_2(0,T;H)} \le C_{0,\varphi}\omega(f,\delta), \tag{9.22}$$

$$\|\Delta f_\delta'\|_{L_2(0,T;H)} \le \frac{1}{\delta}C_{1,\varphi}\omega(f,\delta), \tag{9.23}$$

where

$$C_{0,\varphi} = \sqrt{2}\left(\int_{-1}^{1}(\varphi(\tau))^2\,d\tau\right)^{1/2},\ \ C_{1,\varphi} = \sqrt{2}\left(\int_{-1}^{1}(\varphi'(\tau))^2\,d\tau\right)^{1/2}$$

(for example, for $\varphi(t) = 1 - |t|$, $C_{0,\varphi} = 2\sqrt{3}/3$, $C_{1,\varphi} = 2$).

We consider an investigated family of Cauchy problems with free elements $f_\delta \in W_2^1(0,T;H)$, indexed by the parameter $\delta > 0$:

$$\begin{cases} u_\delta''(t) + A(t)u_\delta(t) + B(t)u_\delta(t) = f_\delta(t), \\ u_\delta(0) = u_\delta'(0) = 0. \end{cases} \tag{9.24}$$

The approximate solution of (9.24) obtained by the Bubnov–Galerkin method is denoted by $u_{\delta,n}$. The following notation is also introduced:

$$\Delta u_\delta(t) = u(t) - u_\delta(t),$$

$$\Delta u_{\delta,n}(t) = u_n(t) - u_{\delta,n}(t),$$

$$\Delta u_{n,\delta}(t) = u_{\delta,n}(t) - u_\delta(t).$$

Assuming that the function Δu_δ is the exact solution of the Cauchy problem of type (9.1) with free element Δf_δ, and that the function $\Delta u_{\delta,n}$ is the approximate solution of this problem obtained by the Bubnov–Galerkin method, we can write down prior estimates of type (9.10), (9.11) as follows:

$$\left[\|u'_{\delta,n}(t)\|_H^2 + \|A_0^{1/2} u_{\delta,n}(t)\|_H^2\right]^{1/2} \le C_5 \|f_\delta\|_{L_2(0,T;H)}, \tag{9.25}$$

$$\left[\|u'_\delta(t)\|_H^2 + \|A_0^{1/2} u_\delta(t)\|_H^2\right]^{1/2} \le C_5 \|f_\delta\|_{L_2(0,T;H)}. \tag{9.26}$$

(almost always ($t \in [0,T]$)).

By applying Theorem 9.1 to (9.24) and taking into account the fact that $f_\delta(0) = 0$, which is true in the case of an odd extension $\tilde{f}$ of the free element of (9.1), we obtain the following inequality:

$$\begin{aligned}\left[\|\Delta u'_{n,\delta}(t)\|_H^2 + \|A_0^{1/2} \Delta u_{n,\delta}(t)\|_H^2\right]^{1/2} &\le C_9^{1/2}\Big(C_6 \|f_\delta\|^2_{L_2(0,T;H)} \\ +C_7 \|f'_\delta\|^2_{L_2(0,T;H)}\Big)^{1/4} &\Big[C_{10}\|f_\delta\|_{L_2(0,T;H)} + C_{11}\big(C_6\|f_\delta\|^2_{L_2(0,T;H)} \\ &+C_7\|f'_\delta\|^2_{L_2(0,T;H)}\big)^{1/2}\Big]^{1/2} \tilde{\lambda}_{n+1}^{-1/4},\end{aligned} \tag{9.27}$$

where $0 \le t \le T$.

The error of the Bubnov–Galerkin method when applied to (9.1), (9.2) can be represented in the following form:

$$\begin{aligned}\left[\|\Delta u'_n(t)\|_H^2 + \|A_0^{1/2}\Delta u_n(t)\|_H^2\right]^{1/2} &\le \Big[\|\Delta u'_\delta(t)\|_H^2 \\ + \|A_0^{1/2}\Delta u_\delta(t)\|_H^2\Big]^{1/2} &+ \left[\|\Delta u'_{\delta,n}(t)\|_H^2 + \|A_0^{1/2}\Delta u_{\delta,n}(t)\|_H^2\right]^{1/2} \\ + \left[\|\Delta u'_{n,\delta}(t)\|_H^2 + \|A_0^{1/2}\Delta u_{n,\delta}(t)\|_H^2\right]^{1/2}&.\end{aligned}$$

Hence, applying the inequalities (9.25)–(9.27), (9.22) and (9.23), we obtain

$$\begin{aligned}\left[\|\Delta u'_n(t)\|_H^2 + \|A_0^{1/2}\Delta u_n(t)\|_H^2\right]^{1/2} &\le 2C_5\|\Delta f_\delta\|^2_{L_2(0,T;H)} \\ + C_9^{1/2}\Big(C_6\|f_\delta\|^2_{L_2(0,T;H)} + C_7\|f'_\delta\|^2_{L_2(0,T;H)}\Big)^{1/4} &\Big[C_{10}\|f_\delta\|_{L_2(0,T;H)} \\ + C_{11}\Big(C_6\|f_\delta\|^2_{L_2(0,T;H)} + C_7\|f'_\delta\|^2_{L_2(0,T;H)}\Big)^{1/2}\Big]^{1/2} &\tilde{\lambda}_{n+1}^{-1/4}\end{aligned}$$

$$\leq 2C_5C_{0,\varphi}\omega(f,\delta) + C_9^{1/2}\Big[\Big(C_6(\|f\|_{L_2(0,T;H)} + C_{0,\varphi}\omega(f,\delta)\Big)^2 + \frac{1}{\delta^2}C_7C_{1,\varphi}^2(\omega(f,\delta))^2\Big]^{1/4}\Big\{C_{10}(\|f\|_{L_2(0,T;H)} + C_{0,\varphi}\omega(f,\delta)) + C_{11}\Big[C_6(\|f_\delta\|_{L_2(0,T;H)} + C_{0,\varphi}\omega(f,\delta))^2 + \frac{1}{\delta^2}C_7C_{1,\varphi}^2 \times (\omega(f,\delta))^2\Big]^{1/2}\Big\}^{1/2}\tilde{\lambda}_{n+1}^{-1/4}.$$

In the above inequality, δ is any positive number. Let $\delta = \tilde{\lambda}_{n+1}^{-1/4}$; we now obtain

$$\begin{aligned}\Big[\|\Delta u_n'(t)\|_H^2 + \|A_0^{1/2}\Delta u_n(t)\|_H^2\Big]^{1/2} &\leq 2C_5C_{0,\varphi}\omega(f,\tilde{\lambda}_{n+1}^{-1/4}) \\ &+ C_9^{1/2}\Big[C_6\Big(\|f\|_{L_2(0,T;H)} + C_{0,\varphi}\omega(f,\tilde{\lambda}_{n+1}^{-1/4})\Big)^2 + C_7C_{1,\varphi}^2 \\ &\times \Big(\omega(f,\tilde{\lambda}_{n+1}^{-1/4})\Big)^2\Big]^{1/4}\Big\{C_{10}\Big(\|f\|_{L_2(0,T;H)} + C_{0,\varphi}\omega(f,\tilde{\lambda}_{n+1}^{-1/4})\Big) \\ &\times \tilde{\lambda}_{n+1}^{-1/4} + C_{11}\Big[C_6\|f\|_{L_2(0,T;H)} + (C_{0,\varphi}\omega(f,\tilde{\lambda}_{n+1}^{-1/4}))^2\tilde{\lambda}_{n+1}^{-1/2} \\ &+ C_7C_{1,\varphi}^2(\omega(f,\tilde{\lambda}_{n+1}^{-1/4}))^2\Big]^2\Big\}^{1/2}\end{aligned} \tag{9.28}$$

(almost always ($t \in [0,T]$)). By introducing the notation

$$D_1 = 2C_7C_{1,\varphi}^2, \quad D_2 = C_7C_{1,\varphi}^2, \tag{9.29}$$

$$e(f,\delta) = \|f\|_{L_2(0,T;H)} + C_{0,\varphi}\omega(d,\delta), \tag{9.30}$$

the (9.28) can be rewritten as (9.21). Therefore Theorem 9.2 has been proved.

Remark 9.2 *The full continuity of the operator A_0^{-1}, $\lambda_i \to \infty$ for $i \to \infty$, implies that the estimates of error of the Bubnov–Galerkin method (9.14) and (9.21) are concised in the sense that the right-hand sides approach zero as $n \to \infty$.*

As an example of the application of the above estimates, we discuss the Cauchy–Dirichlet problem for a hyperbolic equation of second order:

$$\begin{gathered}\frac{\partial^2 u}{\partial t^2} - \sum_{i,j=1}^{P}\frac{\partial}{\partial x}\left(a_{ij}(x,t)\frac{\partial u}{\partial x_j}\right) + \sum_{i=1}^{P} b_i(x,t)\frac{\partial u}{\partial x_i} + c(x,t)u = f(x,t), \\ u|_{t=0} = 0, \quad \left.\frac{\partial u}{\partial t}\right|_{t=0} = 0, \quad u|_{x\in S} = 0,\end{gathered} \tag{9.31}$$

where $u = u(x,t)$ is a soughtfunction defined on $Q = \Omega \times [0,T]$; $x = (x_1,\dots,x_p) \in \Omega$; $t \in [0,T]$, $T > 0$; and Ω is a p-metric space, surrounded by a surface S, such that the boundary functions $y_p = \Psi(y_1,\dots,y_{p-1})$ describing the equations of the surface S in the local coordinate systems have uniformly bounded partial derivatives over all y_k up to the second order inclusive. The eigenfunctions of the Laplace operator ∇^2, equal to zero on S, are written for the space Ω in the explict form (see, for example [159]); and a_{ij}, b_i, c are given functions, satisfying the following conditions:

$$a_{ij}, \frac{\partial a_{ij}}{\partial x_k}, \frac{\partial a_{ij}}{\partial t}, \frac{\partial^2 a_{ij}}{\partial t^2}, b_i, \frac{\partial b_i}{\partial t} \in L_\infty(Q),$$

$$\frac{\partial c}{\partial t} \in L_\infty(0,T; \in L_2(\Omega)),\ a_{ij} = a_{ji}(i,j,k = 1,2,\dots,p),$$

$$\sum_{i,j=1}^{P} a_{ij}(x,t)\xi_i\xi_j \geq \alpha \sum_{i=1}^{P} \xi_i^2. \tag{9.32}$$

Here $\xi = (\xi_1, \dots, \xi_p)$ is any set of p real numbers; α is a positive constant, not depending on t or ξ, $f = f(x,t)$ is a given rational function defined on Q.

Let us convert the problem (9.31) into the form (9.1), (9.2). We introduce the following notation:

$$A(t)u = -\sum_{i,j=1}^{P} \frac{\partial}{\partial x_i}\left(a_{ij}(x,t)\frac{\partial u}{\partial x_j}\right), \quad D(A) = W_2^2(\Omega) \cap \overset{\circ}{W}{}_2^1(\Omega),$$

$$B(t)u = \sum_{i=1}^{P} b_i(x,t)\frac{\partial u}{\partial x_i} + c(x,t)u, \quad D(B) = \overset{\circ}{W}{}_2^1(\Omega),\ H = L_2(\Omega).$$

Here $\overset{\circ}{W}{}_2^1(\Omega)$ depicts the subspace of the Sobolev space $W_2^1(\Omega)$ of functions equal to zero on S. In [140] it is stated that when the conditions on the space Ω and the coefficients a_{ij} are recalculated, the following inequality can be derived for an arbitrary function $u \in W_2^2(\Omega) \cap \overset{\circ}{W}{}_2^1(\Omega)$:

$$m\|u\|^2_{W_2^2(\Omega)} \leq -\int_\Omega A(t)u \nabla^2 u\, dx + \|u\|^2_{\overset{\circ}{W}{}_2^2(\Omega)}, \tag{9.33}$$

where the positive constant m depends only on α and $\max\left\{|a_{ij}|, \left|\frac{\partial a_{ij}}{\partial x_k}\right|, \left|\frac{\partial^2 \Psi}{\partial y_l \partial y_m}\right|\right\}$. Consequently, this constant can be chosen set independently of t. From (9.33), using (9.32), we obtain the inequality

$$\int_\Omega A(t)uA_0u\, dx \geq m\|u\|^2_{W_2^2(\Omega)},$$

where $A_0u = -\nabla^2 u + \frac{1}{\alpha}u$.

The last inequality and the conditions on the space Ω and coefficients a_{ij}, b_i, c given above ensure a uniform upper limit of the norms $\left\|A(t)A_0^{-1}\right\|_{H\to H}$, $\left\|A_0^{1/2}(A(t))^{-1/2}\right\|_{H\to H}$ and also ensure that the inequalities (9.3)–(9.7) are satisfied. We can consider the application of the Bubnov–Galerkin method to the problem (9.31), using the eigenfunctions of the Laplace operator for the space Ω as the basis functions $\nu_i(x)$. If $f \in L_2(0,T;L_2(\Omega)) = L_2(Q)$, an estimate of the form (9.21) exists for the problem (9.31) with $\lambda_n = -\mu_n + \dfrac{1}{\alpha}$, where $\{\mu_n\}_{n=1}^{\infty}$ is a series of positive eigenvalues of the Laplace operator for the space Ω. If $f \in W_2^1(0,T;L(\Omega))$ (this means also if $\dfrac{\partial f}{\partial t} \in L_2(Q)$), an estimate of the form (9.14) also exists.

References

1. Kikoin I.K. (ed.), *Tables of Physical Properties.* Atomizdat, Moscow, 1976 (in Russian).
2. Andrianov I.V., Awrejcewicz J., New trends in asymptotic approaches: summation and interpolation methods, *Appl. Mech. Rev.*, **54**, No. 1, 2001, 69–92.
3. Andrianov I.V., Awrejcewicz J., Numbers or understanding: analytical and numerical methods in the theory of plates and shells, *Facta Universitis, Mech., Automat. Control and Robot.*, **2**, No. 10, 2000, 1319–1327.
4. Andronov A.A., Vitt A.A., Khaikin S.E., *Theory of Oscillations.* Dover, New York, 1987.
5. Awrejcewicz J., Krys'ko V.A., *Techniques and Methods of Plate and Shell Analysis.* Lodz Technical University Press, Lodz, 1996 (in Polish).
6. Awrejcewicz J., Krys'ko V.A., *Numerical Analysis of Shells Oscillations With Thermal Load.* WNT, Scientific Book Foundation, Warsaw, 1998 (in Polish).
7. Awrejcewicz J., Krys'ko V.A., *Dynamics and Stability of Shells With Thermal Excitations.* WNT, Scientific Book Foundation, Warsaw, 1999 (in Polish).
8. Awrejcewicz J., Krys'ko V.A., *Oscillations of Lumped Systems.* WNT, Warsaw, 2000 (in Polish).
9. Awrejcewicz J., Andrianov I.V., *Plates and Shells in Nature, Mechanics and Biomechanics.* WNT, Fundacja Ksiazka Naukowo-Techniczna, Warsaw, 2001 (in Polish).
10. Awrejcewicz J., Strange nonlinear behaviour governed by a set of four averaged amplitude equations, *Meccanica*, 1996, 347–361.
11. Awrejcewicz J., Krys'ko V.A., Dynamical stability of thin shells with thermal convection, in *Proceedings of the International Symposium on Trends in Continuum Physis* (eds. B. T. Maruszewski, W. Muschik, A. Radowicz), Poznan, Poland, 17–20 August, 1998. World Scientific, Singapore, 1999, pp. 35–45.
12. Awrejcewicz J., Krys'ko V.A., Vibration analysis of the plates and shells of moderate thickness, *Journal of Technical Physics*, 40, 3, 1999, 277–305.
13. Awrejcewicz J., Krys'ko V.A., 3D theory versus 2D approximate theory of the free orthotropic (isotropic) plates and shells vibrations. Part 1, derivation of governing equations, *J. of Sound Vib.*, **226**, No. 5, 1999, 807–829.
14. Awrejcewicz J., Krys'ko V.A., 3D theory versus 2D approximate theory of the free orthotropic (isotropic) plates and shells vibrations. Part 2, numerical algorithms and analysis, *J. of Sound Vib.*, **226**, No. 5, 1999, 831–871.
15. Awrejcewicz J., Krys'ko V.A., Kutsemako N., Free vibrations of doubly curved in-plane non-homogeneous shells, *J. Sound Vib.*, **225**, No. 4, 1999, 701–722.

16. Awrejcewicz J., Krys'ko V.A., Period doubling bifurcation and chaos exhibited by an isotropic plate, *Z. Angew. Math. Mech.* **80**, 2000, S267–S268.
17. Awrejcewicz J., Krys'ko V.A., Feigenbaum scenario exhibited by thin plate dynamics, *Nonlinear Dynam.*, **24**, 2001, 373–398.
18. Awrejcewicz J., Krys'ko V.A., Coupled thermoelasticity problems of shallow shells, *J. Systems Anal. Modell. Simulat.* (to appear).
19. Awrejcewicz J., Krys'ko V.A., Krys'ko A.V., Spatial–temporal chaos and solitions exhibited by von Kármán mode, *Int. J. Bifurcation Chaos* (to appear).
20. Awrejcewicz J., Krys'ko V.A., Dynamics of a shell with the added masses, in *Proceedings of the 4th Conference on Dynamical Systems–Theory and Applications*, (eds. J. Awrejcewicz, J. Grabski, J. Mrozowski), Lodz, December 8–9, 1997. Technical University Press, Lodz, 1997, pp. 163–168.
21. Awrejcewicz J., Krys'ko V.A., Three dimensional problem of the orthotropic plate vibrations with the added masses, *Proceedings of the 4th Conference on Dynamical Systems–Theory and Applications* (eds. J. Awrejcewicz, J. Grabski, J. Mrozowski), December 8–9, Lodz, Poland, 1997. Technical University Press, Lodz, 1997, pp. 169–176.
22. Awrejcewicz J., Krys'ko V.A., Krys'ko A.V., Period doubling Hopf bifurcation of thin flexible isotropic plates with an impact load, in *Proceedings of the 4th Conference on Dynamical Systems–Theory and Applications* (eds. J. Awrejcewicz, J. Grabski, J. Mrozowski), Lodz, Poland, 1997. Technical University Press, Lodz, 1997, pp. 63–68.
23. Awrejcewicz J., Krys'ko V.A., Abstract thermoelasticity problems of shallow shells, in *Proceedings of the 5th Conference on Dynamical Systems–Theory and Applications* (eds. J. Awrejcewicz, J. Grabski, J. Mrozowski), Lodz, Poland, December 6–8, 1999. Technical University Press, Lodz, 1999, pp. 89–94.
24. Awrejcewicz J., Krys'ko V.A., Coupled thermomechanical problems in the shell theory, in *Proceedings of the 3rd International Congress on Thermal Stresses '99* (eds. J.J. Skrzypek, R.B. Hetnarski), Cracow, Poland, June 13–17, 1999, pp. 53–56.
25. Awrejcewicz J., Krys'ko V.A., Optimization of plate and shell surfaces, in *Proceedings of the ASME Design Engineering Technical Conferences*, Las Vegas, Nevada, September 12–16, 1999, DETC99/VIB-8139, (published on CD-ROM).
26. Awrejcewicz J., Krys'ko V.A., Optimization of plate and shell surfaces, in *Proceedings of the 10th World Congress on the Theory of Machines and Mechanisms*, Oulu, Finland, June 20–24, 1999, pp. 2128–2133.
27. Awrejcewicz J., Krys'ko V.A., Krys'ko A.V., Non-symmetric oscillations and transition to chaos in freely supported flexible plate sinusoidally excited, *Proceedings of the 5th Conference on Dynamical Systems–Theory and Applications* (eds. J. Awrejcewicz, J. Grabski, J. Mrozowski), Lodz, Poland, December 6–8, 1999. Technical University Press, Lodz, 1999, pp. 95–102.
28. Awrejcewicz J., Krys'ko V.A., Krys'ko A.V., Symmetric oscillations and transition to chaos in a freely supported sinusoidally excited flexible shell, in *Proceedings of the 5th Conference on Dynamical Systems–Theory and Applications* (eds. J. Awrejcewicz, J. Grabski, J. Mrozowski), Lodz, Poland, December 6–8, 1999. Technical University Press, Lodz, 1999, pp. 103–112.

29. Awrejcewicz J., Krys'ko V.A., Lamarque C.-H., Optimal vibroisolation problem of plates and shells, in *Proceedings of the 5th Conference on Dynamical Systems-Theory and Applications* (eds. J. Awrejcewicz, J. Grabski, J. Mrozowski), Lodz, Poland, December 6–8, 1999. Technical University Press, Lodz, 1999, pp. 113–120.
30. Awrejcewicz J., Krys'ko V.A., Mrozowski J., Analysis of free vibrations of non-homogeneous plates using finite element method, in *Proceedings of EURODYN'99, Fourth Conference of the European Association for Structural Dynamics* (eds. L. Fryba, J. Naprstek), Prague, June 7–10, 1999, Structural Dynamics, Vol. I. Balkema, Rotterdam, 1999, pp. 415–420.
31. Awrejcewicz J., Krys'ko V.A., Krys'ko A.V., Symmetric and non-symmetric oscillations and bifurcations of periodically excited plates with non-homogeneous boundary conditions, in *IASS-IACM 2000 Fourth International Colloquium on Computation of Shell & Spatial Structures* (eds. M. Papadrakakis, A. Samartin, E. Onat, Chania, Grecee, June 4–7, 2000. ISASR-NTUA, Athens, 2000, pp. 1–9 (CD-ROM).
32. Awrejcewicz J., Krys'ko V.A., Krys'ko A.V., Vakakis A., Solitons and chaos exhibited by flexible plates sinusoidally excited, in *Nonlinear Dynamics, Chaos, Control and Their Applications to Engineering Sciences*, Vol. 5, *Chaos Control and Times Series* (eds. J.M. Balthazar, P.B. Goncalvez, R. M.Brasil, I.L. Caldas, F.B. Rizatto). Brazilian Society of Mechanical Sciences, Rio de Janeiro, 2000, pp. 258–267.
33. Awrejcewicz J., Krys'ko V.A., Complex parametric oscillations of flexible rectangular plates, in *Proceedings of the 6th Conference on Dynamical Systems–Theory and Applications* (eds. J. Awrejcewicz, J. Grabski, J. Nowakowski), Lodz, Poland, December 10–12, 2001. Technical University Press, Lodz, 2001, pp. 153–164.
34. Awrejcewicz J., Krys'ko V.A., Krys'ko A.V., Regular and chaotic behaviour of flexible plates, in *Proceedings of the Third International Conference on Thin-Walled Structures* (eds. J. Zaras, K. Kowal-Michalska, J. Rhodes), Cracow, Poland, June 5–7, 2001. Elsevier Science, 2001, pp. 349–356.
35. Awrejcewicz J., Krys'ko V.A., Krys'ko A.V., Regular and chaotic behaviour of flexible plates, in *European Congress on Computational Methods in Applied Sciences and Engineering, ECCOMAS Computational Fluid Dynamics Conference 2001*, Swansea, UK, September 4–7, 2001 (CD ROM).
36. Awrejcewicz J., Krys'ko V.A., Krys'ko A.V., Solitons exhibited by the von Kármán equations, in *Applied Mechanics in the Americas, Proceedings of the Seventh Pan-American Congress on Applied Mechanics* (eds. P. Kittl, G. Diaz, D. Mook, J. Geer), Temuco, Chile, January 2–4, 2002, Vol. 9, pp. 653–9 to 653–12.
37. Awrejcewicz J. (ed.), *Bifurcation and Chaos: Theory and Application.* Springer, Berlin, Heidelberg, 1995.
38. Awrejcewicz J., Andrianov I.V., Manevitch L.I., *Asymptotic Approach in Nonlinear Dynamics: New Trends and Applications.* Springer, Berlin, Heidelberg, 1998.
39. Awrejcewicz J., Andrianov I.V., *Asymptotic Methods and Their Applications in Shell Theory.* WNT, Fundacja Ksiazka Naukowo-Techniczna, Warsaw, 2000 (in Polish).

40. Bacinov C.D., Nonlinear oscillations of plates under simultaneous static and vibrating loads, *Appl. Mech.*, **7**, No. 10, 1971, 126–130 (in Russian).
41. Bauer H., Nonlinear response of elastic plates to pulse excitation, *Trans. ASME*, **E35**, No. 1, 1968, 47–52.
42. Beam R.M., Warming R.F., *Numerical Calculating of Two-Dimensional Unsteady Transonic Flows with Circulation.* NASA, TND-76–05, 1974.
43. Beam R.M., Warming R.F., An implicit finite-difference algorithm for hyperbolic system in conservation law form, *J. Comput. Phys.*, **22**, No. 1, 1976, 87–110.
44. Bell J.F., *Experiment Fundamentals of Mechanics of Elastic Solids*, Vol. 2, *Small Deformations.* Nauka, Moscow, 1984 (in Russian).
45. Bellman R., Angel E., *Dynamic Programming and Partial Differential Equations.* Academic Press, New York, 1972.
46. Bernadou M., Oden J.T., An existence theorem for a class of nonlinear shellow shell problem, *J. Math. Pure Appl.*, **60**, 1981, 1–24.
47. Biot M.A., Thermoelasticity and irreversible thermodynamics, *J. Appl. Phys.*, **27**, No. 3, 1956, 240–253.
48. Birger I.A., *Circular Plates and Shells of Revolution.* Oborongiz, Moscow, 1971 (in Russian).
49. Bogaryan K.O., On convergence of errors of the Bubnov–Galerkin and Ritz methods, *Trans. Acad. Sci. SSSR*, **191**, No. 2, 1961 (in Russian).
50. Bollhaus U.F, Lomex G., Numerical modelling of low-frequency nonstationary transonic flows, in *Numerical Solutions of Hydromechanics Problems* (eds. A.Yu. Ishlinskiy, G.G. Tchiorny), News in Foreign Sciences, Mekhanika, No. 14. Mir, Moscow, 1977 (in Russian).
51. Bolotin V.V., *The Dynamic Stability of Elastic Systems.* Holden-Day, San Francisco, 1964.
52. Bolotin V.V., *Nonconservative Problems of the Theory of Elastic Stability.* Pergamon, Oxford, 1963.
53. Borisyuk A.I., Motovilovitz I.A., On temperature fields in a variable thickness shell, *Appl. Mech.*, **3**, No. 12, 1961 (in Russian).
54. Bradley M.E., Lasiecka I., Global stabilization of a von Kármán plate without geometric conditions, in *Identification and Control in Systems Governed by Partial Differential Equations* (eds. H.T. Banks, R.H. Fabiano, K. Ito), Proceedings in Applied Mathematics, No. 68. SIAM, Philadelphia, 1993.
55. Budiansky B., Roth B.S., *Axisymmetric Dynamic Buckling of Clamped Shallow Spherical Shells.* NASA, TN D-1510, 1962, pp. 379–606.
56. Buslov E.P., Experimental investigations on shallow behaviour under dynamic load, in *Theory of Plates and Shells* (ed. L.I. Lur'e). Sudostroenie, Leningrad, 1995, pp. 363–365 (in Russian).
57. Chen L.-W., Hwang J.-R., Axisymmetric dynamic stability of transversely isotropic Mindlin circular plates, *J. Sound Vib.*, **121**, No. 2, 1988, 307–315.
58. Chrzeszczyk A., Generalized solutions of dynamical equations in nonlinear theory of thin elastic shells, *Archiwum Mechaniki Stosowanej*, **35**, No. 5–6, 1983, 555–566.
59. Chrzeszczyk A., On the regularity, uniqueness and continuous dependence for generalized solutions of some coupled problems in nonlinear theory of thermoelastic shells, *Archiwum Mechaniki Stosowanej*, **38**, No. 1–2, 1986, 97–102.

60. Ciarlet P.G., *Mathematical Elasticity*, Vol. 2, *Theory of Plates.* Elsevier Science, Amsterdam, 1997.
61. Ciarlet P.G., *Plates and Junctions in Elastic Multi-Structures: an Asymptotic Analysis*, Vol. 14. Masson, Paris, 1990.
62. Ciarlet P.G., Rabier P., *Les Equations de von Kármán*, Springer, Berlin, Heidelberg, 1980.
63. Cole I.D., On a quasi-linear parabolic equation occuring in aerodynamics. *Q. Appl. Math.*, No. 9, 1951, 226–238.
64. Crocker M.J., Response of panels to oscillating and to moving shock waves. *J. Sound Vib.*, **6**, No. 1, 1967, 38–58.
65. Dafermos K.H., Quasi-linear hyperbolic systems derived from conservation principles, in *Nonlinear Waves* (eds. S. Leibovich, A.R. Seebass). Cornell University Press, Ithaca and London, 1974, pp. 91–112.
66. Day W.A., Cesaro means and recurrence in dynamic thermoelasticity. *Mathematika (London)*, **28**, No. 2, 1981, 211–230.
67. Day W.A., On the status of the uncoupled approximations within quasi-static thermoelasticity. *Mathematika (London)*, **28**, No. 2, 1981, 286–294.
68. Destuynder C., An existence theorem for a nonlinear shell model in large displacements analysis, *Math. Meth. Appl. Sci.*, **5**, 1983, 68–83.
69. Dowell E.H., *Aeroelasticity of Plates and Shells.* Nordhoff, Leiden, 1974.
70. Duhamel J., Second memoire sur les phénomenes thermo-méchanique-nes. *J. Ecole Polytechn.*, No. 15, 1937, 1–15.
71. Dzhishkaryani A.V., Convergence speed of approximate Ritz method, *J. Comput. Math. Math. Phys.*, **3**, No. 4, 1963, 654–663 (in Russian).
72. Dzhishkaryani A.V., On the convergence speed of the Bubnov–Galerkin method, *J. Comput. Math. Math. Phys.*, **4**, No. 2, 1964, 343–348 (in Russian).
73. Ehlers F.E., Weatherill W.H., Sebastian J.D., *The Practical Application of a Finite Difference Method for Analyzing Transonic Flow Over Oscillating Airfoils and Wings.* National Aeronautics and Space Administration, Scientific and Technical Information Office, Springfield, VA, 1978.
74. El-Ghazdly H.A., Sherboune A.N., Deformation theory for elastic-plastic buckling analysis of plates under nonproportional planar loading, *Comput. Struct.*, **22**, No. 2, 1986, 131–149.
75. Faedo S., Un nuovo metodo per l'analisi esistenziale e duantativa dei problemi di propagazione, *Ann. Scuola Norm. Super., Pisa*, No. 1, 1949, 1–40.
76. Fedos'ev V.I., On a certain method of solution to the nonlinear problem of stability of deformable systems, *Appl. Math. Mech.*, **27**, 1963, 265–274 (in Russian).
77. Fischer G., Existence theorems in elasticity, in *Handbuch der Physik*, Bd. 6a/s2, 1972, pp. 347–389.
78. Filippov A.P., *Vibrations of Deformable Bodies.* Mashinostroenie, Moscow, 1970 (in Russian).
79. Fomin V.G., Modification of Godunov method for asymptotic transonic flows, in *Applied Mathematics and Mechanics*, issue 2 (ed. A.P. Khromov). Saratov University Press, Saratov, 1983, pp. 61–66 (in Russian).
80. Galerkin B.G., Rods and plates, in *Some Problems on Elastic Equilibrium of Rods and Plates*, Engineering News (Vestnik Inzhinierov), **1**, No. 19, 1915, pp. 897–908 (in Russian).

81. Galimov K.Z., General theory of plates and shells with finite displacements and deformations, *Appl. Math. Mech.*, **15**, No. 6, 1951, 723–742 (in Russian).
82. Gawinecki J., Existence, uniqueness and regularity of the first boundary–initial value for thermal stress equations of classical and generalized thermomechanics, *J. Techn. Phys.*, **24**, No. 4, 1983, 467–479.
83. Gawinecki J., On the first initial–boundary value problem for the equations of thermal stresses, *Bull. Acad. Pol. Sci. Ser. Sci. Technol.*, **33**, No. 2, 1984, 17–34.
84. Gelos R., Dominiques H., Laura P.A.A., Application of the optimized Galerkin method to the determination of the fundamental frequency of a vibrating circular plate subjected to non-uniform in-plane loading, *J. Sound Vib.*, **114**, No. 3, 1987, 598–600.
85. Germain P., Bader R., *Unicite des ecoulement avec chocs dans la mecanique de Burgers.* Office National d'Etudes et de Recherches Aeronautiques, Paris, 1953, pp. 1–13.
86. Godunov S.K., Riabenkiy V.S., *Difference Schemes.* Elsevier Science, Amsterdam, 1987.
87. Godunov S.K., Zabrodin A.V., Ivanov M.Ya., Krayko A.N., Prokopov G.P., *Numerical Solutions of Multi-Dimensional Problems in Gas Dynamics.* Nauka, Moscow, 1976 (in Russian).
88. Godunov S.K., Zabrodin A.V., Prokopov G.P., Difference scheme for two-dimensional, nonstationary problems of gas dynamics and calculations of a shock wave flow, *Comput. Math. Tech. Phys.*, **1**, No. 6, 1961, 1020–1050 (in Russian).
89. Goldenweiser A.L., *Theory of Thin Elastic Shells.* Gostekhizdat, Moscow, 1953 (in Russian).
90. Gorshkov A.G., Nonstationary interaction of plates and shells with continuous systems. *Proc. Acad. Sci. SSSR, MTT*, No. 4, 1981, 177–189 (in Russian).
91. Gould P.L., *Analysis of Shells and Plates.* Prentice Hall, New Jersey, 1999.
92. Gould P.L., *Introduction to Linear Elasticity*, 2nd ed., Springer, Berlin, Heidelberg, 1999.
93. Gribanov V.F., Panichkin N.G., *Coupled and Dynamical Problems of Thermoelasticity.* Mashinostroenie, Moscow, 1984 (in Russian).
94. Grigolyuk E.I., Kabanov V.V., *Stability of Shells.* Nauka, Moscow, 1978 (in Russian).
95. Grigolyuk E.I., Shalashilin V.I., *Problems of Nonlinear Deformation.* Kluwer Academic, Dordrecht, 1991.
96. Gurov N.P., *Phenomenological Thermodynamics of Nonconvertible Processes.* Nauka, Moscow, 1978 (in Russian).
97. Gvozdev A.A., *Calculations of Load Limit of Construction by Limited Equilibrium Method.* Strojizdat, Moscow, 1949 (in Russian).
98. Hartmann F., *Ordinary Differential Equations.* Birkhäuser, Boston, 1982.
99. Hinton E., *Numerical Methods and Software for Dynamics Analysis of Plates and Shells.* Pineridge, Swansea, 1987.
100. Hopf E., The partial differential equation, *Commun. Pure Appl. Math.*, No. 3, 1950, 201–230.
101. Horn M.A., Lasiecka I., Nonlinear boundary stabilization of parallelly connected Kirchhoff plates, *Dynam. Control*, **6**, No. 3, 1996, 263–292.

102. Huang H.-C., *Static and Dynamic Analyses of Plates and Shells. Theory, Software and Applications.* Springer, Berlin, Heidelberg, 1989.
103. Ignatiev V.A., Sokolov O.L., *Thin-Walled Cellular Structures. Methods for Their Analysis.* Balkema, Rotterdam, 1999.
104. Il'yushin A.A., *Plasticity.* Gostekhizdat, Moscow, 1948 (in Russian).
105. Ivanov G.V., About solution of two-dimensional and spatial problems of transonic flow around bodies, *Comput. Math. Tech. Phys.*, **15**, No. 5, 1975, 1222–1240 (in Russian).
106. Ivanov M.Ja., Ryl'ko O.A., Calculations of transonic flow in spatial nozzles, *Comput. Math. Tech. Phys.*, **12**, No. 5, 1972, 1280–1291 (in Russian).
107. Jawad M., *Theory and Design of Plate and Shell Structures.* Kluwer Academic, Dordrecht, 1994.
108. Kaczkowski Z., On variational principles in thermoelasticity, *Bull. Acad. Pol. Sci. Technol.*, **30**, No. 5–6, 1982, 81–86.
109. Kaplunov J.D., Kossovitch L.Yu., Nolde E.V., *Dynamics of Thin Walled Elastic Bodies.* Academic Press, San Diego, 1997.
110. Karnaukhov V.G., Kirichok I.F., *Coupled Problems in Theory of Elasto-Plastic Plates and Shells.* Naukova Dumka, Kiev, 1986 (in Russian).
111. Karsloy G., Eger D., *Heat Conductivity of Solids.* Nauka, Moscow, 1964 (in Russian).
112. Kauderer H., *Nonlinear Mechanics.* Springer, Berlin, 1958 (in German).
113. Keldysh M.V., On the Galerkin method for solving boundary problems, *Trans. Acad. Sci. SSSR Ser. Math.*, **6**, No. 6, 1942, 309–330 (in Russian).
114. Kelly J.M., Wierzbicki T., Motion of a circular viscoplastic plate subject to projectile impact, *Z. Angew. Math. Phys.*, **18**, 1967, 236–246.
115. Kil'chinskaya G.A., The smallest excitation force principle for generalized thermomechanics, *Trans. Acad. Sci. USSR*, No. 2, 1997, 1092–1095 (in Russian).
116. Kirichenko V.F., Krys'ko V.A., Khametova N.A., On the influence of the coupling effect of temperature and deformation fields on the dynamic stability of shallow shells, *Appl. Mech.*, **24**, No. 11, 1998, 46–50 (in Russian).
117. Kirichenko V.F., Krys'ko V.A., On the existence of solutions to a certain nonlinear coupled problem of thermoelasticity, *Diff. Equat.*, **20**, No. 9, 1984, 1583–1588 (in Russian).
118. Kleiber M., Woźniak C., *Nonlinear Mechanics of Structures*, Vol. 8. Kluwer Academic, Dordrecht, 1991.
119. Kochin N.E., Kibel I.A., Roze N.V., *Theoretical Hydromechanics.* Interscience, New York, 1964.
120. Kolyano Yu.M., Shter E.I., Thermoelasticity of nonhomogeneous bodies, *Eng. Phys. J.*, **3**, No. 6, 1980, 1111–1114 (in Russian).
121. Kovalenko A.D., *Collected Works.* Naukova Dumka, Kiev, 1976 (in Russian).
122. Kovalenko A.D., *Thermoelasticity. Basic Theory and Applications.* Wolters-Noordhoff, Groningen, 1969.
123. Kovalenko A.D., *Thermoelasticity.* Vysshaia Shkola, Kiev, 1975 (in Russian).
124. Kowalski T., Litevska K., Piskorek A., Uniqueness and regularity of the solution of the first initial–boundary value problem in linear thermoelasticity, *Bull. Acad. Pol. Sci. Technol.*, **30**, No. 3–4, 1982, 171–175.

125. Kowalski T., Piskorek A., Existence of solution of boundary-value problem in linear theory of thermoelasticity, *Z. Angew. Math. Mech.*, **61**, No. 5, 1981, T250–T252.
126. Kozdoba L.A., *Solution Methods for Nonlinear Problems of Thermoconductivity.* Nauka, Moscow, 1975 (in Russian).
127. Kozlov V.I., Coupling effect in a problem of thermal impact on rod surface, in *Thermal stresses in structures (collection of articles)* (ed. A.D. Kovalenko). Kiev, 1971, issue 11, pp. 97–102 (in Russian).
128. Kozlov V.I., Temperature oscillations of rectangular plate, *Appl. Mech.*, **8**, No. 4, 1972, 123–127 (in Russian).
129. Kozlov V.I., Thermal impact on circular plate surface taking into account coupling of deformation fields, *Trans. Acad. Sci. USSR, Ser. A*, No. 10, 1971, 923–927 (in Russian).
130. Krasnosel'skiy M.A., Convergence of Galerkin method for nonlinear equations, *Trans. Acad. Sci. SSSR*, **23**, No. 6, 1950, 1121–1124 (in Russian).
131. Krylov K.M., Bogolyubov N.N., On some theorems concerning the behaviour of integrals of partial differential equations of hyperbolic type, *Trans. Acad. Sci. SSSR*, No. 3, 1931, 323–344 (in Russian).
132. Krys'ko V., Awrejcewicz J., Bruk V., The existence and uniqueness of solution of one coupled plate thermomechanics problem, *J. Appl. Anal.* (to appear).
133. Krys'ko V.A., *Nonlinear Statics and Dynamics of Nonhomogeneous Shells.* Saratov University Press, Saratov, 1976 (in Russian).
134. Krys'ko V.A., Dynamic buckling of shells, rectangular in plan, with finite deflection, *Appl. Mech.*, **15**, No. 11, 1979, 58–62 (in Russian).
135. Krys'ko V.A., Kutsemako A.N., *Stability and Oscillations of Nonuniform Shells.* Saratov Technical University, Saratov, 1999 (in Russian).
136. Krys'ko V.A., Mishnik M.P., Calculation of coupled physically nonlinear three-dimensional plates in a temperature field, *Trans. VUZ, Ser. Civ. Eng. Architect.*, No. 9, 1984, 33–37 (in Russian).
137. Kupradze V.D., Gegelija T.G., *Three-Dimensional Problems of The Mathematical Theory of Elasticity and Thermoelasticity.* North-Holland, Amsterdam, 1979.
138. Kutateladze S.S., *Fundamentals of Heat Transfer.* Academic Press, New York, 1963.
139. Ladyzhenskaya O.A., *The Boundary Value Problems of Mathematical Physics.* Springer, New York, 1985.
140. Ladyzhenskaya O.A., *The Mathematical Theory of Viscous Incompressible Flow.* Gordon and Breach, New York, 1969.
141. Lanczos C., *Variational Principles of Mechanics*, 2nd ed. Toronto University Press, Toronto, 1964.
142. Landau L.D., Lifshitz E.M., *A Shorter Course of Theoretical Physics.* Pergamon, Oxford, 1972.
143. Lepik Yu.A., Dynamics of circular and toroidal plates made of rigid–plastic material sensitive to deformation speed, *Appl. Mech., Acad. Sci. USSR*, **5**, No. 1, 1969, 35–41 (in Russian).
144. Libai A., Simmonds J.G., *The Nonlinear Theory of Elastic Shells.* Cambridge University Press, Cambridge, 1998.
145. Liew K.M., Wang C.M., Xiang Y., Kitipornchai S., *Vibration of Mindlin Plates.* Elsevier, Amsterdam, 1998.

146. Lin C.C., Reissner E., Tsien H.S., On two-dimensional nonsteady motion of a slender body in a compressible fluid, *J. Math. Phys.*, **27**, No. 3., 1948. Russian translation in *Gas Dynamics.* Moscow, 1955, pp. 183–196.
147. Lions J.L., *Problèmes aux limites dans les équations aux dérivées partielles.* 2 éd. Presses de l'Université de Montréal, Montréal, 1967 (in French).
148. Lions J.L., Magenes E., *Non-homogeneous Boundary Problems and Their Applications.* Springer, Berlin, 1972.
149. Lord H.W., Shulman Y., A generalized dynamical theory of thermoelasticity. *J. Mech. Phys. Sol.*, **15**, No. 5, 1967, 299–309.
150. Lykov A.V., *Analytical Heat Diffusion Theory.* Academic Press, New York, 1968.
151. Krasnosel'skiy M.A., Baynikko G.M., Zabreyko P.P., *Approximate Solution of Operator Equations.* Nauka, Moscow, 1969 (in Russian).
152. Magnus R., Yoshihara H., Calculations of transonic flow over an oscillating airfoil, *AIAA J.*, No. 1, 1975, 75–98.
153. Malkin Ja.F., On problems of distribution of temperature in flat plates, *Appl. Math. Mech.*, **2**, No. 3, 1939, 317–330 (in Russian).
154. Mansfield E.H., *The Bending and Stretching of Plates.* Cambridge University Press, Cambridge, 1989.
155. Marguerre K., Temperature changes and temperature stresses in plates and shallow bodies, *Ing. Arch.*, **8**, No. 3, 1937, 216–228.
156. Mikhaylov V.P., *Partial Differential Equations.* Mir, Moscow, 1978 (in Russian).
157. Mikhaylovskaya I.E., Novik O.B., Cauchy problems in a class of increasing functions for nonhyperbolic evolutionary systems which are not parabolic, *Siber. Math. J., Novosibirsk*, 1979 (in Russian).
158. Mikhlin S.G., *The Minimum Problem of a Square Functional.* Gostekhizdat, Moscow, 1952 (in Russian).
159. Mikhlin S.G., *The Numerical Performance of Variational Methods.* Wolters-Noordhoff, Groningen, 1971.
160. Mikhlin S.G., *Some Problems in Error Theory.* Leningrad University Press, Leningrad, 1988 (in Russian).
161. Mikhlin S.G., On the convergence of the Galerkin method, *Trans. Acad. Sci. SSSR*, **61**, No. 2, 1948, 197–199 (in Russian).
162. Mikhlin S.G., *Variational Methods in Mathematical Physics.* Pergamon, Oxford, 1964.
163. Mikhlin S.G., On the Ritz method, *Trans. Acad. of Sci. SSSR*, **106**, No. 3, 1956, 391–394 (in Russian).
164. Morozov N.F., *Collected Two-Dimensional Problems of Theory of Elasticity.* Leningrad University Press, Leningrad, 1978 (in Russian).
165. Morozov N.F., On nonlinear oscillations of thin plates with consideration of moment of inertia, *Trans. Acad. Sci. SSSR*, **176**, No. 3, 1967, 522–525 (in Russian).
166. Moskvitin V.V., *Periodic Loads on Structural Elements.* Nauka, Moscow, 1981 (in Russian).
167. Moskvitin V.V., *Plasticity Under Alternating Loads.* MGU, Moscow, 1965 (in Russian).
168. Motovilovitz I.A., Kozlov V.I., *Mechanics of Coupled Fields in Structural Elements*, Vol. 1. Naukova Dumka, Kiev, 1987 (in Russian).

169. Mrozowski J., Awrejcewicz J., Flow-induced chaotic oscillations, in *Sixth International Conference on Flow Induced Vibrations* (ed. P.W. Bergman), London, April 10–12, 1995. Balkema, Rotterdam, 1995, pp. 557–564.
170. Mukhopandhayay M., Free vibration of rectangular plates with edges having different degrees of rotational restraint, *J. Sound Vib.*, **67**, No. 4, 1979, 459–468.
171. Murman E.M., Cole I.D., Calculation of plane steady transonic flows, *AIAA J.*, **9**, No. 1, 1971, 114–121.
172. Mushtari Kh.M., Galimov K.Z., *Non-Linear Theory of Elastic Shells.* Tatknigoizdat, Academy of Sciences, USSR, Kazan' Branch, 1957.
173. Nashed M.Z., The convergence of the method of steepest descents for nonlinear equations with variational or quasi-variational operators, *J. Math. Mech.*, **13**, 1964, 765–794.
174. Nezlin M.V., Snieshkin E.N., Trubnikov A.S., Kelvin–Helmholtz instability and Great Red Spot of Jupiter, *ZETF*, **32**, 1982, 190–193.
175. Nickell R.E., Sackman J.L., Variational principles for linear coupled thermoelasticity, *Q. Appl. Math.*, **26**, No. 1, 1968, 11–26.
176. Nowacki W., *Dynamical Problems in Thermoelasticity.* PWN, Warsaw, 1966 (in Polish).
177. Nowacki W., Thermal stresses in orthotropic plates, *Bull. Acad. Pol. Sci. Ser. Sci. Techol.*, **7**, No. 1, 1959, 1–6.
178. Nowacki W., *Thermal Stresses in Shells. General Report on Non-Classical Shell Problem.* IASS, Warsaw, 1962.
179. Nowacki W., *Thermoelasticity.* Pergamon, London, 1962.
180. Nowacki W., Some dynamic problems of thermoelasticity, *Archiwum Mechaniki Stosowanej*, **11**, No. 2, 1959, 259–283.
181. Osaka H., Fujita M., Hanasaki K., Fujinaka R., A numerical analysis of plates under transverse impact loading, *J. Japan Soc. Technol. Plast.*, **27**, No. 301, 1986, 288–294.
182. Parton V.Z., Perlin P.I., *Mathematical Methods of the Theory of Elasticity.* Mir, Moscow, 1984 (in Russian).
183. Petrashvili V.I., Multidimensional solitons, in *Nonlinear Waves* (ed. Gaponov). Nauka, Moscow, 1979, pp. 5–20 (in Russian).
184. Petryshyn W.V., Direct and iterative methods for the solution of linear operator equations in Hilbert space, *Trans. Am. Math. Soc.*, **105**, No. 1, 1962, 136–175.
185. Pobedrya B.E., Numerical methods in viscoelasticity *Polym. Mech.*, No. 3, 1973, 417–428 (in Russian).
186. Podstrigach Ya.S., Kolyano Yu.M., *Generalized Thermomechanics.* Naukova Dumka, Kiev, 1976 (in Russian).
187. Podstrigach Ya.S., Lapakin V.A., Kolyano Yu.M., *Thermoelasticity of Nonhomogeneous Bodies.* Nauka, Glav. red. fiziko-matematicheskoi lit-ry, Moscow, 1984 (in Russian).
188. Podstrigach Ya.S., Shvetz R.N., *Thermoelasticity of Thin Shells.* Naukova Dumka, Kiev, 1978 (in Russian).
189. Rabinovich M.I., Trubetskov D.I., *Oscillations and Waves in Linear and Nonlinear Systems.* Kluwer Academic, Dordrecht, 1989.
190. Ritz W., Uber eine neue Methode zur Losung gewisser Variationsprobleme der mathematischen Physik, *J. Reine Angew. Math.*, **135**, No. 1, 1909, 1–61.

191. Rozhdestvenskiy B.L., Janenko N.N., Systems of quasilinear equations and their applications to gas dynamics, *Am. Math. Soc.*, **55**, 1983, 0065–9282.
192. Ryzhov O.S., *Transonic Flow in Laval Nozzles.* Computer Centre, Academy of Sciences of the SSSR, Moscow, 1965 (in Russian).
193. Ryzhov O.S., Shefter G.M., On the influence of viscosity and heat conductivity on the structure of compressible flow, *Appl. Math. Mech.*, **28**, No. 6, 1964, 996–1007 (in Russian).
194. Godunov S.K., Zabrodin A.V., Ivanov M.Ja., Krayko A.N., Prokopov G.P., *Numerical Solutions of Multi-Dimensional Problems of Gas Dynamics.* Nauka, Moscow, 1976 (in Russian).
195. Sathyamoorthy M., *Nonlinear Analysis of Structures.* CRC, Boca Raton, FL, 1998.
196. Sedov L.I., *Mechanics of Continuous Media.* World Scientific, Singapore, 1997.
197. Sevost'janov G.D., Fomin V.G., On nonstationary axisymmetric ideal gas flows, *Aerodynamics* (Saratov University Press, Saratov), **5**, No. 8, 1976, 27–32 (in Russian).
198. Shalov V.M., Solution of non-self-adjoint equations by variation method, *Trans. Acad. of Sci. SSSR*, **151**, No. 3, 1963, 511–512 (in Russian).
199. Shiam A.C., Soong I.T., Roth R.S., Dynamic buckling of conical shells with imperfections, *AIAA J.*, **12**, No. 6, 1974, 262–271.
200. Shvetz R.N., Flachok V.M., Equations of mechanical thermal diffusion for anisotropic shells with consideration of transverse deformations, *Math. Meth. Phys. Mech. Fields*, No. 20, 1984, 54–61 (in Russian).
201. Shvetz R.N., Lopat'ev A.A., On properties of dynamic processes in deformable solids with consideration of finite speed of heat transfer, *J. Eng. Phys.*, **25**, No. 4, 1978, 705–712 (in Russian).
202. Shvetz R.N., Lun' E.I., Some aspects of the theory of thermoelasticity of an orthotropic shell with consideration of moments of inertia and transverse shear, *Appl. Mech.*, **7**, No. 10, 1971, 121–125 (in Russian).
203. Shvetz R.N., Variational theorem for mutually coupled thermoelasticity problems of thin shells, *Math. Phys.*, No. 28, 1980, 104–108 (in Russian).
204. Skrypnik I.V., *Nonlinear Elliptic Equations of High Order.* Naukova Dumka, Kiev, 1973 (in Russian).
205. Skurlatov E.D., Experimental investigations of behaviour of shells under dynamic loads, *Elast. Problems*, No. 9, 1972, 79–83 (in Russian).
206. Skurlatov E.D., On the behaviour of cylindrical panels and shells subjected to an incoming pressure wave, in *Theory of Plates and Shells* (ed. L.I. Lur'e). Nauka, Moscow, 1971, pp. 256–261 (in Russian).
207. Skurlatov E.D., Solonenko V.R., Experimental and theoretical investigation of oscillations of cylindrical shells under moving, pulsing pressure jumps, in *State Symposium on Distribution of Elastic and Elasto-Plastic Waves.* Nauka, Moscow, 1978 (in Russian).
208. Smith I.M., Griffiths D.V., *Programming The Finite Element Method*, 3rd ed. Wiley, Chichester, 1998.
209. Sobolev C.L., *Some Application of Functional Analysis in Mathematical Physics.* Leningrad University Press, Leningrad, 1950 (in Russian).

210. Sobolevskiy P.E., On the Bubnov–Galerkin method for parabolic equations in Hilbert space, *Trans. Acad. Sci. SSSR*, **178**, No. 3, 1968, 548–551 (in Russian).
211. Sobotka Z., *Theory of Plasticity and Limit Design of Plates*, Vol. 18. Elsevier, Amsterdam, 1989.
212. Soedel W., *Vibrations of Shells and Plates*, Vol. 10. Marcel Dekker, New York, 1981.
213. Spreiter I.R., Aerodynamics of wings and bodies at transonic speeds, *J. Aero/Space Sci.*, **26**, No. 8, 1959, 465–486. Russian translation in *Mechanics*, No. 3, 1960, 3–50.
214. Stepanov G.B., Kovalenko A.V., Bending of slender plate loaded by short pressure impact, *Strength Problems*, No. 3, 1986, 40–46 (in Russian).
215. Svirskiy I.V., *Methods of the Bubnov-Galerkin Type and Successive Approximations.* Nauka, Moscow, 1968 (in Russian).
216. Szilard R., *Theory and Analysis of Plates, Classical and Numerical Methods.* Prentice-Hall, Englewood Cliffs, NJ, 1974.
217. Thompson D.E., *Design Analysis. Mathematical Modeling of Nonlinear Systems.* Cambridge University Press, Cambridge, 1999.
218. Timashev S.A., *Stability of Reinforced Shells.* Stroyizdat, Moscow, 1974 (in Russian).
219. Traci R.M., Farr I.L., Albano E., Perturbation method for transonic flow about oscillating airfoils, *AIAA J.*, June 1975, 75–877.
220. Ugural A.C., *Stresses in Plates and Shells*, 2nd ed. McGraw-Hill, New York, 1999.
221. Ugural A.C., Fenster S.K., *Advanced Strength and Applied Elasticity.* Prentice Hall, New Jersey, 1995.
222. Vaynikko G.M., Oya P.E., On convergence and convergence velocity of Galerkin method for abstract of evolutionary equations, *Diff. Equations*, **11**, No. 7, 1975, 1269–1277 (in Russian).
223. Vakhlaeva L.F., Krys'ko V.A., Stability of thin shallow shells in a temperature field, *Appl. Mech.*, **19**, No. 1, 1983, 16–23 (in Russian).
224. Valishvilli I.E., *Methods of Computer Calculation of Axisymmetric Shells.* Mashinostroenie, Moscow, 1976 (in Russian).
225. Vashakmadze T.S., *The Theory of Anisotropic Elastic Plates.* Kluwer Academic, Dordrecht, 1999.
226. Vasil'kovskiy S.N., Theorem about uniqueness of solution of coupled thermoelasticity dynamics in stress equations, *Trans. VUZ, Math.*, No. 9, 1984, 21–24 (in Russian).
227. Villagia P., *Mathematical Models for Elastic Structures.* Cambridge University Press, Cambridge, 1997.
228. Vinson J.R., *The Behavior of Thin Walled Structures: Beams, Plates and Shells*, Vol. 8. Nijhoff, Dordrecht, 1988.
229. Vlasov V.Z., *General Shell Theory and Its Technical Applications.* Gostekhizdat, Moscow, 1949 (in Russian).
230. Vlasov V.Z., *Thin-Walled Structures.* Gosstrojizdat, Moscow, 1958 (in Russian).
231. Volmir A.S., *Survey of Investigations on The Theory of Flexible Plates and Shells (Covering The Period From 1941 to 1957).* National Aeronautics and Space Administration, Washington, DC, 1963.

232. Volmir A.S., *Nonlinear Dynamics of Plates and Shells.* Nauka, Moscow, 1972 (in Russian).
233. Volmir A.S., *Shells in Liquid and Gas Flow. Hydroelasticity Problems.* Nauka, Moscow, 1979 (in Russian).
234. Volmir A.S., *Shells in Liquid and Gas Flow.* Nauka, Moscow, 1976 (in Russian).
235. Vorovich I.I., *Nonlinear Theory of Shallow Shells.* Springer, New York, 1999.
236. Vorovich I.I., Lebedev L.P., On the existence of solutions in the nonlinear theory of elastic shells, *Appl. Math. Mech.*, **36**, No. 4, 1972, 691–704 (in Russian).
237. Vorovich I.I., On some simple methods in the nonlinear theory of oscillations of elastic shells, *Appl. Math. Mech.*, **21**, No. 6, 1957, 747–784 (in Russian).
238. Vorovich I.I., Lebedev L.P., Shlafman S.M., On certain direct methods and existence of solutions in nonlinear theory of elastic non-shallow rotational shells, *Appl. Math. Mech.*, **38**, No. 2, 1974, 339–348.
239. Waszczyszyn Z., Cichon C., Radwanska M., *Stability of Structures By Finite Element Method.* Elsevier, Amsterdam, 1994.
240. Weatherill W.H., Sebastian J.D., Ehlers F.E., *The Practical Application of a Finite Difference Method for Analyzing Transonic Flow over Oscillating Airfoils and Wings.* National Aeronautics and Space Administration, Scientific and Technical Information Office, Springfield, VA, 1978.
241. Wierzbicki T., Finite deflection of a circular viscoplastic plate subject to projectile impact, *Int. J. Solids Struct.*, **4**, 1968, 1081–1092.
242. Wierzbicki T., Impulsive loading of rigid viscoplastic plates, *Int. J. Solids Struct.*, **3**, 1967, 635–647.
243. Zarubin A.G., Investigation of Galerkin–Petrov projection procedure by small-steps method, *Trans. Acad. Sci. SSSR*, **297**, No. 4, 1987, 780–784 (in Russian).
244. Zarubin A.G., On the convergence speed of the Faedo–Galerkin method for linear nonstationary equations, *Diff. Equtions*, **8**, No. 4, 1982, 639–645 (in Russian).
245. Zarubin A.G., Tiunchik M.F., On approximate solutions of a certain class of nonlinear nonstationary equations, *Diff. Equations*, **9**, No. 11, 1973, 1966–1974 (in Russian).
246. Zino I.E., Trepp E.A., *Asymptotic Methods in Problems of Theory of Thermal Conductivity and Thermoelasticity.* Leningrad University Press, Leningrad, 1978 (in Russian).
247. Zubrikhin O.A., Gribanov V.F., Skurlatov E.D., Experimental investigation of oscillations of shells in a transonic gas flow, in *Oscillations of Elastic Structures in Liquids* (ed. A.S. Volmir). Nauka, Moscow, 1976, pp. 109–115 (in Russian).

Index

Scientific Computation

A Computational Method in Plasma Physics
F. Bauer, O. Betancourt, P. Garabechan

Implementation of Finite Element Methods for Navier-Stokes Equations
F. Thomasset

Finite-Different Techniques for Vectorized Fluid Dynamics Calculations
Edited by D. Book

Unsteady Viscous Flows
D. P. Telionis

Computational Methods for Fluid Flow
R. Peyret, T. D. Taylor

Computational Methods in Bifurcation Theory and Dissipative Structures
M. Kubicek, M. Marek

Optimal Shape Design for Elliptic Systems
O. Pironneau

The Method of Differential Approximation
Yu. I. Shokin

Computational Galerkin Methods
C. A. J. Fletcher

Numerical Methods for Nonlinear Variational Problems
R. Glowinski

Numerical Methods in Fluid Dynamics
Second Edition M. Holt

Computer Studies of Phase Transitions and Critical Phenomena O. G. Mouritsen

Finite Element Methods in Linear Ideal Magnetohydrodynamics
R. Gruber, J. Rappaz

Numerical Simulation of Plasmas
Y. N. Dnestrovskii, D. P. Kostomarov

Computational Methods for Kinetic Models of Magnetically Confined Plasmas
J. Killeen, G. D. Kerbel, M. C. McCoy, A. A. Mirin

Spectral Methods in Fluid Dynamics
Second Edition C. Canuto, M. Y. Hussaini, A. Quarteroni, T. A. Zang

Computational Techniques for Fluid Dynamics 1 Fundamental and General Techniques Second Edition
C. A. J. Fletcher

Computational Techniques for Fluid Dynamics 2 Specific Techniques for Different Flow Categories Second Edition
C. A. J. Fletcher

Methods for the Localization of Singularities in Numerical Solutions of Gas Dynamics Problems
E. V. Vorozhtsov, N. N. Yanenko

Classical Orthogonal Polynomials of a Discrete Variable
A. F. Nikiforov, S. K. Suslov, V. B. Uvarov

Flux Coordinates and Magnetic Filed Structure: A Guide to a Fundamental Tool of Plasma Theory
W. D. D'haeseleer, W. N. G. Hitchon, J. D. Callen, J. L. Shohet

Monte Carlo Methods in Boundary Value Problems
K. K. Sabelfeld

The Least-Squares Finite Element Method
Theory and Applications in Computational Fluid Dynamics and Electromagnetics
Bo-nan Jiang

Computer Simulation of Dynamic Phenomena
M. L. Wilkins

Grid Generation Methods
V. D. Liseikin

Radiation in Enclosures
A. Mbiock, R. Weber

Large Eddy Simulation for Incompressible Flows An Introduction Second Edition
P. Sagaut

Higher-Order Numerical Methods for Transient Wave Equations
G. C. Cohen

Fundamentals of Computational Fluid Dynamics
H. Lomax, T. H. Pulliam, D. W. Zingg

The Hybrid Multiscale Simulation Technology An Introduction with Application to Astrophysical and Laboratory Plasmas A. S. Lipatov

Computational Aerodynamics and Fluid Dynamics An Introduction J.-J. Chattot

Nonclassical Thermoelastic Problems in Nonlinear Dynamics of Shells Applications of the Bubnov–Galerkin and Finite Difference Numerical Methods J. Awrejcewicz, V. A. Krys'ko

Series homepage – http://www.springer.de/phys/books/sc/

Printing: Mercedes-Druck, Berlin
Binding: Stein+Lehmann, Berlin